WATER QUALITY TRENDS
AND
GEOCHEMICAL MASS BALANCE

ADVANCES IN HYDROLOGICAL PROCESSES

Series Editors   M. G. Anderson
D. E. Walling
N. E. Peters

WATER QUALITY TRENDS AND GEOCHEMICAL MASS BALANCE

Edited by

N. E. PETERS

US Geological Survey, Atlanta, GA, USA

O. P. BRICKER

US Geological Survey, Reston, VA, USA

and

M. M. KENNEDY

US Geological Survey, Reston, VA, USA

JOHN WILEY & SONS

Chichester · New York · Weinheim · Brisbane · Singapore · Toronto

The papers in this volume were originally published in
Hydrological Processes—An International Journal, part **10**(2):
125–335 (1996) and **11** (7): 643–816 (1997).

Other Wiley Editorial Offices

John Wiley & Sons, Inc., 605 Third Avenue,
New York, NY 10158-0012, USA

WILEY–VCH Verlag GmbH, Pappelallee 3,
D-69469 Weinheim, Germany

Jacaranda Wiley Ltd, 33 Park Road, Milton,
Queensland 4064, Australia

John Wiley & Sons (Asia) Pte Ltd, 2 Clementi Loop #02–01,
Jin Xing Distripark, Singapore 129809

John Wiley & Sons (Canada) Ltd, 22 Worcester Road,
Rexdale, Ontario M9W 1L1, Canada

British Library Cataloguing in Publication Data

A catalogue record for this book is available from the British
Library

ISBN 0-471-97868-X

Typeset in 10/12 Times
Printed and bound in Great Britain by BPC Wheatons Ltd, Exeter

CONTENTS

PREFACE

Water is, perhaps, the single most important substance on earth. It forms the basis for life as we know it. Organisms, including man, can only survive if there is an adequate supply of water available. In addition, the water must be of the proper quality to support life. The living world has evolved and adjusted over hundreds of millions of years to variation in the natural distribution and chemical composition of water at the earth's surface. Although 98% of the earth's water is sea water, humans are most concerned with the remaining 2% of fresh water that supports most terrestrial life on earth. In the last several hundred years, man has increasingly impacted on both the distribution and the quality of fresh water by building dams, draining wetlands, re-routing water for irrigation and water supply, discharging industrial and municipal wastes and applying fertilizer and pesticides that find their way into streams, rivers, lakes and groundwater. The ever-growing demand for water, coupled with the rate at which much of the earth's fresh waters are being adversely affected by man's activities, portends a developing crisis in the not-too-distant future if environmental resources are not appropriately managed. Rational management of the quality of our water resources requires a basic understanding of the natural processes that affect water quality and human impacts on those processes. We need to know how the composition of water evolves under natural conditions as a baseline against which to compare the effects of human interferences. In addition, we must be able to observe how water quality changes with time as a function of both natural and anthropogenic stresses. This requires tracking of water quality by systematic measurements or monitoring for trend detection.

This volume proves an overview of the tracking of water-quality changes, through both trend analysis and geochemical mass balances. The term 'trend', as used herein, captures the spirit of the word as (1) 'a prevailing tendency or inclination'; or more rigorously as (2) 'a straight line or other statistical curve showing a tendency of some function to grow or decline over a period of time'. The mass balances approach is simply tracking of the changes of mass of specific elements or compounds in the water–landscape system. Water passing through a watershed interacts with watershed materials – the vegetation, soils, regolith, and bedrock – and reflects these interactions by changes in chemical composition and in other characteristics. The flux of water and accompanying dissolved and suspended materials into and out of the watershed provides a frame of reference within which biotic and abiotic information can be interwoven to understand watershed ecosystem processes and how they affect the quality of water.

Human interferences in the environment, either through physical alteration of the landscape or through emissions from industrial or municipal sources, fertilizer application and liquid and solid waste disposal, can accelerate trends in water-quality parameters. Natural variations in climate and vegetational succession also can affect trends. It is important to establish the processes controlling water-quality evolution under natural conditions as well as to identify human-induced changes. As water-quality degradation continues, the threat to human and ecosystem health increases. Consequently, there is an increasing need to assess water-quality trends and mass transfers to provide not only a warning of the rates of water-resource degradation, but also to provide knowledge of causal relationships to manage these precious resources effectively.

The first half of this volume provides a perspective on trends in water quality, particularly temporal aspects. This includes an overview of techniques used to evaluate temporal changes in water quality. Case studies are given of solute trends in precipitation and streamwater from small catchments in the United States and the United Kingdom, trends in temperatures of surface waters, trends in heavy metals, trends in nutrient concentrations and loadings, trends in eutrophication research and control, trends in pathogenic agents in fresh waters and trends in groundwater quality.

The second half of the volume focuses on geochemical mass balance including the current status of the method and how it has been applied in some recent studies. Topics include the assessment of atmospheric deposition inputs to watersheds, methods used to quantify solute exports from watersheds, rates of mineral weathering and its effect on stream chemistry in forested systems and on bare rock, biogeochemical cycling and fluxes of nitrogen and sulfur in watershed systems, and application of the use of geographic information systems in watershed mass balance studies.

The collection of papers in this volume, although only a small sample, gives a sense of the variety of information on water-quality trends in various aquatic environments and geochemical mass balance applications in watershed and catchment studies. The combination of approaches attests to their primary importance in elucidating the complex biogeochemical processes that govern the functioning of these systems. The composition of natural waters and the impact of human activities on water quality are strongly dependent on watershed processes. We hope that this volume will serve to extend our understanding of those processes and encourage innovative applications of these data analysis techniques to the study of environmental systems. Also, one should not discount the continuing need for quality environmental monitoring data as these data provide the basic building blocks used for the trend and mass balance assessments presented herein.

Special Issue Editors
Norman E. Peters
Owen P. Bricker
Margaret M. Kennedy

SECTION I

WATER QUALITY TRENDS

1

REVIEW OF METHODS FOR THE DETECTION AND ESTIMATION OF TRENDS WITH EMPHASIS ON WATER QUALITY APPLICATIONS

SYLVIA R. ESTERBY

National Water Research Institute, PO Box 5050, Burlington, Ontario, Canada L7R 4A6

ABSTRACT

Methods for the detection and estimation of trends which are suitable for the type of data sets available from water quality and atmospheric deposition monitoring programmes are considered. Parametric and non-parametric methods which are based on the assumption of monotonic trend and which account for seasonality through blocking on season are described. The topics included are heterogeneity of trend, missing data, covariates, censored data, serial dependence and multivariate extensions. The basis for the non-parametric methods being the method of choice for current large data sets of short to moderate length is reviewed. A more general definition of trend as the component of gradual change over time is consistent with another group of methods and some examples are given. Spatial temporal data sets and longer temporal records are also briefly considered. A broad overview of the topic of trend analysis is given, with technicalities left to the references cited. The necessity of defining what is meant by trend in the context of the design and objectives of the programme is emphasized, as is the need to model the variability in the data more generally.

INTRODUCTION

The methods which are suitable for the detection and estimation of trends in a particular situation will be determined by: (1) the definition of trend; (2) models of trend and other components of variability; (3) data set characteristics; and (4) programme objectives for which data were collected.

The term 'trend' is widely used in the vernacular to mean general direction or tendency. It is used in this general way when trend assessment is given as an objective of a water quality monitoring programme and in reports where undocumented graphical summaries are given as evidence of trends. When used in the vernacular as an objective of a monitoring programme, the reader must draw upon a personal perception of trend to make the statement more precise. In conclusions based on descriptive methods, this perception may not be explicitly stated, but it may be deducible from the graphic used or the statements made.

In contrast, a numerical analysis of water quality data to detect or estimate trend involves a much more specific definition of trend, although it may not be stated explicitly, and this definition is motivated by the objectives of the data collection programme and characteristics of the data set. An objective to assess trends in the ambient concentration of a pollutant will require a different analysis from one to assess trends in the anthropogenic component of the load of the pollutant to a system. Similarly, the type of changes over time which can be detected by a programme of monthly samples for a 10 year period may be very different from that with biweekly sampling for 25 years or daily sampling for five years. Each situation requires a model that represents particular features of the data set and, in the model, the definition of trend is irrevocably linked to the way other sources of variability are included. Esterby (1993) compared the link between the way trend and seasonality are modelled in several trend analysis methods used for water quality data to show what is meant by trend in each method. Brillinger (1994) also addressed the question of what a trend is through the analysis of environmental data sets using time series and point process methods and concluded that the definition depends on the circumstances.

This paper is based on the view that trend analysis is an analysis of the variation in data collected over time, with the major objective of the analysis being the extraction of information about changes over time and the expression of this information quantitatively. Because the term 'trend' is used so pervasively, but generally is defined so imprecisely, the definition of trend will be made explicit for the methods considered here. Thus methods are broadly grouped according to the type of trend: trend as a monotonic change or more general forms of change. Types of methods which are dictated by the nature of the data set and objectives of the data collection programme are then separated within these broad topics. Descriptive methods will be discussed first because they are essential to all methods of analysis, and in some situations that is all that has been used. A river water quality data set is used to give examples of some useful plots and to illustrate types of assumptions about variability due to trend and other sources of variability that underlie different methods of analysis. The applicability of methods in the presence of missing observations, serial dependence and censored data (some observations below the analytical detection limit) and methods for longer data records and spatial/temporal data will also be considered.

The papers in this issue on *Trends in Water Quality* cover many aspects of water quality and a comprehensive review of methods is not practicable. Further, trend analysis is currently a high priority for many programmes, and an extensive literature exists. The intent herein is to show the breadth of the topic, the type of methods which are available and areas where work is continuing. Examples have been chosen which accommodate many features of water quality data sets and it is acknowledged that other examples are available which would have been equally illustrative. The important features of the methods will be covered, with details left to the references cited. Throughout this paper, it will be assumed that consistent sampling, laboratory and reporting procedures have been maintained over the period considered. Emphasis will be on data sets from programmes with a fixed sampling interval, moderate number of years sampled and multiple water quality indicators sampled at several stations within a monitoring network.

DESCRIPTIVE METHODS

Descriptive methods can be defined as methods which do not involve either formal hypothesis testing or estimation and include graphical methods and summary statistics. They are mentioned here separately to emphasize their importance, but they also form part of any numerical analysis, from initial inspection of the data to presentation of the results.

Sometimes graphical methods or summary statistics are all that is possible or appropriate. Two examples follow. A report on preliminary environmental indicators (OECD, 1991) cites increasing trends in the nitrate content of river waters and gives a supporting figure consisting of a bar graph of the annual mean concentration of nitrate for rivers in 14 countries based on data from five or fewer years . Data in global monitoring programmes lack the consistency with respect to sampling and analytical techniques possible in national programmes, and, apparently for the data displayed in this example, very few years of data are available. Thus the records may not be considered long enough to warrant further analysis. It is left to the reader to assess the plot using his or her personal notion of trend. Conclusions about changes over time in the past are drawn from plots of metal concentrations in samples of sediment cores versus estimated time and similarly from plots of inferred pH versus estimated time (Charles and Norton, 1986). There are many assumptions that go into such reconstructions and it is difficult to assess the magnitude of uncertainty due to the assumptions relative to the variability of points on the plot. The evaluation of sources of variability in the interval estimation of past pH from diatom frustules (Oehlert, 1988) provides an example of the types of considerations involved.

There are two aspects to good graphical representations. The first is the choice of data and presentation in a manner that is consistent with the design of the data collection and method(s) of analysis. Secondly, the actual form and specific details of the graphic determines how clearly information in the data is conveyed, and whether it is done without distortion. Cleveland (1985) deals with graphical methods and principles and with graphical perception, the scientific basis for these principles. Helsel and Hirsch (1992) discuss the subject of good presentation graphics for water resource applications. These two references should be consulted for this second aspect. Examples of plots that can be used to initially examine a data set are given here to illustrate the first aspect, namely, plots that represent data in a form which corresponds to the method being used, where the methods are some of those reviewed in later sections.

Notation and an example

Data from a programme where observations for a number of water quality indicators are collected from one location, with multiple observations within the year over several years, can be represented by (y_{ij}, t_{ij}, x_{ij}), where y_{ij} and x_{ij} are, respectively, the vector of p water quality indicators and the vector of q ancillary variables observed at the jth sample collection in year i, and t_{ij} is the day of the jth sampling in year i for $j = 1, 2, \ldots, m_i$ and $i = 1, 2, \ldots n$. This notation accounts for either unequal spacing between observations or event sampling. The case of one water quality indicator, one ancillary variable and monthly sampling reduces to the three scalars (y_{ij}, t_{ij}, x_{ij}) for $j = 1, 2, \ldots, 12$, $i = 1, 2, \ldots, n$ and $t_{ij} = j$.

Mean monthly total phosphorus (TP) concentrations (mg l^{-1}), nitrate nitrogen (NO$_3$-N) concentrations (mg l^{-1}) and discharge (m s^{-1}) in the Niagara River at Niagara-on-the-Lake (NOTL) measured from 1976 to 1992 provide the example. In the above notation, (y_{ij}, j, x_{ij}) represent the concentration of the water quality variable, the month j and the discharge in month j and year i for $i = 1, 2, \ldots, 17$. Plots of these data are given which: (1) illustrate the general features of the data (Figures 1–3); (2) display the data in a manner consistent with the treatment of season as a blocking variable (Figure 4); and (3) display the data for methods which model the seasonal cycle as a smooth curve (Figures 1 and 5).

An overview of the features of the TP and discharge time series, including the relationship between them, is given by plotting the data in the order of observation, where vertical lines are used to separate the years so that the seasonal pattern can be discerned (Figure 1). There are several features, salient to modelling the variability, which can be observed from this plot. The variability within each year is large and the differences between years are marked for both sets of measurements. The general form of the change within year, i.e. seasonal cycle, consists of a maximum for discharge and a minimum for TP in the middle months. The scatter about the seasonal pattern for TP due to year to year variability is shown in Figure 2.

A preliminary examination of the relationship between TP concentrations and discharge can be obtained from a plot of all the data (Figure 3). From Figure 2, where a distinct symbol is used for each year, it can be seen that high values, which lie above the broad horizontal scatter in Figure 3, occur between March and October or December 1990 or 1991, where the latter month is on the increasing arm of the seasonal cycle. Thus further plots of these variables for individual months or for groups of months representing phases of the hydrological cycle or some other suitable separation are indicated if the relationship between discharge and TP is to be better characterized.

In the next section, parametric and non-parametric methods for monotonic trend under seasonal blocking are discussed. These methods involve an assumption that the seasonal component for a specific season, e.g. January, if seasons are taken to be months, can be represented by an additive component that is constant over years. The methods use the data, not in the order of observation as shown in Figure 1, but rather as displayed in Figure 4, where for month j, y_{1j}, $y_{2j}, \ldots, y_{17j}$ are plotted against year $1, 2, \ldots, 17$. The trend over years can be assessed by examining the observations for all years made in a specific month and, assuming homogeneity of trend in all months (i.e. in each of the plots in Figure 4), the overall assessment is some consensus from these individual assessments. Thus Figure 4 shows the change over time in the subsets of data that are used to calculate the statistics for trend, and facilitates the detection of non-linear changes and heterogeneity of trend. The data for all years with observations in a particular month, as the vertical band at the location of that month, is shown in Figure 2. However, Figure 2 shows more about scatter and the central value for the month, which is represented as the seasonal component for model 3 considered in the next section, than about changes over time.

To visualize the data for the component models that are discussed under the section on trend as change over time, the data would be dispayed as in Figure 1. The component models considered formulate trend as the slowly varying change over years, the seasonal component as a more rapidly varying change and the remainder of the variation is treated as random variation. A first look at such components can be obtained

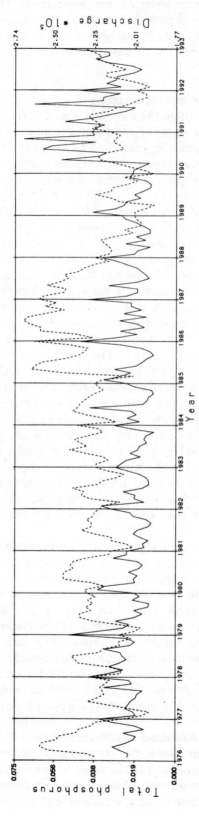

Figure 1. Temporal variability in total phosphorus concentrations, mg l^{-1}, (solid line) and discharge, m s^{-1} (broken line); the observations are joined by lines to aid in visualizing temporal changes

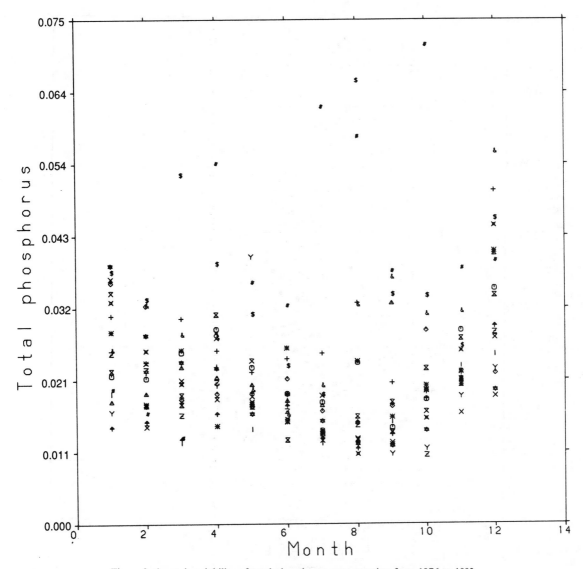

Figure 2. Annual variability of total phosphorus concentration from 1976 to 1992

by different degrees of smoothing. An example of this is shown for NO_3-N concentrations (Figure 5). The *lo*cally *we*ighted regression and *s*moothing *s*catterplot technique (LOWESS) of Cleveland and Grosse (1991) was applied to the original data (top plot) and to the difference between data and smooth (middle plot). The remainder after smoothing twice is given in the bottom plot. The middle plot shows that there is year to year variability, above that of the gradual change over time represented by the curve in the top plot, as well as a variable seasonal component.

TREND AS MONOTONIC CHANGE

The assumption of monotonic change is consistent with the general understanding of trend and provides results that are readily understood. For example, the statement that, for the River Thames, an upward trend is evident over the period 1928 to 1978 when the mean annual NO_3-N concentrations rose from $2 \cdot 5$ to $8 \cdot 0 \, \text{mg} \, \text{l}^{-1}$ (Meybeck *et al.*, 1989) implies a monotonic change. The statement gives complete information about the change in level in a qualitative way, i.e. without a measure of uncertainty, if the rate of

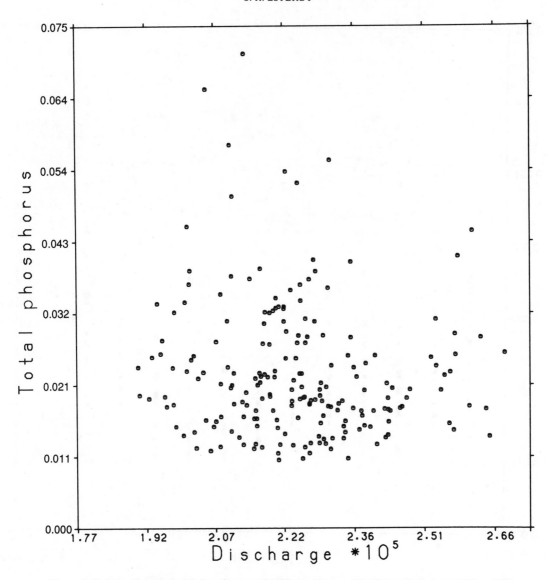

Figure 3. Relation between total phosphorus and discharge for monthly observations from 1976 to 1992

change is constant, i.e. linear trend. Detection methods that are based on the assumption of monotonic change are often chosen. If the magnitude of change is to be estimated, then further specification of the form of change is required. The possibilities are linear change, step change or multiple step changes in the same direction and a curve with a first derivative that does not change sign. The topics to be considered are listed in Table I.

DETECTION: NON-PARAMETRIC TESTS

Non-parametric methods have been favoured for the analysis of the large data sets from several national monitoring programmes. These programmes have now been operating long enough to allow an assessment of temporal trends. The dimensions of the data sets which are large are the number of water quality parameters measured and number of locations sampled, not the number of years. This has prompted a search for a method which is easy to use and widely applicable, so that the same method can be applied

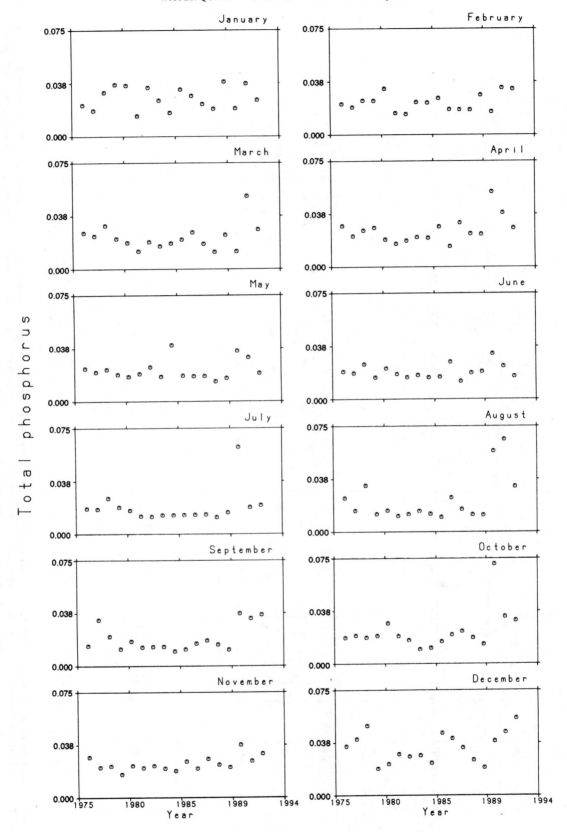

Figure 4. Variations in total phosphorus by month and year

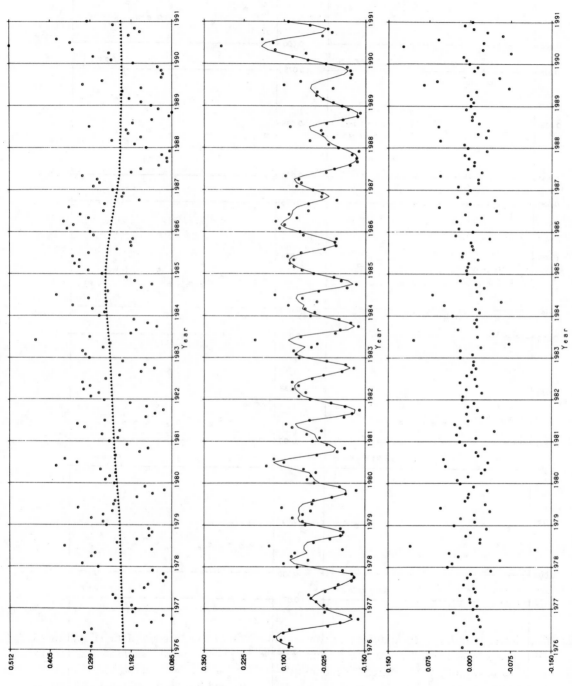

Figure 5. Observed nitrate nitrogen (○) and the LOWESS smoothed values (△) from low frequency smoothing are shown in the top panel. The difference (○) between the observations and the smoothed values shown in the top panel and the smooth of these differences (line), from moderate frequency smoothing, are shown in the middle panel. The remainder, i.e. differences between points and curve in the middle panel, are shown in the bottom panel

Table I. Topics related to methods for monotonic trends

Hypothesis testing for trend detection
    Monotonic trend in non-parametric methods
    Specific form of monotonic trend in regression
Estimation of change over time
    Specified form: linear, step change, curvature
    Time of step change
Testing homogeneity of trend across seasons
Performance in the presence of
    Serial correlation
    Missing data
    Values below the detection limit
Multivariate test for detection

---

to all data sets in a programme. Non-parametric methods involve fewer assumptions and are simpler to apply than parametric methods, as the latter require more assumption checking. There has been widespread use of the Mann–Kendall statistics and many papers investigating the properties and extensions following the introduction of a seasonal version of the statistic to the water quality literature by Hirsch *et al.* (1982). Methods based on the Spearman correlation coefficient have also been used (Lettenmaier, 1976; El-Shaarawi *et al.*, 1983; McLeod *et al.*, 1991). These non-parametric methods have high power (or efficiency) under both normality and departures from normality.

The focus of this section is on trend as a monotonic change and hypothesis testing as the method for trend detection. Consider the set of data consisting of the observations $y_{1j}, y_{2j}, \ldots, y_{nj}$ for season j in years 1 to n; for example, the data for TP in January shown in the first panel of Figure 4. The Mann–Kendall statistic for season j is:

$$S_j = \sum_{i < k} sgn(y_{kj} - y_{ij}) \tag{1}$$

where

$$sgn(x) = \begin{cases} 1 & x > 0 \\ 0 & \text{if} & x = 0 \\ -1 & x < 0 \end{cases} \tag{2}$$

S_j will be positive if the y_{ij} increase with i and negative if they decrease with i. The hypothesis tested is that $y_{1j}, y_{2j}, \ldots, y_{nj}$ are a sample of n independent and identically distributed random variables. The alternative hypothesis is that y_{ij}, y_{kj} are not independently and identically distributed for all i, k where $i \neq k$ (two-sided test). It is used as a test for trend because the statistic is powerful for departures from randomness in the form of monotonic change over time. It will be a test for trend free of seasonality if $y_{1j}, y_{2j}, \ldots, y_{nj}$ can be described by a model of the form

$$y_{ij} = \alpha_j + f(i) + \epsilon_{ij} \tag{3}$$

where the effect due to month j, α_j, is additive and constant over years 1 to n, because under Equation (3) the term α_j disappears from the difference in Equation (1). The trend component is indicated as $f(i)$ for generality as it is assumed only that the trend is monotone.

The test is also sensitive to serial correlation in the $y_{1j}, y_{2j}, \ldots, y_{nj}$ and has been evaluated as a test for such. Through a Monte Carlo experiment, Hipel *et al.* (1986) compared Kendall's τ [τ_j is equivalent to S_j, as $\tau_j = 2S_j/n(n-1)$] and the lag-1 serial correlation coefficient, r_1, as tests for trend against both purely deterministic and purely stochastic trends. Although r_1 was more powerful for the stochastic trends, τ

had a power as high as 0·80 for one type of stochastic trend. These results are worth noting because they emphasize the nature of the hypothesis test, i.e. a test of independence which is sensitive to departures in the form of monotonic change or serial dependence. They apply to non-seasonal data, such as $y_{1j}, y_{2j}, \ldots, y_{nj}$ and for this instance, independence will often be tenable as observations are a year or more apart. The effect of serial correlation between observations within a year will be considered in the following.

Seasonal test

The extension to seasonal data, known as the seasonal Kendall trend test (Hirsch *et al.*, 1982) uses the statistic

$$S = \sum_{j=1}^{m} S_j \tag{4}$$

that is, the sum of the statistic for individual seasons over all seasons and which, in terms of Figure 4, is the sum of the S_j over all panels in the figure. The hypothesis being tested includes the assumption that for season j, $y_{1j}, y_{2j}, \ldots, y_{nj}$ are independent and identically distributed, where $j = 1, 2, \ldots, m$, but also that y_{ij} for all i and j is an independent sample. In practical terms, this means that adjacent observations are assumed to be serially uncorrelated. This may be tenable for sampling intervals of two weeks or longer (Lettenmaier, 1976).

For data sets as small as $n = 2$ and $m = 12$, the normal approximation is adequate and thus the test is easy to use. The method also accommodates both (1) a moderate number of missing observations, in which instance the number of years with an observation in season j, n_j, is not equal to n for at least one j and (2) values below the detection limit, as the latter are treated as ties. For details on computation, performance and strategies when there is more than one observation per season in a year, see Hirsch *et al.* (1982).

The statistic S is an example of a weighted sum of Kendall τ's under blocking, where seasons form blocks of data. Alternatively, Spearman's correlation coefficient, r_s, could be used as the test statistic to calculate within blocks. The tests based on τ and r_s have been shown to have essentially the same power for $n_j = n$ (see van Belle and Hughes, 1984 for a summary) and $n_j \neq n$ (Taylor, 1987). van Belle and Hughes have also compared blocking with the alternative method of accounting for seasonality in which the seasonal mean is subtracted before calculating the test statistics (i.e. use $y_{ij} - \bar{y}_{.j}$, where $\bar{y}_{.j}$ is the mean for season j). However, this method has not been used extensively as missing data are difficult to handle.

Heterogeneity of trend

If there is heterogeneity of trend between seasons, because S is the sum of the S_j, the positive and negative values will tend to cancel and can result in a conclusion of no trend. Under such heterogeneity, the use of S is inappropriate. van Belle and Hughes (1984) show that the sum

$$\sum_{j=1}^{m} \left\{ \frac{S_j^2}{\mathrm{var}(S_j)} \right\} \tag{5}$$

can be partitioned into two components so that heterogeneity of trend can be tested before testing the hypothesis of monotonic trend in all seasons. Separate conclusions about trend in individual seasons or groups of seasons are appropriate under heterogeneity.

Serial correlation

The seasonal Kendall test for trend, modified to include an estimate of the covariance between the statistics for different seasons, S_j and S_k for $j \neq k$ (Dietz and Killeen, 1981), was given by Hirsch and Slack (1984). The performance of the original and modified seasonal Kendall tests was evaluated by a Monte Carlo study assuming autoregressive moving average processes with a range of parameter values and nominal significance levels. The modified test was shown to provide more accurate significance levels for $n \geq 10$

and short-term serial correlation typical of that found in an examination of the sample autocorrelation function for many water quality and flow data sets. Hirsch and Slack (1984) also showed that the improvement in the significance level was accompanied by a loss in power under independence. These properties were further documented by Taylor and Loftis (1989) in an extensive Monte Carlo study.

Zetterqvist (1988) and El-Shaarawi and Niculescu (1992) obtained the variance of the Kendall trend statistic and proved asymptotic normality of the test statistic for a strictly stationary process with continuous marginal distribution function under several different models for the dependence. Zetterqvist considered the seasonal Kendall trend statistic, S, and an m-dependent process for m smaller than half the number of seasons. This retains the assumption of independence of observations within a season, but allows for dependence between consecutive observations within a year for a lag $\leqslant m$. El- Shaarawi and Niculescu (1992) considered two cases. The first is the non-seasonal Mann–Kendall trend statistic and a moving average process of order 1 or 2. An example would be the observations within a season with serial correlation between these observations of the assumed form. The second case is the seasonal trend statistic with dependence between adjacent observations in a year and between observations in the same season which are one year apart. Through calculation of the exact variance in each case, assuming Gaussian moving average processes of the appropriate lag (Zetterqvist considered lags up to 3), both studies showed that substantial under-estimation or over-estimation of variance of the statistic occurs if independence is assumed in the presence of positive or negative serial correlation, respectively. El-Shaarawi and Niculescu (1992), by using Monte Carlo simulation to estimate the terms in the variance expressions for the Laplace and Cauchy distributions, also showed that the distribution had much less effect on the variance than did serial correlation.

In the presence of serial correlation of a particular form, as in the above three cases, the Kendall trend test is no longer non-parametric as the parameters defining the serial dependence must be estimated. The form of the monotonic trend must also be specified so that residuals can be obtained and the dependency parameters estimated from them.

Missing data

From the practical point of view, missing data have been handled by ignoring the missing data and calculating the test statistic as if the sample is complete. Alvo and Cabilio (1994) note that this is equivalent to defining a new time-scale on which the observed data is equally spaced. The trend tests based on the Kendall and Spearman statistics are obtained by taking the time ordering as the second variable. If a record extends over n years, but observations are available for only k years, the time vector forms a complete ranking of $1, 2, \ldots, n$, but there are not enough observations to provide a complete ranking $1, 2, \ldots, n$, for the variable of interest. Performing the test on the time vector $1, 2, \ldots, k$ ignores the missing data. Alvo and Cabilio define statistics that account for the missing data. Based on the asymptotic relative efficiency of the new Spearman statistic relative to that obtained by ignoring the missing data for testing the hypothesis of zero slope in a normal regression model, the new statistic is always more efficient. Performance depends on the number and location of missing data and further evaluation for particular examples is required.

Multivariate tests

As noted earlier, data sets typically include many water quality parameters and indications that water quality has changed may come simultaneously from some or all available parameters. This has lead to the investigation of multivariate techniques. Dietz and Killeen (1981) describe an extension of the Mann–Kendall statistic for multivariate data where the trend for individual variables may be in different directions. An example of the type of data assumed is given by the matrix $Y_j = \{y_{ijl}\}$ for season j, where the element y_{ijl} is the observation for water quality parameter l in season j of year i and $i = 1, 2, \ldots, n$ and $l = 1, 2, \ldots, p$. The test statistic is the quadratic form $S'V^-S$ of the vector, $S = (S_1, S_2, \ldots, S_p)$, of the Mann–Kendall statistics for the p parameters and V, the covariance matrix of S. Lettenmaier (1988), noting that the above test has poor power for record lengths typical of stream quality data, proposed as a test statistic the sum of the squares of the S_l for $l = 1, 2, \ldots, p$, which he called the covariance eigenvalue (CE) method. To account for the additional dimension of season, the Mann–Kendall statistics are calculated separately for each season within variable and then summed over season for each variable.

The performance of these multivariate tests has been evaluated for situations typical of river monitoring programmes (Lettenmaier, 1988; Loftis *et al.*, 1991a) and lake monitoring programmes (Loftis *et al.*, 1991b). Assuming a linear trend, normal or log-normal error distributions and between-variable and between-season correlations, the first two studies showed that the CE method was more powerful than the method of Deitz and Killeen [called covariance inversion (CI) by Lettenmaier] and, although both had empirical significance levels lower than the nominal, the CE method was less conservative than the CI method. Loftis *et al.* (1991b) evaluated modifications which retained dependence between variables, but assumed independence between seasons, the latter assumption having been shown to be tenable for the lake acidification data sets which motivated their study. The major result of the study was to show that these multivariate tests are more powerful than individual tests with nominal significance levels modified by a Bonferroni inequality.

The modified seasonal Kendall trend test given by Hirsch and Slack (1984) can also be viewed as a multivariate test and has been called the covariance sum test. It was evaluated in the above studies, but, as it does not allow for trends of different direction, it would not always be applicable. For example, Loftis *et al.* (1991b) report an increase in acid neutralizing capacity and decrease in sulphate in some lakes.

Smith *et al.* (1993) give a general framework for multivariate trend tests and show how the tests discussed here fit into this framework, as do tests based on the Spearman trend statistic. Further results are also given on testing the heterogeneity of trend, the null distribution of the CE statistic and the identification of variables that are important contributors to overall trend through canonical analysis.

ESTIMATION: NON-PARAMETRIC METHODS

First consider non-seasonal data such as the observations over years in season j given by $y_{1j}, y_{2j}, \ldots, y_{nj}$. The Mann–Kendall statistic, S_j, defined by Equation (1), is based on the sign of the differences $y_{kj} - y_{ij}$. Estimates are available for the slope parameter of a linear trend (Theil, 1950; Sen, 1968) and the magnitude of the step in a step change, based on these differences, but now the magnitude of the difference is taken into account. The slope estimator, B_j, for season j is the median of the $n(n-1)/2$ quantities $(y_{kj} - y_{ij})/(k - i)$ for $i < k$ and $i, k = 1, 2, \ldots n$. This can be extended to seasonal data (Hirsch *et al.*, 1982) by calculating these quantities for each season, i.e. for $j = 1, 2, \ldots, m$, and then taking the median, B, over all $mn(n-1)/2$ quantities (number of quantities equals $\sum n_j(n_j - 1)/2$ for $j = 1, 2, \ldots, m$ if $n_j \neq n$ for all j). Gilbert (1987) gives an approximate confidence interval for the true slope.

Non-parametric estimation of the magnitude of a step change in a water quality variable has been considered by Hirsch (1988) and the seasonal Hodges–Lehman estimator, Δ_{SHL}, which is the analogue of B, was found to perform well. Suppose that a step change occurred after year c. Then the objective is to estimate the change in level between the two periods $i = 1, 2, \ldots, c$ and $i = c + 1, \ldots, n$. For season j, all possible differences $(y_{kj} - y_{ij})$ are calculated, where $i = 1, 2, \ldots, c$ and $k = c + 1, \ldots, n$, and this is done for $j = 1, 2, \ldots, m$. Δ_{SHL} is the median of these differences, with $mc(n - c)$ differences if $n_j = n$ for all j. Helsel and Hirsch (1992) give methods for calculating the confidence interval for the change in level.

A corresponding non-parametric method of testing for the existence of a step change and estimating the point of this change is given by Pettitt (1979). Although this has not been applied to water quality variables, it has been used to test for changes in diatom concentrations in sediment cores (Esterby *et al.*, 1986). Even when the time of an intervention is known, it may be appropriate to estimate the time at which a change occurs in the variable being followed because there may be a lag in the response to the intervention.

Censored data, missing data, non-monotonic change

It was noted earlier that, in the statistics calculated for hypothesis testing, values below the analytical detection limit can be treated as ties and thus pose no added problem. In estimation, this is not the case because the numerical differences are used. For a small number of observations reported as below the detection limit, Gilbert (1987) suggests assigning these observations a value equal to half the detection limit. Helsel and Hirsch (1992) suggest calculating B by setting observations below the detection limit to zero,

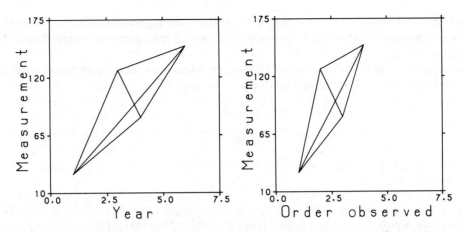

Figure 6. The left-hand panel shows all possible terms contributing to the non-parametric slope estimator, B, with individual terms represented by lines joining the observation pairs (t_i, y_i) (modified from Helsel and Hirsch, 1992); $t = (1, 2, 3, 4, 5, 6)$ and $y = (28, -, 127, 82, -, 150)$ with dashes indicating missing data. The right-hand panel shows the six terms when missing data are ignored; $t = (1, 2, 3, 4)$ and $y = (28, 127, 82, 150)$

then by setting them to the detection limit and using this method only if the two values of B do not differ substantially. This also applies to calculation of Δ_{SHL}. Bias induced in the estimation of means by replacing values below the detection limit by a constant between zero and the detection limit has been evaluated both by Monte Carlo methods (e.g. Gilliom and Helsel, 1986) and analytically (El-Shaarawi and Esterby, 1992) and similar considerations apply here.

Missing data are accounted for directly in the trend estimators since the differences between a pair of observations is divided by the difference in the time of observation, i.e. the individual terms (slopes) are given by $(y_{kj} - y_{ik})/(k - j)$. The left-hand panel in Figure 6 shows all possible slopes which contribute to the slope estimator, B, for a set of observations where data are missing at two of six observations. In the right-hand panel, observations on either side of a data gap are treated as contiguous, as in the trend statistic S_l, to illustrate how this alters the time-scale. This was discussed earlier under the topic of missing data as it affects hypothesis tests.

Non-linear monotonic change could be detected by the non-parametric tests of the hypothesis of randomness. El-Shaarawi and Niculescu (1993) provide a test for detecting non-linearity. The hypothesis that the form of change is linear is tested with a statistic obtained by calculating the Mann–Kendall statistic on the first differences of the observations. Thus for season j, the first differences $z_{ij} = y_{i+1j} - y_{ij}$ for $i = 1, 2, \ldots, n - 1$ are used to calculate the S_l in Equation (1). The variance of S_l is that of a stationary moving average process of order one (El-Shaarawi and Niculescu, 1992) and thus a statistic of the usual form is available. Normality of the test statistic is proved and, by Monte Carlo simulations, the approximation is shown to be good for $n \geqslant 20$ and the test is shown to be moderately sensitive to non-normality.

A non-parametric estimate of the magnitude of change, analogous to the estimates for linear and step changes, could be obtained when the change is in the form of a non-linear curve. A reasonable alternative is to characterize the change by a non-parametric smoothing procedure such as LOWESS (Cleveland and Grosse, 1991).

PARAMETRIC METHODS

Normal least-squares regression is the primary focus of this section, particularly models restricted to monotonic trend and to those which could also be dealt with by the above non-parametric methods. Some more general situations that arise naturally from the regression method being discussed are also included. The seasonal Kendall test for trend is free of the effects of seasonality if the data can be described by a model such as Equation (3). Thus only regression models with such an additive seasonal effect are considered. Detection and estimation are not separated, as was done for the non-parametric methods, because the form of the trend component must be specified to perform the regression analysis and thus testing the hypothesis about the existence of trend is not based on a more general definition of trend than the definition used in trend estimation.

For season j, the monotonic trends considered here, i.e. linear, non-linear and step change, can be modelled, respectively, as the following modifications to Equation (3)

$$
\begin{aligned}
y_{ij} &= \alpha_j + \beta_j i + \epsilon_{ij} \\
y_{ij} &= \alpha_j + \beta_{j1} i + \beta_{j2} i^2 + \epsilon_{ij} \\
y_{ij} &= \alpha_j + \Delta + \epsilon_{ij} \ i \leqslant c \\
&= \alpha_j + \epsilon_{ij} \ i > c
\end{aligned}
\tag{6}
$$

where the second equation will be monotonic only for restricted values of i and of the parameters. Homogeneity of trend can be tested by partial F-tests (Draper and Smith, 1981) for the model with seasonal trend parameters relative to a model with the same trend parameter(s) for all seasons. Under homogeneity, a common linear trend for all seasons would be obtained from the model

$$
y_{ij} = \alpha_j + \beta i + \epsilon_{ij}
\tag{7}
$$

Alternatively, $m(i - 1) + j$ may be used instead of i as the time variable for the trend components in the first two equations in Equation (6) and in Equation (7).

Under the null hypothesis for the seasonal Kendall trend test the $y_{1j}, y_{2j}, \ldots, y_{nj}$ are independent and identically distributed, and y_{ij}, for all i, j, are independent. The assumption here is that the ϵ_{ij}, for all i and j, are independent and normally distributed with 0 mean and constant variance. This involves the additional assumptions of normality and homogeneity of variance over all seasons. Techniques which are useful when these assumptions are violated are weighted regression and transformations. Serial correlation can be accommodated by weighted least squares or a correction to the F statistic, but as with the non-parametric methods, is more problematic than other data features. Moderate missing data do not pose a computational problem and unequal spacing is incorporated directly as the actual spacings on the time-scale, i.e. t_{ij}, are used in the regression computations. In the presence of values below the detection limit (type I censoring in the statistical literature), regression parameters are estimated by iterative maximum likelihood methods (Aitken, 1981). Multivariate regression can be used for trends and Smith *et al.* (1993) show that the non-parametric multivariate test for trend can be thought of as an extension of the multivariate trend test under normality. An alternative to the simultaneous estimation of seasonal and trend parameters is to deseasonalize the data by subtracting the season mean from the data for that season before estimating trend. This has been used most frequently when serial dependence is present.

A method for estimating the point of change in a regression relationship using likelihood methods (Esterby and El-Shaarawi, 1981) has been shown to be useful for environmental variables. Kawamura *et al.* (1986) derived a method of detecting an abrupt change and estimating the state parameters using the adaptive Kalman filter. MacNeill *et al.* (1991) considered methods for detecting change-points in both mean level and serial correlation structure. MacLeod *et al.* (1983) illustrate the use of intervention analysis to estimate the magnitude of the change in the mean level of a water quality variable when an intervention such as clear-cutting has occurred at a known time in a river basin.

Taylor and Loftis (1989) provide a good description of the analysis of covariance, give a method using t-statistics which accommodates seasonal differences in variance and suggest performing the regression analysis on the ranks of the water quality variable. Through Monte Carlo simulations, regression methods have been shown to perform poorer than non-parametric tests under conditions that violate the assumptions of regression (e.g. Hirsch and Slack, 1984; Taylor and Loftis, 1989). However, under the conditions of the simulation of Taylor and Loftis, neither parametric nor non-parametric methods performed well in the presence of serial correlation. In particular cases, regression methods may be satisfactory. Esterby (1993) gives an example where the same conclusions were drawn from regression, assuming a linear trend and blocking on season, as from the seasonal Kendall trend test and very similar parametric and

Table II. Models for trend methods with data blocked by season and assuming monotonic trend

Feature	Description	
Seasonality	Additive seasonal effect, constant for season over all years	
Trend	Step change or step changes in the same direction, linear change or limited curvature	
	Non-parametric	Parametric
Distributional assumptions	Observations identically distributed within season, independent overall	Errors independent and normally distributed with mean zero and constant variance
Serial dependency	Results available for some models	General structure through weighted least squares

non-parametric confidence intervals for the slope were obtained. Stensland *et al.* (1986) report linear trends in ions measured in bulk precipitation samples collected at Hubbard Brook, where the analysis was performed by least-squares regression on seasonally adjusted values (monthly mean subtracted) and the standard errors were corrected for autocorrelation. An example of regression analysis on the ranks of a water quality variable is given by the analysis of Stoddard (1991), who estimated time trends in water quality variables in New York streams which were potentially affected by acidic deposition.

Generally, regression methods have not been the method of choice for the large data sets from monitoring programmes where the number of years is the smallest dimension and the type of method sought has been one that is easy to use and of general applicability. A complete regression analysis includes diagnostic tests to check model assumptions and model reformulation, if necessary, and will often be impractical for the routine analysis of large data sets. However, the flexibility of regression models potentially make them a better tool for the detailed analysis of data and they will be considered again in the following. A brief summary of the models is given in Table II

TREND AS CHANGE OVER TIME

With seasons treated as blocks as above, the statistics for the individual seasons were calculated from the n observations in season j, $y_{1j}, y_{2j}, \ldots, y_{nj}$, $j = 1, 2, \ldots, m$, and, under the assumption of homogeneity of trend over seasons, the statistics were combined to give a summary for all seasons. Consider now the data in the order of observation (Figure 1), i.e. $y_{11}, y_{12}, \ldots, y_{1m}, y_{21}, y_{22}, \ldots, y_{2m}, \ldots, y_{n1}, \ldots, y_{nm}$. Decomposition models more general than Equation (3) can be represented by

$$y_{ij} = T_{ij} + S_{ij} + R_{ij} \tag{8}$$

where y_{ij} can be thought of as a sum of contributions due to various factors, here trend, T_{ij}, season, S_{ij}, and remainder or residual, R_{ij}. The trend and seasonal components are usually taken as deterministic components and the remainder as random, either independent or serially dependent. Parametric models may be assumed for the trend and seasonal components, or they may be represented by curves from smoothing procedures where the separation of the components arises through different degrees of smoothness. Trend is the slowly varying component and thus the smoothest, the seasonal component is the next smoothest and the remainder is rough. There is also an assumption that the seasonal component is cyclical and has a period of one year. Thus the trend is defined relative to the other components. An explicit form for the trend may not be specified at the outset or that specified may be modified as need is indicated from diagnostics in the method of fitting the model. If models for serial dependence are to be fitted, equally spaced observations will be needed for most methods.

There are many situations where a flexible approach to assessing change is appropriate. A non-monotonic change may have occurred in a data set, which will not be detected by methods sensitive only to monotonic change, and although only a few years of data are available, it is important to know if any change appears to be occurring. When a long record is available, it is reasonable to expect that it contains more information about change over time than a shorter record and more effort may need to be expended to extract the information. Methods which model each of the components in Equation (8) are capable of providing more information and are thus preferable when they are applicable.

Several examples are given here of the application of methods which are more general in one or more ways than those considered in the previous section. The estimation of a curvilinear trend over years within the framework of blocking is a simple extension. Liu (1992) fitted a quadratic (U-shaped) model for the time trend in monthly volume-weighted means of sulphate concentrations in wet deposition samples. The seasonal component was removed rather than modelled as the data was first deseasonalized by subtracting the mean for the appropriate season. Autoregressive moving average models were fitted to the residuals.

A smooth seasonal component, as provided by harmonic terms, may be more realistic than an additive constant, i.e. α_i, for season j in Equation (3), and has been used in the analysis of physical measurements such as temperature (McMichael and Hunter, 1972; Neilson and Hsieh, 1982) and chemical water quality parameters (El-Shaarawi *et al.*, 1983). An additive model for monthly data, which assumes a stable annual cycle and a linear trend, could be represented by

$$y_{ij} = \mu + \beta_1 \cos\frac{2\pi j}{12} + \beta_2 \sin\frac{2\pi j}{12} + \omega\left(i + \frac{j}{12}\right) + \eta_{ij} \tag{9}$$

where the second and third terms provide for a cyclic seasonal component, ω is the slope and η_{ij} are the residual terms. Additional harmonic terms, other forms for non-seasonal deterministic terms and independent or dependent errors can be accommodated in such models. Reinsel and Tiao (1987) found that a model with annual and semi-annual seasonal components, a two-phase trend component and η_{ij} modelled as an autoregressive process provided a good description of the variability in monthly averages of stratospheric total ozone. Esterby *et al.* (1992) showed that, although a significant linear trend was found in specific conductance at several stations on a river system when the non-parametric and regression methods of the previous section were used, the data could be better fitted by a model with yearly mean and yearly harmonic seasonal components. The residual diagnostics indicated that there were marked departures from linearity in the trend over years. The final conclusion was that several of the early years had significantly higher levels than the other years in the period of observation. Multiple comparison procedures were used on the estimates of the yearly mean parameters in the model.

There are water quality parameters for which harmonic components will not provide an adequate fit to the seasonal component and a more general method for modelling seasonal variation will be needed. One solution is to use a smoothing procedure such as LOWESS (Cleveland and Grosse, 1991) or moving averages. Esterby (1993) showed that a seasonal trend decomposition procedure, STL, based on LOWESS (Cleveland *et al.*, 1990), provided a reasonable representation of the seasonal and trend components for the specific conductance data of one of the stations mentioned earlier (Esterby *et al.*,1992). The trend component, although smooth in this analysis, was consistent with the previous finding of several years being higher than the others. More smoothing of the trend led to behaviour in the residuals similar to that observed when a linear trend, instead of yearly means, was used in the regression analysis. Quantification of the magnitude of change relative to the residual component will involve modelling of the latter component, and although this was not done for the specific conductance data, the feasibilty of using STL together with a standard ARIMA model was noted by Cleveland *et al.,* (1990).

Longer records, with more frequent sampling (typically daily), are available for variables such as temperature, discharge and biological oxygen demand (BOD). Such records have been found to be well characterized by Fourier harmonics for the cyclical deterministic components and ARIMA models for the random components (McMichael and Hunter, 1972; Mehta *et al.*, 1975; Neilson and Hseih, 1982). The seasonal (12-month) component was dominant and the combination of deterministic and random components were able to account for in excess of 80% of the variation. BOD *loading* could be modelled by this approach, but not *concentration*. The purpose of these studies was characterization for forecasting and simulation. However, the results provide insights useful in component modelling for trend analysis.

Measurements or observations which do not require modern technology may provide longer records that can be used to infer changes in environmental conditions. An example is river height as a response to deforestation. Brillinger (1989) used a 100 year record of daily stage on the Amazon River to illustrate a test for monotonic trend suitable for a long series with stationary autocorrelated error component. A model for the response at time t of the form

$$Y(t) = S(t) + E(t) \qquad (10)$$

is assumed, where $S(t)$ is a deterministic component and $E(t)$ is a zero mean stationary error series. A test statistic and asymptotic distribution of the statistic are provided for a test of the hypothesis that $S(t)$ is constant against the alternative $S(t)$ that is monotonically increasing. The statistic used is the linear combination $\sum c(t)Y(t)$ where $c(t)$ is defined such that it is relativley large and negative at the beginning of the series and large and positive at the end of the series. The error series is estimated by the residuals $\hat{E} = Y(t) + \hat{S}(t)$, where $\hat{S}(t)$ is a running mean and the residuals are used to calculate an estimate of the error power spectrum and the variance of the statistic. The analysis of the Amazon River stage data was further considered in Brillinger (1994).

Unequal spacing and variable seasonal components

The issue of equal spacing of observations was mentioned under methods for serially dependent data and under the topic of missing data in the non-parametric trend analysis methods. It is clear that missing data haphazardly change the scale of observation. What is not so obvious is the effect of arbitrarily taking samples equally spaced in time as being on a meaningful scale when hydrological events occur at different times in different years. Esterby (1993) discusses this with respect to treating season as a block and shows that one feature of the yearly variation, the month of minimum specific conductance, varies from April to July over the 15 year record. If the seasonal effect for season j varies from year to year, the differences, $y_{ij} - y_{kj}$, will include differences due to season. Regression and smoothing methods which use the day of observation were shown to be an alternative to using blocking on season when data are collected monthly, but the interval between samples is not exactly the same. The advantage to such an approach is modelling the seasonal cycle for each year as this incorporates the variability of the hydrological cycle. The major disadvantage is complication of the statistical analysis, particularly if serial dependence exists.

This topic extends to sampling design and the methods of fixed-interval and event-based sampling. Consider an analogous, but simpler, situation. In sampling from two media of unequal homogeneity, more samples would be taken from the more heterogenous medium to obtain estimates of the mean level of each medium with equal precision. Fixed-interval sampling, unless it is of adequate frequency, will result in a poor representation of the changes in concentration at the time of year when large changes happen quickly. Thus for parameters where this situation occurs there are several considerations, including logistical and statistical, which need to be balanced.

INCLUSION OF CONCOMITANT INFORMATION

It is often important to include other variables in the analysis, e.g. when it is the anthropogenic component of the change over time that is to be estimated or when variability due to other factors needs to be removed from the error to detect small changes. Examples of the relationship between the variable of interest and another variable are the relationship between ion concentration in precipitation and the amount of precipitation, and the concentration of a water quality parameter and streamflow. To illustrate what is to be accomplished, consider a simple example. Denote by X the variable to be followed for change over time, where X is measured in a system with different states. The states are such that the same value of X would appear to be different in the various states. If measurements $x_1, x_2, \ldots, x_n$ are made at times $t_1, t_2, \ldots, t_n$, then to obtain an estimate of change over time, free from bias due to modifying factors, the natural state of the system being monitored should be the same at each time t_i. If the state of the system cannot be

Table III. Methods for including covariates in trend analysis

---

Direct inclusion in analysis, e.g. multiple regression
Adjusting variables before analysis for trend
Choice of basis for variable, e.g. flux instead of concentration
Constant value of concomitant variable through
    Choice of time of sampling
    Choice of observations used in analysis

---

controlled, as is the case here, adjustments can be made to attempt to approximate this condition of constant state.

Various techniques of accounting for such relationships are available (Table III). Of the trend methods considered here, LOWESS and the parametric methods allow the inclusion of covariates directly as independent variables. Adjustment of variables is required before calculating the Mann–Kendall or seasonal Kendall statistics or the slope estimators, B or B_j. Alley (1988) and Smith and Rose (1991) compare procedures where adjustment for the covariate was (1) to the dependent variable only and (2) to both dependent variable and time. Through Monte Carlo experiments, Alley (1988) showed that method (1) underestimates the magnitude of a linear time trend and is thus less powerful than method (2) for both regression and the application of Kendall's test to residuals. Smith and Rose (1991) noted that Alley's Monte Carlo experiments were for the case where the covariate was not correlated with time and, through further Monte Carlo experiments, showed that the power to detect a linear time trend in the dependent variable decreased for both stepwise and stagewise regression as the correlation of the covariate with time increased, but the reduction in power was greater for stagewise regression. They also showed, by example, that the Kendall correlation under (2) need not always be greater than under (1). McLeod et al. (1991) illustrated the use of the Spearman partial rank statistic, which provides the analogue of the Pearson partial correlation, but applied to the ranks.

For the case of trends in river water quality, although discharge has been used extensively as a covariate, there is no simple solution because the relationship between concentration and streamflow varies from stream to stream, season to season and parameter to parameter. Such variability has been studied and two examples are Teti (1984) and Mohaupt (1986). Hirsch et al. (1991) discuss this question in some detail and advise that flow adjustment should not be used where human activity has altered the probability distribution of discharge. Zetterqvist (1991) considers the problem of separating the effect of a number of factors on the single observable series $\{x_t\}$ and describes a component model. The observable series is taken as the sum of unobserved series which describe the factors, natural, man-made and random, operating on the water quality variable to produce the observed series. Interventions, activities of a different nature which affect the random error component (e.g. activities which give permanent change and activities whose effect disappears rapidly) are included and the model is applied to phosphorus concentrations in the Ljungbyån River, Sweden. In another approach to the problem of several factors affecting the variable of interest, Peters and Turk (1981) used the criterion that a major change in an individual ion should indicate a specific cause, whereas a change in all ions at once suggests some general factor, such as climate. This was implemented by fitting the relationship between an individual ion and an index variable, specific conductance, using linear regression and then looking for time changes in the relationship for the individual ions.

Similar considerations apply when the concentration of ions in deposition samples are to be used to follow deposition changes over time. Stein et al. (1993), in an analysis to estimate the effect of emission changes on the sulphate concentration in precipitation, use a regression model of the sulphate concentration on a particular day as a function of the time of year and amount of precipitation. Eynon and Switzer (1983), in modelling rainfall acidity using daily pH values for days with measurable rainfall, make a correction for rainfall volume that is based on a simple scavenging mechanism. Switzer (1988) further considers models for the scavenging mechanism.

Table IV. Summary by location-wise trend statistics versus trend in year-wise summary of amount (or concentration) over the region of interest

	Location				Amount for year i	Statistic for trend in amount over time
	1	2	$...j$	$...p$		
Year 1	y_{11}	y_{12}	$...y_{1j}$	$...y_{1p}$	$\hat{a}_1$	
2	y_{21}	y_{22}	$...y_{2j}$	$...y_{2p}$	$\hat{a}_2$	
.	.	.	...	...	.	tr_a Calculated from
i	y_{i1}	y_{i2}	$...y_{ij}$	$...y_{ip}$	$\hat{a}_i$	statistics for year
.	.	.	...	...	.	in the previous column
n	y_{n1}	y_{n2}	$...y_{nj}$	$...y_{np}$	$\hat{a}_n$	
Result of trend test or estimate	tr_1	tr_2	$...tr_j$	$...tr_p$		

Summary over all locations:
(a) $\bar{tr}$ obtained by summing over locations
(b) Number of locations
$n_p = $ # positive trend
$n_n = $ # negative trend
$n_0 = $ # no trend

SPATIAL–TEMPORAL DATA

In monitoring networks, data are collected at several locations over time. Although the primary focus is time trend, at some level of reporting the spatial position will enter. There are two situations to consider (Table IV). The first is the case where a summary is to be made of the trend analyses performed separately at each sampling location. Typically, the types of changes over time are classified and the number of locations exhibiting each type of change is reported. The second situation arises when sampling over time is performed at several locations in a continuous medium, such as the atmosphere or a large body of water, and areal or volume estimates of the environmental parameter at a given time are required. There are occasions where summaries of trend analyses at individual locations may be all that is required in the second situation. In determining the procedure for the collection of an individual sample — for example, a sample of water from a river — spatial and temporal variability must be considered, but this is at another scale and is not the topic here. It is assumed that the sampling procedure has been designed to account for small-scale temporal and spatial heterogeneity, e.g. cross-sectional and depth heterogeneity within a river.

Examples of summaries of temporal trends at individual locations over a geographical region (location-wise statistics in Table IV) are given by Alexander and Smith (1988) and Lettenmaier et al. (1991) by showing the significance level for a Kendall trend test on the data from a particular location at that location on a map using different symbols for different levels of significance. Lettenmaier et al. (1991) also tabulate the the proportion of locations showing positive, negative or no trend, and Alexander and Smith (1991) sum statistics across stations in a spatially aggregated moving window seasonal Kendall test. Sandén and Rahm (1993) report the results of the analysis of nutrient concentrations measured at nine locations in the Baltic Sea between 1968 and 1990. The seasonal Kendall test for trend was applied separately for each sampling depth at each station and the degree of significance by depth is shown at the location of sampling on a map of the Baltic Sea. The estimates of slopes from linear time-trend analysis for ions in bulk deposition samples obtained by Barnes et al. (1982) are given in tabular form by Stensland et al. (1986).

Regional estimates are important for environmental variables when effects are expected to be exhibited on a regional basis. Thus more is needed than the estimates at the monitoring locations. The results of many studies involving regional estimates have been given as hand-drawn isopleths or computer-interpolated maps. Bilonick (1985) gives several examples of such methods for sulphate concentrations and deposition in North America. A common difficulty when these methods are applied is the failure to report the method used for obtaining the estimates and the lack of a measure of reliability of the maps. Methods which model the spatial-temporal correlation structure of the data have the potential to overcome these inadequacies. Several examples showing progress in handling this type of data are given below.

Rainfall–event pH measurements were modelled as the sum of spatial–temporal stochastic components and deterministic components for seasonal variation and rainfall washout by Eynon and Switzer (1983). The model was used to obtain contour maps of seasonal and rainfall-adjusted average pH over the monitoring region in the eastern USA. Bilonick (1985) applied three-dimensional universal kriging (time and two space dimensions) to two USGS sulphate deposition data sets collected in 1965 to 1979 and 1980 and 1981. Maps were generated for sulphate deposition, standard deviation by year and yearly average deposition for New York State for 1966 to 1979. The spatial–temporal semi-variogram exhibited a seasonal time component and an isotropic spatial component. Examples of contributions to the understanding of the components in spatial–temporal models and to the method development are given by Egbert and Lettenmaier (1986) and Haas (1990), respectively. Egbert and Lettenmaier (1986) analysed 1980–1981 NADP data and found substantial spatial correlation at weekly and longer time-scales which was anisotropic, but found little or no temporal correlation. Haas (1990) proposed a moving window procedure for ordinary kriging, which implicitly accounts for spatial covariance non-stationarity and evaluated the method for NADP data. Although the last three examples deal with spatial and short-term temporal structure, they contribute to the base of knowledge required to model long-term regional time trends.

DISCUSSION

Many of the references cited here are concerned with providing the environmental scientist with methods suitable for a particular problem. Several books (Gilbert,1987; Helsel and Hirsch, 1992; Hipel and McLeod, 1994) and papers (e.g. Berryman et al., 1988; Hirsch et al., 1991; McLeod et al., 1991) provide advice on the choice of methods as well as describing the methods. Some of the papers deal with the evaluation of methods under assumptions chosen to represent a particular application and others provide new methodological developments motivated by environmental applications. Table V gives an overview of the features of the methods discussed herein.

In the introduction, two inter-related issues were raised, specifically the need to be precise when using the term trend and the view that trend analysis involves a broader assessment of variability. Throughout the paper the methods have been grouped on the basis of the model for trend and other forms of variability. These are important considerations. There are implications in terms of scientific integrity and the economics of monitoring programmes. With respect to the first, it is important to report defensible conclusions, i.e. conclusions that will not be shown to be wrong if more data are obtained, although they may be shown to be incomplete. More data may be in the form of either longer records or other variables.

Currently, with records of 10–20 years, it is possible to model short-term trends and seasonal components within the year. With longer records, additional scales of temporal variation may need to be modelled — for example, those induced by long-term climatic cycles. Hirsch and Peters (1988) discuss the difficulty of interpreting trends detected from short records. By subsampling five year intervals of a 17 year record of sulphate deposition data, they illustrate that it is possible to detect trends of the opposite direction in the short term from the trend determined from a longer data set. The difficulty illustrated by this example can be overcome to some degree by clearly reporting the specific form of trend assumed, the assumption about what constitutes a measure of variability for judging whether the magnitude of the trend is important, and the characteristics of the data set. Further, the general understanding of the word 'trend' tends to imply something long term or continuing, and the wrong conclusion maybe drawn more by implication if specifics are not given.

The problem of what constitutes a measure of variability for judging whether the magnitude of the trend is important is inherent to observational data sets such as those obtained from monitoring programmes. It stems from the inability to vary some factors while holding others constant, as can be done in controlled experiments. Answers will be specific to particular problems, but this is an area which warrants further

Table V. Summary of the major strengths and limitations of methods for temporal trends

Method	Features	Strengths	Limitations
Seasonal Kendall trend test, S, and associated estimators, B and Δ_{SHL}*		Robust against departures from normality Resistant to outliers Easy to use and summarize On the basis of above, provide widely applicable procedure	Cannot simultaneously include covariates Cannot accommodate complex situations
S, B and Δ_{SHL}*	Assumption of independence		Only some results available. Methods lose advantageous features under dependence
S, B and Δ_{SHL}* and regression from model (3)	Block by season	Separate results available for each season plus over all seasons (latter under homogeneity)	Seasonal effect not removed if it varies year to year
B and Δ_{SHL}*	Use differences of observations	Robust to extremes (due to use of median)	Must provide a number for non-detects. Not a limitation for S which uses ranks
Decomposition models; see Equation (8)		Flexibility with respect to forms of trend, seasonality and inclusion of covariates	More effort required than for S, B and, Δ_{SHL}
Regression	Distributional assumptions	Many diagnostics available for model checking Methods available for handling dependency	More effort required to meet assumptions: not always possible
Smoothing		Flexibility	Still largely descriptive

* Procedures based on Spearman rank correlation have similar properties to S, B and Δ_{SHL}

work. Understanding how the model underlying the data analysis apportions variability to different components is a good start.

Progress towards answering the question of how to interpret trends from data records of the moderate length currently available is possible by modelling the anthropogenic component of the record with the help of covariates or by being able to relate the observed quality variable levels to known emissions, or both. This takes the analysis beyond simple hypothesis testing to more complex modelling, hypothesis testing and estimation. The chance of omitting a factor, the inclusion of which would lead to a different conclusion about changes over time, will be lessened by performing an analysis which accounts for as many sources of variability as possible. However, it will often be the case that important variables are identified only after the analysis of data collected over several years.

A comprehensive analysis of the data is also consistent with obtaining the maximum information from the data, which is an important consideration when the amount of resources invested in collecting the data is taken into consideration. Such analyses require that the appropriate data have been collected. Some of the recommendations made by Stensland *et al.* (1986: 195), with some modification to make them more generic, apply to monitoring programmes in general, i.e. monitoring of variables that will characterize climatologically or other naturally induced variation in the parameter of interest, an emphasis on network design that will ensure the capability of detection of changes of the type and size that are of interest, devotion of more effort to the analysis of available data and, as an integral part of the monitoring programme, establishment of a mechanism that ensures, not only the collection of quality assured data, but also the analysis and reporting of the results.

ACKNOWLEDGEMENTS

The author thanks K. Kuntz for helping to make the Niagara River data available, V. Bhavsar for preparing the plots, the staff of the Water Quality Branch — Ontario Region and National Water Quality Laboratory involved with collecting and analysing the samples and two referees for their helpful comments.

REFERENCES

Aitken, M. 1981. 'A note on the regression analysis of censored data', *Technometrics*, **23**, 161–163.

Alexander, R. B., and Smith, R.A. 1988. 'Trends in lead concentrations in major U.S. rivers and their relation to historical changes in gasoline-lead consumption', *Wat. Resour. Bull.*, **24**, 557–569.

Alley, W. M. 1988. 'Using exogenous variables in testing for monotonic trends in hydrologic times series', *Wat. Resour. Res.*, **24**, 1955–1961.

Alvo, M., and Cabilio, P. 1994. 'Rank test of trend when data are incomplete', *Environmetrics*, **5**, 21–27.

Barnes, C. R., Schroeder, R. A., and Peters, N.E. 1982. 'Changes in the chemistry of bulk precipitation in New York State, 1965–1978', *Northeast. Environ. Sci.*, **1**, 187–197.

Berryman, D., Bobée, B., Cluis, D., and Haemmerli, J. 1988. 'Nonparametric tests for trend detection in water quality time series', *Wat. Resour. Bull.*, **24**, 545–556.

Bilonick, R. A. 1985. 'The space–time distribution of sulphate in the northeastern United States', *Atmos. Environ.*, **19**, 1829–1845.

Brillinger, D. R. 1989. 'Consistent detection of a monotonic trend superposed on a stationary time series', *Biometrika*, **76**, 23–30.

Brillinger, D. R. 1994. 'Trend analysis: time series and point process problems', *Environmetrics*, **5**, 1–19.

Charles, D. F., and Norton, S.A. 1986. 'Paleolimnological evidence for trends in atmospheric deposition of acids and metals' in *Acid Deposition, Long-Term Trends*. National Academy Press, Washington. pp. 335–431.

Cleveland, R. B., Cleveland, W. S., McRae, J. E., and Terpenning, I. 1990. 'STL: A seasonal-trend decomposition procedure based on loess', *J. Off. Stat.*, **6**, 3–73.

Cleveland, W. S. 1985. *The Elements of Graphing data*. Wadsworth Advanced Books and Software, Monterey, p. 323.

Cleveland, W. S., and Grosse, E. 1991. 'Computational methods for local regression', *Statistics Computing*, **1**, 47–62.

Dietz, E. J.. and Killeen, T. J. 1981. 'A non-parametric multivariate test for monotone trend with pharmaceutical applications', *J. Am. Stat. Assoc.*, **76**, 169–174.

Draper, N. R., and Smith, H. 1981. *Applied Regression Analysis*. Wiley, New York. 709 pp.

Egbert, G. D., and Lettenmaier, D.P. 1986. 'Stochastic modeling of the space–time structure of atmospheric chemical deposition', *Wat. Resour. Res.*, **22**, 165–179.

El-Shaarawi, A. H., and Esterby, S. R. 1992. 'Replacement of censored observations by arbitrary values: an evaluation', *Wat. Res.*, **26**, 835- 844.

El-Shaarawi, A. H., and Niculescu, S. 1992. 'On Kendalls tau as a test for trend in time series data', *Environmetrics*, **3**, 385–411.

El-Shaarawi, A. H., and Niculescu, S. 1993. 'A simple test for detecting non-linear trend', *Environmetrics*, **4**, 233–242.

El-Shaarawi, A. H., Esterby, S. R., and Kuntz, K. W. 1983. 'A statistical evaluation of trends in the water quality of the Niagara River', *J. Great Lakes Res.*, **9**, 234–240.

Esterby, S. R. 1993. 'Trend analysis methods for environmental data', *Environmetrics*, **4**, 459–481.

Esterby, S. R., and El-Shaarawi, A. H. 1981. 'Inference about the point of change in a regression model', *Appl. Statis.*, **30**, 277–285.

Esterby, S. R., Delorme,L. D., Duthie, H., and Harper, N. S. 1986. 'Analysis of multiple depth profiles in sediment cores: an application to pollen and diatom data from lakes sensitive to acidification', *Hydrobiologia*, **141**, 207–235.

Esterby, S. R., El-Shaarawi, A. H., and Block, H. O. 1992. 'Detection of water quality changes along a river system', *Environ. Monit. Assess.*, **23**, 219–242.

Eynon, B. P., and Switzer, P. 1983. 'The variability of rainfall acidity', *Can. J. Statis.*, **11**, 11–24.

Gilbert, R. O. 1987. *Statistical Methods for Environmental Pollution Monitoring*. Van Nostrand Reinhold, New York. 320 pp.

Gilliom, R. J., and Helsel, D. R. 1986. 'Estimation of distributional parameters from censored trace level water quality data 1. Estimation techniques', *Wat. Resour. Res.*, **22**,135–146.

Haas, T. C. 1990. 'Kriging and automated variogram modelling within a moving window', *Atmos. Environ.*, **24**, 1759–1769.

Helsel, D. R., and Hirsch, R. M. 1992. *Statistical Methods in Water Resources*. Elsevier, Amsterdam. 522 pp.

Hipel, K. W., and McLeod, A. I. 1994. *Time Series Modelling of Environmental and Water Resources Systems*. Elsevier, Amsterdam.

Hipel, K. W., McLeod, A. I., and Fosu, P. K. 1986. 'Empirical power comparisons of some tests for trend' in El-Shaarawi, A. H. and Kwiatowski, R.E. (Eds), *Statistical Aspects of Water Quality Monitoring*. Elsevier, Amsterdam, 347–362.

Hirsch, R. M. 1988. 'Statistical methods and sampling design for estimating step trends in surface water quality', *Wat. Resour. Bull.*, **24**, 493–503.

Hirsch, R. M., and Peters, N. E. 1988. 'Short-term trends in sulphate deposition at selected bulk precipitation stations in New York', *Atmos. Environ.*, **22**, 1175–1178.

Hirsch, R. M., and Slack, J. R. 1984. 'A non-parametric trend test for seasonal data with serial dependence', *Wat. Resour. Res.*, **20**, 727–732.

Hirsch, R. M., Slack, J. R., and Smith, R. A. 1982. 'Nonparametric tests for trend in water quality', *Wat. Resour. Res.*, **18**, 107–121.

Hirsch, R. M., Alexander, R. B., and Smith, R. A. 1991. 'Selection of methods for the detection and estimation of trends in water quality', *Wat. Resour. Res.*, **27**, 803–813.

Kawamura, A., Jinno, K., Ueda, T., and Medina, R. R. 1986. 'Detection of abrupt changes in water quality time series by the adaptive Kalman filter' in Lerner, D. (Ed.), *Monitoring to Detect Changes in Water Quality Series. Int. Assoc. Hydrol. Sci.*, **157**, 285–296.

Lettenmaier, D. P. 1976. 'Detection of trends in water quality data from records with dependent observations', *Wat. Resour. Res.*, **12**, 1037–1046.

Lettenmaier, D. P. 1988. 'Multivariate non-parametric tests for trend in water quality', *Wat. Resour. Bull.*, **24**, 505–512.

Lettenmaier, D. P., Hooper, E. R., Wagoner, C., and Faris, K.B. 1991. 'Trends in stream quality in the continental United States, 1978–1987', *Wat. Resour. Res.*, **27**, 327–339.

Liu, J. 1992. 'Global trend of acid wet deposition — a parametric modelling of sulphate concentrations', *Environmetrics*, **3**, 413–429.

Loftis, J. C., Taylor, C. H., and Chapman, P. L. 1991a. 'Multivariate tests for trend in water quality', *Wat. Resour. Res.*, **27**, 1419–1429.

Loftis, J. C., Taylor, C. H., Newell, A. D., and Chapman, P. L. 1991b. 'Multivariate trend testing of lake water quality', *Wat. Resour. Bull.*, **27**, 461–473.

MacNeill, I. B., Tang, S. M., and Jandhyala, V. K. 1991. 'A search for the source of the Niles change-points', *Environmetrics*, **2**, 341–375.

McLeod, A. I., Hipel, K. W., and Comancho, F. 1983. 'Trend assessment of water quality time series', *Wat. Resour. Bull.*, **19**, 537–547.

McLeod, A. I., Hipel, K. W., and Bodo, B. A. 1991. 'Trend analysis methodology for water quality time series', *Environmetrics*, **2**, 169–200.

McMichael, F. C., and Hunter, J. S. 1972. 'Stochastic modelling of temperature and flow in rivers', *Wat. Resour. Res.*, **8**, 87–98.

Mehta, B. M., Ahlert, R. C., and Yu, S. L. 1975. 'Stochastic variation of water quality of the Passaic River', *Wat. Resour. Res.*, **11**, 300–308.

Meybeck, M., Chapman, D. V., and Helmer, R. 1986. *Global Freshwater Quality, A First Assessment*. World Health Organization and United Nations Environment Programme, Basil Blackwell, Oxford. 306 pp.

Mohaupt, V. 1986. 'Nutrient–discharge relationships in a flatland river system and optimization of sampling' in Lerner, D. (Ed.), *Monitoring to Detect Changes in Water Quality Series. Int. Assoc. Hydrol. Sci.*, **157**, 297–304.

Neilson, B. J., and Hsieh, B. B. 1982. 'Analysis of water temperature records using a deterministic-stochastic model' in El-Shaarawi, A. H. and Esterby, S. R. (Eds), *Time Series Methods in the Hydrosciences*. Elsevier, Amsterdam. pp. 465–473.

OECD 1991. *Environmental Indicators, A Preliminary Set*. Organisation for Economic Co-operation and Development, Paris.

Oehlert, G. W. 1988. 'Interval estimates for diatom-inferred lake pH histories', *Can. J. Statis.*, **16**, 51–60.

Peters, N. E., and Turk, J. T. 1981. 'Increases in sodium and chloride in Mohawk River, New York, attributed to road salt', *Wat. Resour. Bull.*, **17**, 586–598.

Pettitt, A. N. 1979. 'A non-parametric approach to the change-point problem', *Appl. Statis.*, **23**, 126–135.

Puri, M. L., and Sen, P. K. 1971. *Nonparametric Methods in Multivariate Analysis*. Wiley, New York.

Reinsel, G. C., and Tiao, G. C. 1987. 'Impact of chlorofluoromethanes on stratospheric ozone', *J. Am. Statis. Assoc.*, **82**, 20–30.

Sandén, P., and Rahm, L. 1993. 'Nutrient trends in the Baltic Sea', *Environmetrics*, **4**, 75–103.

Sen, P. K. 1968. 'Estimates of the regression coefficient based on Kendalls tau', *J. Am. Statis. Assoc.*, **63**, 1379–1389.

Smith, E. P., and Rose, K. A. 1991. 'Trend detection in the presence of covariates: stagewise versus multiple regression', *Environmetrics*, **2**, 153–168.

Smith, E. P., Rheem, S., and Holtzman, G. I. 1993. 'Multivariate assessment of trend in environmental variables' in Patil, G. P., and Rao, C. R. (Eds), *Multivariate Environmental Statistics*. Elsevier, Amsterdam. pp. 491–507.

Stein, M. L., Shen, X., and Styer, P. E. 1993. 'Applications of a simple regression model to acid rain data', *Can. J. Statis.*, **21**, 331–346.

Stensland, G. J., Whelpdale D. M., and Oehlert, G. 1986. 'Precipitation chemistry' in *Acid Deposition, Long-Term Trends*. Committee on Monitoring and Assessment of Trends in Acid Deposition, National Research Council, National Academy Press, Washington. pp. 128–199.

Stoddard, J. L. 1991. 'Trends in Catskill Stream water quality: evidence from historical data', *Wat. Resour. Res.*, **27**, 2855–2864.

Switzer, P. 1988. 'An analysis of hourly deposition data', *Can. J. Statis.*, **16**, 39–50.

Taylor, C. H., and Loftis, J. C. 1989. 'Testing for trend in lake and ground water quality time series', *Wat. Resour. Bull.*, **25**, 715–726.

Taylor, J. M. G. 1987. 'Kendalls and Spearmans correlation coefficients in the presence of a blocking variable', *Biometrics*, **43**, 409–416.

Teti, P. 1984. 'Time-variant differences in chemistry among four small streams', *Wat. Resour. Res.*, **20**, 347–359.

Theil, H. 1950. 'A rank-invariant method of linear and polynomial regression analysis, 1, 2, and 3', *Ned. Akad. Wetensch Proc.*, **53**, 386–392, 521–525, 1397–1412.

van Belle, G., and Hughes, J. P. 1984. 'Nonparametric tests for trend in water quality', *Wat. Resour. Res.*, **20**, 127–136.

Zetterqvist, L. 1988. 'Asymptotic distribution of Manns test for trend for m-dependent seasonal observations', *Scand. J. Statis.*, **15**, 81–95.

Zetterqvist, L. 1991. 'Statistical estimation and interpretation of trends in water quality time series', *Wat. Resour. Res.*, **27**, 1637–1648.

2

TRENDS IN THE CHEMISTRY OF PRECIPITATION AND SURFACE WATER IN A NATIONAL NETWORK OF SMALL WATERSHEDS

BRENT T. AULENBACH* AND RICHARD P. HOOPER

US Geological Survey, 3039 Amwiler Road, Suite 130, Atlanta, GA 30360, USA

AND

OWEN P. BRICKER

US Geological Survey, National Center, MS 432, 12201 Sunrise Valley Drive, Reston, VA 22092, USA

ABSTRACT

Trends in precipitation and surface water chemistry at a network of 15 small watersheds (< 10 km^2) in the USA were evaluated using a statistical test for monotonic trends (the seasonal Kendall test) and a graphical smoothing technique for the visual identification of trends. Composite precipitation samples were collected weekly and surface water samples were collected at least monthly. Concentrations were adjusted before trend analysis, by volume for precipitation samples and by flow for surface water samples. A relation between precipitation and surface water trends was not evident either for individual inorganic solutes or for solute combinations, such as ionic strength, at most sites. The only exception was chloride, for which there was a similar trend at 60% of the sites. The smoothing technique indicated that short-term patterns in precipitation chemistry were not reflected in surface waters. The magnitude of the short-term variations in surface water concentration was generally larger than the overall long-term trend, possibly because flow adjustment did not adequately correct for climatic variability. Detecting the relation between precipitation and surface water chemistry trends may be improved by using a more powerful sampling strategy and by developing better methods of concentration adjustment to remove the effects of natural variation in surface waters.

INTRODUCTION

Monitoring the effects of large-scale environmental disturbances, such as acid precipitation, is challenging both because of the cost of sampling a large area and because of the difficulty in detecting small changes in the face of large natural variability (Turk, 1983). Regional surveys, such as the Norwegian 1000-lake survey (Rosseland and Henriksen, 1990), the US Eastern and Western Lake Surveys (Landers *et al.*, 1986; Linthurst *et al.*, 1986), or the US National Stream Survey (Herlihy *et al.*, 1991; Kaufmann *et al.*, 1991) provide a 'snapshot' of water quality at a particular time and, in the case of the American surveys, were carried out within a random sampling framework to allow population inferences to be made. The results of the Norwegian survey, performed in 1986, were compared with an earlier survey conducted in 1974–1975 to infer population trends in lake water quality and fisheries status.

Fixed-station networks provide a complementary approach to regional surveys. Although population

---

* To whom correspondence should be addressed.
E-mail: btaulenb@fs1dgadrv.er.usgs.gov. Tel.: 770-903-9148. Fax.: 770-903-9199.

inferences cannot be made from these networks, additional sampling better characterizes the variability of surface water quality than the single 'index' samples on which surveys rely. Previous networks have monitored precipitation (Peters *et al.*, 1982; Hirsch and Peters, 1988) or streamwater (Steele, *et al.*, 1974; Hirsch *et al.*, 1982; Smith and Alexander, 1983; Smith *et al.*, 1987; Murdoch and Stoddard, 1993; Stoddard and Kellogg, 1993) through time to determine the impacts of acid deposition. Peters *et al.* (1982) compared precipitation and stream water trends in sulphate and hydrogen ions on a regional scale. In this study, precipitation and surface water trends are compared for individual watersheds to permit a more process-oriented assessment.

Fifteen watersheds (Figure 1) were chosen to form a national network of small watersheds in which to study trends in the chemistry of precipitation and surface water and the relation between precipitation and surface water chemistry. These watersheds represent a broad range of geology among generally sensitive rock types (ones which would have a minimal influence on surface water chemistry), a variety of surface water systems (drainage lakes, seepage lakes and streams), a wide geographical coverage and minimal anthropogenic disturbance. Watershed characteristics are summarized in Table I. Ten sites are located

Table I. Site characteristics of sampled watersheds

Watershed	Physiographic province*	Area (ha)	Elevation range (m asl)	Bedrock geology	Watershed type	Surface water sampling frequency
Biscuit Brook, New York	Appalachian Plateaus	995	628–1130	Sandstone and shale	Stream	Weekly + events
North Fork Bens Creek, Pennsylvania	Appalachian Plateaus	894	408–835	Sandstone, shale and limestone	Stream	Monthly
Hauver Branch, Catoctin Mountain, Maryland	Blue Ridge Province	550	314–570	Greenstone	Stream	Weekly to 1982; bi-weekly to 1984; monthly thereafter
Hunting Creek, Catoctin Mountain, Maryland	Blue Ridge Province	1040	314–573	Greenstone	Stream	Weekly to 1982; bi-weekly to 1984; monthly thereafter
Mill Run, Massanutten Mountain, Virginia	Valley and Ridge	303	366–671	Sandstone and shale	Stream	Bi-weekly to 1983; monthly thereafter
Shelter Run, Massanutten Mountain, Virginia	Valley and Ridge Province	36	366–579	Sandstone and shale	Stream	Bi-weekly to 1983; monthly thereafter
Old Rag Mountain, Virginia	Blue Ridge Province	260	488–975	Granite	Stream	Bi-weekly to 1983; monthly thereafter
Jordan Creek, North Carolina	Piedmont Province/ Coastal Plain	93	117–135	Unconsolidated sand	Stream	Monthly
Panola Mountain, Georgia	Piedmont Province	41	224–276	Granodiorite and gneiss	Stream	Weekly + events
Lake Lucerne, Florida	Coastal Plain	67	40–54	Unconsolidated sand and clay	Seepage lake	Monthly
Lake Clara, Wisconsin	Superior Upland	120	460–490	Till to sandy outwash	Drainage lake	Monthly
Vandercook Lake, Wisconsin	Superior Upland	290	495–510	Sand and gravel outwash	Seepage lake	Monthly
Loch Vale, Colorado	Southern Rocky Mountains	660	3100–4000	Granite	Alpine drainage lake	Weekly + events
Ned Wilson Lake, Colorado	Southern Rocky Mountains	50	3389–3436	Basalt	Drainage lake	Bi-weekly
Goat Lake, Washington	Cascade-Sierra Mountains	777	960–2160	Quartz diorite and pelitic schist	Alpine drainage lake	Monthly

*After Fenneman (1946)

on the east coast from New York to Florida and fall along a gradient in acidic deposition. Two sites are located in Wisconsin and represent two different types of lakes (seepage and drainage) receiving the same quality of atmospheric deposition. Two high elevation lakes in the Rocky Mountains, one on the eastern flank and one on the western flank of the continental divide, are also in the network. One site in the Cascade Mountains is located downwind of Seattle/Tacoma, WA. Results from the input–output monitoring programme are reported in this paper, although more detailed process research was conducted at a subset of sites (Figure 1).

The network design assumed that trends in precipitation chemistry would be reflected in trends in surface water chemistry and that these trends could be detected in volume-weighted or flow-adjusted concentration of weekly precipitation and monthly surface water samples, respectively (Hirsch et al., 1982). Sampling was not sufficiently frequent at all sites to allow for a reliable estimation of mass flux, which has also been used in trend analysis (Driscoll et al., 1989). It was assumed that such trends would be interpretable with only a general understanding of acid–base reactions in a watershed, such as primary mineral weathering and sulphate adsorption by the soils, without site-specific information. Similar trends in unreactive solutes, such as chloride, or in related reactive solutes, such as hydrogen ion in precipitation and silica in surface water (related by primary weathering reactions), would suggest that precipitation influences surface–water chemistry.

METHODS

Precipitation and surface water samples were collected between 1982 and 1993, with sampling periods at individual sites ranging from three to ten years. Trend analyses were performed on both precipitation and surface water samples for calcium, magnesium, sodium, potassium, sulphate, nitrate, chloride and usually ammonium. Trend analyses were also performed for hydrogen ions in precipitation samples and for alkalinity and dissolved silica in surface water samples. In addition to these measured solutes, trend analyses were performed for ionic strength, which was calculated for both precipitation and surface water

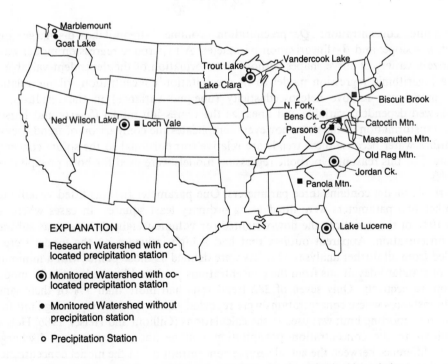

Figure 1. Location of watersheds and precipitation stations used in this analysis

samples. These trend analyses were performed at all 15 watersheds sites resulting in 282 separate trend tests. The sample size for each test ranged from 32 to 1281.

Sampling frequency

Weekly composite precipitation samples were collected using Aerochem Metrics (the use of trade, product or firm names is for descriptive purposes only and does not imply endorsement by US Geological Survey) samplers, which automatically expose the collection bucket only during rainfall, precluding contamination of the sample by dry deposition. These samples were collected within or very near each watershed, except for three watersheds where data from the nearest National Acid Deposition Program/National Trends Network (NADP/NTN) site were used to provide atmospheric deposition inputs (Figure 1). Surface water samples were collected at fixed intervals ranging from weekly to monthly at all sites (Table I). At Loch Vale, CO and Ned Wilson Lake, CO samples could not be collected during one or more winter months. In addition, at three of the sites — Biscuit Brook, NY; Panola Mountain., GA; and Loch Vale, CO — event-based sampling was performed during some part of the record (Table I).

Concentration adjustment for precipitation volume and discharge

Before performing any trend analysis, sample concentrations were adjusted to account for different hydrological conditions. Precipitation concentrations were adjusted using a concentration–precipitation volume model because large storms tend to have lower solute concentrations than small storms (Peters *et al.*, 1982). This relation, although weakened by the weekly composite sampling method, was observed in these data. Stream and lake water concentrations were adjusted similarly using a concentration–discharge model (or a concentration–stage model for seepage lakes) to remove any systematic variation with respect to these variables. By adjusting concentrations in this manner, the effects of temporal patterns in discharge or precipitation volume on concentration trends can be reduced.

Concentration adjustments were made using a hyperbolic regression model (Johnson *et al.*, 1969)

$$C = b + \frac{m}{(1 + \beta Q)} \tag{1}$$

where C = solute concentration; Q = precipitation volume, stream/lake discharge or lake stage; b = intercept; m = slope; and β = linearization parameter. A hyperbolic regression model was used because it can linearize a variety of curve shapes without transformation of the dependent variable. Other studies have used a logarithmic regression model to fit precipitation concentration/volume relations because a logarithmic model can remove heteroscedasticity (unequal variance) in the residuals, normalize the positively skewed dependent variable and linearize the regression model (Shertz and Hirsch, 1985). The logarithmic transformation of the data, however, complicates the comparison of trends between precipitation and surface water because these trends are relative (proportional to the solute concentration) rather than absolute. For this reason, hyperbolic regression models were used for both precipitation and surface water samples.

The hyperbolic model contains three parameters. One parameter (β) was fitted visually to linearize the data; the other two parameters were fitted using ordinary least squares. In cases where a model could not explain 10% of the variation, the flow-weighted (or volume-weighted) mean was subtracted from the measured concentration. Apparent outliers that had a high influence on the model were removed and were excluded from all further analyses. Outliers were defined as those points with a minimum of 2.6 standard errors or standard deviations from the concentrations predicted by the hyperbolic model or the mean concentration, respectively. Only seven of 282 trend tests had more than 5% of their samples excluded (Table II). In instances where concentrations were reported as less than the reporting limit for the chemical analysis, half the reporting limit was used in the calculations (Gilliom and Helsel, 1986; Helsel and Gilliom, 1986). Residuals to the concentration–precipitation volume and concentration–discharge models are defined as the difference between the actual sample concentration and the model concentration. These concentration residuals were then used in trend analyses.

Table II. Trend tests for which more than 5% of points were excluded as outliers

Site	Solute	Data points excluded	Total number of data points	Percentage of data points excluded
N. Fork Bens Creek, PA	Streamwater SO_4^{2-}	5	97	5·2
Mill Run, VA	Streamwater Si	11	164	6·7
Lake Lucerne, FL	Lake water alkalinity	7	51	14·
Trout Lake, WI	Precipitation Na^+	10	172	5·8
Lake Clara, WI	Lake water SO_4^{2-}	2	33	6·1
Lake Clara, WI	Lake water NO_3^-	2	32	6·3
Vandercook Lake, WI	Precipitation Na^+	3	60	5·0

Seasonal Kendall test for trends

The seasonal Kendall test, a non-parametric technique for detecting monotonic trends (Hirsch *et al.*, 1982), compares a concentration residual with concentration residuals of the same season in each successive year over the length of the record to determine whether there was a change in concentration. The seasonal Kendall test is particularly useful for monitoring data because the test is not influenced by missing values and is insensitive to outliers. Seasons are defined as calendar months for this study. If multiple samples were taken within a month, the median concentration residual was used for comparison. Each comparison was assigned a value: 1 for an increase in concentration, −1 for a decrease and 0 for no change in concentration residual (a tie). Previous studies using this test typically did not account for analytical uncertainty in the measurement of water chemistry. To account for this, a comparison was considered a tie if the difference varied less than a certain amount. Each test was run three times using different levels of analytical uncertainty (0, 5 and 10% of the absolute value of the sample concentration). Zero percent represents the typical way the test has been performed whereas 5 and 10% represent reasonable values for analytical uncertainty.

The test statistic, S_i, is defined as the sum of the assigned values from the comparisons

$$S_i = \sum_{k=1}^{n_i-1} \sum_{j=k+1}^{n_i} \text{sgn}(x_{ij} - x_{ik}) \tag{2}$$

where x = median concentration residual; i = season; j,k = year; n = total number of years; and

$$\text{sgn}(\theta) = \begin{cases} 1 & \text{if } \theta > 0 \\ 0 & \text{if } \theta = 0. \\ -1 & \text{if } \theta < 0 \end{cases}$$

This test statistic S_i is approximately normally distributed under the null hypothesis of no trend. Two-sided trend analyses were conducted at the $\alpha = 0·10$ level. The ability to detect a monotonic trend is improved as the magnitude of the trend increases, the standard error or standard deviation of the concentration–precipitation volume/discharge model decreases and/or the period of record increases. We cannot determine the magnitude of the trend that can be detected based on the standard error or standard deviation of the concentration–precipitation volume/discharge model and period of record alone because the magnitude of the trend and the standard error or standard deviation of the model are related. An increase in the magnitude of the trend may result in an increase in the standard error or standard deviation of the model because a larger trend represents a greater excursion from the model prediction. The slope of the trend is given by the seasonal Kendall slope estimator (Hirsch *et al.*, 1982), defined as the median of slopes calculated between each concentration-residual comparison.

Smoothed concentration residuals

To examine more complex concentration patterns than monotonic trends detected by the seasonal Kendall test, residuals to the concentration–precipitation volume/discharge models also were plotted against time. Because the scatter in these concentration residuals make trends difficult to observe, a smoothing technique, called *lo*cally *w*eighted regression and *s*moothing *s*catterplots or LOWESS (Cleveland, 1979) was applied to make long-term trends more apparent.

The value of the smooth curve for every observation is determined by a robust, weighted least-squares model fit to the observations that fall within a range from that point, known as the window. Each observation that falls within the window is weighted by the product of the concentration-weighting function and a time-weighting function. Weights are chosen to reduce the influence of outliers and emphasize the central tendency of the data. A bi-square weighting function was used for the concentration weighting function having a window size defined as six times the median value of the absolute differences between the predicted and observed concentration residuals at each point in time and a tri-cube weighting function was applied to the data along the time axis, as suggested by Cleveland (1979). As LOWESS calculates a smoothed value for each observation, it does not require values to be evenly spaced through time, but enough values must be contained within the chosen window to calculate a weighted least squares.

A window size of 1·5 years was chosen for the time-weighting function. Smoothed data with a window size of one year retains the strong seasonal variation in the data (Figure 2). When a window size of three years is used, all seasons have approximately equal weights and, hence, the smoothed data are unaffected by seasonal variations. However, this window size eliminates some variations that have a duration of longer than a season and is a substantial portion of the average record of approximately eight years. Therefore, a window size of 1·5 years was used because it removed much of the seasonal variation while tracking the concentration residuals more closely than a larger window size. Note that seasonal variability at the beginning and end of the record has a large influence on the smoothed curve for the smaller time windows, resulting in a pronounced upturn or downturn that is an artifact of the LOWESS technique. These initial or terminal 'hooks' should be ignored when evaluating trends.

Normalized concentration ratios

To determine the likelihood of detecting a direct influence of precipitation on surface water, a normalized concentration ratio was constructed to compare precipitation and surface water solute concentrations. Because evapotranspiration will influence the surface water concentration, evapotranspiration was calculated at each site using precipitation volume and discharge records for the period of the study (when available) and assuming that evapotranspiration is the difference between precipitation and runoff. The ratio of the surface water mean concentration to the precipitation mean concentration was normalized by dividing by the expected concentration ratio due to evapotranspiration. A concentration ratio of unity would indi-

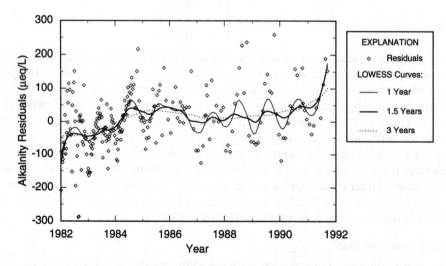

Figure 2. Example of the effect of window size on locally weighted scatter plot smoothing (LOWESS) for streamwater alkalinity concentration residuals at Hunting Creek, Catoctin Mountain, Maryland

Table III. Average precipitation, discharge and solute concentration in precipitation and surface water samples

Site	Type*	Precipitation volume/discharge (cm/y)	H⁺/ANC (μequiv/l)	Ca²⁺ (μequiv/l)	Mg²⁺ (μequiv/l)	Na⁺ (μequiv/l)	K⁺ (μequiv/l)	NH₄⁺ (μequiv/l)	SO₄²⁻ (μequiv/l)	NO₃⁻ (μequiv/l)	Cl⁻ (μequiv/l)	Dissolved Si (μM/l)	Ionic strength (μequiv/l)
Biscuit Brook, NY	Precipitation	129·2	57·1	3·8	1·7	3·0	0·5	11·3	46·1	24·7	4·2	—	104·7
	Weekly stream	91·5	18·6	132·4	46·7	16·3	6·8	—	131·1	28·4	15·7	31·7	359·6
	All stream	91·5	3·1	112·3	38·7	14·1	9·0	—	111·4	36·3	12·8	23·5	354·2
Parsons WV N Fork Bens Creek, PA	Precipitation	126·3	65·1	7·6	1·8	2·2	0·6	12·4	60·8	25·9	3·6	—	125·0
	Stream	—	93·7	160·0	76·7	79·2	16·2	1·6	204·5	40·0	106·3	60·5	619·5
Catoctin Mtn, MD	Precipitation	112·1	69·1	9·2	2·6	5·6	1·9	15·3	52·2	22·8	11·0	—	121·1
Hauver Branch	Stream	51·3	250·2	262·6	198·8	103·5	6·4	—	187·0	31·3	119·3	186·6	910·0
Hunting Creek	Stream	74·5	306·5	340·7	207·6	252·7	17·0	—	149·7	50·5	293·8	224·9	1166·1
Massanutten Mtn, VA	Precipitation	86·9	52·3	7·0	1·7	4·0	1·7	16·2	53·6	24·7	8·5	—	108·0
Mill Run	Stream	36·0	2·6	62·3	92·8	27·8	27·1	0·8	221·7	1·9	33·4	90·1	424·2
Shelter Run	Stream	—	526·3	446·9	195·5	35·2	31·8	1·3	152·6	1·3	31·5	102·4	1121·8
Old Rag Mtn, VA	Precipitation	111·6	48·2	7·4	2·1	7·6	2·8	15·1	51·2	24·7	13·6	—	109·7
	Stream	40·2	53·6	52·7	27·4	58·6	12·4	—	40·1	0·6	27·7	173·1	215·2
Jordan Creek, NC	Precipitation	119·0	37·9	3·6	2·5	8·2	0·6	8·7	34·5	15·7	9·4	—	78·1
	Stream	24·9	—	12·4	23·7	38·1	8·3	2·5	31·0	2·2	40·0	67·0	112·4
Panola Mtn, GA	Precipitation	115·6	28·5	3·6	2·1	5·4	1·1	11·9	42·5	17·1	7·5	—	90·0
	Weekly stream	30·1	179·7	83·7	57·2	115·5	22·2	0·7	38·4	1·2	44·6	242·2	360·2
	All stream	30·1	58·2	71·3	42·6	66·2	23·2	0·2	84·3	3·1	31·2	126·3	288·1
Lake Lucerne, FL	Precipitation	128·6	31·2	15·7	5·5	15·2	1·6	10·4	26·9	16·1	16·2	—	93·7
	Seepage lake†	—	83·4	354·3	378·1	653·4	103·6	1·5	700·2	0·1	670·5	4·6	2193·3
Trout Lake, WI	Precipitation	92·1	18·4	10·4	2·9	1·7	0·6	17·9	31·4	18·0	2·2	—	67·4
Lake Clara, WI	Drainage lake†	—	—	92·3	48·6	119·6	18·4	3·4	90·2	1·9	131·8	1·4	134·2
Vandercook Lake, WI	Precipitation	87·4	20·5	25·8	8·4	12·1	3·4	36·5	40·7	25·0	4·0	—	—
	Seepage lake†	—	—	62·9	31·4	25·3	6·2	3·7	83·6	2·1	7·9	1·4	—
Loch Vale, CO	Precipitation	100·7	12·3	8·4	2·1	0·9	1·8	6·4	12·8	11·1	2·4	—	39·5
	Weekly lake	72·5	34·7	60·7	15·8	17·9	4·2	1·1	29·6	16·3	4·6	27·2	146·7
	All lake	72·5	35·7	63·0	16·7	18·3	4·6	1·6	30·7	17·8	5·1	28·9	152·2
Ned Wilson Lake, CO	Precipitation	140·7	7·6	15·0	3·5	3·4	1·4	0·8	13·5	3·2	3·2	—	38·9
	Drainage lake	—	51·3	43·9	16·5	6·3	4·1	—	10·1	0·3	3·0	3·4	104·7
Marblemount, WA	Precipitation	198·8	12·0	1·6	2·2	6·8	0·3	1·6	7·4	4·8	8·1	—	28·1
Goat Lake, WA	Drainage lake	363·9	121·1	92·2	35·2	19·6	7·2	1·6	27·7	4·2	8·8	44·8	236·1

* Solute concentrations are volume weighted or flow weighted means except where noted
† Solute concentrations are means

cate that the difference between precipitation and surface water solute concentration could be explained solely by evapotranspiration; a concentration ratio greater than unity indicates that there is a net export of that solute; and a concentration ratio less than unity indicates a net retention of that solute.

A direct influence might be expected for solutes that have a normalized concentration ratio around unity, because those solutes seem to be conservative. A direct influence is less likely to be detected when the normalized concentration ratio differs from unity, because this is an indication that processes other than evapotranspiration are influencing the concentration of solutes in surface waters. If these other processes are variable, then the ability to detect a small trend in surface water chemistry as a result of a trend in precipitation chemistry is diminished. Even in instances when the normalized concentration ratio is around unity, the ability to detect direct influences can be difficult because of time lags in which a trend in precipitation chemistry does not immediately affect surface water chemistry due to watershed processes.

RESULTS

Volume-weighted mean concentrations for precipitation, flow-weighted mean concentrations for streams and drainage lakes and mean concentrations for seepage lakes are listed in Table III. Deviations from the mean concentrations are shown as LOWESS smoothed concentration residuals at each of the 15 watersheds (Figure 3a–o). The top panels in the two columns are the precipitation quantity and stream/lake discharge (or lake stage for seepage lakes) at each site (no smoothing has been performed on these data). Missing values for precipitation quantity are indicated by crosses below the plot, indicating that at least one weekly precipitation volume was missing from that month. Missing values for discharge or stage are indicated by a discontinuity in the line, indicating an incomplete record for that month. There may be no corresponding gap in the smoothed concentration residuals because instantaneous stage values were available at the time of sample collection.

Each panel has a series of three symbols which are the results of the seasonal Kendall test (+ for a significant uptrend, − for a significant down trend and 0 for no significant trend; significant trend determined using a two-sided test with an α level of 0·10) at three levels of analytical uncertainty (0, 5, and 10%, respectively) for the solutes, or measurement uncertainty in the case of precipitation and discharge time series. At those sites where event samples were collected, a separate set of seasonal Kendall test results are shown in the surface water solute panels. Each panel has a minimum range of 40 μequiv l^{-1}. Where a larger range is needed to display the curves, a grid line is drawn at every 20 μequiv l^{-1} to highlight the larger range for that panel.

Trend analysis methods

The results of seasonal Kendall test are generally consistent with the visual impression of the monotonic trend given by the smooth curves. As noted earlier, when the solute has a strong seasonal variation, the beginning and end of the smoothed line will curve sharply [e.g. precipitation sulphate concentration at Hauver Branch, MD (Figure 3c)]. Aside from this artifact of the LOWESS technique, a significant trend can result unexpectedly from a small absolute change if the variability in the concentration residuals is low. Examples of this are the significant downtrends in calcium, magnesium and sodium concentration in precipitation at Biscuit Brook, NY (Figure 3a). Because only monotonic trends are detected by the seasonal Kendall test, the timing of deviations in the record is important. Deviations that occur in the middle of the record, regardless of size, will not result in a significant trend, as observed for the lake water sulphate concentration at Lake Lucerne, FL (Figure 3j).

The LOWESS technique provides additional information for trend identification and interpretation than the seasonal Kendall test and slope estimator. This is because the LOWESS curves, although not a statistical test for trends, provide a more detailed view of the data than the seasonal Kendall test, which identifies only monotonic trends, and the slope estimator, which approximates the magnitude of the trend for the period of record. For example, at Biscuit Brook, NY (Figure 3a), although there is no trend in the concentration of ammonium, sulphate or nitrate in precipitation, the similarity of the LOWESS curves indicates that these solutes are highly correlated in precipitation. Conversely, at Lake Clara, WI (Figure 3k), although chloride is significantly decreasing in both precipitation and lake water, the pattern of the LOW-ESS curve for lake water has structure that is not present in the LOWESS curve for precipitation, and the magnitude of the decrease is greater in the lake water than in precipitation. This indicates that factors other than precipitation chloride concentrations are controlling the concentrations of chloride in the lake water at Lake Clara. The LOWESS curves can also be used to identify time lags between precipitation chemistry and its effect on surface water chemistry. An example of this is sodium at Old Rag Mountain, VA (Figure 3g), where the patterns of the LOWESS curves indicate that there is an apparent 1·5 year lag between precipitation and stream water concentration residuals.

At alpine lake sites, in which snowmelt dominates the annual hydrograph, simple flow adjustment may not be adequate to remove concentration patterns induced by the preferential elution of solutes from the snowpack in which the first snowmelt has a higher solute concentration than later meltwater. This is evident at Loch Vale, CO, the most intensively sampled alpine site. Most lake outlet solutes had patterns in LOW-ESS curves in which there was a seasonal decrease followed by an increase in concentration residuals around the time of snowmelt (Figure 3m). Because of preferential elution, the concentration–discharge model underestimates the concentration at the beginning of snowmelt, followed by an overestimation in concentration late in the snowmelt when the meltwater concentration is lower. This pattern is equally evident in the raw data as in the LOWESS curves, so it is not an artifact of the smoothing technique that could result from the discontinuous record. The other two alpine lakes, Ned Wilson Lake, CO (Figure 3n) and Goat Lake, WA (Figure 3o), do not show similar patterns, perhaps because of less intensive or incomplete sampling of the snowmelt period.

The use of analytical uncertainty levels of 0, 5 and 10% did not change the results of the seasonal Kendall test in 248 instances (88% of the tests). In 22 instances (7·8% of the tests), significant trends became insignificant as the uncertainty level increased. In 12 instances (4·3% of the tests), insignificant trends became significant. This counter-intuitive result arose because an increased number of ties reduced the variance of the test statistic and, thus, increased the standardized z-score (the ratio of the test statistic to its standard deviation).

The use of event-based surface water samples influenced the results of the seasonal Kendall test for at least one of the analytical uncertainty levels for chloride at Biscuit Brook, NY (Figure 3a) for five solutes/solute combinations at Panola Mountain, GA (calcium, sodium, sulphate, dissolved silica and ionic strength; Figures 3i and 6), and for two solutes at Loch Vale, CO (calcium and alkalinity; Figure 3m). Comparisons of weekly versus weekly plus event surface water samples indicate that the inclusion of event-based samples made insignificant trends become significant in six instances (19% of the 32 possible comparisons) and significant trends become insignificant in two instances (6·2% of the comparisons). If event samples helped to better define the concentration–discharge relation, false negative tests would be expected. However, the occurrence of false positive tests indicates that the flow adjustment may not have adequately captured the hydrological influences on the concentrations of these solutes. The smoothed curves were not different for most solutes at most sites, although the inclusion of event samples increased the variability of concentration residuals for some solutes at both Biscuit Brook and Panola Mountain.

Relation between precipitation and surface water chemistry

A direct influence of precipitation on surface water solute trends was explored by comparing the results of the seasonal Kendall test and LOWESS curves for individual sites and solutes/solute combinations between precipitation and surface water. Of the 137 possible test comparisons, 22 surface water trends (16%) were in the opposite direction from the precipitation trend, 13 (9·5%) had trends in the same direction, 46 (43%) had trends in neither precipitation nor surface water and 52 (42%) exhibited a trend in either precipitation or surface water, but no trend in the other water type. In general, a direct relation between precipitation and surface water solutes might be difficult to detect, but the number of opposite trends between precipitation and surface water was striking. The magnitude of trends in surface waters was generally equal to or greater than the magnitude of precipitation trends. Across all solutes and all sites, the precipitation slope estimator varied between −3 and 3 μequiv l^{-1} y^{-1}, whereas the surface water slope estimator varied between −22 and 25 μequiv l^{-1} y^{-1} (Figure 4). The LOWESS curves indicate a large

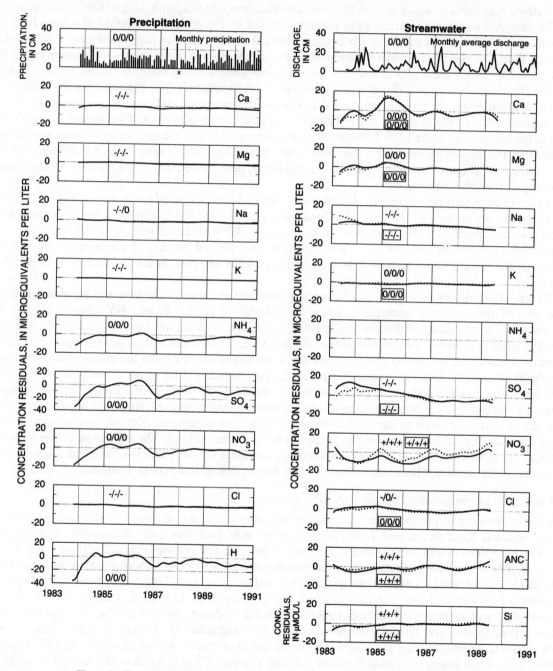

EXPLANATION

-/-/- RESULTS OF SEASONAL KENDALL TEST FOR TREND . Plus, zero, and minus indicate a significant uptrend, no significant
⌐-/-/-⌐ trend, and a significant downtrend, respectively, at a 10% (two-sided) confidence level. The first, second, and third symbols
are the results at 0, 5, and 10% levels of analytical uncertainty, respectively. Plain symbols are results for weekly to monthly
samples; outlined symbols are results for weekly and event samples.
— LOWESS-SMOOTHED LINE THROUGH CONCENTRATION RESIDUALS FROM WEEKLY TO MONTHLY SAMPLES.
.... LOWESS-SMOOTHED LINE THROUGH CONCENTRATION RESIDUALS FROM WEEKLY AND EVENT SAMPLES.

Figure 3a. Trends in precipitation and streamwater chemistry at Biscuit Brook, New York

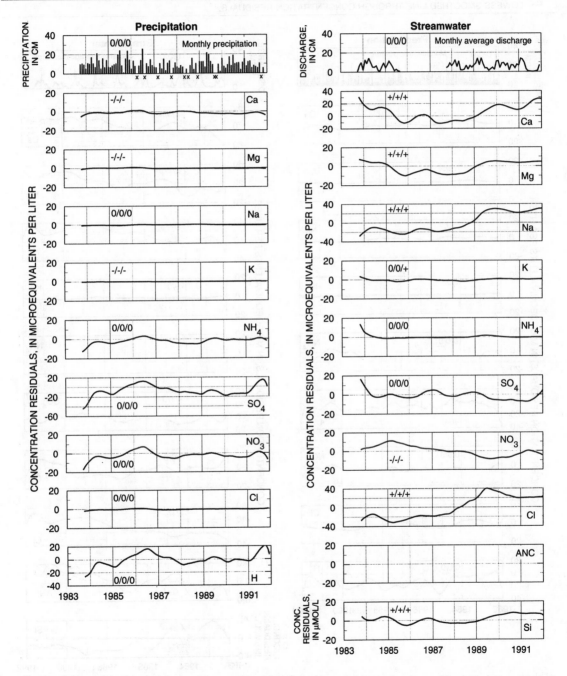

EXPLANATION

-/-/- RESULTS OF SEASONAL KENDALL TEST FOR TREND . Plus, zero, and minus indicate a significant uptrend, no significant trend, and a significant downtrend, respectively, at a 10% (two-sided) confidence level. The first, second, and third symbols are the results at 0, 5, and 10% levels of analytical uncertainty, respectively.
— LOWESS-SMOOTHED LINE THROUGH CONCENTRATION RESIDUALS.

Figure 3b. Trends in precipitation chemistry at Parsons, West Virginia and in streamwater chemistry at North Fork Bens Creek, Pennsylvania

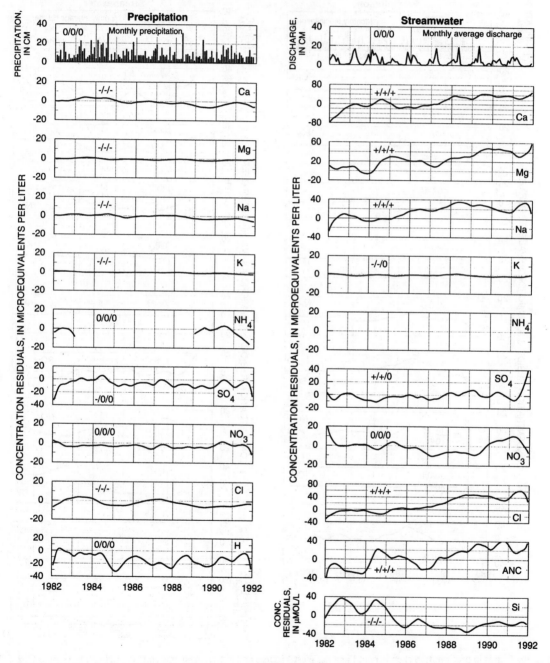

Figure 3c. Trends in precipitation and streamwater chemistry at Hauver Branch, Catoctin Mountain, Maryland

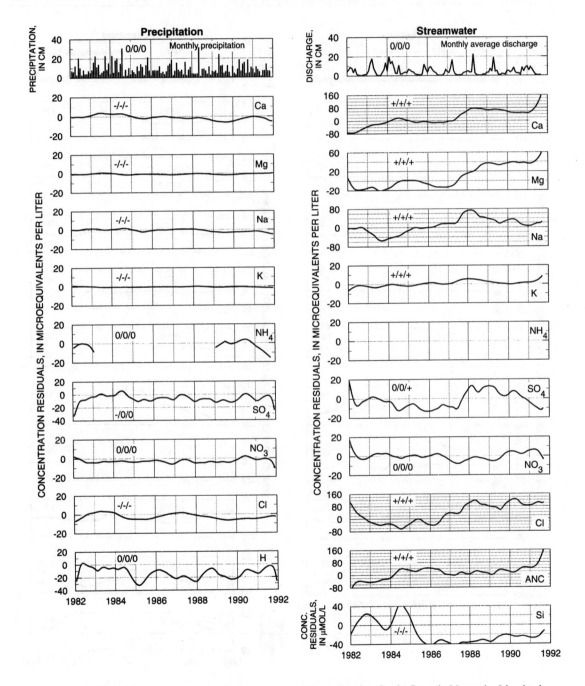

Figure 3d. Trends in precipitation and streamwater chemistry at Hunting Creek, Catoctin Mountain, Maryland

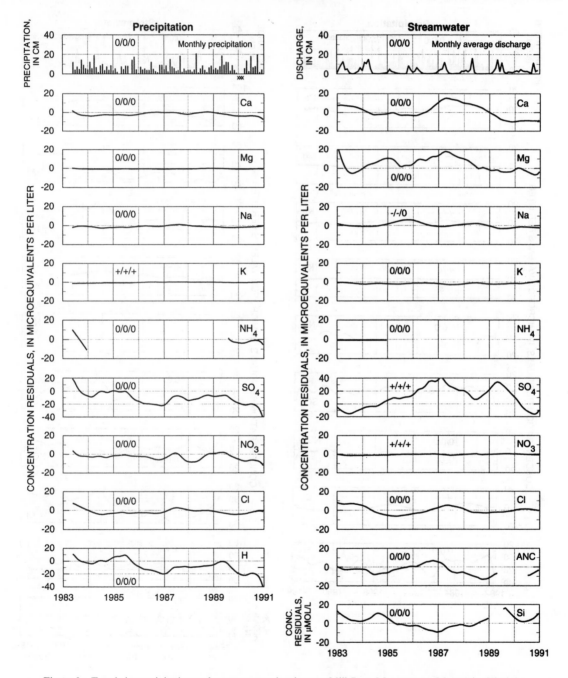

Figure 3e. Trends in precipitation and streamwater chemistry at Mill Run, Massanutten Mountain, Virginia

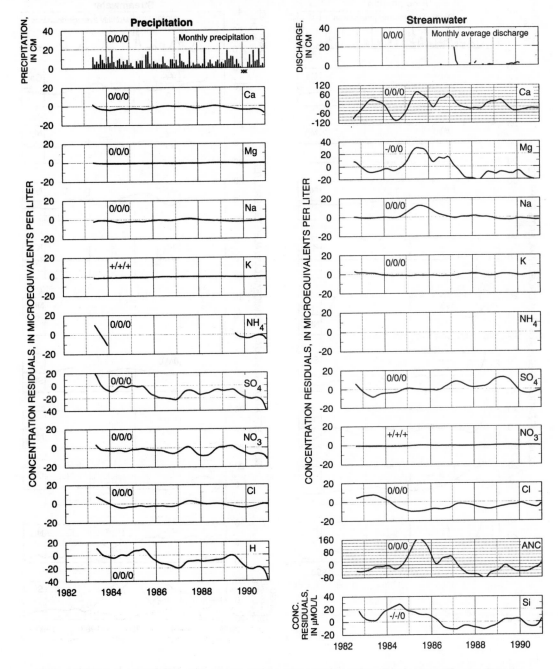

EXPLANATION

-/-/- RESULTS OF SEASONAL KENDALL TEST FOR TREND . Plus, zero, and minus indicate a significant uptrend, no significant trend, and a significant downtrend, respectively, at a 10% (two-sided) confidence level. The first, second, and third symbols are the results at 0, 5, and 10% levels of analytical uncertainty, respectively.

— LOWESS-SMOOTHED LINE THROUGH CONCENTRATION RESIDUALS.

Figure 3f. Trends in precipitation and streamwater chemistry at Shelter Run, Massanutten Mountain, Virginia

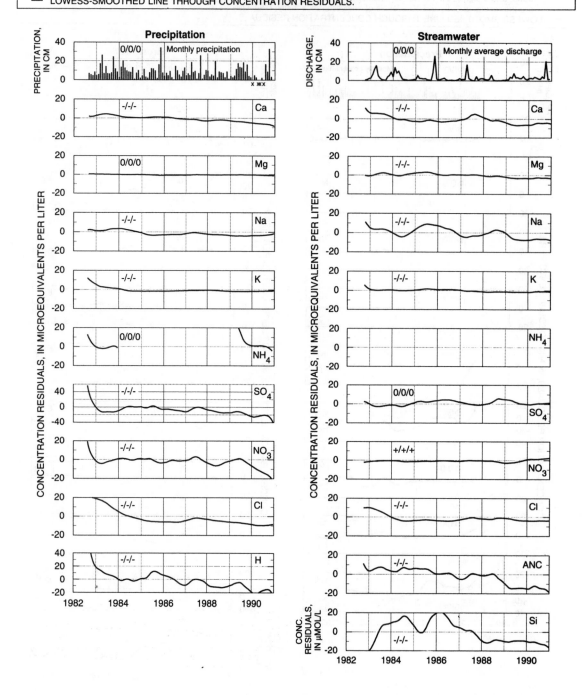

Figure 3g. Trends in precipitation and streamwater chemistry at Old Rag Mountain, Virginia

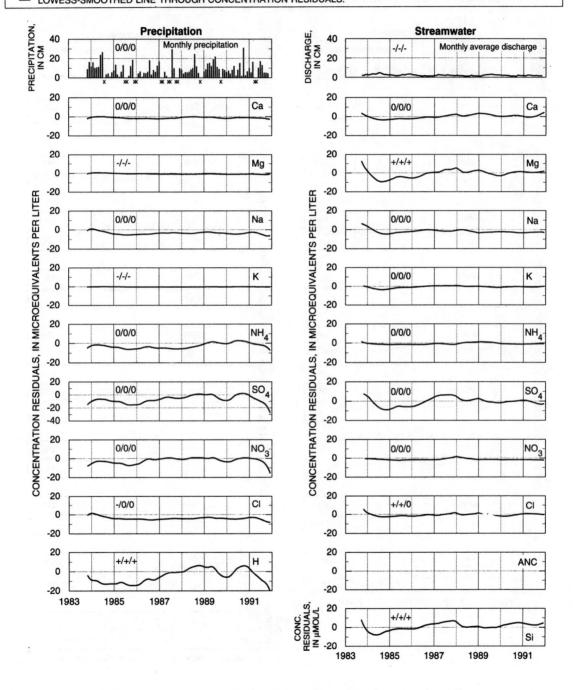

Figure 3h. Trends in precipitation and streamwater chemistry at Jordan Creek, North Carolina

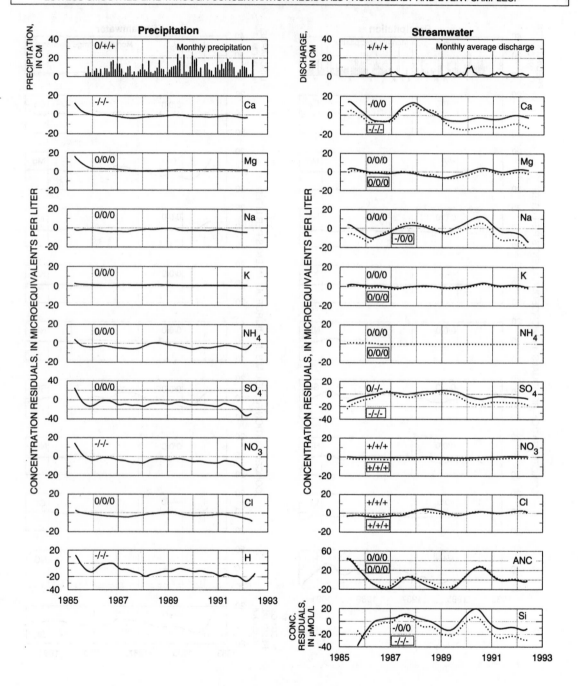

Figure 3i. Trends in precipitation and streamwater chemistry at Panola Mountain, Georgia

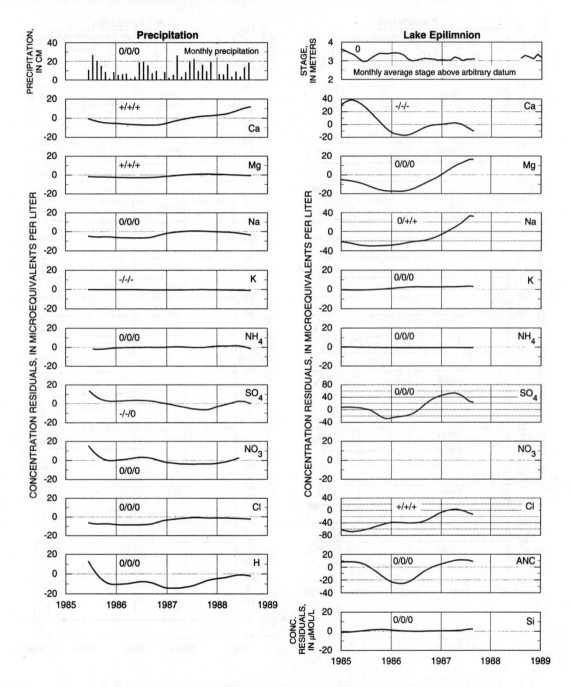

Figure 3j. Trends in precipitation and lake water chemistry at Lake Lucerne, Florida

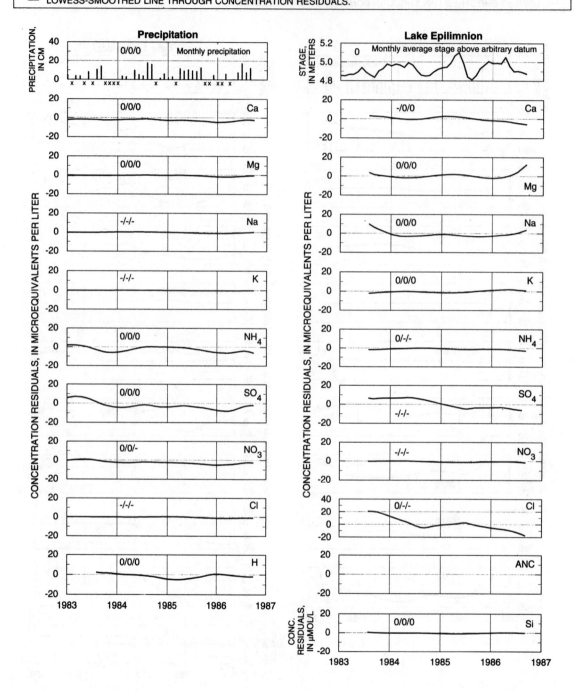

Figure 3k. Trends in precipitation chemistry at Trout Lake, Wisconsin and in lake water chemistry at Lake Clara, Wisconsin

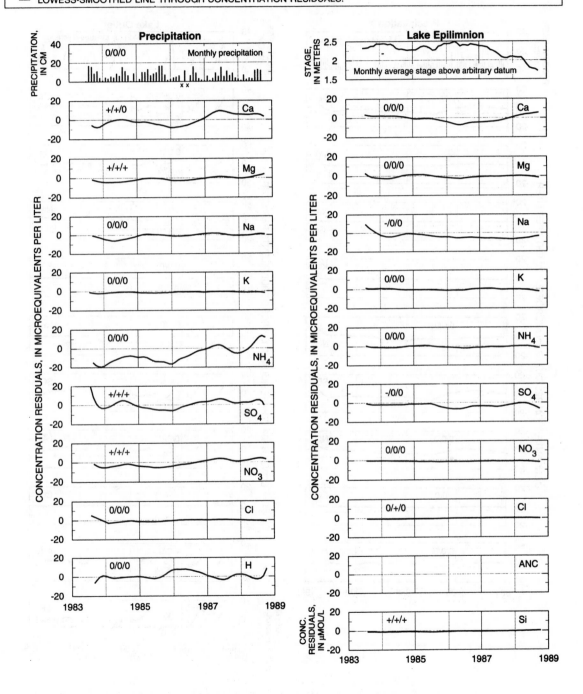

Figure 31. Trends in precipitation and lake water chemistry at Vandercook Lake, Wisconsin

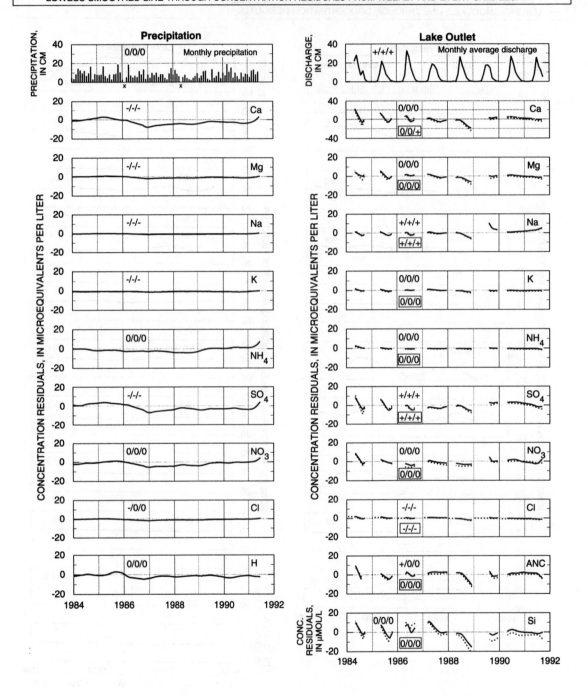

Figure 3m. Trends in precipitation and lake outlet water chemistry at Loch Vale, Colorado

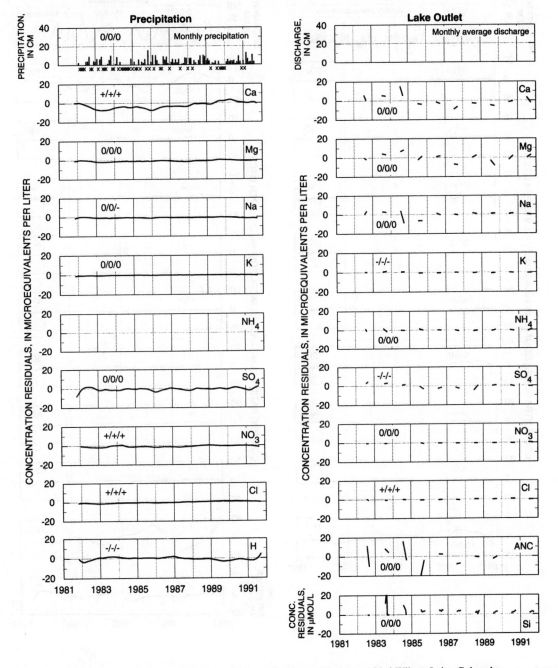

Figure 3n. Trends in precipitation and lake outlet water chemistry at Ned Wilson Lake, Colorado

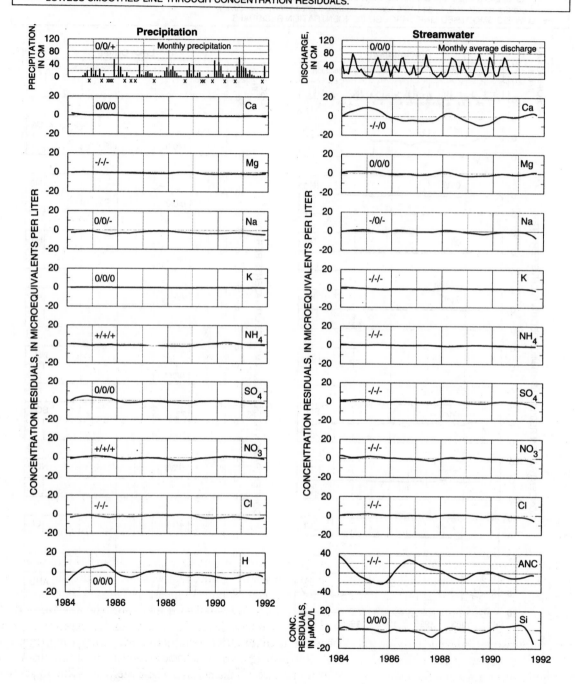

Figure 3o. Trends in precipitation chemistry at Marblemount, Washington and in streamwater chemistry at Goat Lake, Washington

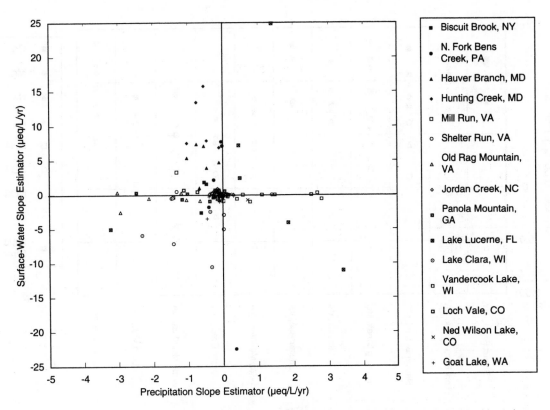

Figure 4. Surface water slope estimators versus precipitation slope estimators for watersheds used in this study

amount of short-term variations in surface water trends, despite flow adjustments to remove hydrological effects. These short-term trends were often of similar magnitude to the long-term trends.

Normalized concentration factors indicated that at most sites only sulphate, nitrate and chloride have concentration factors near unity, indicating that a direct influence is most likely to be identified for these solutes (Figure 5). Of these three solutes, only chloride, perhaps the most unreactive solute measured, had similar precipitation and surface water LOWESS curves and the same trends as indicated from the seasonal Kendall test at nine of the 15 sites. For four of these sites, however, there was little variation in chloride in precipitation and surface water samples. In some instances, the decoupling of the precipitation and surface water trends for chloride can be explained. At Hunting Creek and Hauver Branch, MD (Figure 3b and 3c), road salt is known to be a contaminant of streamwater. At North Fork Bens Creek, PA, the similar pattern for sodium and chloride trends in streamwater is most likely controlled by the leakage of saline water from an abandoned gas well within the watershed (Witt and Bikerman, 1991), not atmospheric inputs of sodium and chloride.

The lack of a direct influence of precipitation chemistry on surface water chemistry for sulphate and nitrate may be due to the dry deposition of sulphur and nitric oxides that was not measured in this study. Dry deposition can be a significant portion of the total deposition of sulphate and nitrate. For example, in one study at a site in northern Connecticut, Geigert et al. (1994) indicated that approximately one-half the total nitrate deposition and one-quarter of the total sulphate deposition was from dry deposition. If dry deposition varies independently of wet deposition, it could affect trend detection. A direct influence may also not be apparent for nitrate because of biological uptake, which may control the concentration of nitrate in surface waters. Sulphate may not exhibit a direct influence at south-eastern sites due to sulphate retention in soils. At Panola Mountain, GA sulphate retention was 83% of the total deposition of sulphate (Huntington et al., 1994).

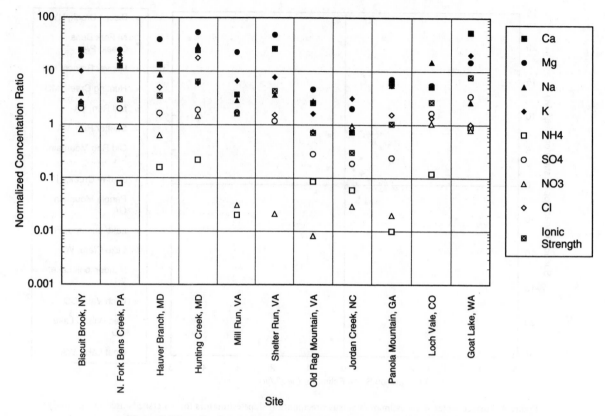

Figure 5. Normalized concentration ratios for watersheds used in this study

Indirect influences between precipitation and surface water chemistry were also explored by examining solutes or solute combinations related by chemical reactions. For example, a change in ionic strength in precipitation would be expected to be reflected in surface water for most catchments. However, only four of the 13 sites which could be compared have similar LOWESS curves and seasonal Kendall test trends in ionic strength (Figure 6). Similarly, we would expect long-term trends in hydrogen ions in precipitation to be associated with trends in surface water dissolved silica concentrations due to the effects of acidity on mineral weathering reactions, at least in biologically unproductive waters. Five of the 15 sites exhibit this relation (Figure 7).

Intersite trend relations

Atmospheric deposition exhibits sufficient spatial coherence to allow interpolation between monitoring sites (Wilson and Mohnen, 1982; Seilkop and Finkelstein, 1987; Zemba *et al.*, 1988). Stream water concentration patterns for six sites in Pennsylvania, Maryland and Virginia, however, exhibited little similarity (Figure 3b–g), except for the adjacent paired watersheds in Maryland (Hauver Branch and Hunting Creek). This lack of spatial coherence indicates either that terrestrial processes may control surface water chemistry and these processes are substantially different among these sites, or that the trends are the artifact of inadequate sampling. There were some adjacent watersheds which did exhibit similarities in surface water between some solutes. Jordan Creek, NC (Figure 3h) and Panola Mountain, GA (Figure 3i) had similar LOWESS patterns in sulphate concentration residuals. The Wisconsin lakes (Lake Clara and Vandercook Lake; Figure 3k and 3l) had very similar LOWESS patterns between each of the basic cations (calcium, magnesium, sodium and potassium). Surprisingly, however, in these two instances there was less coherence between sites in precipitation for the same solutes. This suggests that in each of these two instances, similar terrestrial processes may be controlling the surface water chemistry for the similar solutes.

DISCUSSION

Despite the fact that precipitation chemistry must ultimately influence surface water chemistry, the relation between these time series is apparent only for chloride in this data set, even though the sampled watersheds were small and generally had unreactive bedrock. Assuming that the concentration patterns adequately capture these trends, continued monitoring would be necessary to detect the surface water responses to the declines in precipitation sulphate concentrations that have been reported (Baier and Cohn, 1993)

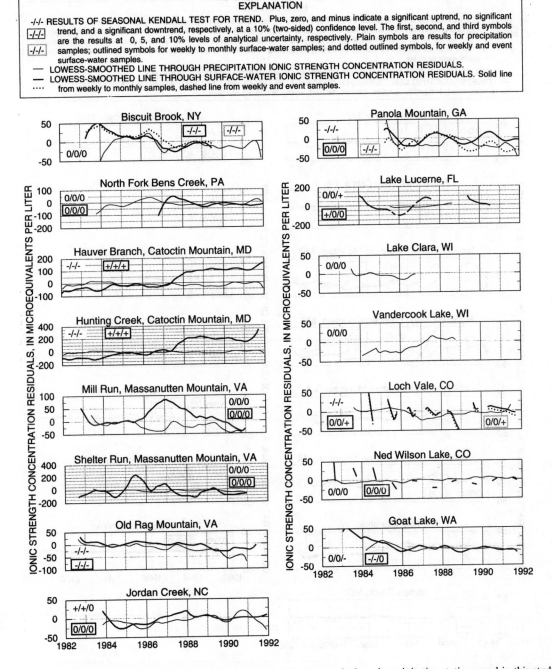

Figure 6. Trends in precipitation and surface water ionic strengths at watersheds and precipitation stations used in this study

and immediate improvement in surface water quality from declining acid precipitation should not be expected.

If a relation exists between precipitation chemistry and surface water chemistry for this period of record, an increase in the power of the statistical test by better filtering of natural variability would be required to detect it. Although surface water samples were flow-adjusted, the residual variation in surface water concentrations remained large relative to variations in the precipitation concentration residuals. The LOWESS

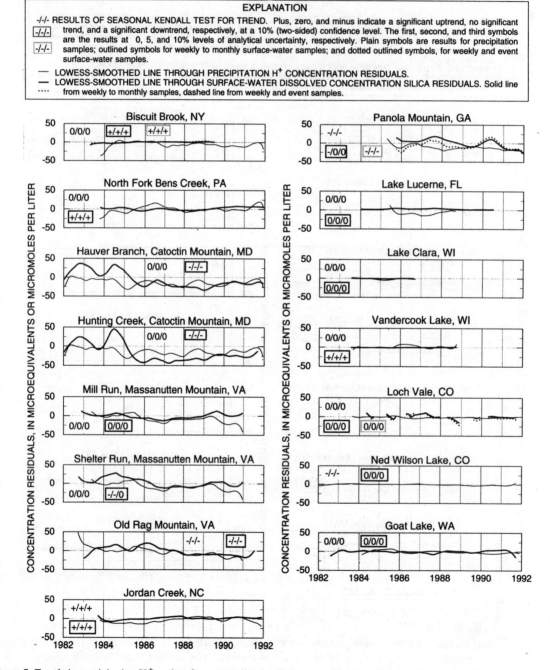

Figure 7. Trends in precipitation H$^+$ and surface water dissolved silica at watersheds and precipitation stations used in this study

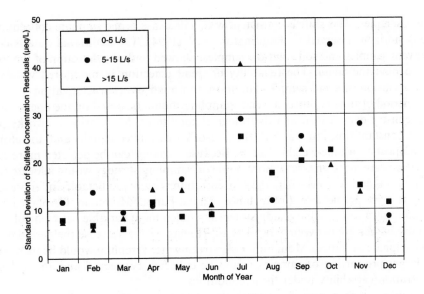

Figure 8. Standard deviation of sulphate concentration residuals at Panola Mountain, GA for different seasons and discharge classes

curves also indicated large short-term variation in surface water concentration residuals that were unrelated to seasonal cycles. One possible cause is the short-term climatic variations, which result in changes in hydrological, geochemical and biological processes which affect surface water chemistry. Simple flow adjustment does not adequately account for these changes.

The weak relation between precipitation and surface water trends and the large natural variability in surface water chemistry observed in this study using a fixed-station network indicates that detecting and interpreting trends using regional surveys may also have similar problems. Fixed-station networks define surface water chemistry over different seasons and hydrological conditions by collecting many samples per year at each site. We would expect it to be even more difficult to relate precipitation and surface water trends using a regional survey which typically uses one index sample per year for each site. This would be especially true if the surface water trend was related to a specific season or hydrological condition which may not be well represented by the index sample. The large natural variability observed in surface water chemistry over short periods in time indicates that obtaining representative index samples may be more problematic than previously assumed. It is important that the variability in surface water chemistry during the index period be assessed in regional surveys. Such variation would add to the 'noise' in the data set and, thereby, weaken the relation between precipitation and surface-water chemistry. This noise need not be white noise. If the natural variability of index samples is associated with regional climatic variations, trends could be induced that may be wrongly associated with depositional trends. If this is so, it would be important to remove climatic variation from the index samples. Removing climatic variations from regional survey data would be even more difficult than for a fixed-station network because there is less data at individual sites in regional surveys to relate with climatic factors.

Because some trend tests were sensitive to the inclusion of event samples, the adequacy of weekly to monthly sampling for characterizing the surface water concentration of these sites is questionable. In their analysis of the mass flux of sulphate from Panola Mountain, GA, one of the sites included in this study, Huntington *et al.* (1994) found a decline in sulphate transport between 1987 and 1992 that resulted in a trend both from a shift in the concentration–discharge relation for sulphate and from a change in hydrological flow paths resulting from different rainfall amounts and distribution between the years. In a further analysis of this data set, Hooper and Aulenbach (1993) demonstrated that the change in the concentration–discharge relation did not occur uniformly across all discharges, but is greater at higher discharges. Therefore, it would be more difficult to detect a trend with a weekly to monthly sampling interval which would contain few event samples. This shift is consistent with a decrease in the sulphate concentration of the soil solution in the A horizon observed by Huntington *et al.* (1994). The additional data collected at Panola Mountain, the event samples and soil solution samples, allowed both the detection of the trend and insight into the causes behind the trend. The reliability of trend detection in monitored watersheds would be increased by the inclusion of event samples and more thorough site characterization.

To reduce the period of time to detect a trend, sampling should be based on the unexplained variation in surface water chemistry over season and hydrological conditions. Monitoring networks typically sample surface water at a constant time interval such that seasons are sampled equally and, therefore, hydrological conditions are sampled on a time-weighted basis. Because a change can be detected more easily when concentration predictions are more precise, a more powerful sampling strategy would allocate fewer samples to hydrological conditions and seasons where the concentration is more predictable and more samples to those conditions where it is less predictable. For instance, data at Panola Mountain, GA indicated that residuals to the sulphate concentration–discharge adjustment are much more variable between July and November, especially when discharges are between 5 and $15\,\mathrm{l\,s^{-1}}$ (Figure 8). To decrease the period of time needed to detect a trend in sulphate at Panola Mountain, GA, streamwater sampling would be increased during these months and a larger proportion of samples would be collected between 5 and $15\,\mathrm{l\,s^{-1}}$ to better define sulphate concentration variability under those conditions.

Is a fixed-station network approach adequate for detecting regional precipitation and surface water trends? This study has shown that flow adjustment alone is not adequate to remove all natural variability in surface water concentrations and that constant time interval sampling with a weekly to monthly interval may not be adequate. However, ongoing watershed monitoring studies such as United Nations Economic Commission of Europe's Integrated Monitoring program on the effects of air pollution use similar sampling protocols (Pylvänäinen, 1993). Our results indicate that we might improve the ability to detect trends in surface water with a fixed-station network approach by using a stratified sampling scheme and improved data analysis techniques. Firstly, we would characterize each site for unexplained variability in surface water chemistry versus hydrological status and seasonality at the beginning of the study. Then sampling would be defined by this variability with appropriate coverage to be able to detect trends over all discharge classes and seasons. These samples would have to be integrated into a single representative concentration for each season of each year using a weighting scheme. This approach would probably detect a trend in fewer years, but requires more sampling during that period of time. Although solute concentrations can still be affected by long-term trends in climate, this approach can better account for natural variability in surface waters.

REFERENCES

Baier, W. G. and Cohn, T. A. 1993. 'Trend analysis of sulfate, nitrate, and pH data collected at National Acid Deposition Program/ National Trends Network stations between 1980 and 1991', *US Geol. Surv. Open-File Rep.*, **93–56**, 13 pp.

Cleveland, W. S. 1979. 'Robust locally weighted regression and smoothing scatterplots'. *J. Am. Stat. Assoc.*, **74**, 829–836.

Driscoll, C. T., Likens, G. E., Hedin, L. O., Eaton, J. S., and Bormann, F. H. 1989. 'Changes in the chemistry of surface waters'. *Environ. Sci. Technol.*, **23**, 137–143.

Fenneman, N. M. 1946. *Physical Divisions of the United States*. Reference Code **38077-H1-UG-07M-11**. U.S. Geological Survey, Reston.

Geigert, M. A., Nikolaidis, N. P., Miller, D. R., and Heitert, J. 1994. 'Deposition rates for sulfur and nitrogen to a hardwood forest in northern Connecticut, U.S.A.', *Atmos. Environ.*, **28**, 1689–1697.

Gilliom, R. J. and Helsel, D. R., 1986. 'Estimation of distributional parameters for censored trace level water quality data. 1. Verification and applications', *Wat. Resour. Res.*, **22**, 147–155.

Helsel, D. R. and Gilliom, R. J. 1986. 'Estimation of distributional parameters for censored trace level water quality data. 2. Estimation techniques', *Wat. Resour. Res.*, **22**, 135–146.

Herlihy, A., Kaufmann, P., and Mitch, M. 1991. 'Stream chemistry in the Eastern United-States. 2. Current sources of acidity in acidic and low acid-neutralizing capacity streams', *Wat. Resour. Res.*, **27**, 629–642.

Hirsch, R. M. and Peters, N. E. 1988. 'Short-term trends in sulfate deposition at selected bulk precipitation stations in New York', *Atmos. Environ.*, 22, 1175–1178.

Hirsch, R. M., Slack, J.R., and Smith, R.A. 1982. 'Techniques of trend analysis for monthly water quality data', *Wat. Resour. Res.*, **18**, 107–121.

Hooper, R. P. and Aulenbach, B. T. 1993. 'The role of sampling frequency in determining water-quality trends' [abstract], *Eos, Trans. Am. Geophys. Union*, **74**, 279.

Huntington, T. G., Hooper, R. P., and Aulenbach, B. T. 1994. 'Hydrologic processes controlling sulfur mobility: a small watershed approach', *Wat. Resour. Res.*, **30**, 283–295.

Johnson, N. M., Likens, G. E., Bormann, F. H., Fisher, D. W., and Pierce, R. S. 1969. 'A working model for the variation in stream-water chemistry at the Hubbard Brook Experimental Forest, New Hampshire', *Wat. Resour. Res.*, **5**, 1353–1363.

Kaufmann, P., Herlihy, A., Mitch, M., Messer, J., and Overton, W. 1991. 'Stream chemistry in the Eastern United-States 1. Synoptic survey design, acid–base status, and regional patterns', *Wat. Resour. Res.*, **27**, 611–627.

Landers, D. *et al.* 1986. 'Characteristics of lakes in the Western United States. Vol. I. Population descriptions and physico-chemical relationships', *US Environmental Protection Agency Rep*, **EPA/600/3–86/054a**, US Environmental Protection Agency, Washington, 176 pp.

Linthurst, R., Landers, D., Eilers, J., Brakke, D., Overton, W., Meier, E., and Crowe, R. 1986. 'Characteristics of lakes in the Eastern United States. Vol. I. Population descriptions and physico-chemical relationships', *US Environmental Protection Agency Rep*, **EPA/600/4–86/007a**, US Environmental Protection Agency, Washington, 136 pp.

Murdoch, P. S., and Stoddard, J. L. 1993. 'Chemical characteristics and temporal trends in eight streams of the Catskill Mountains, New York', *Wat. Air Soil Pollut.*, **67**, 367–395.

Peters, N. E., Schroeder, R. A., and Troutman, D. E. 1982. 'Temporal trends in the acidity of precipitation and surface waters of New York', *US Geol. Surv. Wat. Supply Pap.*, **2188**, 35 pp.

Pylvänäinen, M. (Ed.) 1993. *Manual for Integrated Monitoring: Programme Phase 1993–1996*. Environmental Data Centre, Helsinki. 114 pp.

Rosseland, B. and Henriksen, A. 1990. 'Acidification in Norway—loss of fish populations and the 1000-lake survey 1986', *Sci. Total Environ.*, **96**, 45–56.

Seilkop, S. K. and Finkelstein, P. L. 1987. 'Acid precipitation patterns and trends in eastern North America, 1980–1984', *J. Climate Appl. Meteorol.*, **26**, 980–994.

Shertz, T. L. and Hirsch, R. M. 1985. 'Trend analysis of weekly acid rain data 1978–1983', *US Geol. Surv. Wat. Resour. Invest. Rep.*, **85–4211**, 64 pp.

Smith, R. A. and Alexander, R. B. 1983. 'Evidence for acid-precipitation-induced trends in stream chemistry at hydrologic bench-mark stations', *US Geol. Surv. Circ.*, **910**, 12 pp.

Smith, R. A., Alexander, R. B., and Wolman, M. G. 1987. 'Analysis and interpretation of water-quality trends in major U.S. rivers, 1974–81', *US Geol. Surv. Wat. Supply Pap.*, **2307**, 25 pp.

Steele, T. D., Gilroy, E. J., and Hawkinson, R. O. 1974. 'An assessment of areal and temporal variations in streamflow quality accounting network', *US Geol. Surv. Open-File Rep.*, **74–217**, 210 pp.

Stoddard, J. L. and Kellogg, J. H. 1993. 'Trends and patterns in lake acidification in the state of Vermont: evidence from the long-term monitoring project', *Wat. Air Soil Pollut.*, **67**, 301–317.

Turk, J. T. 1983. 'An evaluation of trends in the acidity of precipitation and the related acidification of surface water in North America', *US Geol. Surv. Wat. Supply Pap.*, **2249**, 18 pp.

Wilson, J. W. and Mohnen, V. A. 1982. 'An analysis of spatial variability of the dominant ions in precipitation in the eastern United States', *Wat. Air Soil Pollut.*, **18**, 199–213.

Witt, E. C. III and Bikerman, M. 1991. 'Geochemical mass-balance in a small forested watershed in southwestern Pennsylvania' in *Proceedings, 1991 National Conference on Irrigation and Drainage, Honolulu, Hawaii, 22–26 July, 1991*. American Society of Civil Engineers, Irrigation and Drainage Division, New York. pp. 516–523.

Zemba, S. G., Golomb, D., and Fay, J. A. 1988. 'Wet sulfate and nitrate patterns in eastern North America', *Atmos. Environ.*, **22**, 2751–2761.

3

WATER QUALITY TRENDS AT AN UPLAND SITE IN WALES, UK, 1983–1993

A. J. ROBSON AND C. NEAL

Institute of Hydrology, Crowmarsh Gifford, Wallingford, OX10 8BB, UK

ABSTRACT

Ten years of detailed upland stream and bulk deposition water quality data from Plynlimon, mid-Wales, are examined for trend. A robust statistical test (the seasonal Kendall test) is applied and data are presented graphically. Smoothing techniques are used to highlight the patterns of change which underlie high data scatter. The graphs show long-term cycles within the data which violate the assumptions of common statistical tests for trend. These cycles relate to fluctuations in the weather patterns at Plynlimon. Even though the seasonal Kendall test is significant for some determinands, the evidence from the graphs suggests that many of these 'trends' are unlikely to continue.

For solutes in rainfall, there is no convincing long-term trend. There is a possible increase in ammonium concentrations, which may indicate an increasing atmospheric source generated by farming activities, but this will require a longer data series for confirmation. Several trace metal concentrations increased significantly part way through the study period, but later returned to the original levels. The bulk precipitation sea salt input has been uneven over the 10-year sampling period, with the highest inputs occurring during the wetter winters.

For solutes in streamwaters, there are clear trends in dissolved organic carbon (DOC), iodine and bromide, which increase over time and may be attributed to an increase in organic decomposition in the catchment. Previous studies in Wales have shown similar behaviour for colour, which is related to DOC, but the corresponding changes for bromide and iodine are new. For most other streamwater determinands, any changes are masked by the effects of year to year variations in the quality and quantity of rainfall. For example, zinc and chromium variations parallel the corresponding rainfall quantity variations. The effect of rainfall quality variation is marked for marine-derived elements such as chloride. For sulphate, streamwater variations are inverted relative to chloride. This suggests that dry deposition may vary with weather conditions: high when the wind direction is from the land and low when weather systems are predominantly frontal and laden with sea salts. Alternatively, high sea salt rainfall may be affecting absorption/solubility reactions in the soils.

There are four main conclusions. Firstly, there is no indication of changing acid deposition inputs or changing acidity within the runoff, despite a decline in UK sulphur dioxide emissions. Secondly, streamwater DOC has shown an increase over time, but there is no clear corresponding decrease in pH as might be expected from acidification theory. Thirdly, there are cyclical variations in bulk precipitation inputs and in streamwater quality, which mean that trends cannot be established even with 10 years of data. Long-term cycles are likely to exist in other environmental data and extreme care is required for the interpretation of trend, especially if data sets are short. This aspect strongly supports the continuation of long-term monitoring programmes over several decades. Finally, the graphical application strongly enhances data analysis and should be considered an essential component of trend investigation.

INTRODUCTION

There continues to be much concern over the long-term impacts of atmospheric pollution on acidic and acid-sensitive environments. Even in areas where acid deposition has been reduced, there is evidence of continued soil acidification and loss of exchangeable base cations (Christophersen *et al.*, 1990). This may well mean that proposed reductions in acidic oxide emissions across Europe will be insufficient to improve stream ecology for many areas. Indeed, water quality may continue to deteriorate if current long-term acidification models are correct. Such models rely on theories which have not yet been verified and which, in some instances, have been invalidated in the field. Because of this, it is important that environmental policies are not based solely on modelled results. Data from long-term catchment monitoring studies must be examined, so that the true situation can be identified. Analyses of long-term data may provide clues about what will occur in the future.

In the UK, there is evidence of surface water acidification, but detailed mechanisms are poorly understood. Present theory dictates that the input of acidic oxides to acidic upland systems leads to a loss of base cations from the soil and increases in the acidity of soil and streamwaters. Soil acidity may also be further increased by planting conifers, a factor of concern because of the extensive upland areas afforested since the war (Neal *et al.*, 1992). Increased acidity, coupled with the resulting mobilization of toxic forms of aluminium, threatens fish and leads to a deterioration in aquatic species diversity. The time-scale of concern is potentially fairly long. For example, a typical plantation forest cycle is around 40 years.

Because of these environmental concerns, and given the limited information on detailed hydrochemical mechanisms, several UK catchment studies have been initiated to try and discern the effects of acidification on surface water quality. In this paper, trend analysis is performed for the Plynlimon catchment in the Welsh uplands. In comparison with most sites, Plynlimon data provide an unusually long and unusually detailed record (Hornung *et al.*, 1990; UK Acid Waters Monitoring Network, 1991). An intensive sampling programme at this typical upland site has weekly hydrochemical data for more than a decade. Detailed chemical analysis is available for a wide range of determinands. The prime concern is for changes in the main pollutant species such as aluminium, hydrogen and sulphate, although trends are considered for trace metals (e.g. manganese and zinc) and organic material, which also are affected by acidification (Hemond, 1994; Mulder and Cresser, 1994; Vesely, 1994). This study provides a trend assessment for both the standard determinands and components not usually determined at other sites. Emphasis is placed on examining the observed streamwater changes in relation to bulk precipitation inputs, flow variations and catchment processes. An advanced graphical presentation of the data, not commonly used in trend analysis of water quality, is used to smooth the data, highlighting the underlying patterns of change. The graphical approach is used to interpret the results of a standard statistical test, the seasonal Kendall test, for trend in environmental data.

Trend, time-scale and process

In this study, the aim is to detect trends which are likely to continue over many years and which indicate internal catchment change. Because catchment processes are complex, a trend in streamwater chemistry can arise for multiple reasons, each of which may have different implications for the future. For example, a decline in base cations and an increase in hydrogen and aluminium concentrations may indicate catchment acidification (UK Acid Waters Review Group, 1988). However, such a base cation decrease could also be observed if there was decreased atmospheric deposition because fewer ions would then be moving through the system. Examining streamwater trends in isolation is not sufficient. Rather, it is essential to examine the observed trends in the context of within-catchment chemical and hydrological processes as well as bulk precipitation inputs.

Although the time-scale of interest is a decade or more, changes in streamwater quality can be observed at many scales. Such changes can relate to storm episodes, lasting hours to days (Davies *et al.*, 1992), or they can relate to seasonal features. On a slightly longer time-scale, water quality may show long-term cycles caused by year to year variations in weather patterns. All of the above need to be identified so that the underlying trends, if any, can be distinguished. Apparent, but false, long-term trends can easily result if short-term, seasonal and long-term cycles are not taken into account. A decade-long trend seen today may later merely turn out to be part of a longer term fluctuation. The more that is understood about processes, and the longer the data series, the better the chance of correctly detecting trend.

PLYNLIMON CATCHMENTS

The Plynlimon site is located 38 km from the west coast of mid-Wales and is typical of much of upland Britain. The bedrock consists of Lower Palaeozoic mudstones, shales and grits. The soils are acidic and range from peats and podzols to gleys. The Afon Hafren (catchment area 347 ha; altitude 350–690 m) is the main headwater tributary to the River Severn. The upper half of the Hafren catchment drains acidic semi-natural moorlands and peats. The lower half, which was once semi-natural moorland, is densely planted with conifers, which date from 1937 to 1964. Plynlimon has a cool, wet maritime climate: the annual rainfall averages 2400 mm and the annual evapotranspiration loss ranges from 400 to 600 mm. More extensive descriptions of the Plynlimon area are given by Kirby *et al.* (1991) and in Newson (1976).

Chemical sampling programme and analysis

As part of the Plynlimon programme, rainfall and streamwater samples from the upper part of the river Severn (including the Afon Hafren, of concern in this paper) have been collected at approximately weekly intervals since 1983. Rainfall samples were bulked weekly averages and streamwater samples were instantaneous grab samples. A wide range of chemical analyses have been carried out on these samples, including pH, conductivity and major, minor and trace elements. Samples are filtered (0·45 μm) and stored in acid-washed polypropylene bottles. All samples were pre-concentrated 20-fold and analysed using inductively coupled plasma atomic emission spectrometry and inductively coupled plasma emission mass spectrometry for major, minor and trace metals. Automatic colorimetric techniques were used to determine F, Cl, Br, I, NO_3, NH_4 and Si for 2 μm filtered samples which had been stored in chromic acid-washed bottles. Electrometric techniques were used to determine pH and alkalinity, the latter by Gran titration, for unfiltered samples at field temperature. Dissolved organic carbon (DOC) was determined using a TOCsinII aqueous carbon analyser. Rainfall DOC is not presented because of problems with DOC leaching from aging plastic funnels. Streamwater was collected in glass bottles and is not affected by this problem. Early in the study period, silica concentrations in rainfall were suspiciously high (around 2 mg l^{-1}), whereas later on silica decreased to very low values. Rainfall silica data are not used in the following analysis because of possible contamination by the rainfall collector. For all determinations, quality control checks were made throughout the sampling period and instruments have been recalibrated if the specified accuracy limits (usually 2%, but up to 10% for some trace metals) are exceeded.

The detection limits varied for the different determinands. For major components, the lowest quotable values (in mg l^{-1}) are about 0·1 for Na, Ca and Mg, 0·04 for K and SO_4, 0·1 for Si and NO_3 and 0·2 for DOC and Cl. Throughout the text, concentrations of SO_4 and NO_3 are given as mg l^{-1} of the species (rather than as S and N, respectively). For the trace constituents, the lowest quotable values (in μg l^{-1}), are 0·2 for Sr, Ba, Co, Mn, Y and I, 0·4 for Li, 1·5 for B, Cu, Fe, Zn and Br, and 0·05 for other components determined by inductively coupled plasma emission mass spectrometry.

Further methodological details are presented elsewhere (Neal *et al.*, 1992; 1986). For several minor elements, the method of analysis was changed in 1992 and the last year of data has not been used in examining trends.

Primary catchment chemical processes

To interpret trends for a determinand, it is important that the trend is related to the atmospheric deposition and hydrochemical processes within the catchment. Here, a brief overview is provided and further information can be found in Neal *et al.* (1986; 1990a; 1992) and Robson (1993).

Wet and dry deposition are important chemical sources for the Plynlimon catchments. The chemistry of wet deposition is variable and is linked to meteorological conditions (UK Review Group on Acid Rain, 1990). Marine salts are brought in by Atlantic frontal systems, which are common during the winter. Pollutant components (e.g. ammonium, nitrate and heavy metals) derive from long-range transport from a variety of sources (e.g. from industry and agriculture; UK Review Group on Acid Rain, 1990) and are often highest in summer and in small storms. Dry deposition for some constituents is high, enhanced by the rough forest vegetation and high altitude. For example, 33% of the total sulphate deposition to the catchment is estimated to be dry deposition which is not collected by the bulk precipitation collectors.

The substantial variations in rainfall chemistry are typically damped by the catchment, yielding a longer term alteration of stream chemistry (Reynolds and Pommeroy, 1988; Neal and Rosier, 1990). This reflects the importance of chemical storage within the catchment (Robson, 1993) and the modification of wet deposition by chemical interaction with the soils. There are marked differences in the chemical characteristics of upper and lower soil solutions and groundwater (Reynolds *et al.*, 1986). The main hydrochemical processes occurring in these zones are summarized in the following.

Upper soils. These soils have a high organic content. They are strongly acidic (pH values from 3·7 to 4·2) and are enriched in humic acids. Biological activity, adsorption–desorption and organic decomposition processes are important in determining the mobility of hydrogen ion, nutrients (phosphate, nitrate and potassium), bromide, iodine and sulphate. Cation-exchange processes involving the organic materials also are important: hydrogen and aluminium ions in the soil are exchanged for base cations. The pH and aluminium concentration and speciation are further altered by reactions with humic acids.

Lower soils. The lower soils are mainly formed from weathering of inorganic aluminosilicate bedrock and from residual bedrock materials themselves. Acidic waters from the upper soils partly dissolve the weathering products, producing solutions bearing aluminium and transition metals. The soil solutions typically range in pH from 4·2 to 4·9 and have much lower humic acid contents than the upper soils due to both flocculation and decomposition reactions. Sulphate adsorption also can occur.

Groundwater/bedrock zone. In this zone, aluminosilicate bedrock minerals are present and weathering reactions predominate. Base cations (calcium and some magnesium) are released and the pH increases so that aluminium and transition metals are precipitated: some sulphate may also be precipitated.

Streamwater chemistry. Streamwater chemistry typically is flow-dependent (Neal *et al.*, 1992) and is strongly influenced by both the current and antecedent hydrological conditions (Durand *et al.*, 1994). This dependency reflects the marked chemical gradients described earlier and the alteration of hydrological pathways during storm events. The discharge of waters from the soils is largest during storm flow. Thus storm flow waters are acidic and enriched in transition metals and aluminium. In contrast, baseflow waters, derived from groundwater, are calcium- and hydrogencarbonate-rich, and provide the most ecologically favourable conditions in the stream.

Overall, streamwater chemistry may be viewed as the result of rainfall which is chemically modified by the soils and bedrock, stored in the catchment for variable periods, and is finally released to the stream via episodically altering flow pathways.

Trend analysis methods

In this paper, two approaches are taken. Firstly, robust statistical tests for trend have been applied. Secondly, advanced graphical presentation has been used to give further insight into the data and to check out the assumptions made in the statistical tests. The two approaches are detailed in the following.

Statistical tests for trend. The non-parametric seasonal Kendall test was used to test for trend. This test is commonly used for testing for monotonic trend in environmental data (Christophersen *et al.*, 1990; Walker, 1991; Waters and Jenkins, 1992; Kahl *et al.*, 1993; Newell, 1993). The seasonal Kendall test is an adaptation of the non-parametric Kendall test which can be used for data with seasonality and serial correlation (Hirsch *et al.*, 1982; Hirsch and Slack, 1984). The test is suited to non-normal data because it uses ranked data in its calculations. It can also handle data with missing values. It has been widely recommended for use with water quality data (e.g. Berryman *et al.*, 1988; Taylor and Loftis, 1989).

A seasonal Kendall slope estimator (based on the Sen estimate of slope; Sen, 1968) is used to estimate the annual rate of change. The estimate is relatively robust with respect to extreme values as it is constructed by calculating all pairwise slopes within a season and taking the median of all these values over all the seasons. For comparison, a robust technique, least-trimmed squares regression, was also used to estimate the slope. The estimated slope is that slope which minimizes the sum of the smallest half of the squared residuals. Note that this method does not allow for serial correlation in the data.

Although the seasonal Kendall test is robust to several factors, it is based on the assumption that concentrations increase or decrease monotonically or remain stable over long periods of time (Sirois, 1993). This assumption is not necessarily valid for environmental data (Sirois and Summers, 1989; UK Review Group on Acid Rain, 1990). Graphical presentation is one means of checking whether the test assumptions hold.

Graphical presentation. Graphical presentation provides a means of examining data and visually summarizing change, without making restrictive assumptions about the nature of this change. In this paper, the time series of concentrations is overlain with a smoothed curve, which is used to aid identification of the broader characteristics of change. The graphs allow cross-comparisons to be made between the different determinands and between rain and streamwaters.

The method selected for smoothing the data is a *locally weighted* regression *smoothing scatterplot* technique, called LOWESS (Cleveland, 1979; Chambers and Hastie, 1992), which was used within the Splus programming environment (both *loess* and *lowess* functions exist in this program; the newer and more general function, *loess*, was used for the analysis herein; Becker *et al.*, 1988). The smoothed curve at any point is derived by fitting a local quadratic polynomial between the predictors and the dependent variable. For any value of the predictors, a weighting function is applied to nearby points, and least-squares optimization is used to fit the local curve. In this application, a robust bilinear weight function was used, which is suited to cases where errors are symmetrically distributed. Flow and streamwater quality data are highly skewed at Plynlimon and, in common with much environmental data, have a distribution which is approximately log-normal. For positive data series, a log transformation was applied. This improved the symmetry of the data considerably, reducing the influence of the extreme points. The effect of these extreme values on the fitted LOWESS curves was investigated and was found to make only a slight difference to the fit. Log-transformed data were back-transformed before plotting. Values less than the lowest quotable value (*lqv*) were assigned a value of half the *lqv* for these log transformations. For several graphs, a further LOWESS curve is presented in addition to the LOWESS curve fitted to the concentration data. This is weighted for flow (or volume in the case of rainfall) and is used to emphasize changes at high flow, and thus changes in load. A LOWESS curve weighted towards low flow data was also investigated, but generally paralleled the other curves and is therefore not shown. The degree of smoothness used in fitting the curve may be adjusted and here it is selected to emphasize the general drift in the data rather than the seasonal or short-term changes. It should also be noted that the accuracy of the smoothed curve is less at the start and end of each data series because fewer data points can be used to produce it. When examining the LOWESS curves, allowance should be made for lower certainty at the start and end of the records.

The LOWESS plots were produced for all determinands and were examined to see if change had been monotonic throughout the period. The results were used to assess the validity of the assumption of monotonic change. For determinands with a significant seasonal Kendall statistic, the data were investigated further. The time series were decomposed into seasonal, trend and residual components using the LOWESS based STL routine of Splus (Cleveland *et al.*, 1990; Statistical Sciences, 1993). This technique uses smoothing techniques to estimate the frequency components of the data set. A smoothed seasonal component is extracted first, followed by a low frequency component (the trend). The remainder corresponds to the other higher frequency variations caused by factors such as rainfall, climate and flow. Note that, as with the LOWESS curves, the smoothing applied in this technique requires a subjective choice of parameters. The same smoothing parameters were used for all determinands; they were chosen to try and emphasize the underlying variations. In this decomposition, no transformation of the data was used because of the difficulty of subsequently interpreting the components. However, the robust bilinear weight function was still used in the curve fitting.

For streamwater quality determinands, the effect of flow-induced variation was also considered by looking at the trends at low, medium and high flows. Each flow grouping was selected to contain the same number of sample points, with 25% of the points in one regime also overlapping into the next regime. Scatter plots and LOWESS curves were used to see how the trends varied over the 10 year period at different flow levels. This is important because almost all determinands show some form of relationship with flow due to the chemical gradients in the catchment (see earlier section). The exercise was aimed at detecting whether any of the 'trends' are an artefact of uneven flow distribution.

APPLICATION TO PLYNLIMON

The seasonal Kendall test was applied to all determinands which were detected in the rainfall and stream-waters, except for rainfall DOC and Si (Table I). Time series plots showing LOWESS curves were also examined. For sodium, calcium, magnesium and sulphate, excess concentrations (the non-marine component as estimated from the chloride concentration) also were examined. In the case of rainfall, these excesses represent the components associated with pollutant sources. Correspondingly, for the streamwaters, the excesses relate to a combination of sources including weathering, atmospheric deposition and cation

Table I. Seasonal Kendall test results and estimated slopes for Plynlimon rainfall and streamwaters. The estimated slopes have been multiplied by 10 to show the change in concentration over a decade. Statistically significant seasonal Kendall p-values are shown in bold

Determinand	Rainfall				Streamwaters			
	Seasonal Kendall p-value	Seasonal Kendall slope (×10)	Robust regression slope (×10)	Mean concentration	Seasonal Kendall p-value	Seasonal Kendall slope (×10)	Robust regression slope (×10)	Mean concentration
Flow (mm h^{-1})	NA	NA	NA	NA	0·32	0·02	0·04	0·21
Cl (mg l^{-1})	0·81	−0·26	−0·73	4·52	0·23	1·03	1·09	7·24
Na (mg l^{-1})	0·69	0·14	−0·29	2·24	**0·09**	0·55	0·51	4·05
K (mg l^{-1})	0·36	−0·02	0	0·13	**0·01**	−0·02	−0·04	0·14
Ca (mg l^{-1})	0·42	0·04	0	0·3	0·16	−0·07	−0·07	0·89
Mg (mg l^{-1})	0·87	0	−0·04	0·28	**0·04**	−0·07	−0·07	0·78
SO$_4$ (mg l^{-1})	0·8	0·05	−0·04	2·18	0·27	−0·35	−0·33	4·1
NO$_3$ (mg l^{-1})	0·5	0·16	0·18	1·68	**0·06**	−0·16	−0·18	1·23
NH$_4$ (mg l^{-1})	**0·07**	0·18	0·22	0·51	0·3	0	0	0·01
DOC (mg l^{-1})	NA	NA	NA	NA	**<0·005**	0·7	0·66	1·25
Br (μg l^{-1})	0·11	−3·33	−3·39	17·45	**<0·005**	4	4·45	21·44
I (μg l^{-1})	0·35	−0·08	0	1·66	**<0·005**	0·33	0·33	0·98
pH	0·49	−0·03	0·04	4·92	0·82	0·05	−0·04	5·39
H (μequiv. l^{-1})	0·52	1·51	−0·95	23·16	0·7	−0·35	0·11	8·06
Alkalinity (μequiv. l^{-1})	0·2	−7·61	−0·84	−20·13	0·8	0·61	1·31	−2·49
Si (mg l^{-1})	NA	NA	NA	NA	0·29	−0·14	−0·15	1·71
Sr (μg l^{-1})	0·65	0·01	0	1·98	0·27	−0·02	−0·02	5·33
Li (μg l^{-1})	NA	NA	NA	0·07	0·69	0	0	2·03
Y (μg l^{-1})	NA	NA	NA	0·01	0·83	0	0·04	0·26
Ba (μg l^{-1})	**0·09**	−0·66	−0·99	3·26	0·56	−0·08	0·07	2·91
Mn (μg l^{-1})	0·16	−0·35	−0·26	1·93	0·96	0·12	0·18	34·37
Cu (μg l^{-1})	0·13	−0·92	−1·02	2·18	**0·05**	−0·75	−0·66	1·57
Fe (μg l^{-1})	**0·05**	−1·7	−0·04	9·64	0·24	7·05	11·86	73·08
Co (μg l^{-1})	NA	NA	NA	0·02	**0·05**	−0·25	−0·22	1·86
Zn (μg l^{-1})	**0·02**	−5·74	−5·07	13·61	**0·03**	−5·44	−3·43	13·43
Al (μg l^{-1})	0·71	−0·92	−0·91	10·98	0·71	−16·25	0·15	184·75
Cr (μg l^{-1})	0·56	0·17	−0·22	2·65	**0·04**	−2·93	−1·82	3·46
Excess Na (mg l^{-1})	0·28	0·09	−0·04	−0·27	0·6	−0·2	−0·18	0·02
Excess K (mg l^{-1})	0·94	−0	0	0·04	**0·01**	−0·06	−0·07	−0·01
Excess Ca (mg l^{-1})	**0·1**	0·05	0·04	0·17	**0·04**	−0·09	−0·11	0·73
Excess Mg (mg l^{-1})	0·13	−0·02	−0·04	−0·03	**0·01**	−0·16	−0·18	0·29
Excess SO$_4$ (mg l^{-1})	0·82	−0·07	0·11	1·4	0·17	−0·49	−0·47	3·08

exchange. Several data series had a significant test statistic but, as will be seen below, many of these also showed a non-monotonic trend, suggesting long-term cycles in the data. Many of these long-term cycles appeared to be rooted in the year by year variations in the marine rainfall component. This variation is propagated through to the stream and affects many streamwater anions and cations (not just the marine species). The length of the cycles indicate that much longer data series would be required before this component could be estimated and separated from any underlying trend.

Cycles in the data are important when interpreting the seasonal Kendall test results. An artificial data set can be readily constructed that is trend-free and contains seasonal and long-term cycles, but results in a statistically significant trend according to the seasonal Kendall test (see Figure 1). Thus the seasonal Kendall statistic cannot reliably give an accurate test of trend when the underlying changes are non-monotonic. For this reason, determinands, for which the seasonal Kendall statistic is significant, are examined in more detail to see whether the assumption of monotonic trend is reasonable. In cases where the seasonal Kendall is significant but trend is non-monotonic, the length of data series is probably insufficient for conclusive proof.

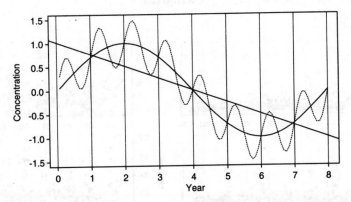

Figure 1. Artificial data containing long-term and annual cycles, but without any underlying trend. The seasonal Kendall test for these data (dotted line) gives a significant result ($p < 0.043$) which is misleading. The straight line shows the apparent trend that is estimated for this data series

Trends in rainfall chemistry

Fluctuations in rainfall volume are important because of their effect on stream flow. Rainfall chemistry also is linked to volume because very wet periods are usually associated with Atlantic frontal systems. At Plynlimon, the annual rainfall for the study period ranged from 2117 mm in 1984 to 2897 mm in 1983, the year during which sampling began. 1990 was also wetter (2839 mm) than average (2400 mm) and was followed by a dry year in 1991 (2251 mm). Annual and monthly rainfall totals and a smoothed rainfall curve, which highlights differences between the seasons, are shown in Figure 2. Rainfall during the summer of 1984 stands out as being exceptionally dry.

Yttrium, cobalt and lithium, which are detectable in streamwaters, were at or near the lowest quotable value for rainfall and are not included in the rainfall trend analysis.

Rainfall species showing possible trend. The following species showed a possible trend: ammonium*, barium*, iron, zinc, excess calcium* (*denotes only significant at a 90% level; the remainder are significant at a 95% level).

Several pollutant components had a significant seasonal Kendall test statistic (Table I). Of these, the heavy

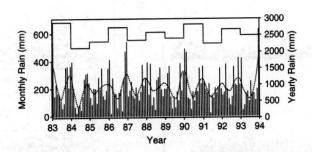

Figure 2. Monthly and annual rainfall for Plynlimon. Vertical lines show the monthly precipitation totals. The dotted line shows a smoothed LOWESS curve fitted through the monthly data which highlights periods which are unusually dry or wet. The stepped solid line shows the annual rainfall totals for the period

Figure 3. Rainfall time series plots. For each chemical species, a pair of graphs is shown; the left-hand graph shows the entire data set and the right-hand graph shows an enlargement after removing the top and bottom 5% of the data. The thick vertical mark on the left hand axes marks the extent of the axes for the right-hand graphs. LOWESS curves are shown; the solid line is fitted to unweighted data whereas the dotted lines have been volume-weighted

Figure 3A and B over page

Rainfall

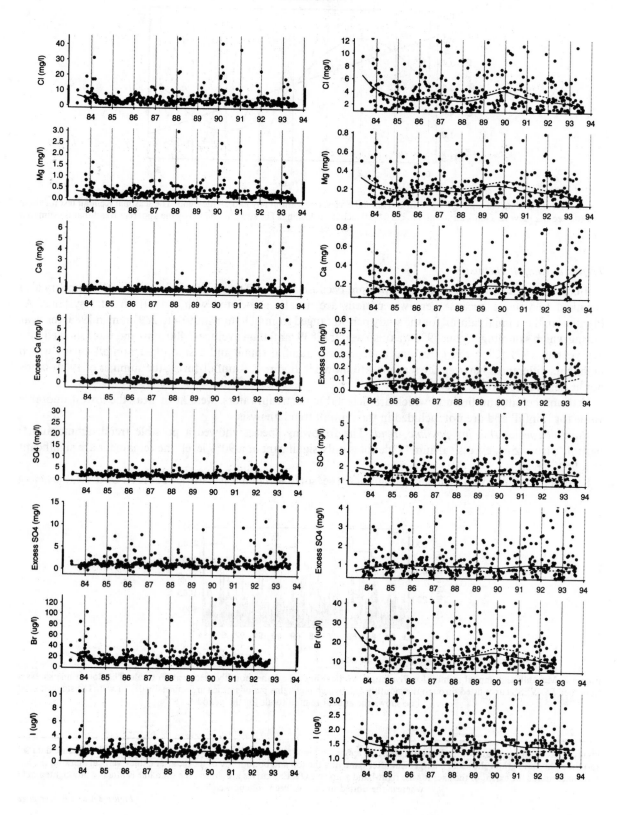

Rainfall

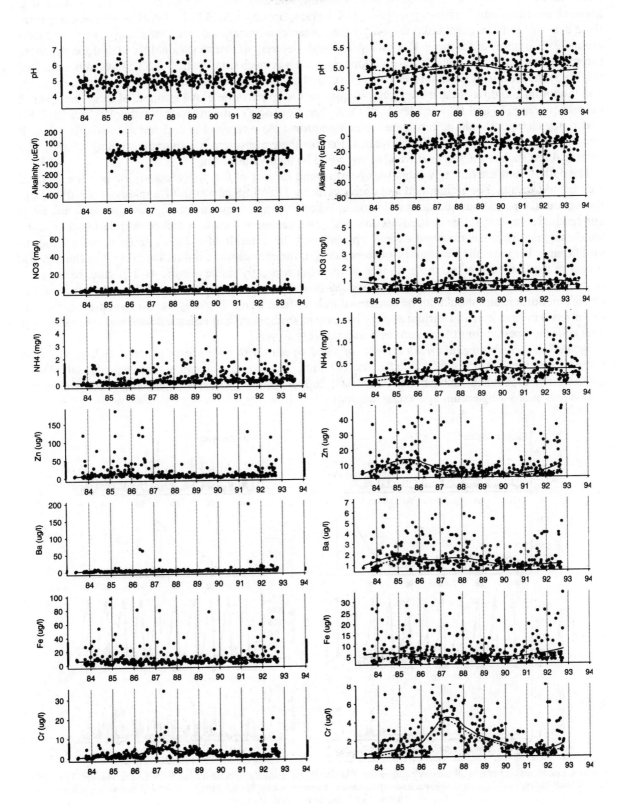

metals and excess calcium are probably from industry, and ammonium is probably from agriculture (e.g. livestock emissions and fertilizer application; UK Review Group on Acid Rain, 1990). Regression slopes for these determinands are substantial relative to their mean concentrations; often the change over 10 years is around 25% or more of the mean. Robust regression slopes were similar to seasonal Kendall slope estimates, except for iron where the robust regression slope estimate was much smaller. LOWESS curves were examined (Figure 3) and a complex situation is evident. The underlying trends for some of the determinands are strongly non-monotonic, and for these the seasonal Kendall test is inappropriate. An example of this is zinc, which bulges between 1984 and 1987 and levels off thereafter. For some of the others it is rather less clear whether underlying changes should be considered to be monotonic. As a result, any decision as to whether there is trend becomes very subjective. What is clear, is that for these measurements an unquestioning reliance on a statistical test such as the seasonal Kendall test could be highly misleading.

The data do not convincingly exhibit long-term trend for any of the determinands, despite the seasonal Kendall test results. For ammonium and excess calcium, LOWESS curves show a relatively steady change throughout the period but questions still remain. When the time series are decomposed into seasonal, trend and residual components (see methods section; Figure 4), ammonium shows a decline over the last four years, which suggests that the observed overall uptrend may be a result of a high annual variation and an insufficiently long sampling period. For excess calcium, the change was non-uniform. Most of the increase occurs in the last three years, the first seven years being relatively stable (Figure 4). For zinc, barium and (arguably) iron, LOWESS curves (Figure 3) and seasonal decomposition indicate that the underlying changes are not monotonic and should not be considered as long-term trend. The only way to establish whether the downtrends in zinc, iron and barium and the increase in ammonium and excess calcium real is to extend the data collection period for at least another 5 and possibly 10 years.

Rainfall species showing no overall evidence of trend. Species with no evidence of trend include: chloride, sodium, calcium, strontium, sulphate, magnesium, potassium, bromide, iodine, pH, alkalinity, conductivity, aluminium, manganese, chromium, copper, nitrate, excess sodium, excess sodium, excess potassium, excess magnesium and excess sulphate.

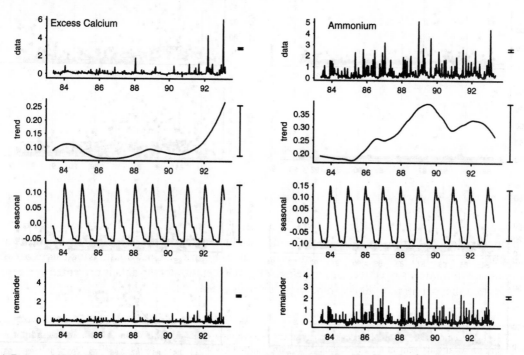

Figure 4. Decomposition of rainfall time series into a trend, seasonal and residual component using LOWESS techniques. The vertical bars on the right represent the same distance and can be used to compare scales. In each instance, the trend and seasonal components are small relative to the residual noise

Most components showed no evidence of trend (the seasonal Kendall statistic was not significant). These components include sea salt sources (Na, Cl, Mg, Br, I), pollutant sources (many trace metals, excess components and alkalinity) and a combination of these sources (Ca, Sr, SO_4, K and conductivity). Despite the lack of significant trends, many display noticeable shifts over the 10 year period. The long-term changes in these determinands are briefly examined because they are relevant to the interpretation of streamwater data.

The time distribution of sea salt input from rainfall has been very uneven over the 10 year sampling period (e.g. chloride, Figure 3). The highest inputs occurred in the wettest winters of 1983–1984 and 1989–1990, but there is no evidence of overall trend. Sulphate, calcium, magnesium, potassium, strontium, bromide, iodine and conductivity all show a temporal variation which roughly follows chloride, although calcium increases towards the end of the record. Excess sulphate is not as variable as the marine components.

Rainfall aluminium and manganese show similar changes to barium and zinc, suggesting that they have common sources. The pattern of fluctuation for these species is distinct from the marine variations, but LOWESS curves do not suggest trend. Chromium has a very distinctive temporal variation with a large, short-lived and unexplained 'bulge' between 1986 and 1989 (Figure 3). More recently, chromium concentrations have returned to much lower levels. Fluctuations in pH, alkalinity, nitrate and iodine are observed, but show no clear relation with any other constituents.

Summary. Rainfall data indicate that there have been some marked changes between 1983 and 1993. The most uniform changes are for ammonium and excess calcium, but it is not clear that these changes will continue. The longer term temporal variations in many of the determinands suggest long-term cycles in the data. In part, these relate to fluctuations in weather conditions, particularly for marine inputs. Given these large variations, the assumption of monotonic trend required for tests, such as the seasonal Kendall test, is dubious. The rainfall data show no significant changes for the 10 year period either in the wet-deposited pollutants usually associated with acid deposition (sulphate and nitrate), or in rainfall acidity, as represented by pH and alkalinity.

Trends in streamwater chemistry

In this section, water quality parameters for Hafren streamwaters are examined for trend (Table I). Streamwater data are presented in Figure 5 using a similar format to that used for the rainfall data. As with rainfall, both marine and excess concentrations are examined.

Note that very low flows occurred during the first summer of the data record and have not occurred since then. This may account for some of the distinctive behaviour seen in the early years, so that changes seen at this time are not necessarily evidence of long-term trend.

Species showing possible trend. Species showing a possible trend include: sodium, excess calcium*, magnesium, excess magnesium, potassium, excess potassium, DOC, iodine, bromide, copper*, chromium, zinc, cobalt*(* denotes only significant at a 90% level; the remainder are significant at a 95% level).

The components having possible trends include some associated with sea salt sources (Na, Mg, I, Br), pollutant sources (Cu, Cr, Zn, Co) and those from a combination of these sources (excess Ca, excess Mg). Excess cation concentrations in the streamwater are also partly the result of cation-exchange processes. The estimated slopes of change for these species are generally proportionally smaller than those in rainfall. Most robust regression slopes are similar to the seasonal Kendall slopes, the exceptions being for chromium, zinc and potassium. As with the rainfall, the LOWESS curves have been examined (Figure 5) and the time series have been decomposed into trend, seasonal and residual components (see Figure 6 for examples). Furthermore, the effect of flow regime on trend is examined (Figure 7). Once again, conclusions are difficult to draw and are ultimately subjective. Nevertheless, our interpretation of the data is presented and results are graded between the extremes of 'likely trend' and 'unlikely trend'.

Figure 5. Streamwater time series plots. For each measurement two graphs are shown; the left-hand plot shows the full range and the right-hand plot shows an enlargement after removing the top and bottom 2% of the data. The thick vertical mark on the left-hand axes shows the extent of the axes used in the right-hand graph. LOWESS curves are shown; the solid line is fitted to the unweighted data whereas the dotted lines have been flow-weighted

Figure 5A, B and C over page

Afon Hafren

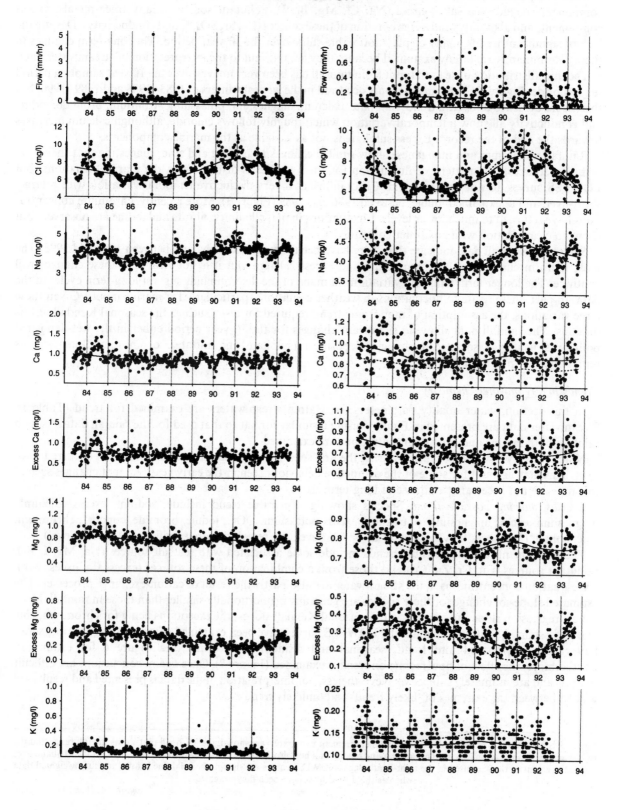

Afon Hafren

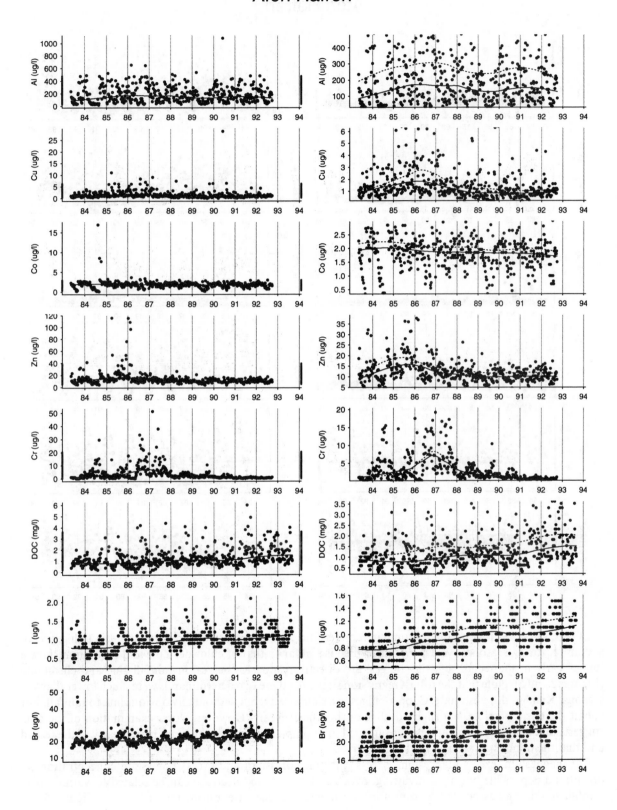

Afon Hafren

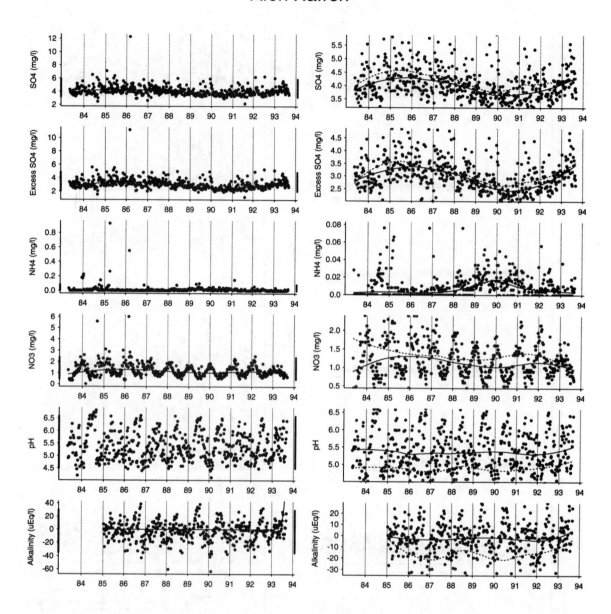

LOWESS curves for sodium, chromium, copper and zinc suggest that long-term trend is unlikely. For sodium, a strong seasonal cycle is evident, which is similar to that of chloride. Zinc and chromium show a 'bulge' early in the record and the 'bulges' are directly related to 'bulges' in rainfall zinc and chromium (Figures 3 and 5). Note that the robust regression slopes for these two determinands were much lower than the seasonal Kendall slope estimate. Copper bulges in a similar way, but without any obvious link to rainfall. For cobalt there is a possible, but slight, decline during the 10 years. However, closer examination of the data suggests that cobalt concentrations were affected in a complex way by the long dry period early in the record and that the trend is an artefact of this dry period.

For magnesium, excess magnesium, excess calcium, potassium and excess potassium concentrations, a downtrend is possible but not convincing. Much of the observed decrease can be related to variations in rainfall volume and associated concentrations. For example, decreases during the early years (Figures 5 and

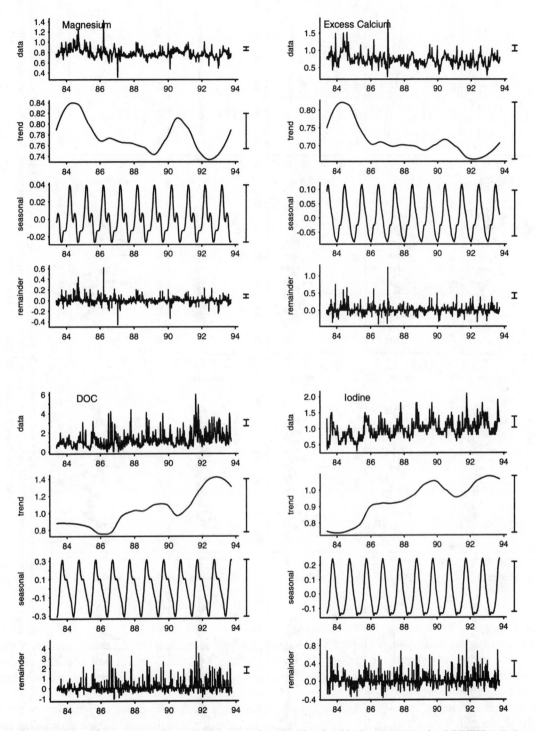

Figure 6. Decomposition of streamwater time series into a trend, seasonal and residual component using LOWESS techniques. The vertical bars on the right represent the same distance and can be used to compare scales. In each instance the trend and seasonal components are small relative to the residual noise

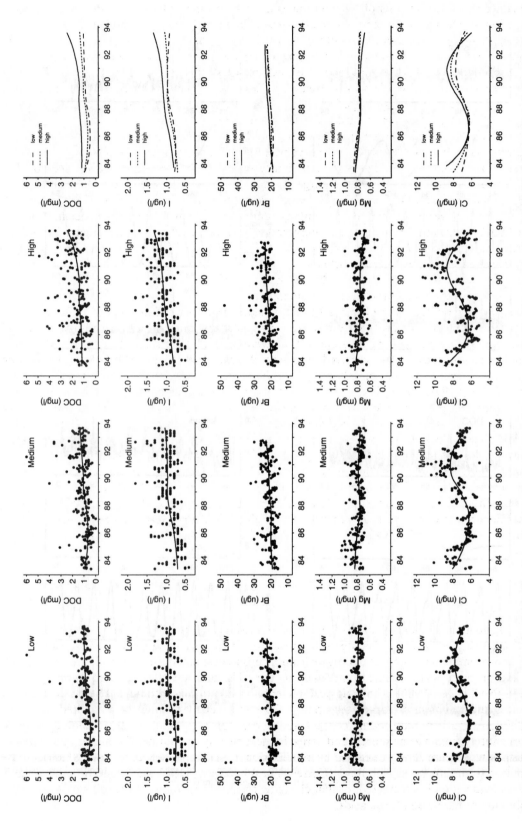

Figure 7. Effect of flow class on streamwater quality. For each determinand, four graphs are shown. The first three graphs show LOWESS curves and data for low, medium and high flows (see text). The fourth graph combines the three LOWESS curves. Magnesium and chloride graphs are shown for comparison. Note that for chloride the variation becomes progressively damped and delayed at lower flows

6) can be explained by the dry summer of 1984, during which base cation-rich baseflow waters were sampled. A high weathering source during this period is also suggested by several other determinands, e.g. strontium and silica. Changes in other base cation concentrations are similar to the cycling of chloride, e.g. magnesium and excess magnesium (for excess magnesium, the changes are inverted relative to chloride). Consequently, broad changes in magnesium (Figure 6) and excess magnesium are strongly non-monotonic. For potassium, excess potassium and excess calcium (Figure 6), trends are nearer to being monotonic. However, most of the overall downtrend is due to the early dry period, after which the pattern is similar to that of chloride. Had sampling begun 18 months later, trend would probably not have arisen.

The only components for which the trend is convincing are DOC, iodine and bromide (Figures 5 and 6), i.e. determinands associated with reactions involving organic matter. For each, the trend is relatively steady and near monotonic throughout the period and is not obviously associated with trends in rainfall. The slopes of the trends for these determinands are relatively large compared with the mean concentration (Table I). For example, the estimated slope for DOC is $0.07 \text{ mg l}^{-1} \text{ year}^{-1}$ compared with a mean value of 1.25 mg l^{-1}, which when evaluated for the 10 year study period represents a change of more than 50% in DOC concentrations. In addition, for DOC, iodine and bromide, increasing trend is observed at all flow levels (Figure 7) and, for iodine, the trend may be the greatest at high flows.

Streamwater species showing no overall evidence of trend. Species showing no evidence of trend include: chloride, calcium, sulphate, conductivity, pH, alkalinity, ammonium, nitrate, aluminium, lithium, strontium, yttrium, barium, manganese, iron and aluminium.

Streamwater components showing no overall trend (i.e. the seasonal Kendall test was not statistically significant) include some associated with sea salt sources (Cl), pollutant sources (NH_4), metals (Al, Fe, Li, Y, Ba and Mn), nitrate and acidity (pH and alkalinity). The variation for chloride concentration is very strongly sinusoidal, and concentrations are less scattered than other determinands (Figure 5). Streamwater chloride variations are lagged relative to the rainfall inputs, e.g. compare the peak rainfall chloride in 1990 with streamwater chloride in 1991. The lag is greatest at low flows and least at high flows (Figure 7). Several other streamwater determinands, such as conductivity, calcium, magnesium, potassium and strontium, vary sinusoidally, in a similar way to chloride, although effects of the early dry period also are apparent for some. Total and excess sulphate also vary sinusoidally, but the variation is out of phase with chloride. Many other streamwater determinands, including lithium, barium, manganese, iron, yttrium, aluminium, nitrate, ammonium and pH, show variation, but show no evidence of trend and no obvious links with rainfall contributions or with chloride. Of these, only aluminium and nitrate showed similarities in their temporal variability. Ammonium concentrations show a 'bulge' part way through the period (Figure 5), but there is no obvious link with rainfall except that the streamwater 'bulge' occurs simultaneously with a peak in the rainfall LOWESS curve.

DISCUSSION

Process interpretation

Concurrent uptrends in streamwater DOC, bromide and iodine suggest a common source. All of these determinands are strongly affected by biological processes (e.g. bromide and iodine are essential nutrients). Also, iodine and bromide have similar hydrogeochemical properties; inputs from rainfall are biologically assimilated and/or react with organic compounds in the soils, and are later released during organic decomposition (Neal *et al.*, 1990b). The biological influence on stream concentrations is very strong for all three substances and results in a marked seasonality. Observed increases cannot be attributed simply to changes in rainfall inputs; rainfall iodine concentrations were fairly constant, whereas rainfall bromide derives almost entirely from marine sources and varied in the same way as chloride. The higher rate of iodine increase at high flows (Figure 7) may be caused by mobilization from organic-rich upper soils, which increase their contribution to the stream at high flows. Uptrends in streamwater DOC, iodine and bromide suggest that organic decomposition has increased, or that changing chemical conditions of the soil have lead to the increased solubility of humic and fulvic acids.

Increases in streamwater colour, which is related to DOC, have been reported in other UK studies (McDonald *et al.*, 1989; Kay *et al.*, 1989) and Plynlimon results add to this evidence for a general DOC increase in upland streams. Streamwater DOC increase may well be of concern to the water industry because of increased treatment costs for drinking water. Previously reported uptrends have been attributed to a variety of factors. For example, dry summers can result in increased erosion of peat hags on hill tops, with surface decomposition resulting in increased colour. However, it is unclear for the present study as to why the changes have taken place. Furthermore, trends in bromide and iodine concentrations have not been reported previously. The trends indicate that there has probably been some change with respect to biological processes in the soils. It may well prove that DOC trends are directly related to these processes, but further research is required to identify the actual mechanisms. Within the context of upland streamwater acidification, increases in DOC have not lead to decreases in pH. This is contrary to expectation because DOC contains humic and fulvic acids.

There is a marked inverse relationship between streamwater sulphate and chloride over the 10 year period. When sea salt inputs are high, wet deposition of sulphate increases, and streamwater sulphate would be expected to increase (in parallel to chloride). Instead, the opposite is observed. Two possible explanations for this discrepancy are as follows. The first is that dry and gaseous deposition of sulphate/sulphur dioxide varies substantially with weather conditions. This is reasonable as the dry deposition of sulphate and sulphur dioxide might be expected to be higher during drier periods when the wind direction is mainly from the land (pollution sources) and to be lower during years when weather systems are predominantly frontal (UK Review Group on Acid Rain, 1990). An alternative explanation is that the high sea-salt rainfall affects the nature of the adsorption and solubility processes which control soil water sulphate concentration. However, the detailed mechanisms for this are complex and very sophisticated thermodynamic modelling techniques are needed to test this hypothesis. For us, the former hypothesis seems most likely. If this is the case, then annual variations in the dry deposition of sulphate and sulphur dioxide are marked and future hydrochemical models may need to take such variations into account.

For the base cations, stream concentrations roughly parallel chloride. However, excess concentrations of sodium, magnesium and potassium, which would be expected to change steadily and to be unaffected by variations in marine source contributions, are instead inverted relative to chloride. Excess cation concentrations are often used as an indicator of change in cation-exchange stores, but this may not be appropriate at Plynlimon. The inverse relationship may result from base cations moving through the system more slowly than chloride; a slight attenuation in the base cation signal, relative to chloride, would cause a decrease in the excess component when sea salts are high. If sodium, magnesium and potassium are exchanged onto the soils when sea salt contributions are high, then calcium and/or hydrogen ions would be released simultaneously. Stored ions would be subsequently released to streamwaters during succeeding events, effectively damping the rainfall signal. Unlike the other base cations, excess streamwater calcium remains relatively constant. This may be due to the large calcium weathering component, which could dominate controls by cation exchange, or to calcium inputs from dry deposition, which offset the cation-exchange source.

Metal concentrations in streamwater are closely linked to rainfall inputs. The data series show some of these metals behaving in a near-conservative way. Rainfall is the dominant source of zinc and chromium (mineralization in the catchment is low) and the rainfall signal is quickly passed through the catchment. There are clearly multiple sources of metal pollution as evidenced by variable timing of peak concentrations for different metals.

Trend analysis

The use of both statistical and graphical techniques for examining trend highlights difficulties of trend analysis for water quality data. Standard statistical tests, such as the seasonal Kendall test, do not allow for long-term cycles in the data; they assume an underlying monotonic trend. However, smoothed fits of the data indicate that long-term cycles are present for many of the determinands studied herein. These long-term cycles may be caused by cycles in weather as have been observed elsewhere (e.g. South Africa; du Plessis and van Veelen, 1991) and which are not unique to Plynlimon. Likewise air quality and precipitation data are known to have long-term cycles, which also make it difficult to identify trends (UK Review on Acid Rain, 1990;

Sirois, 1993). The experience from Plynlimon suggests that caution should be used when examining streamwater quality data for trend.

Variations in sea salt inputs are seen here as system noise because they are damped markedly by the catchment as observed in streamwater. However, this noise is considerable. Streamwater quality trends of *real* interest, e.g. evidence of long-term acidification, must be distinguished from this background noise, making it difficult to identify trends with only 10 years of data. It is important to examine streamwater data allowing for climatological variations and with knowledge of rainfall inputs. For determinands which are affected by cycles, much longer data sets are required to identify trends, except when trend is very pronounced and cyclic components are small. *Artificial* trends will be erroneously detected if shorter time series data are examined. If UK sites are to provide useful information on acidification trends, they must be maintained for much longer than is currently proposed.

Several important features of the data have been identified. The data vary markedly, having large residual scatter, particularly for rainfall, even after allowing for seasonal effects (Figures 4 and 6). Where trends are detected, they are small relative to this residual scatter. Even so, in most instances the estimated trend slopes are large when compared with the mean concentrations. The data show serial correlation, non-normality and complex interdependencies. A more detailed investigation which incorporates some of these properties is clearly possible. Ideally, factors such as rainfall dependency, flow dependency and the impact of long meterological variation should be included. A step in this direction would be to incorporate some flow variation by further use of LOWESS smoothing techniques (Hirsch et al., 1991). To include more explicit accounting for rainfall and flow dependency is highly complex and probably requires the use of hydro-chemical catchment models. Sophisticated time series techniques such as fractional differencing might help to describe the nature of the variations induced by the long-term weather variations (Haslett and Rafferty, 1989). Given the inherent problems arising from the year to year and shorter term fluctuations in the atmospheric inputs to the catchments, it is unclear how much can be gained towards process understanding from this type of analysis.

CONCLUSIONS AND FUTURE CONSIDERATIONS

The 10 years of rainfall data showed no significant trend in pollutants in bulk precipitation usually associated with acid deposition (sulphate and nitrate). The absence of a sulphate trend in streamwaters suggests that there has been no long-term trend in the dry deposited component either, despite a general decrease in UK sulphate and sulphur dioxide emissions. Consequently, a decrease in sulphur emissions and an increase in NO_x emissions over the last 10 years have had little effect on acidification inputs to Plynlimon, and presumably to the Welsh uplands in general. Instead, the most dominant changes in rainfall inputs over the last 10 years are related to variations in weather.

In streamwater, the most pronounced changes are uptrends for DOC, iodine and bromide concentrations. These trends are consistent with general patterns of increasing colour in upland streams. There have been possible declines in some base cations, but these have probably occurred because of a dry period at the start of the sampling period. Trends observed at Plynlimon contrast with several studies showing a downtrend in base cations. For these sites it has been inferred that continued acidification is taking place in acid-sensitive European catchments (Christophersen et al., 1990; Kirchner, 1993; Moldan and Cerny, 1994).

Analysis of the Plynlimon data shows the need for much longer data sets for trend analysis and the need to maintain long-term monitoring programmes. This aspect of the study is very important in that financial support for long-term monitoring is dwindling in favour of short-term, regional networks which are sampled at a much lower frequency. Although in principle such a change is attractive to funding agencies, the results are unsound and ultimately such endeavours represent poor value for the money expended.

With regard to assessing the trends in environmental data, we recommend that graphical techniques are more widely used when analysing trends in the chemical quality of rainfall and streamwater. Graphical smoothing strongly enhances data analysis, especially when the data are highly variable, and should be an essential component of trend analysis. Furthermore, graphs should be given considerable weight along with the standard statistical information, and complete data series should be more widely shown in publications.

ACKNOWLEDGEMENTS

Data collection was carried out by the staff at Plynlimon. Samples were analysed by the analytical chemists at Wallingford. Helpful suggestions were received from Dr R. J. Gibbens.

REFERENCES

Becker, R. A., Chambers, J. M., and Wilks, A. R. 1988. *The New S Language*. Wadsworth and Brooks/Cole Computer Science Series, Pacific Grove. 1702 pp.

Berryman, D., Bobee, B., Cluis, D., and Haemmerli, J., 1988. 'Nonparametric tests for trend detection in water quality time series', *Wat. Resour. Bull.* **24**, 545–556.

Chambers, J. M. and Hastie, T. J. 1992. *Statistical Models in S*. Wadsworth and Brooks/Cole Advanced Books and Software, Pacific Grove. 608pp.

Christophersen, N., Robson, A., Neal, C., Whitehead, P. G., Vigerust, B., and Henriksen, A., 1990. 'Evidence for long term deterioration of stream water chemistry and soil acidification at the Birkenes site', *J. Hydrol.*, **116**, 63–76.

Cleveland, W. S. 1979. 'Robust locally weighted regression and smoothing scatterplots', *J. Am. Statis. Soc.*, **74**, 386; 829–836.

Cleveland, R. B., Cleveland, W. S., McRae, J. E., and Terpening, I., 1990. 'STL: a seasonal-trend decomposition procedure based on Loess', *J. Off. Statis.*, **6**, 3–73.

Davies, T. D., Tranter, M., Wigington Jr, P. J., and Eshleman, K. N., 1992. 'Acidic episodes in Europe', *J. Hydrol.*, **132**, 25–70.

Durand, P., Neal, C., Jeffery, H. A., Ryland, G. P., and Neal, M. 1994. 'Minor, major and trace element mobility in the Plynlimon afforested catchments (Wales): general trends, and effects of felling and climate variations', *J. Hydrol.*, **157**, 139–156.

Haslett, J. and Rafferty, A. E. 1989. 'Space–time modelling with long memory dependence: assessing Ireland's wind power resource', *Appl. Statis.*, **38**, 1–50.

Hemond, H. F. 1994. 'Role of organic acids in acidification of fresh waters' in Steinberg, C. E. W. and Wright, R. F. (eds), *Acidification of Fresh Waters; Implications for the Future*. Wiley, Chichester. pp. 103–115.

Hirsch, R. M. and Slack, J. R. 1984. 'A nonparametric text for seasonal data with serial dependence', *Wat. Resour. Res.*, **20**, 727–732.

Hirsch, R. M., Slack, J. R., and Smith, R. A. 1982. 'Techniques of trend analysis for monthly water quality data', *Wat. Resour. Res.*, **18**, 107–121.

Hirsch, R. M., Alexander, R. B., and Smith, R. A. 1991. 'Selection of methods for the detection and estimation of trends in water quality', *Wat. Resour. Res.*, **27**, 803–813.

Hornung, M., Roda, F., and Langan, S. J. 1990. 'A review of small catchments in Western Europe producing hydrochemical budgets', *Air Pollut. Res. Rep. 28*. Commission of the European Communities, Brussels.

Kahl, J. S., Haines, T. A., Norton, S. A., and Davis, R. B. 1993. 'Recent trends in the acid base status of surface waters in Maine, USA', *Wat. Air. Soil Pollut.*, **67**, 281–300.

Kay, D., Boon, R., and Crowther, J. 1989. 'Coloured waters in Wales: spatial and temporal trends' in *Second National Hydrological Symposium, August 1989*. Institute of Hydrology, Wallingford. pp. 1.49–1.57.

Kirby, C., Newson, M. D., and Gilman, K. 1991. 'Plynlimon research, the first two decades', *Inst. Hydrol. Rep.* **109**, 1–187.

Kirchner, J. 1993. 'Chronic base depletion revealed by long term monitoring from Norwegian catchments' in Cerny, J. (Ed.) *Abstract, BIOGEOMON Symposium on Ecosystem Behaviour, Prague, September 1993*. Czech Geological Survey, Prague. pp. 146–147.

Moldan, B. and Cerny, J. 1994. 'Small catchment research' in Moldan, B. and Cerny, J. (Eds). *Biogeochemistry of Small Catchments: a Tool for Environmental Research*. Wiley, Chichester. pp. 1–29.

McDonald, A. T., Edwards, A. M. C., Naden, P. S., Martin, D., and Mitchell, G. 1989. 'Discoloured runoff in the Yorkshire Pennines' in *Second National Hydrological Symposium, August 1989*. Institute of Hydrology, Wallingford. pp. 1.49–1.57.

Mulder, J. and Cresser, M. S. 1994. 'Soil and soil solution chemistry' in Moldan, B. and Cerny, J. (Eds). *Biogeochemistry of Small Catchments: a Tool for Environmental Research*. Wiley, Chichester. pp. 107–131.

Neal, C. and Rosier, P. T. W. 1990. 'Chemical studies of chloride and stable oxygen isotopes in two conifer afforested and moorland sites in the British uplands', *J. Hydrol.*, **15**, 269–283.

Neal, C., Walls, J., and Dunn, C. S. 1986. 'Major, minor and trace element mobility in the acidic forested upland catchment of the upper River Severn, Mid-Wales', *Q. J. Geol. Soc. London*, **143**, 635–648.

Neal, C., Smith, C. J., and Hill, S., 1992. 'Forestry impact on upland water quality', *Inst. Hydrol. Rep.*, **119**, 1–50.

Neal, C., Smith, C. J., Walls, J., Billingham, P., Hill, S., and Neal, M. 1990a. 'Hydrogeochemical variations in Hafren forest streams, Mid-Wales', *J. Hydrol.*, **116**, 185–200.

Neal, C., Smith, C. J., Walls, J., Billingham, P., Hill, S., and Neal, M. 1990b. 'Comments on the hydrochemical regulation of the halogen elements in rainfall, stemflow, throughfall and stream waters at an acidic forested area in mid-Wales', *Sci. Total Environ.*, **91**, 1–11.

Newell, A. D. 1993. 'Inter-regional comparison of patterns and trends in surface water acidification across the United States', *Wat. Air. Soil. Pollut.*, **67**, 257–280.

Newson, M. D. 1976. 'The physiography, deposits and vegetation of the Plynlimon catchments', *Inst. Hydrol. Rep.*, **30**, 1–59.

Plessis du, H. M. and Veelen van, M. 1991. 'Water quality: salinization and eutrophication time series and trends in South Africa', *South Afr. J. Sci.* **87**, 11–16.

Reynolds, B. and Pommeroy, A. B. 1988. 'Hydrogeochemistry of chloride in an upland catchment in mid-Wales', *J. Hydrol.*, **99**, 19–32.

Reynolds, B., Neal, C., Hornung, M., and Stevens, P. A. 1986. 'Baseflow buffering of stream water acidity in five mid-Wales catchments', *J. Hydrol.*, **87**, 167–185.

Robson, A. J. 1993. 'The uses of continuous measurement in understanding and modelling the hydrochemistry of the uplands', *PhD Thesis*, Univ. Lancaster, 278 pp.

Sen, P. K. 1968. 'Estimates of the regression coefficient base on Kendall's Tau', *J. Am. Statis. Assoc.*, **63**, 1379–1389.

Sirois, A. 1993. 'Temporal variation of sulphate and nitrate concentration in precipitation in eastern north America, 1979–1990', *Atmos. Environ.* **27A**, 945–963.

Sirois, A. and Summers, P. W. 1989. 'An estimation of the atmospheric input of sulphur and nitrogen oxides to the Kejimkujik Watershed: 1979–1987', *Wat. Air. Soil Pollut.*, **46**, 29–43.

Statistical Sciences 1993. *S-PLUS Reference Manual, Version 3.2.* StatSci, a division of MathSoft, Seattle.

Taylor, C. H. and Loftis, J. C. 1989. 'Testing for trend in Lake and ground water quality time series', *Wat. Resour. Bull.* **25**, 715–726.

UK Acid Waters Monitoring Network 1991. *Site Descriptions and Methodology.* Ensis, London.

UK Acid Waters Review Group, 1988. *Acidity in the United Kingdom Fresh Waters. Second Report.* HMSO, London. 61pp.

UK Review Group on Acid Rain 1990. *Acid Deposition in the United Kingdom, 1986–1988. Third Report.* Department of the Environment, London.

Vesely, J. 1994. 'Effect of acidification on trace metal transport in fresh waters' in Steinberg, C. E. W. and Wright, R. F. (Eds), *Acidification of Fresh Waters; Implications for the Future.* Wiley, Chichester. pp. 141–151.

Walker, W. W. 1991. 'Water quality trends at inflows to Everglades National Park', *Wat. Resour. Bull.*, **27**, 59–72.

Waters, D. and Jenkins, A. 1992. 'Impacts of afforestation on water quality trends in two catchments in Mid-Wales', *Environ. Pollut.*, **77**, 167–172.

4

TRENDS IN STREAM AND RIVER TEMPERATURE

B. W. WEBB

Department of Geography, University of Exeter, Exeter, EX4 4RJ, UK

ABSTRACT

Information on past and likely future trends in water temperature from different parts of the world is collated. The potential causes of trends in the thermal regimes of streams and rivers are many, but the existing database of water temperature information is inadequate to provide a global perspective on changes during the recent, let alone the more remote, past. Data from Europe suggest that warming of up to ca. 1°C in mean river temperatures has occurred during the 20th century, but that this trend has not been continuous, is distorted by extreme hydrological events, is not correlated with simple hydrometeorological factors and has been influenced by a variety of human activities. Predictive studies indicate that an accelerated rise in stream and river temperatures will occur during the next century as a consequence of global warming. However, forecasts must be tentative because future climatic conditions are uncertain and interactions between climate, hydrological and vegetation changes are complex.

INTRODUCTION

Water temperature is arguably the most important physical property of streams and rivers. Temperature exerts a strong influence on many physical and chemical characteristics of water, which include vapour pressure, surface tension, density and viscosity (Stevens *et al.*, 1975), the solubility of oxygen and other gases (e.g. ASCE Committee on Sanitary Engineering Research, 1961), sediment concentrations and transportation (e.g. Lane *et al.*, 1949), chemical reaction rates (e.g. Brezonik, 1972) and the presence or absence of pathogens (e.g. Hendricks, 1972).

The ecology of streams and rivers is also fundamentally affected by the thermal regime (e.g. Macan, 1974). An enormous amount of research has been devoted to the thermobiology of freshwater organisms and this has demonstrated that water temperature can moderate many different aspects of stream and river biota. These include the geographical distribution, growth and metabolism, food and feeding habits, reproduction and life histories, movements and migrations, behaviour and tolerance to parasites, diseases and pollution (e.g. Brett, 1956; Ordal and Pacha, 1963; Rose, 1967; Anderson, 1969; Magnuson *et al.*, 1979; Ward and Stanford, 1979; Holtby *et al.*, 1989; Pöckl, 1992; Ward, 1992; Crisp, 1993, Elliott, 1994).

Human use of river water may also be affected by river temperature. The efficiency of water purification and treatment methods and the palatability of domestic supplies, the effectiveness of irrigation, the economics of commercial aquaculture and of industrial processes requiring cooling water, and the suitability of water courses for recreation, including swimming and angling, are related to river temperatures (e.g. Clark and England, 1963; Raney, 1963; Hoak, 1965; Devik, 1975; North, 1980). Furthermore, controlling the thermal regime may be an important goal in the human management of river systems (e.g. Churchill, 1965; Croley *et al.*, 1981; Theurer *et al.*, 1985; Cassidy and Dunn, 1987; Bartholow, 1991).

The importance of water temperature as a parameter of stream and river quality is enhanced by its considerable sensitivity to modification by natural factors and human activities. Spatial variations in thermal regime are apparent at global, national, catchment and channel scales, and predominantly reflect the influence of latitude, altitude and continentality at the macro-scale, climate and hydrology at the meso-scale and insolation receipt and substratum conditions at the micro-scale (e.g. Smith, 1968; Smith and Lavis, 1975; Mosely, 1982; Steele, 1982; Bilby, 1984; Ward, 1985; Webb and Walling, 1986; Ozaki, 1988; Crisp, 1990; Rutherford *et al.*, 1992). Water temperature may also exhibit marked annual, diurnal and storm-period fluctuations which follow seasonal and daily rhythms in the amount and type of heat energy gained and lost by a water course, and reflect the volume and sources of runoff contributing to river flow (Smith, 1975; Webb and Walling, 1985; Shanley and Peters, 1988; Rutherford *et al.*, 1993).

Human activities may strongly modify water temperature in space and time. The discharge of heated effluents, especially in the form of the cooling water used in the generation of electrical power, results in a direct alteration of river thermal regime (e.g. Parker, 1974; Zaric, 1978; Langford, 1983). Stream and river temperatures may also be modified in an indirect way through changes in catchment land use and channel condition, which alter the energy budget and the thermal capacity of the water course. A wide variety of indirect impacts associated with agriculture, channelization, flow modification and urbanization may modify stream and river temperatures (e.g. Stoeckeler and Voskuil, 1959; Sylvester, 1963; Pluhowski, 1970; Parrish *et al.*, 1978; Hockey *et al.*, 1982; Dymond, 1984; Shields and Sanders, 1986; Quigley *et al.*, 1989; Quinn *et al.*, 1992), but the effects of river impoundment and forestry practices have received most attention (e.g. Greene, 1950; Brown and Krygier, 1967; Jaske and Goebel, 1967; Nishizawa and Yambe, 1970; Ward, 1974; Graynoth, 1979; Walker *et al.*, 1978; Mackie *et al.*, 1983; Petts, 1986; Crisp, 1987; Beschta and Taylor, 1988; Holtby, 1988; Byren and Davies, 1989; O'Keeffe *et al.*, 1990; Weatherley and Ormerod, 1990; Webb and Walling, 1993a; Rowe and Taylor, 1994).

Given the importance of water temperature to the quality, ecology, utility and sensitivity of streams and rivers, it is of considerable interest to understand how thermal regimes have changed in the past and how they may be modified in the future. Long-term trends in water temperature, however, have been one of the least intensively studied aspects of thermal regime. The present paper collates the available information on the nature, magnitude and controls of trends in water temperature, including the documentation of past changes and the prediction of future evolution. Attention will be given to the potential causes of changes in stream and river thermal regime and to the data and techniques for the identification of water temperature trends. Several case studies will be presented from different areas of the world to illustrate trends in water temperature that have occurred in the past and may take place in the future.

POTENTIAL CAUSES OF CHANGE IN THERMAL REGIME

The fundamental cause of changes and trends in water temperature can be traced to the modification of the energy budget and/or the thermal capacity of a water course by natural processes or human activities. The energy or heat budget of a stream or river reach is comprised of the following major components

$$Q_n = \pm Q_r \pm Q_e \pm Q_h \pm Q_{hb} + Q_{fc} \pm Q_a$$

where Q_n = total net heat exchange; Q_r = heat flux due to net radiation; Q_e = heat flux due to evaporation and condensation; Q_h = heat flux due to sensible transfer between air and water; Q_{hb} = heat flux due to bed conduction; Q_{fc} = heat flux due to friction; and Q_a = heat flux due to advective transfer in precipitation, groundwater, tributary inflows, streamflow and effluent discharges.

This budget determines the amount of energy available to modify the stream or river water temperature. When the river temperature is in an unsteady, non-equilibrium state, the sensitivity of a water course to changing inputs and outputs of heat energy will reflect its thermal capacity. This depends, in turn, on the volume of water to be heated or cooled. The larger the water volume, the greater the capacity for heat storage and the less responsive the stream or river will be to alterations in the energy budget.

Trends in water temperature can be caused by the permanent modification of any component of the river heat budget. For example, natural changes in climate may alter the amount of radiation reaching the river channel or modify the temperature of the air in contact with the water surface. Consequently, radiative and sensible heat fluxes affecting the water course are altered and the water temperature may change. Human activities may also disrupt the energy budgets of streams and rivers. The shortwave solar radiation reaching the stream surface may be dramatically increased if forestry practices remove riparian tree cover (e.g.

Pluhowski, 1972; Beschta *et al.*, 1987). For example, Binkley and Brown (1993) demonstrate, from a synthesis of studies in North America, that summer temperature maxima are typically increased by 2–6°C if, during forest harvesting, a buffer strip is not left along the riparian zone to retain stream shading. The cumulative effects of forest harvesting may promote long-term trends in water temperature. Beschta and Taylor (1988) have shown that average daily maximum and minimum stream temperatures generally increased due to forest cutting in the Salmon Creek watershed, Oregon, USA, during the period 1955–1984. However, they also caution that water temperature trends in forested catchments will reflect the complex interaction of harvesting extent, management practices and the occurrence of major geomorphic events. In the latter context, it was demonstrated that the greatest increases in water temperature recorded during the 30 year study period tended to follow the occurrence of large peak flows and associated mass soil failures, which caused channel scouring and the removal of riparian vegetation. Water temperature trends may also occur as catchments recover from the effects of forest harvesting. For example, Hostetler (1991) demonstrated, by graphical and time series analysis, that water temperatures gradually declined over a 20 year period following the 1969 clear-cut logging in the Steamboat Creek basin, Oregon, USA, as riparian vegetation regrew and a shading canopy became re-established.

Increases in advected heat into river systems from effluent discharges is also potentially a very important cause of long-term trends in water temperature. In particular, the dramatic growth in electrical power generation during this century (Parker, 1974; Langford, 1983) has greatly increased the quantity of surface water abstracted for cooling purposes, which is ultimately returned to river systems as heated effluent. For example, thermoelectric power generation in 1985 accounted for freshwater withdrawals of ca. $6 \times 108 \, \text{m}^3$ day^{-1} in the USA (Carr *et al.*, 1987). The effect of heated effluents in raising the temperature of major water courses is illustrated by River Rhine in Germany. For the Rhine below the confluence with the River Voerde, Zimmermann and Geldner (1978) calculated that discharges from power stations and other industries in 1970 raised summer and autumn temperature maxima by more than 2 and 3°C, respectively.

The heat advected into a reach from upstream may also be modified by dam construction and river regulation. The thermal characteristics of water issuing from a reservoir typically differ from ambient conditions in an unregulated water course because of the greater thermal inertia of the impounded water mass, thermal stratification of the reservoir, or alteration of groundwater circulation downstream of the dam (e.g. Sylvester, 1963; Lavis and Smith, 1972; Stanford and Ward, 1979; Ward, 1982; Cowx *et al.*, 1987; Webb and Walling, 1988; Palmer and O'Keeffe, 1989). Modification of the discharge regime in the regulated river may also influence water temperature through the impact on thermal capacity. Although impoundment may have an immediate effect on downstream water temperature, by eliminating freezing conditions, depressing summer maxima, delaying the annual cycle of variation and reducing diel fluctuation, the impact of regulation may show considerable variability from year to year over the long term. For example, an evaluation of the effects of a reservoir in the headwaters of the River Exe, Devon, UK during a 10 year period following the attainment of top water level (Webb and Walling, 1993a) revealed significant inter-annual contrasts in the impact of the impoundment on mean and extreme values, as well as the characteristics of daily temperature fluctuation (Table I).

Abstraction of water for industrial, agricultural and domestic purposes, or diversion of flow in water transfer schemes may also cause changes in water temperature. These activities alter the volume of discharge in a stream or river and thereby affect the thermal capacity, without necessarily modifying the components of the heat budget. A decrease in thermal capacity will encourage higher water temperature maxima and lower minima below the point of diversion or abstraction (e.g. Dymond, 1984) and this effect may be exacerbated by reduced water velocity and increased residence time associated with lower flow volumes. Furthermore, river diversion may impact water temperatures over long distances of river channel. For example, McLeod (1979) predicted that diverting the McGregor River, with a mean annual flow of $240 \, \text{m}^3 \, \text{s}^{-1}$, away from the Fraser River in British Columbia, Canada, would increase the summer water temperatures by 0·7–1·1°C in the mainstream of the Fraser for a distance of 720 km below the diversion. Although water temperature changes for some streams and rivers may be attributed to a single cause, such as forestry practices, it is likely, in many instances, that long-term trends result from a combination of impacts which affect several components of the river heat budget and the thermal capacity of the watercourse. In particular, this situation will be true of water temperature trends detected for major river systems that have experienced a variety of climatic and human modifications during the past century.

DATA AVAILABILITY AND TREND IDENTIFICATION

Water temperature can be measured with relative ease (e.g. Stevens *et al.*, 1975; Bartholow, 1989) and data are currently collected in a wide range of contexts, varying from the routine monitoring and surveillance activities of statutory authorities responsible for water resources management to specialized research investigations that seek to elucidate the thermal characteristics of streams and rivers for a variety of applied and academic purposes. However, data that are both detailed and long term enough for rigorous assessment of trend are rare. For example, hourly river temperature data collected for periods of 10 or more years are restricted to a handful of studies (e.g. Vannote and Sweeney, 1980; Hostetler, 1991; Webb and Walling, 1992). As with many water quality parameters, a general dearth of long, reliable and unbroken river temperature records reflects the later development of interest in quality, compared with quantity, monitoring for rivers of many countries.

Some world-wide information on river water temperatures is available from the Global Environmental Monitoring System water quality monitoring project (GEMS/Water) (Barabas, 1986; Meybeck *et al.*, 1989) and is published as triennial data summaries (UNEP/WHO/UNESCO/WMO, 1983; 1987; 1990). These data, however, can only provide a broad perspective on spatial contrasts and temporal trends in water temperature at the global level because the GEMS/Water programme is based on infrequent monitoring ($\leqslant$ 24 samples per year). GEMS sampling also is biased to rivers of the northern hemisphere and those affected by industrial and urban development, and some of the world's major water courses are deleted from the GEMS/Water programme (Meybeck, 1985). Water temperature is not reported for all river sites in the GEMS/Water network. Although information was available from 49 countries in the period 1979–1987, the number of sites having temperature measurements has varied between triennial periods for some countries, and the number of sites and countries reporting data have decreased from the 1979–1981 to the 1985–1987 triennium (Table II). Sites in Africa are particularly sparse, where information on river temperatures may be further restricted to one or two observations within a triennial period.

Water temperature monitoring has been most intensive in Europe, North America and Japan, and the longest tradition of measurement is found in central, eastern and northern European countries, where there is a practical need to monitor conditions in rivers subject to winter freezing (Smith, 1972). Water temperature data extending back more than 100 years have been reported for some rivers in the former USSR (Lemmelä *et al.*, 1990). Daily measurements began for many of the major rivers of Austria in the mid-1890s (Nobilis, 1978; Dokulil *et al.*, 1993). Systematic monitoring of river temperatures commenced during the early years of the present century in several Nordic countries (Nybrant, 1954; Thendrup, 1985).

Table I. Difference in water temperature statistics between the Upper Haddeo (UH), a tributary regulated by the Wimbleball Reservoir in the north−east of the Exe Basin, Devon, UK and the River Pulham (PU), a neighbouring, but unregulated, water-course. Differences are expressed as UH−PU in °C

Year	Annual value			Diel fluctuation		
	Maximum	Mean	Minimum	Maximum	Mean	Minimum
1981	−0·6	+0·4	+3·0	−4·1	−0·8	0·0
1982	−3·7	−0·3	+1·1	−4·7	−1·3	−0·2
1983	−3·3	+0·4	+3·2	−0·6	−1·1	−0·1
1984	−3·7	+0·2	+2·0	−6·0	−1·9	−0·3
1985	−2·5	+1·1	+3·3	−6·6	−1·7	−0·3
1986	−3·5	+0·4	+3·0	−3·0	−1·3	−0·1
1987	−1·7	+0·5	+3·7	−4·4	−1·5	−0·1
1988	−0·1	+0·6	+3·9	−3·9	−1·3	−0·1
1989	−0·3	+1·2	+3·9	−4·9	−1·7	−0·1
1990	−1·6	+1·0	+4·2	−6·1	−1·8	−0·1

Table II. Number of river sites reporting water temperature data in three triennial periods of the GEMS/Water programme (based on UNEP/ WHO/UNESCO/WMO 1983; 1987; 1990)

Country	Triennium		
	1979–1981	1982–1984	1985–1987
Africa			
Kenya	—	5	6
Mali	—	—	2
Senegal	—	—	1
United Republic of Tanzania	3	2	—
The Americas			
Argentina	6	5	3
Brazil	—	6	8
Canada	9	8	9
Chile	2	2	1
Columbia	—	2	1
Ecuador	1	2	1
Guatemala	3	—	—
Mexico	9	9	9
Panama	2	2	2
Peru	1	—	—
Uruguay	1	4	4
Eastern Mediterranean			
Egypt	7	—	—
Iran	5	5	—
Morrocco	—	—	1
Pakistan	4	4	4
Sudan	1	1	1
Tunisia	1	1	—
Europe			
Belgium	9	9	9
Denmark	4	—	—
Federal Republic of Germany	12	—	—
Finland	3	3	3
France	16	—	—
Grand Duchy of Luxembourg	1	—	—
Hungary	2	2	2
Ireland	4	—	—
Italy	5	—	—
Netherlands	5	5	—
Norway	2	3	3
Portugal	1	1	1
Spain	5	—	—
Turkey	3	3	3
UK	8	9	9
South-East Asia			
Bangladesh	5	5	—
India	24	23	31
Indonesia	6	—	6
Sri Lanka	2	—	—
Thailand	2	2	2
Western Pacific			
Australia	8	5	5
China	3	3	3
Fiji	1	1	1
Japan	9	9	9
Malaysia	6	6	5
New Zealand	3	3	2
Philippines	2	2	1
Republic of Korea	—	1	1
Total	206	153	149

Several approaches have been used to identify past and future trends in water temperature. In the absence of long-term and high quality data, changes in thermal characteristics have sometimes been inferred from the comparison of data collected in periods separated by many years. Blakey (1966), for example, compared records for the Mississippi River in 1923 and in 1962 and found that the mean annual water temperature was 1·6°C higher in 1962, despite the mean annual air temperature being 2·2°C lower. However, because of the climatic vagaries of the individual years studied and differences in the measurement techniques between the two periods, it was not possible to conclude definitively that the Mississippi had experienced a systematic increase in water temperature between 1923 and 1962.

Continuous river temperature data may be subjected to statistical analysis to identify temporal trends. Ludwig et al. (1990) used linear regression to analyse April and October mean water temperatures and mean annual water temperatures for the River Danube at the Reichsbruecke in Vienna during the 1968–1987 period. Increases in temperature were detected and the slope of the upward trend was found to be greater for mean October than mean April water temperatures, although these results were not tested for statistical significance.

In recent years, a growing interest in water quality trends (e.g. Smith et al., 1987) has led to the development of special statistical techniques for the identification of significant temporal changes in water quality (e.g. Montgomery and Loftis, 1987; Berryman et al., 1988; Uri, 1991). A detailed discussion of these methods is provided in the paper in this issue by Esterby. The seasonal Kendall test and associated seasonal Kendall slope estimator are particularly useful for the detection of trends in water temperature (Hirsch et al., 1982; Hirsch and Slack, 1984). These methods cope with a strong seasonality in the water quality time series and problems of data non-normality, flow relatedness, missing values and serial correlation.

Future changes in stream and river temperatures may be predicted in several ways. One approach is to extrapolate past trends into the future (e.g. Webb and Nobilis, 1994). This method, however, is limited by the small number of rivers for which good quality data on past changes in water temperature exist, and by the assumption that past trends will continue unaltered in direction and magnitude in the future.

Another approach is to use relationships between air and water temperature (e.g. Smith, 1981; Crisp, 1992; Jeppesen and Iversen, 1987; Stefan and Preud'homme, 1993) that may be applied to future changes in climate (e.g. Mackey and Berrie, 1991; Webb, 1992). This approach assumes that air–water temperature relationships will remain unchanged in the future and that they can be used to extrapolate water temperature. Non-linearities in the relationship at low air temperatures and increased scatter and lower sensitivity when air and water temperatures averaged over shorter time periods (e.g. days and hours) are related (Crisp and Howson, 1982; Stefan and Preud'homme, 1993) limit the potential of this method for predicting future temperature in detail. Furthermore, this approach relies on the air–water temperature relationship to provide a reasonable surrogate for the complex heat exchanges. In the latter context it has been acknowledged that predictions of future temperatures in UK chalk streams on the basis of air–water temperature relationships are likely to be too low, because the effects of future changes in groundwater temperatures are not included (Mackey and Berrie, 1991).

A third approach uses more sophisticated models of actual or equilibrium water temperature based on energy budgets and heat transport to predict future water temperatures (e.g. Voos et al., 1987; Cooter and Cooter, 1990; Meisner, 1990; Bartholow, 1991; Sinokrot and Stefan, 1993; Stefan and Sinokrot, 1993). Physically based models provide a powerful and flexible tool for estimating future trends in stream and river temperatures. However, the success of this approach depends not only on the extent to which modelled temperatures can be made to fit the observed temperatures for present thermal regimes, but also on the reliability of estimates of future meteorological and hydrological conditions which are required for simulation purposes. Predictions of the effects of global warming on river temperatures, for example, can vary appreciably depending on the scenarios for changes in air temperature, relative humidity, cloud cover and solar radiation, which are derived from different global climate models for use in the water temperature simulation (e.g. Cooter and Cooter, 1990).

TRENDS IN WATER TEMPERATURE BEHAVIOUR

Results from several studies in different countries are used to illustrate the nature and controls of trends in

Table III. Characteristics of tributaries in the Exe Basin, Devon, UK from which monthly time series of water temperature have been analysed for trend over the period 1977–1990

Catchment	Drainage area (km²)	Elevation (mOD)	Aspect	Geology	Soils	Land use	Channel character
Black Ball Stream	2·1	286·5	E-facing	Massive sandstones and interbedded shales	Cambic stagnohumic gleys on interfluves and ferric stagnopodzols on slopes	Stable heather moorland	Not significantly shaded and stream bed of silt and gravel
River Pulham	19·1	183·0	S-facing	Slates and sandstones	Brown earths and brown podzolics	Cattle pasture	Partly shaded by trees and silty stream bed
Iron Mill Stream	33·5	95·4	E-facing	Sandstones, siltstones and shales	Brown podzolics on valley floor, brown earths on slopes and seasonally waterlogged surface gleys on interfluves	Riparian woodland and cattle pasture	Significant shading and stream bed of silt, sand and rock

water temperature, which were reported for a range of catchments of different scale and character. These trends relate to time periods varying from less than one decade to much of the present century.

British rivers

Results from detailed river temperature monitoring in the Exe Basin, Devon, UK show: (1) significant recent increases in water temperatures; (2) the changes in some aspects of water temperature are greater than others; (3) the impacts of local human activities may equal climatic variables in causing stream temperature increases; and (4) statistical significance of short-term temperature trends may be distorted by extreme hydrological events (Webb and Walling, 1992; 1993b). Continuous temperature data from 1977 to 1990, collected in three small tributaries of the Exe Basin (Table III), were used to derive monthly values of maximum, mean maximum, mean, mean minimum and minimum water temperatures. These time series and equivalent monthly values of air temperature recorded at Exeter Airport were evaluated for trend significance and slope using the seasonal Kendall test and seasonal Kendall slope estimator, respectively. The results (Table IV) revealed statistically significant temperature increases ($p < 0.05$) for nine of the 15 water temperature trends tested. Significant increases varied in magnitude from 0·5 to 1·3°C over the 14 year study period.

All three rivers exhibited significant increases in monthly mean maxima, which reflected a strong increase of 1·4°C in the same air temperature statistic as recorded at Exeter Airport. Increases were also significant for monthly mean air temperature and water temperature in the River Pulham and the Iron Mill Stream. Trends in water temperatures were least strongly developed for minima and for the upland moorland catchment of the Black Ball Stream, where temperature increases during the study period were small and not statistically significant except monthly mean maxima (Table IV). In contrast, all water temperature statistics for the forested Iron Mill Stream exhibited a significant increase over the study period. At this station, increases in monthly maximum, mean and mean minimum water temperatures exceeded the equivalent increases in the Exeter Airport air temperature. Upward trends in water temperature for the Iron Mill Stream have been locally enhanced by tree removal adjacent to the monitoring station during the last two years of the study. In particular, increased exposure of the channel results in greater increases in summer maxima during 1989 and 1990 for the Iron Mill Stream compared with the other two catchments.

Table IV. Trends in water and air temperatures at sites in the Exe Basin, Devon, UK over the period 1977–1990, based on the application of the seasonal Kendall test and slope stimator to monthly values

Monthly temperature parameter and station*	Significance level (p)†	Slope of trend‡
Maxima		
BBS	0·271	0·020
IMS	*<0·001*	0·092
RP	*0·039*	0·050
EA	0·082	0·073
Mean maxima		
BBS	*0·030*	0·039
IMS	*<0·001*	0·075
RP	*0·050*	0·040
EA	*0·002*	0·100
Means		
BBS	0·187	0·017
IMS	*<0·001*	0·073
RP	*0·048*	0·029
EA	*0·048*	0·050
Mean minima		
BBS	0·749	0·000
IMS	*<0·001*	0·058
RP	0·116	0·025
EA	0·342	0·022
Minima		
BBS	0·646	0·000
IMS	*0·025*	0·050
RP	0·689	0·013
EA	0·150	0·050

* BBS = Black Ball Stream; IMS = Iron Mill Stream; RP = River Pulham; and EA = Exeter Airport (air temperature). † Significance is two-tailed and an italic value denotes a trend significant at the $p < 0.05$ level. ‡ Slope of trend is expressed as °C yr^{-1}.

Analysis of water temperature from the Black Ball Stream also revealed significant increases in monthly mean maxima for 1977–1990, but the increase was not significant for 1976–1989 (Webb and Walling, 1992). This difference is attributed to the occurrence of exceptionally low flows and attendant high water temperatures during the drought of summer 1976 (Walling and Carter, 1980), which has had a distorting effect on water temperature trends over the last two decades in many British rivers.

US rivers

The spatial variation in water temperature trends over the conterminous USA is provided by Smith and Alexander (1983) through an analysis of river temperatures during 1974–1981 at 364 river sites in the 'Benchmark' and 'NASQAN' monitoring networks of the US Geological Survey. From 70 to 90 observations were typically available at each site and the statistical significance and slope of trends were tested using the seasonal Kendall test and slope estimator for both raw and flow-adjusted temperatures. The latter were calculated as a residual between actual temperatures and those estimated from a 'best fit' relationship between temperature and discharge at each site (Hirsch *et al.*, 1982).

The results show that significant changes in temperature over the relatively short study period have been neither universal nor systematic for rivers across the USA. Significant trends were detected at 90 sites, or ca. 25% of the rivers investigated (Figure 1). However, at 29 of these sites a significant trend was evident for the raw temperatures, but not for the flow-adjusted temperatures. Water temperature increases occurred at most

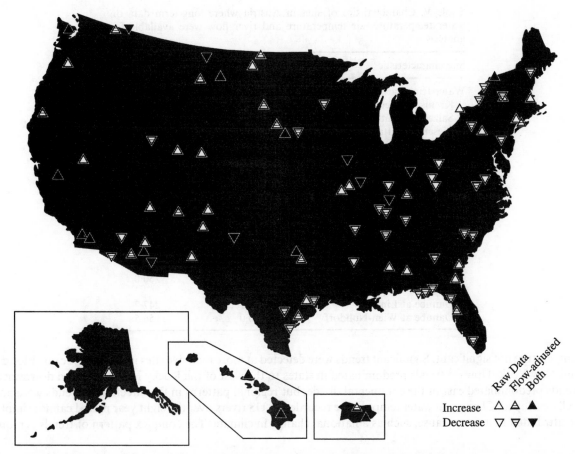

Figure 1. Trends in water temperature in rivers of the USA from 1974 to 1981 based on information in Smith and Alexander (1983). Trends were identifed using the seasonal Kendall test and slope estimator and stations with statistically significant changes ($p < 0.05$) in raw and/or flow-adjusted temperature data are depicted

of these sites (22); these averaged 2·2% of the site-mean temperature, but exceeded 5% in some instances. An absence of a significant trend in flow-adjusted temperatures suggests that raw temperature increases were promoted by decreasing river flows. Similarly, an interpretation that increasing flows were responsible for declining water temperatures applies to the small number of sites (seven) where a significant decrease in raw temperatures, but not in flow-adjusted temperatures, was recorded. Decreases at these sites did not exceed 3.4% of the site-mean temperatures.

Significant trends in flow-adjusted, but not in raw, temperatures were detected at 40 sites. At these sites, changes for raw temperatures have been obscured by fluctuations in river discharge, and trends only emerge when the influence of flow variations is removed. The majority of sites in this category (26) exhibited a significant decrease in water temperature, which averaged 4% of the site-mean temperature, but exceeded 10% in some instances. Significant increases in flow- adjusted temperatures occurred at the remaining 14 sites and averaged 3·5%, but never exceeded 7%, of the site-mean temperature. Significant trends were recorded for both raw and flow-adjusted water temperatures at 21 sites. Trends were in the same direction at these sites for all but one case, and uptrends or downtrends were encountered in almost equal proportions. Flow adjustment improved the significance of the trend for most sites (14), but the magnitude of the change over the study period did not exceed 7% of the site-mean temperatures.

Sites associated with significant changes in water temperature during the period 1974–1981 were found in most parts of the USA (Figure 1). Only in Kansas, Wisconsin, Michigan, Louisiana, Mississippi, North Carolina and some small states located along the eastern seaboard and in New England were trends in river

Table V. Characteristics of sites in Austria where long-term data on water temperature, air temperature and river flow were available for analysis

Site/characteristic	Drainage area (km^2)	Elevation (m)
Water temperature		
Krems at Kremsmünster	142·4	338·8
Salzach at Mittersill	591·2	783·0
Lieser at Spittal	1036·0	549·1
Gail at Federaun	1304·8	499·5
Traun at Wels	3498·6	304·7
Salzach at Salzburg	4427·3	408·1
Mur at Graz	6988·9	340·4
Inn at Schärding	25 663·8	299·8
Danube at Linz	79 490·1	247·7
Danube at Ybbs	92 464·2	212·2
Air temperature		
Kremsmünster	—	390·0
Graz-Universität	—	365·0
Linz-Stadt	—	260·0
River flow		
Danube at Linz	79 490·1	247·7
Danube at Wien-Nußdorf	101 700·0	156·5

temperature not significant. Significant trends were detected at more than five sites in Arizona, Texas, Florida and New York. Upward trends predominated in states to the west of the Rocky Mountains and downward trends predominated east of the continental divide, but regional patterns in the direction of trends were not well developed. Changes in water temperature recorded in US rivers over the eight year period cannot simply be attributed to a single cause, such as a national change in climate. The complex pattern of trends is more

Table VI. Trends in water temperature, air temperature and river flow at sites in Austria over the period 1901–1990 based on the application of the seasonal Kendall test and slope estimator to monthly mean values

Site/characteristic	Slope of trend	Trend	Statistical significance (p)
Water temperature			
Kremsmünster	0·00000	0·00°C	0·303
Mittersill	0·00000	0·00°C	0·961
Spittal	0·00968	0·87°C	0·000
Federaun	0·00526	0·47°C	0·000
Wels	0·01404	1·26°C	0·000
Salzburg	0·00000	0·00°C	0·140
Graz	0·00727	0·65°C	0·000
Schärding	0·00601	0·54°C	0·000
Linz	0·00893	0·80°C	0·000
Ybbs	0·01111	1·00°C	0·000
Air temperature			
Kremsmünster	0·00000	0·00°C	0·444
Graz-Universität	0·00465	0·42°C	0·007
Linz-Stadt	0·00342	0·31°C	0·080
River flow			
Linz	0·21779	19·6 m^3 s^{-1}	0·623
Wien Nußdorf	0·00000	0·00 m^3 s^{-1}	0·994

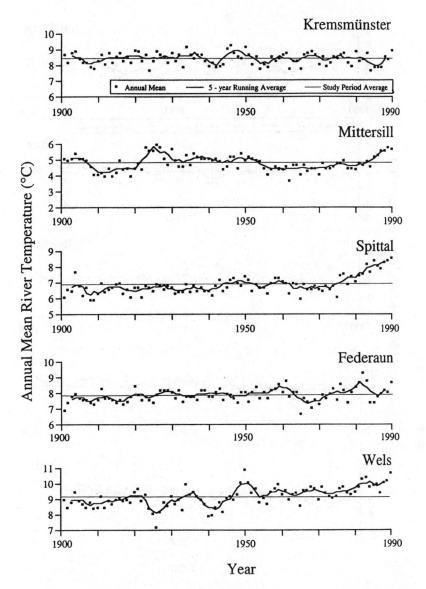

Figure 2. Annual mean, five-year running average and study period average water and air temperatures and river flows in the period 1901–1990 for selected sites in Austria

likely to reflect the local impacts of human activity through changes in heated effluent discharges, modification of river flows and alteration of riparian and catchment land uses. Also, some trends may reflect natural climatological and hydrological fluctuations at regional and local scales.

Austrian rivers

A unique view of water temperature trends over the 20th century is provided by data collected in long-term monitoring programmes for Austrian rivers. Continuous monthly mean water temperatures derived from daily measurements from 1901 to 1990 are available from the Austrian State Hydrographic Service (Hydrographisches Zentralbüro, Bundesministerium für Land- und Forstwirtschaft) for 10 catchments with a range in size and elevation (Table V). Analysis of these data, using the seasonal Kendall test and slope estimator, suggests that a general increase in water temperature has occurred during the present century, but also that the magnitude of increase has varied markedly among rivers (Table VI). A statistically significant

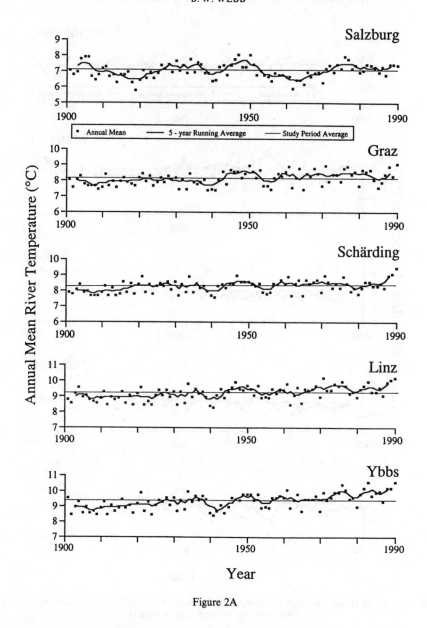

Figure 2A

($p < 0.05$) increase in the monthly mean water temperature was detected for seven of the study sites and ranged from $0.47°C$ in the River Gail at Federaun to $1.26°C$ in the River Wels at Traun.

Mean temperatures at Linz and Ybbs on the mainstream of the Danube increase by at least $0.8°C$ during the 90 year period. An evaluation of five year running average annual mean temperatures (Figure 2) indicates that temperatures in the River Danube were generally below the long-term average for the first half of the study period, but largely above average during the second half, although the significant increase identified by the seasonal Kendall test has been neither steady nor continuous. Analysis of trends for individual months, using linear least-squares regression, revealed that the greatest increases for the mainstream sites on the Inn and the Danube occurred in the autumn and early winter (Table VII). For the site at Ybbs, monthly mean water temperatures recorded in November increased by $> 2.1°C$ during the 90 year period. In contrast, no significant change in water temperature was evident for the months of May, June and July at the mainstream sites on the Inn and the Danube.

A comparison of trends in water temperature with that of monthly mean air temperature and river flow

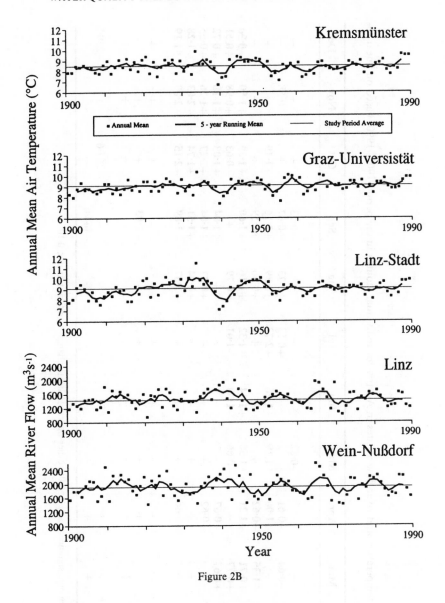

Figure 2B

suggests that water temperature trends for the mainstream of the River Danube cannot be simply attributed to changing climatic and hydrological conditions during the 20th century. Air temperatures at the town of Linz on the banks of the Danube increased by only 0·31°C, which is less than 50% of the increase in water temperature in the adjacent river (Table VI). The air temperature at Graz-Universitat increased less than the water temperature for the River Mur at Graz, and no trend was detected in monthly mean air temperature at Kremsmünster. Furthermore, no air temperature trend was detected for individual months, with the exception of September at Graz-Universität (Table VII). Trends in the five year running averages of mean annual air temperature were generally dissimilar to water temperature trends for the mainstream of the Danube (Figure 2). Furthermore, no statistically significant trend for river flow of the Danube at Linz and Vienna (Wien- Nußdorf) was evident (Table VI). Significant increases and decreases in flow for the months of February and September, respectively, are not consistent with monthly water temperature trends for the Danube (Table VII) and the variations in the running average annual mean temperatures did not correlate with those for water temperature and discharge (Figure 2).

Increasing human impacts have undoubtedly caused water temperatures to increase in the Danube

Table VII. Trends in water temperature, air temperature and river flow at sites in Austria for individual months in the period 1901–1990, based on linear regression analysis of monthly mean values

Site/characteristic	Jan	Feb	Mar	Apr	May	Jun	Jul	Aug	Sep	Oct	Nov	Dec
Water temperature (°C)												
Kremsmünster	−	+	−	−	−0·82	−	+0·53	+	+0·53	+	+	−
Mittersill	+	−	**−0·64**	−0·61	−	−	+	+0·73	+0·67	+1·19	+0·58	−
Spittal	+	+	+	+1·00	+1·19	+1·26	**+2·73**	**+2·89**	**+2·16**	+	+	−
Federaun	+0·88	**+1·84**	**+1·35**	+0·62	+	+	+	+0·56	+0·59	+	−	+
Wels	+1·16	+1·30	+0·71	+1·21	+1·50	+1·14	+1·32	+1·16	+1·60	+1·39	+1·50	+0·90
Salzburg	+0·74	+1·17	+0·77	+	−0·66	−1·00	−1·01	−0·79	+	+0·63	+0·98	+0·51
Graz	+1·02	+1·53	+0·67	+0·67	+	+	−	+	+1·04	+1·03	+1·10	+0·72
Schärding	+	+0·72	+	+0·62	+	−	+	+0·77	+1·04	+1·27	+1·26	+0·55
Linz	+0·76	+1·28	+	+	+	+	+	+0·94	+1·51	+1·74	+2·03	+1·28
Ybbs	+0·70	+1·32	+	+0·66	+	+	+	+1·37	+1·89	+2·15	+2·00	+1·19
Air temperature (°C)												
Kremsmünster	−	+	−	+	−	+	+	+	+	+	+	+
Graz-Universität	+	+	+	+	−	+	+	+	+1·03	+	+	−
Linz	+	+	+	+	+	+	+	+	+	+	+	+
River Flow (m³ s⁻¹)												
Linz	+	+393·9	+	+	−	−	−	−	−	−267·6	+	+239·8
Wien–Nußdorf	−	+398·1	+	+	−	−	−	−	−464·0	−	−	+

+ = Increase; − = decrease; **bold figures** indicate change significant at $p < 0.05$; other figures indicate change significant at $p < 0.10$.

mainstream over the last 90 years. Heated effluent discharges have increased during the 20th century and the Danube has concurrently been markedly affected by canalization and regulation to improve river navigation, reduce overbank flooding and generate electrical power (e.g. Liepolt, 1972; Schiller, 1983; Bacalbasa-Dobrovici, 1989; Dokulil et al., 1993). From 1955 to 1984, nine dams were built across the Danube in the section between the German border and Vienna. These dams created extensive impounded reaches, causing the water velocity to be markedly reduced, the residence time and river depth to be substantially increased and channel banks to be significantly armoured with coarse material (Humpesch, 1992). In the absence of significant stratification of the river, these changes may have led to heating. The greatest increases in water temperature occur in October and November when the flow is lowest. The low flow not only affects the thermal capacity of the water-course, but the impounded reaches are also sensitive to heating because the circulation is reduced.

Table VIII. Studies predicting the effects of global warming on stream and river temperatures

Study area	Approach	Scenario	Forecast	Source
Southern USA	Modelling mean July equilibrium surface water temperatures for 15 m wide streams	Doubled atmospheric CO_2	Increases of 1–7°C, but spatial pattern of rise varying with climatic scenarios from different GCMs	Cooter and Cooter (1990)
UK rivers	Monthly mean air–water temperature relationships	Increased air temperatures of 2–3°C by the year 2050	Increases of 1·0–3·6°C depending on catchment characteristics	Webb (1992)
Five streams in north central USA	Deterministic modelling of long-term average daily water temperature in freely flowing reaches	Doubled atmospheric CO_2	Increases of 2·4–8·4°C with magnitude of rise depending on which GCM climatic scenario used	Stefan and Sinokrot (1993)
River Danube at Linz, Austria	Multiple regression analysis of monthly mean water temperatures	Increases in monthly mean air temperature of 2·3–3·4°C and 15% reduction in flow between June and September by the year 2030	Increase in monthly means of 1·2–2·1°C	Webb and Nobilis (1994)
River Lambourn and Water of Leith, UK	Monthly mean air–water temperature relationships	Increases in monthly mean air temperature of 2–4°C	Increase in monthly means of 1·1–2·2°C in River Lambourn and 1·8–3·6°C in Water of Leith	Mackey and Berrie (1992)
Rouge and Humber rivers, southern Ontario, Canada	Deterministic modelling of daily water temperature in July and August	Increase in mean summer air temperature of 4·1°C and mean groundwater temperature of 4·8°C	Increases of up to 4·5°C	Meisner (1990)

In the snowmelt-dominated flow regime of the Danube, the highest discharges occur in mid-summer and the river at this time is much less sensitive to any human or climatic impacts that typically cause water temperature increases. A study of the River Danube upstream in Germany (Deisenhofer, 1992) also reported a significant increase in annual mean water temperature of ca. 1°C from 1930 to 1990, but no concurrent trend in air temperature. Uptrends in winter were larger than in summer and water temperature increases were attributed to several anthropogenic activities, including canalization and discharges of heated effluent from power stations.

Some Austrian rivers, which are smaller than the Danube and are not regulated by mainstream dams, such as the Mur and the Gail, have smaller increases in water temperature (Table VI, Figure 2), which are less pronounced in autumn and early winter (Table VII). However, trends are large for other small Austrian rivers and can be linked to human impacts. For example, the water temperature generally increased by 0.87°C in the River Lieser at Spittal and uptrends were significant for the months between April and October exceeding 2.5°C for August and September (Tables VI and VII). The river temperature increase at Spittal has been particularly marked in the last decade or so, and may be attributed to human modification of the river channel. Since 1977, flow has been diverted from ca. 130 km^2 (> 12%) of the upstream catchment area. This diversion has not only removed a source of cold-water runoff during the meltwater season, but has also reduced the flow volume and thermal capacity of the channel below the diversion. The Lieser at Spittal has thereby become more sensitive to solar heating during the summer months.

Results from Austrian rivers also suggest that in addition to impacts of human activity the natural catchment characteristics may influence long-term trends in river temperature. The largest overall increase in temperature, having significant uptrends in each month of the year, occurs in the River Traun, which contains large lakes (Attersee and Traunsee) in its drainage. The proportionately large lake area is sensitive to small increases in air temperature, which affects the timing, duration and stability of warm epilimnial layers (Robertson and Ragotzkie, 1990). Temperatures of lake outflows and the river downstream may consequently be affected by changes in the thermal structure of the lakes. No significant trend in water temperature was observed for the River Krems at Kremsmünster and the River Salzach at Mittersill, which were the most upstream sites. In some of the spring months, temperatures at these sites decreased (Table VII). The absence of significant overall trends in water temperature may be attributed to relatively less exposure to increasing human impacts during the 20th century in headwater tributaries, but close proximity to glacial runoff also may be a significant factor at these sites. Any increase in the contribution of cold meltwater to river flow as a consequence of increasing air temperatures in recent years would cause decreasing water temperatures at upstream sites, especially during the spring months. No significant trend in water temperature was observed at Salzburg, which is located further downstream on the River Salzach. This site was unusual in exhibiting significant and relatively strong decreases in monthly water temperature during the study period of ca. 1·0°C for the mid-summer in June and July (Table VII). An explanation of the lack of trends might again be related to trends in the amount of meltwater runoff. However, this site has also been influenced by the discharge of cooling water from a nearby electricity generating plant so that alterations in effluent volumes from this source also may be a significant factor affecting river temperatures.

FUTURE CHANGES IN WATER TEMPERATURE

Many studies have been conducted to predict how water temperature in specific river reaches may change in the future due to various human activities. For example, the US Fish and Wildlife Service has been investigating the potential effects of flow releases from reservoirs, the modification of channel morphology and the alteration of riparian vegetation on water temperatures in a range of US rivers using the stream network temperature model (Theurer et al., 1982; 1985; Lifton et al., 1985; Bartholow, 1991). However, interest has also increased in the factors affecting the general changes in thermal regime and trends in stream and river temperatures that may arise from climate change through global warming.

Table VIII summarizes the results of six studies which were conducted to predict changes in water temperature resulting from climate modification as projected for around the middle of the 21st century by general circulation or global climate models (GCMs). Results from each study suggest that stream and river temperatures will increase during the next century, although the magnitude of the predicted rise varies from 1 to more than 7°C. Differences between GCMs in the projected climatic changes may cause considerable variation in the predictions of future water temperatures for a given area. Stefan and Sinokrot (1993), for example, predicted, by coupling a deterministic river heat transport model with climatic changes projected to follow a doubling of atmospheric CO_2 from the UK Meteorological Office GCM, that the average water temperature would rise by 7·3–8·4°C for five freely flowing water courses in the north-central USA. In contrast, the same model predicted considerably lower increases of 2·4–4·7°C for these water courses when climatic projections from three American GCMs [Goddard Institute for Space Studies, Columbia University (GISS); Geophysical Fluid Dynamics Laboratory, Princeton University (GFDL); Oregon State University (OSU)] were used. Similarly, variations between the three American GCM scenarios resulted in significant differences in the patterns of change in mean July equilibrium water temperature predicted for the southern USA as a consequence of doubling atmospheric CO_2 (Cooter and Cooter, 1990). Maximum increases in river temperature were projected for Virginia and Texas by the GISS model, for Virginia and North Carolina by the GDFL model and for an area in the lower Mississippi River, covering parts of north-east Texas, south-west Oklahoma, southern Arkansas, northern Louisiana and northern and central Mississippi by the OSU model.

The extent to which water temperature may respond to future changes in air temperature and other climatic variables varies between streams and rivers of different character. Mackey and Berrie (1991) used air and water temperature relationships to suggest that increases in water temperature following increases in air temperature would be moderated in chalk streams fed by groundwater compared with rivers which were predominantly fed by surface sources. Monthly mean air temperature increases of 2°C in summer and 4°C in winter were projected to cause monthly mean water temperature increases from 1·1 to 2·2°C for the River Lambourn, which drains an area of chalk in southern England, and from 1·8 to 3·6°C for the Water of Leith in Scotland, which is fed from non-calcareous lithologies. A study of future water temperatures in UK rivers based on relationships between mean monthly air and water temperatures for 36 river sites (Webb, 1992) suggested that air temperature increases from 2 to 3°C by the year 2050 will result in average river temperature increases from 1·8 to 2·7°C. However, this study also indicated that smaller increases of 1·0–1·5°C and 1·4–2·1°C might occur in spring-fed and forested streams, respectively, and larger increases of 2·4–3·6°C might occur in streams and rivers with wider, shallower and more exposed channels. Increases in water temperature due to global warming are likely to vary spatially and the pattern will be complex across any sizable area because it will reflect the superimposition of different responses in catchments of contrasting character on likely regional and national gradients in future climatic change. In the UK, for example, it is predicted that the winter mean air temperature in the year 2050 may be ca. 2·2°C higher than at present in south-west England, but ca. 3·6°C greater in northern Scotland (UKCCIRG, 1991). On the basis of typical air and water temperature relationships for British rivers, the winter mean water temperatures might be expected to increase 2·0°C in the extreme south-west and 3·2°C in the extreme north by the middle of the next century.

Increases in water temperature with future global warming also may not be constant throughout the year. The results of modelling the temperatures of water courses in the north-central USA suggest that long-term trends in temperature increases during seasonal warming from March to July would be less marked than during seasonal cooling from August to November (Stefan and Sinokrot, 1993). Such differences may reflect seasonal variations in the future change of influential climatic variables, such as air temperature, and seasonal contrasts in the sensitivity of rivers to the effects of global warming. These controls may not always operate in the same direction. For example, GCMs predict that the mean water temperature increases by the year 2030 for most of Europe will be significantly larger in the winter than in the summer (Bultot et al., 1988; 1992), whereas analysis of air and water temperature relationships for different months (Webb and Walling, 1993b) suggests that some rivers in south-west England may be more sensitive to air temperature increases that occur in summer. Consequently, future increases in water temperature may not be uniform or follow a simple pattern throughout the year. For example, the predicted monthly mean water temperature for the River Daube at Linz in Austria will increase by 1·2°C in January and June by the year 2030, but by 1·9°C in March and 2·1°C in November (Webb and Nobilis, 1994).

In assessing the effects of global warming on thermal regime, it is important to consider not only the direct effect of changes in air temperature, solar radiation and other climatic factors, but also the indirect impact of modifications to streamflow and riparian vegetation. Several studies have suggested that water temperature is inversely related to river discharge (e.g. Smith, 1975; Smith and Lavis, 1975; Hockey et al., 1982), which reflects the reduction in thermal capacity of a water course as the flow volume decreases. Any decrease in river flow as a consequence of global warming might therefore be expected to exacerbate increases in river temperature. This effect, however, may not be particularly significant, especially in free-flowing streams, where temperatures are already often close to equilibrium. Sinokrot and Stefan (1993) predict that a 50% reduction of flow in the Straight River, Minnesota, USA would not affect the annual thermal regime as reflected in daily average temperatures, whereas regression analysis suggests that the same flow decrease would increase temperatures at a headwater and a mainstream site in the Exe Basin, Devon, UK by no more than 0·8°C. Flow reduction, however, may cause groundwater or reservoir water to warm or cool to an equilibrium level over a shorter distance and may also exaggerate short-term temperature fluctuations in a freely flowing stream (Stefan and Sinokrot, 1993). Increases in groundwater temperature following global warming, however, may contribute significantly to increases in stream and river temperatures. For example, Meisner (1990), using a deterministic river temperature model, predicted that the changes in air temperature following a doubling of atmospheric CO_2 would solely cause increases in water temperature of no more than 2°C for the Rouge and Humber Rivers in southern Ontario, Canada. However, by including the groundwater temperature increases associated with global warming, the model predicted increases in temperature of up to 4·5°C.

Changes in riparian vegetation as a consequence of increased CO_2 levels and a generally warmer atmosphere may be of greater significance to future river temperatures than any alteration of the runoff regime. Riparian tree cover is particularly important for reducing the input of shortwave radiation to the river channel, and Stefan and Sinokrot (1993) predicted that if trees along the Straight River disappeared, as a consequence of future climate changes, then an additional 6°C water temperature increase would occur. Therefore, the interaction between riparian vegetation and climatic change is likely to complicate the spatial pattern of river temperature increases that can be expected during the next century. Cooter and Cooter (1990) showed for the southern USA that projected patterns of forest migration following a doubling of atmospheric CO_2 could increase riparian shading east of 97°W longitude which, in turn, would limit water temperature increases to ca. 1°C. Further to the west, forest migration was projected to be unimportant because of drier conditions and no moderation of water temperature increases following climate change was anticipated.

CONCLUSIONS

Alteration of the energy budget and thermal capacity of a water course, through many different natural and human causes, can cause significant trends in water temperature. Unfortunately, the world database of stream and river temperatures is inadequate to give a reliable global picture of how thermal regimes have changed over the recent, let alone the more distant, past. Data from Europe for this century suggest that mean river temperatures have increased by less than or equal to ca. 1°C. Furthermore, this uptrend has not been monotonic, and is sensitive in magnitude and significance to the time period investigated and especially to the occurrence of hydrological extremes, such as the 1976 drought in the UK.

The factors responsible for past trends in water temperature are likely to be multiple and their interrelationships complex. Rising air temperatures may account for the higher river temperatures, especially during the winter period, which have occurred in recent years. Over the longer term, however, there does not appear to be a close relationship between the trends of water temperature and those of simple hydrometeorological variables. It seems certain that changes in the thermal regime detected during the 20th century have been influenced by human activities, and especially increasing effluent discharges, river impoundment and riparian vegetation removal.

Predictive studies suggest that river temperatures are generally set to increase further during the next century as a consequence of global warming. Also, the rate of rise during the next 50 years is predicted to be equal to, or greater than, that recorded during the last 100 years. Such forecasts, however, need to be used with caution because of uncertainties in the climatic scenarios associated with global warming predictions, and because of interactions and feedbacks between climatic change on the one hand and hydrology and especially riparian vegetation on the other. The importance of even fairly small changes in water temperature to river ecology dictates that further refinements are needed in modelling to better define the nature of stream and river temperature trends for the 21st century, and that more extensive monitoring programmes are required, especially for many parts of the southern hemisphere, so that a better global perspective on current and future water temperature may be achieved.

ACKNOWLEDGEMENTS

The author is grateful to Dr Franz Nobilis of the Hydrographisches Zentralbüro, Bundesministerium für Land- und Forstwirtschaft, Vienna for his generous and substantial help in supplying water temperature and other data for Austrian rivers. A grant from the Department of the Environment to investigate the effects of climate change on thermal regimes (contract PECD/7/7348), which supported research of the UK rivers reported in part of this study, is also gratefully acknowledged. Helpful comments by Professor Des Walling, University of Exeter and skillful editing by Dr Jake Peters, US Geological Survey, with respect to earlier drafts of this paper are also much appreciated.

REFERENCES

Anderson, R. R. 1969. 'Temperature and rooted aquatic plants', *Chesapeake Sci*, **10**, 157–164.

ASCE Committee on Sanitary Engineering Research 1961. 'Effect of water temperature on stream reaeration', *Am. Soc. Civ. Eng. J. Sanitary Engin. Div.* **87**(SA6), 59–71.

Bacalbasa-Dobrovici, N. 1989. 'The Danube River and its fisheries' in Dodge, D.P. (Ed.), *Proceedings of the International Large River Symposium (LARS)*, *Can. J. Fish Aquat. Sci. Spec. Publ.*, **106**, 455–468.

Barabas, S. 1986. 'Monitoring natural waters for drinking-water quality', *World Health Stat. Q.*, **39**, 32–45.

Bartholow, J. M. 1989. 'Stream temperature investigations: field and analytic methods', *U.S. Fish Wildlife Serv. Instream Flow Info. Pap.*, **13**.

Bartholow, J. M. 1991. 'A modeling assessment of the thermal régime for an urban sport fishery', *Environ. Manage.*, **15**, 833–845.

Berryman, D., Bobee, B., Cluis, D., and Haemmerli, J. 1988. 'Nonparametric tests for trend detection in water quality time series', *Wat. Resour. Bull.*, **24**, 545–556.

Beschta, R.L. and Taylor, R.L. 1988. 'Stream temperature increases and land use in a forested Oregon watershed', *Wat Resour Bull.*, **24**, 19–25.

Beschta, R. L., Bilby, R. E., Brown, G. W., Holtby, L. B., and Hofstra, T. D. 1987. 'Stream temperature and aquatic habitat: fisheries and forestry interactions' in Salo, E. F., and Cundy, T. (Eds), *Stream-side Management: An Interdisciplinary Symposium on Forestry and Fisheries Interactions*. College of Fisheries and College of Forest Resources, University of Washington, Seattle, pp. 191–232.

Bilby, R.E. 1984. 'Characteristics and frequency of cool-water areas in a Western Washington stream', *J. Freshwater Ecol.* **2**, 593–602.

Binkley, D. and Brown, T. C. 1993. 'Forest practices as nonpoint sources of pollution in North America', *Wat Resour. Bull.* **29**, 729–740.

Blakey, J. F. 1966. 'Temperature of surface waters in the conterminous United States', *US Geol. Surv. Hydrol. Invest. Atlas*, **HA-235**.

Brett, J. R. 1956. 'Some principles in the thermal requirements of fishes', *Q. Rev. Biol.* **31**, 75–87.

Brezonik, P. L. 1972. 'Chemical kinetics and dynamics in natural water systems' in Ciaccio, L.L. (Ed.), *Water and Water Pollution Handbook*. Vol. 3. Marcel Dekker, New York. pp. 831–913.

Brown, G. W. and Krygier, J. T. 1967. 'Changing water temperatures in small mountain streams', *J. Soil Wat. Conserv.*, **22**, 242–244.

Bultot, F., Coppens, A., Dupriez, G. L., Gellens, D., and Meulenberghs, F. 1988.'Repercussions of a CO_2 doubling on the water cycle and on the water balance—a case study for Belgium', *J. Hydrol.*, **99**, 319–347.

Bultot, F., Gellens, D., Speafico, M., and Schädler, B. 1992. 'Repercussions of a CO_2 doubling on the water balance—a case study in Switzerland', *J. Hydrol.* **137**, 199–208.

Byren, B. A. and Davies, B. R. 1989. 'The effect of stream regulation on the physico-chemical properties of the Palmiet River, South Africa', *Regul. Riv.*, **3**, 107–121.

Carr, J. E., Chase, E. B., Paulson, R. W., and Moody, D. W. 1987. 'National Water Summary 1987—hydrologic events and water supply and use', *US Geol. Surv. Wat. Supply Pap*, **2350**.

Cassidy, R. A. and Dunn, P. E. 1987. 'Water temperature control and areal oxygen consumption rates at a new reservoir, and the effects on the release waters' in Craig, J. F., and Kemper, J. B. (Eds), *Regulated Streams. Adv. Ecol.* Plenum Press, New York. pp. 339–351.

Churchill, M. A. 1965. 'Control of temperature through streamflow regulation' in *Symposium on Streamflow Regulation for Quality Control. US Public Health Serv. Pub.*, **999-WP-30**, 179–201.

Clark, D. and England, G. 1963. 'Thermal power generation' in *Conservation of Water Resources in the UK*. Institution of Civil Engineers, London. pp. 43–51.

Cooter, E. J. and Cooter, W. S. 1990. 'Impacts of greenhouse warming on water temperature and water quality in the southern United States', *Climate Res.*, **1**, 1–12.

Cowx, I. G., Young, W. O., and Booth, J. 1987. 'Thermal characteristics of two regulated rivers in mid-Wales, UK', *Regul. Riv.* **1**, 85–91.

Crisp, D. T. 1987. 'Thermal resetting of streams by reservour releases with special reference to effects on salmonid fishes' in Craig, J. F., and Kemper, J. B. (Eds), *Regulated Streams. Adv. Ecol.* Plenum Press, New York. pp. 163–182.

Crisp, D. T. 1990. 'Water temperature in a stream gravel bed and implications for salmonid incubation', *Freshwater Biol.* **23**, 601–612.

Crisp, D. T. 1992. 'Measurement of stream water temperature and biological applications to salmonid fishes, grayling and dace (including ready reckoners)', *Freshwater Biol. Assoc. Occ. Publ.*, **29**, 1–72.

Crisp, D. T. 1993. 'The environmental requirements of salmon and trout in fresh water', *Freshwater Forum*, **3**, 176–202.

Crisp, D. T., and Howson, G. 1982. 'Effect of air temperature upon mean water temperature in streams in the north Pennines and English Lake District', *Freshwater Biol.*, **12**, 359–267.

Croley, T. E., Giaquinta, A. R., and Woodhouse, R. A. 1981. 'River thermal standards costs in the upper midwest', *Am. Soc. Civ. Engin. J. Ener. Div.* **107**(EY1), 65–77.

Devik, O. 1975. 'Waste-heat and nutrient-loaded effluents in the aquaculture' in *Environmental Effects of Cooling Systems at Nuclear Power Plants.* International Atomic Energy Agency, Vienna. pp. 693–702.

Deisenhofer, E. 'Das Wassertemperaturrégime der deutschen Donau' in *Proceedings XVI. Konferenze der Donauländer über hydrologische Vorhersagen und hydrologisch-wasserwirtschaftliche Grundlagen* (Kelheim, Germany, 18–21 May, 1992). pp. 537–541.

Dokulil, M. T., Humpesch, U. U., Schmidt, R., and Pöckl, M. 1993. 'Limnologie: Auswirkungen geänderter Klimaverhältnisse auf die Ökologie von Oberflächengewässer in Österreich' in *Anthropogene Klimaänderungen: Mîgliche Auswirkungen Auf Österreich. Mîgliche Massnahmen in Österreich,* Österreichische Akademie der Wissenschaften.

Dymond, J. R. 1984. 'Water temperature change caused by abstraction', *Am. Soc. Civ. Engin. J. Hydr. Engin.* **110**, 987–991.

Elliott, J. M. 1994. *Quantitative Ecology and the Brown Trout.* Oxford University Press, Oxford.

Graynoth, E. 1979. 'Effects of logging on stream environments and faunas in Nelson', *N. Z. J. Mar. Freshwater Res.*, **13**, 79–109.

Greene, G. E. 1950. 'Land use and trout streams', *J. Soil Wat. Conserv.* **5**, 125–126.

Hendricks, C. W. 1972. 'Enteric bacterial growth rates in river waters', *Appl. Microbiol.*, **24**, 168–174.

Hirsch, R. M. and Slack, J. R. 1984. 'A nonparametric trend test for seasonal data with serial dependence', *Wat. Resour. Res.*, **20**, 727–732.

Hirsch, R. M., Slack, J. R., and Smith, R. A. 1982. 'Techniques of trend analysis for monthly water quality data,' *Wat. Resour. Res.*, **18**, 107–121.

Hoak, R. D., 1965. 'Hot water—a growing industry concern', *Industr. Wat. Engin.*, **2**, 10–14.

Hockey, J. B., Owens, I. F., and Tapper, N. J. 1982. 'Empirical and theoretical models to isolate the effect of discharge on summer water temperatures in the Hurunui River', *J. Hydrol. (N. Z.)*, **21**, 1–12.

Holtby, L. B. 1988. 'Effects of logging on stream temperatures in Carnation Creek, British Columbia, and associated impacts on the Coho salmon (*Oncorhynchus kisutch*)', *Can. J. Fish. Aquat. Sci.*, **45**, 502–515.

Holtby, L. B., McMahon, T. E., and Scrivener, J. C. 1989. 'Stream temperatures and inter-annual variability in the emigration timing of coho salmon (*Oncorhynchus kisutch*) smolts and fry and chum salmon (*O. keta*) fry from Carnation Creek, British Columbia', *Can. J. Fish. Aquat. Sci.* **46**, 1396–1405.

Hostetler, S. W. 1991. 'Analysis and modeling of long-term stream temperatures on the Steamboat Creek Basin, Oregon: implications for land use and fish habitat', *Wat. Resour. Bull.* **27**, 637–647.

Humpesche, U.U. 1992. 'Ecosystem study Altenwörth: impacts of a hydroelectric power-station on the River Danube in Austria', *Freshwater Forum*, **2**, 33–58.

Jaske, R. T. and Goebel, J. B. 1967. 'Effects of dam construction on temperatures of Columbia River', *J. Am. Waterworks Assoc.* **59**, 935–942.

Jeppesen, E. and Iversen, T. M. 1987. 'Two simple models for estimating daily mean water temperatures and diel variations in a Danish low gradient stream', *Oikos*, **49**, 149–155.

Lane, E. W., Carlson, E. J., and Hanson, O. S. 1949. 'Low temperature increases sediment transportation in Colorado River', *Civ. Engin*, **19**, 45–46.

Langford, T. E. 1983. *Electricity Generation and the Ecology of Natural Waters.* Liverpool University Press, Liverpool.

Lavis, M. E. and Smith, K. 1972. 'Reservoir storage and the thermal régime of rivers, with special reference to the River Lune, Yorkshire', *Sci. Total Environ*, **1**, 81–90.

Lemmelä, R., Liebscher, H., and Nobilis, F. 1990. *Studies and Models for Evaluating the Impact of Climate Variability and Change on Water Resources Within WMO-Regional Association VI (Europe).* World Meteorological Organization Regional Association VI (Europe) Working Group on Hydrology.

Liepolt, R. 1972. 'Uses of the Danube River', in Oglesby, R. T., Carson, C. A., and McCann, J. A. (Eds), *River Ecology and Man.* Academic Press, New York. pp. 233–249.

Lifton, W. S., Voos, K. A., and Gilbert, D. A. 1985. 'Simulation of the Pit 3, 4 and 5 Hydroelectric Project using the USFWS Instream Temperature Model' in *Waterpower 85, Proceedings of an International Conference on Hydropower, Las Vegas, Nevada, September 1985.* Vol. 3. American Society of Civil Engineers. pp. 1805–1814.

Lifton, W. S., Voos, K. A., and Gilbert, D.A. 1987. 'The simulation of variable release temperatures from the Rock Creek-Cresta Project using the USFWS Instream Temperature Model' in *Waterpower 87, Proceedings of an International Conference on Hydropower, Portland, Oregon, August 1987.* American Society of Civil Engineers. pp. 610–619.

Ludwig, Ch., Ranner, H., Kavka, G., Kohl, W., and Humpesch, U. 1990. 'Long-term and seasonal aspects of the quality of the River Danube within the region of Vienna (Austria)' in Miloradov, M. (Ed.), *Water Pollution Control in the Danube Basin.* International Association on Water Pollution Research and Control. pp. 51–58.

Macan, T .T. 1974. *Freshwater Ecology.* 2nd edn. Longman Group, London.

Mackey, A. P. and Berrie, A. D. 1991. 'The prediction of water temperatures in chalk streams from air temperatures', *Hydrobiologia*, **210**, 183–189.

Mackie, G. L., Rooke, J. B., Roff, J. C., and Gerrath, J. F. 1983. 'Effects of changes in discharge level on temperature and oxygen régimes in a new reservoir and downstream', *Hydrobiologia*, **101**, 179–188.

Magnuson, J. J., Crowder, L. B., and Medvick, P. A. 1979. 'Temperature as an ecological resource', *Am. Zool*, **19**, 331–343.

McLeod, G. D. 1979. 'McGregor River diversion Fraser River water temperature study', in *River Basin Management*, Vol. 1, *Proceedings, Fourth National Hydrotechnical Conference, Vancouver, BC, Canada, May 1979*, The Canadian Society for Civil Engineering. pp. 180–195.

Meisner, J. D. 1990. 'Potential loss of thermal habitat for brook trout, due to climatic warming, in two southern Ontario streams', *Trans. Am. Fish. Soc.*, **119**, 282–291.

Mosley, M. P. 1982. 'New Zealand river temperature regimes', *Wat. Soil Misc. Publ.* **36**, Water and Soil Division, Ministry of Works and Development for the National Water and Soil Conservation Organisation, Christchurch.

Meybeck, M. 1985. 'The GEMS/Water program (1978–1983)', *Wat. Qual. Bull.* **10**(4), 167–173; 215.

Meybeck, M., Chapman, D. V., and Helmer, R. 1989. *Global Freshwater Quality. A First Assessment.* Global Environment Monitoring System, World Health Organization and the United Nations Environment Programme, Blackwell Reference, Oxford.

Montgomery, R. H. and Loftis, J. C. 1987. 'Applicability of the t-test for detecting trends in water quality variables', *Wat. Resour. Bull*, **23**, 653–662.

Nishizawa, T. and Yambe, K. 1970. 'Change in downstream temperature caused by the construction of reservoirs', *Part I, Sci. Rep. Tokyo Kyoiku Daigaku*, **10**(100), 237–252.

Nobilis, F. 1978. 'Zur Frage des wahrscheinlichen Auftretens extrem hoher Wassertemperaturen der Donau bei Wien', *Oesterreichische Wasserwirtschaft*, **30**, 125–128.

North, E. 1980. 'The effects of water temperature and flow upon angling success in the River Severn', *Fish. Manage.*, **11**, 1–9.

Nybrant, G. 1954. 'Temperature measurements in lakes and rivers performed by the Meteorological and Hydrological Institute of Sweden', *Int. Assoc. Sci. Hydrol. General Assembly of Rome*, **3**, 62–72.

O'Keeffe, J. H., Palmer, R. W., Byren, B. A., and Davies, B. R. 1990. 'The effects of impoundment on the physicochemistry of two contrasting South African river systems', *Reg. Riv.*, **5**, 97–110.

Ordal, E. J. and Pacha, R. E. 1963. 'The effects of temperature on disease in fish' in Eldridge, E. (Ed.), *Water Temperature. Influences, Effects and Control. Proceedings of the Twelfth Pacific North West Symposium on Water Pollution Research*, US Department of Health, Education and Welfare. pp. 39–56.

Ozaki, V. L. 1988. 'Geomorphic and hydrologic conditions for cold pool formation on Redwood Creek, California', *Redwood Natl Park Res. Dev. Rep.*, **24**.

Palmer, R. W. and O'Keeffe, J. H. 1989. 'Temperature characteristics of an impounded river', *Arch. Hydrobiol.*, **116**, 471–485.

Parker, F. L. 1974. 'Thermal pollution and the environment', in Sax, N. I. (Ed.), *Industrial Pollution*. Van Nostrand Rheinhold, New York. pp. 150–196.

Parrish, J. D., Maciolek, J. A., Timbol, A. S., Hathaway, C. B., and Norton, S. E. 1978. 'Stream channel modification in Hawaii. Part D: summary report', *US Department of the Interior, Fish and Wildlife Service, Biological Services Program*, **FWS/OBS - 78/19**.

Petts, G. E. 1986. 'Water quality characteristics of regulated rivers', *Progr. Phys. Geogr.*, **10**, 492–516.

Pluhowski, E. J. 1970. 'Urbanization and its effects on the temperature of the streams on Long Island, New York', *US Geol. Surv. Prof. Pap.*, **627-D**.

Pluhowski, E .J. 1972. 'Clear-cutting and its effect on the water temperature of a small stream in northern Virginia', *US Geol. Surv. Prof. Pap.*, **800-C**, C257–C262.

Pöckl, M. 1992. 'Effects of temperature, age and body size on moulting and growth in freshwater amphipods *Gammarus fossarum* and *G. roeseli*', *Freshwater Biol*, **27**, 211–225.

Quigley, T. M., Sanderson, H. R., and Tiedemann, A. R. (Eds) 1989. 'Managing interior Northwest rangelands: the Oregon Range Evaluation Project', *USDA Forest Serv. General Tech. Note*, **PNW-GTR-238**, Pacific Northwest Research Station, Portland.

Quinn, J. M., Williamson, R. B., Smith, R. K., and Vickers, M. L. 1992. 'Effects of riparian grazing and channelisation on streams in Southland, New Zealand. 2. Benthic invertebrates', *N. Z. J. Mar. Freshwater Res.*, **26**, 259–273.

Raney, F. C. 1963. 'Rice water temperature', *Calif. Agric.*, **17**(9), 6–7.

Robertson, D. M. and Ragotzkie, R. A. 1990. 'Changes in the thermal structure of moderate to large sized lakes in response to changes in air temperature', *Aqua Sci.*, **52**, 360–380.

Rose, A. H. (Ed.) 1967. *Thermobiology*. Academic Press, London.

Rowe, L. K. and Taylor, C. H. 1994. 'Hydrology and related changes after harvesting native forest catchments and establishing *Pinus radiata* plantations. Part 3. Stream temperatures. *Hydrol. Process.*, **8**, 299–310.

Rutherford, J. C., Williams, B. L., and Hoare, R. A. 1992. 'Transverse mixing and surface heat exchange in the Waikato River: a comparison of two models', *N. Z. J. Mar. Freshwater Res.*, **26**, 435–452.

Rutherford, J. C., Macaskill, J. B., and Williams, B. L. 1993. 'Natural water temperature variations in the lower Waikato River, New Zealand', *N. Z. J. Mar. Freshwater Res.*, **27**, 71–85.

Schiller, G. 1983. 'Die Veränderung der näturlichen Wasserführung durch Speicherkraftwerke', *Mitteilungsblatt des Hydrographischen Dienstes in österreich*, **51**, 1–14.

Shanley, J. B. and Peters, N. E. 1988. 'Preliminary observations of streamflow generation during storms in a forested Piedmont watershed using temperature as a tracer', *J. Cont. Hydrol.*, **3**, 349–365.

Shields, F. D. and Sanders, T. G. 1986. 'Water quality effects of excavation and diversion', *Am. Soc. of Civ. Engin. J. Environ. Engin.* **112**, 211–228.

Sinokrot, B. A. and Stefan, H. G. 1993. Stream temperature dynamics: measurements and modeling. *Wat. Resour. Res.*, **29**, 2299–2312.

Smith, K. 1968. 'Some thermal characteristics of two rivers in the Pennine area of Northern England', *J. Hydrol.*, **6**, 405–416.

Smith, K. 1972. 'River water temperatures: an environmental review', *Scot. Geogr. Mag.*, **88**, 211–220.

Smith, K. 1975. 'Water temperature variations within a major river system', *Nordic Hydrol*, **6**, 155–169.

Smith, K. 1981. 'The prediction of river water temperatures', *Hydrol. Sci. Bull.*, **26**, 19–32.

Smith, K. and Lavis, M. E. 1975. 'Environmental influences on the temperature of a small upland stream', *Oikos*, **26**, 228–236.

Smith, R. A. and Alexander, R. B. 1983. 'A statistical summary of data from the U.S. Geological Surveys National Water Quality Networks', *US Geol. Surv. Open-File Rep.*, **83–533**.

Smith, R. A., Alexander, R. B., and Wolman, M. G. 1987. 'Water quality trends in the nations rivers', *Science*, **235**, 1607–1615.

Stanford, J. A. and Ward, J. V. 1979. 'Stream regulation in North America,' in Ward, J. V. and Stanford, J. A. (Eds), *The Ecology of Regulated Streams*. Plenum Press, New York. pp. 215–236.

Steele, T. D. 1982. 'A characterization of stream temperatures in Pakistan using harmonic analysis', *Hydrol. Sci. J.*, **4**, 451–467.

Stefan, H. G. and Preud'homme, E. B. 1993. 'Stream temperature estimation from air temperature', *Wat. Resour. Bull.*, **29**, 27–45.

Stefan, H. G. and Sinokrot, B. A. 1993. 'Projected global climate change impact on water temperatures in five north central U.S. streams', *Climatic Change*, **24**, 353–381.

Stevens, H. H., Ficke, J. F., and Smoot, G. F. 1975. 'Water temperature, influential factors, field measurement and data presentation', in *US Geol. Surv. Tech of Wat. Resour. Invest*, Book 1, Ch. D1.

Stoeckeler, J. H. and Voskuil, G. J. 1959. 'Water temperature reduction in shortened spring channels of southwestern Wisconsin trout streams', *Am. Fish. Soc. Trans.*, **88**, 286–288.

Sylvester, R. O. 1963. 'Effects of water uses and impoundments on water temperature', in Eldridge, E. (Ed.), *Water Temperature. Influences, Effects and Control*, Proceedings of the Twelfth Pacific North West Symposium on Water Pollution Research. US Department of Health, Education and Welfare. pp. 6–28.

Thendrup, A. (Ed.) 1985 'Present practices in the Nordic countries for estimating temperature and temperature changes in lakes and rivers', *Nordic Hydrol. Progr., NHP Rep. No. 11*.

Theurer, F. D., Voos, K. A., and Prewitt, C. G. 1982. 'Application of IFGs Instream Water Temperature Model in the Upper Colorado River', in *International Symposium on Hydrometeorology*, June 1982. American Water Resources Association. pp. 287–292.

Theurer, F. D., Lines, I., and Nelson, T. 1985. 'Interaction between riparian vegetation, water temperature, and salmonid habitat in the Tucanon River', *Wat. Resour. Bull.*, **21**, 5364.

UKCCIRG, 1991. *The Potential Effects of Climate Change in the United Kingdom*. United Kingdom Climate Change Impacts Review Group, HMSO, London.

UNEP/WHO/UNESCO/WMO 1983. *GEMS/Water Data Summary 1979–1981*. World Health Organization Collaborating Centre for Surface and Groundwater Quality, Canada Centre for Inland Waters, Burlington, Ontario.

UNEP/WHO/UNESCO/WMO 1987. *GEMS/Water Data Summary 1982–1984*. World Health Organization Collaborating Centre for Surface and Groundwater Quality, Canada Centre for Inland Waters, Burlington, Ontario.

UNEP/WHO/UNESCO/WMO 1990. *GEMS/Water Data Summary 1985–1987*. World Health Organization Collaborating Centre for Surface and Groundwater Quality, Canada Centre for Inland Waters, Burlington, Ontario.

Uri, N. D. 1991. 'Detecting a trend in water quality', *Research J. Wat. Pollut. Control Federation*, **63**, 868–872.

Vannote, R. L. and Sweeney, B. W. 1980. 'Geographic analysis of thermal equilibria; a conceptual model for evaluating the effect of natural and modified thermal régimes on aquatic insect communities', *Am. Naturalist*, **115**, 667–695.

Voos, K. A., Lifton, W. S., and Gilbert, D. A. 1987. 'Simulation of the Stanislaus Project: performance of the U.S. Fish and Wildlife Service Instream Temperature model on a complex system' in *Waterpower 87, Proccedings of an International Conference on Hydropower*, Portland, Oregon, August 1987. American Society of Civil Engineers. pp. 746–755.

Walker, K. F., Hillman, T. J., and Williams, W. D. 1978. 'Effects of impoundments on rivers: an Australian case study', *Verh. Int Verein für Theoret. Angew. Limnol.*, **20**, 1695–1701.

Walling, D. E. and Carter, R. 1980. 'River water temperatures', In Doornkamp, J. C. and Gregory, K. J. (Ed), *Atlas of Drought in Britain 1975–76*. Institute of British Geographers, London 49 pp.

Ward, J. V. 1974. 'A temperature-stressed ecosystem below a hypolimnial release mountain reservoir', *Arch. für Hydrobiol.*, **74**, 247–275.

Ward, J. V. 1982. 'Ecological aspects of stream regulation: responses in downstream lotic reaches', *Wat. Pollut. and Manage. Rev. (New Dehli)*, **2**, 1–26.

Ward, J. V. 1985. 'Thermal characteristics of running waters', *Hydrobiologia*, **125**, 31–46.

Ward, J. V. 1992. *Aquatic Insect Ecology, 1. Biology and Habitat*, Wiley, New York.

Ward, J. V. and Stanford, J. A. 1979. 'Ecological factors controlling stream zoobenthos with emphasis on thermal modification of regulated streams' In Ward, J. V. and Stanford J. A. (Eds.), *The Ecology of Regulated Streams*. Plenum Press, New York. pp. 35–55.

Weatherley, N. S., and Ormerod, S. J. 1990. 'Forests and the temperature of upland streams in Wales: a modelling exploration of the biological effects', *Freshwater Biol*, **24**, 109–122.

Webb, B. W. 1992. 'Climate change and the thermal regime of rivers', *Unpublished Report for the UK Department of the Environment*.

Webb, B. W. and Nobilis, F. 1994. Water temperature behaviour in the River Danube during the twentieth century. *Hydrobiologia*, **291**, 105–113.

Webb, B. W. and Walling, D. E. 1985. Temporal variation of river water temperatures in a Devon river system', *Hydrol. Sci. J.*, **30**, 449–464.

Webb, B. W. and Walling, D. E. 1986. Spatial variation of water temperature characteristics and behaviour in a Devon river system. *Freshwater Biology*, **16**, 585–608.

Webb, B. W. and Walling, D. E. 1988. 'Modification of temperature behaviour through regulation of a British river system', *Reg Riv.* **2**, 103–116.

Webb, B. W. and Walling, D. E. 1992. 'Long term water temperature behaviour and trends in a Devon, UK, river system', *Hydrol. Sci. J.*, **37**, 567–580.

Webb, B. W. and Walling, D. E. 1993a. 'Temporal variability in the impact of river regulation on thermal regime and some biological implications', *Freshwater Biol*, **29**, 167–182.

Webb, B. W. and Walling, D. E. 1993b. 'Longer-term water temperature behaviour in an upland stream', *Hydrol. Process.* **7**, 19–32.

Zaric, Z. P. (Ed.) 1978. *Thermal Effluent Disposal from Power Generation*. Hemisphere, Washington.

Zimmerman C. and Geldner, P. 1978. 'Thermal loading of river systems'. In Z. P. Zaric (ed.), *Thermal Effluent Disposal from Power Generation*. Hemisphere, Washington. pp. 175–193.

5

HEAVY METALS IN THE HYDROLOGICAL CYCLE: TRENDS AND EXPLANATION

I. D. L. FOSTER AND S. M. CHARLESWORTH

Centre For Environmental Research and Consultancy, Geography Division, School of Natural and Environmental Sciences, Coventry University, Priory Street, Coventry CV1 5FB, UK

ABSTRACT

This paper reviews the major sources and transport characteristics of heavy metals in the hydrological cycle. It is demonstrated that heavy metal releases to the environment have changed from 19th and early 20th century production-related activities to consumption-oriented factors in more recent times. The relative roles of particle size, sorption and desorption processes, partitioning and the chemical speciation of heavy metals on fine sediments are identified to understand the likely fate of heavy metals released into fluvial systems. It is argued that the spatial and temporal distribution of heavy metals in the river corridor depends not only on an understanding of metal solubility and speciation, but also on an understanding of sediment dynamics which control, for example, floodplain alluviation and the accumulation of metals in the bottom sediments of contaminated rivers, lakes and reservoirs. Existing long- and short-term records are examined to identify the current state of knowledge about the factors which affect heavy metal releases into aquatic environments. With limited exceptions, it is shown that few long-term studies of trends in heavy metal transport are available although, for some major rivers, limited data on trends in metal concentration exists. Palaeo-limnological reconstruction techniques, based on an analysis of lake and reservoir sediments, are identified as a possible means of supplementing monitored records of heavy metal transport. Although numerous studies have suggested that trends in atmospheric contamination, mining and urbanization may be identified in the bottom sediment record, other research has shown that the radionuclide-based chronology and the heavy metal distribution within the sediment are more likely to be a function of post-depositional remobilization than the history of metal loading to the basin. Despite these limitations, it is shown that the incorporation of reservoir bottom sediment analysis into a heavy metal research programme, based in river corridors of Midland England, provides an opportunity to identify and quantify the relative contribution of point and non-point contributions to the heavy metal budget and to relate trends in metal contamination to specific periods of catchment disturbance.

INTRODUCTION

The last 20 years have witnessed a growing interest in the sources, fate and transport of contaminants in fluvial systems. This interest, and the rapidly growing legislative control on contaminant discharges to the hydrosphere (e.g. EC, 1991; Wright, 1992), reflects an increase in our awareness of the impact of contaminants, not only on human health, but also on the health and sustainability of aquatic ecosystems. A wide range of organic and inorganic pollutants have been released directly into fluvial systems from industrial discharges and sewage outfalls. Indirect sources to the fluvial system include atmospheric releases during processing and product use (Meybeck and Helmer, 1989). The latter is exemplified in particular by the emission of gases and particulates from power stations and motor vehicle exhausts.

This paper focuses on the release of heavy metals into terrestrial aquatic systems and is concerned

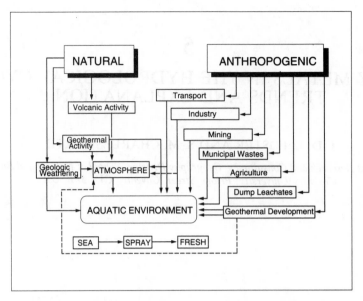

Figure 1. Natural and anthropogenic sources of heavy metals in aquatic environments

with temporal and spatial trends in metal release to the environment. The review is organized into six major sections.

1. An examination of trends in heavy metals production and their points of entry into, and possible pathways through, the hydrological cycle.
2. A consideration of the physical and chemical factors which control the transport and fate of heavy metals in fluvial systems.
3. Identification of those parts of the fluvial system and, in particular, the river corridor, which provide historical evidence of the release of heavy metals into the aquatic environment.
4. A review of documented trends in heavy metal transport in rivers over short and long time-scales.
5. An assessment of the techniques used to reconstruct the intermediate and long-term history of heavy metal contamination in catchments from an analysis of lake and reservoir bottom sediments.
6. A review of lake catchment research undertaken by the authors in Midland England, which has focused on quantifying long-term trends in metal concentrations and on discriminating the relative significance of atmospheric and point sources.

Examples of both global and regional metal pathways will be given to illustrate the scale of the problem and the inherent site-specific nature of contamination history. The overall consequences of increasing heavy metal loadings to fluvial systems through time is emphasized.

HEAVY METAL CONTAMINATION OF THE ENVIRONMENT: A TEMPORAL PERSPECTIVE

Heavy metals, as defined by Davies (1980), are those elements in the periodic table with a density of

Figure 2. Trends in metal production and release into the environment. (A) Annual world-wide production of selected metals from 1700 to 1983. Modified from Brown *et al.* (1990). (B) Concentrations of As, Cd, Pb and Zn in emissions from a range of industrial sources in Europe. Modified from Pacyna and Münch (1987). (C) Average annual enrichment factors related to dry deposition velocities of elements in the air at Styrrup, Nottingham, UK (1972–1981). Modified from Cawse (1987). (D) Long-term trends in global atmospheric metal emissions (Nriagu, 1979). (E) Declining atmospheric metal concentrations in UK air, 1976–1988. Modified from Quality of Urban Air Group (1993). (F) Change from production-associated to consumption- and fossil fuel-associated Pb release into the environment of the Hudson–Rariton River Basin (Tarr and Ayres, 1990)

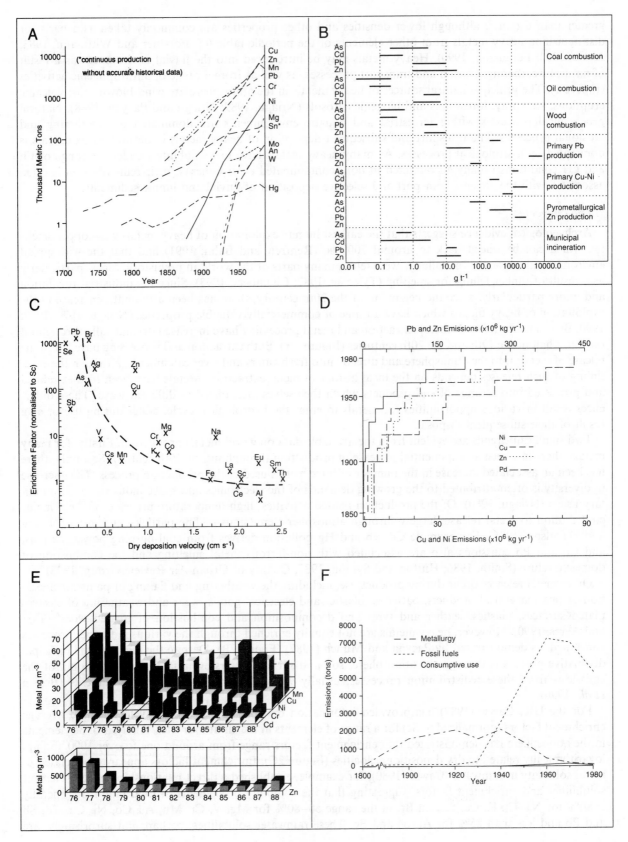

Figure 2

greater than $6\,\mathrm{g\,cm^{-3}}$, although lower densities and other properties are commonly taken as a basis for distinguishing heavy metals from other elements of the periodic table (cf. Förstner and Wittmann, 1983; Jarvis, 1983; Fergusson, 1990). Heavy metals may be introduced into the fluvial environment as a result of natural weathering, erosion and transport processes, as well as from a range of anthropogenic activities (Figure 1). The principal natural sources of heavy metals in the atmosphere are wind-blown soils, volcanic eruptions, sea spray, forest fires and biogenic aerosols (Nriagu, 1989; Nriagu and Pacyna, 1988). Natural geochemical processes within terrestrial and aquatic environments are dominated by weathering and erosional processes. The transport and subsequent deposition of heavy metals in natural environments is dominated by hydrological processes. As many heavy metals have low solubilities under the range of Eh and pH conditions usually encountered in non- contaminated environments, their redistribution is often associated with the erosion, transport and selective deposition of fluvial and limnic sediments.

Human impacts

Archaeological evidence suggests that the earliest human exploitation of heavy metals was copper smelting which can be traced back to around 7000 BC (Renfrew and Bahn, 1991) and that the widespread environmental impacts of metallurgy were felt in many parts of the world in prehistoric times, particularly as a result of timber felling for smelting (Tylecote, 1987; Chambers, 1993). Since the industrial revolution and, more particularly, since the beginning of the 20th century, there has been a dramatic increase in the exploitation of heavy metals which have a range of commercially valuable properties (Nriagu, 1979; 1989; 1990; Brown et al., 1990). Mineral extraction and metal processing have increased dramatically throughout the latter half of the 19th and the 20th centuries (Figure 2A). Both extraction and processing have led to the release of metals into the atmosphere and directly into freshwaters and river catchments, although the exact timing of such releases depends on the local history of metal extraction. Metals have been further refined and processed into a range of end-products which themselves are utilized in different ways. These differences result in various opportunities for metals to enter the hydrological cycle, either during use or as a result of their subsequent disposal.

Two significant trends are evident from the available data on world metals production. Firstly, for many metals, there has been an exponential increase in production throughout the 20th century. Secondly, there has been an associated increase in the number of trace metals used in the production process. This increase in diversity is often attributed to the growing demands of the electronics and space industries and to military usage (Nriagu, 1980). Of the production-related activities, high temperature processes make a major contribution to metal release, largely through atmospheric emissions of fine particles and gases (Nriagu and Davidson, 1986). For example Cd, Pb and Hg pollution derives from coal-burning power stations, and Cd and Pb emissions also are associated with non-ferrous metal smelting and the combustion of domestic refuse (Smith, 1986; Hutton and Symon, 1987; Quality of Urban Air Review Group, 1993).

Other metal releases occur during product use, including the weathering and flaking of paints, incineration of pharmaceutical products, batteries, plastics and electrical goods, wear and weathering of electroplated surfaces, plastics, leather and tyres and decomposition and combustion of treated wood (Tarr and Ayres, 1990). However, heavy metals are not equally enriched through the various industrial processes concerned, as demonstrated by Pacyna and Münch (1987) (Figure 2B). Despite the fact that the principal dissipative process is through the atmosphere, soil and surface waters eventually receive the bulk of the emissions from these redistribution processes, usually via surface runoff and sewage effluents (Brown et al., 1990).

For the UK, Cawse (1987) has provided information on both dry deposition velocities and average enrichment factors (normalized to Sc) for a range of elements in the air at four sites. Analysis of 36 elements in the atmosphere has demonstrated that enrichment factors range from around one to over 1000 and are logarithmically related to dry deposition velocities (Figure 2C). Enrichment factors appear to be related to metal solubility in rainwater. Cawse (1980), for example, established a direct relationship between element solubilities and enrichment factors, suggesting that the solubility of elements in total fallout is generally $>80\%$ for Na, Cl, K, Ca, Se and Br, in the range 50–80% for Mg, V, Cr, Mn, As, Co, Ni, Cu, Cd, Sb and Pb and less than 25% for Al, Sc and Fe. These rainwater solubilities, perhaps not surprisingly, are

Table I. Comparison of metal concentrations ($\mu g\,l^{-1}$) in stormwater, roof and street runoff in part of the Karlsruhe/Waldstadt region

Parameter	Runoff		
	Stormwater	Roof	Street
pH	4·9	6·2	6·4
Pb	5	104	311
Cd	1·5	1·0	6·4
Zn	5	24	603
Cu	1·5	235	108
Ni	5	—	57

Based on data in Xanthopolous and Hahn (1993)

similar to the hydrological solubilities of heavy metals published by Meybeck and Helmer (1989: 288). The relative solubility of various atmospherically derived elements, as well as their grain size and crystallinity, clearly has significant implications for their incorporation into, and mode of transport through, the hydrosphere.

Long-term trends in global atmospheric emissions of heavy metals were calculated by Nriagu (1979) (Figure 2D). In parallel with the metal production data of Figure 2A, emissions increase exponentially in the 20th century. Aggregated data of this type, however, hide significant short-term temporal and spatial variations. The highest concentrations usually occur in winter (Mészáros et al., 1987; Cawse, 1987) and short-term monitoring has detected significant diurnal variations (Hindy et al., 1987). The global increase in emissions identified by Nriagu (1979; Figure 2D) through 1980 has been slowing down or even reversed. The 10 year record of atmospheric heavy metal concentrations in the UK, for example, shows a general decline through time (Figure 2E).

Only limited global data are available on long-term trends in direct metal release to the aquatic environment. Recent calculations by Nriagu (1993), however, have shown the possible magnitude of human impacts of mercury release locally, which include the effects of colonial silver processing in Spanish America. It was calculated that between the years 1570 and 1820, an average of ca. $570\,t\,yr^{-1}$ of Hg were released to the environment, whereas between 1820 and 1900 the average rose to ca. $875\,t\,yr^{-1}$. Of the total output, ca. 10% was estimated to be lost during transport to the silver mines, 25–30% was left behind in tailings or washed into streams and 60–65% was lost to the atmosphere during refinement. Furthermore, Nriagu argues that the current store of Hg in processing wastes and fluvial sediments may experience conditions in which it will continue to be methylated and, therefore, continue to contribute to the global anthropogenic release of Hg to the atmosphere. Similar contemporary problems of Hg release, in this instance during gold ore processing in Brazil, were discussed by Salomons at a meeting in Toronto in 1993 on Heavy Metals in the Environment (A. J. Horowitz, pers. comm.).

Other research has demonstrated long-term trends in Pb release switching from production- to consumption-related processes. In the Hudson–Rariton River Basin, Tarr and Ayres (1990) have shown that the early part of the metal release cycle was associated with mineral exploitation, whereas the latter part of the cycle is associated with consumptive release (Figure 2F).

In addition to point source contamination derived from metal extraction and processing, and the shift from exploitive to process-based contamination sources, significant metal contamination is known to be associated with urban hydrological systems (Ellis et al., 1986; Driver and Lystrom, 1987; Marsalek, 1991). Stormwater discharge includes roof runoff, street runoff and combined sewer flow (Hahn, 1993). Of the three urban sources, street runoff is generally found to be the most contaminated by heavy metals, mineral oils and polycyclic hydrocarbons (Förster, 1993; Xanthopolous and Hahn, 1993; Table I).

Stormwater sewers route large amounts of solids and metals through urban catchments to receiving streams. The majority of the suspended solids from street runoff have particle sizes ranging from 20 to

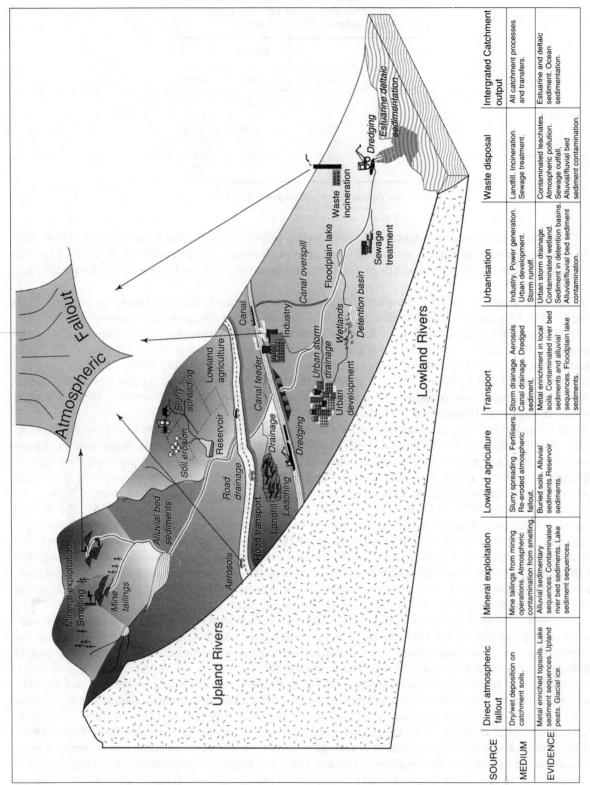

Figure 3. Simple model of metal transfers in a fluvial system due to human activity

SOURCE	Direct atmospheric fallout	Mineral exploitation	Lowland agriculture	Transport	Urbanisation	Waste disposal	Intergrated Catchment output
MEDIUM	Dry/wet deposition on catchment soils.	Mine tailings from mining operations. Atmospheric contamination from smelting.	Slurry spreading. Fertilisers. Re-eroded atmospheric fallout.	Storm drainage. Aerosols. Canal drainage. Dredged sediment.	Industry. Power generation. Urban development. Storm runoff.	Landfill. Incineration. Sewage treatment.	All catchment processes and transfers.
EVIDENCE	Metal enriched topsoils. Lake sediment sequences. Upland peats. Glacial ice.	Alluvial sedimentary sequences. Contaminated river bed sediments. Lake sediment sequences.	Buried soils. Alluvial sediments. Reservoir sediments.	Metal enrichment in local soils. Contaminated river bed sediments and alluvial sequences. Floodplain lake sediments.	Urban storm drainage. Contaminated wetland. Sediment in detention basins. Alluvial/fluvial sediment contamination.	Contaminated leachates. Atmospheric pollution. Sewage outfall. Alluvial/fluvial bed sediment contamination.	Estuarine and deltaic sediment. Ocean sedimentation.

Table IIA. Percentage of particulate-bound metals in total metal discharge of selected polluted rivers

Parameter	Amazon	Mississippi	USA	Germany
Cd	—	88·9	—	30
Zn	—	90·1	40	45
Mn	83	98·5	—	8–97
Cr	83	98·5	76	72
Cu	93	91·6	63	55
Pb	—	99·2	84	79
Fe	99·4	99·9	98	98

After Salomons and Förstner (1984)

Table IIB. Dissolved transport index (DTI), calculated as the percentage dissolved transport divided by total transport

DTI (%)	Characteristic
90–50: Br, I, Sa, Cl, Ca, Na, Sr	High solubility
50–10: Li, N, Sb, As, Mg, B, Mo, F, Cu, Zn, Ba, K	
10–1: P, Ni, Si, Rb, U, Co, Mn, Th, Pb, V, Cs	
<1: Ga, Tm, Lu, Gd, Ti, Er, Nd, Ho, La, Sm, Tb, Yb, Fe, Eu, Ce, Pr, Al	Low solubility ↓

After Martin and Meybeck (1979)

3000 μm. Sediment removal rates of as much as 60 g day^{-1} have been recorded for road surfaces with metal loadings, reaching 22 and 24 mg day^{-1} for Zn and Pb, respectively. The highest metal concentrations are usually associated with median particle sizes of around 300 μm (Ellis and Revitt, 1982; Harrop et al., 1983; Ellis and Harrop, 1984; Beckwith et al., 1986). In many urban environments, the most significant amounts of heavy metals are associated with sewage effluents (often dispersed through barge dumping at sea or land spreading) and through the disposal of municipal wastes to landfills (Förstner, 1986; Fergusson, 1990). For example, Imhoff et al. (1981) calculated that 55% of the heavy metals in the Ruhr River were discharged from municipal and industrial waste water treatment plants and the balance was derived from weathering and erosion. The municipal treatment plants received 31% of their heavy metals from domestic and 69% from industrial effluents. However, the relative proportion varied significantly with different heavy metals, with 90% of Cr, 65% of Ni and 50% of Cu discharged by industry.

For many trace metals, anthropogenic release into the environment from all sources far exceeds natural mobilization via the pathways shown in Figure 1. For example, global anthropogenic enrichment factors for As, Cd, Pb, Se and Hg were calculated by Nriagu and Davidson (1986) to be 3, 7·6, 24·1, 2·8 and 11·3, respectively.

The 20th century increase in metal production, processing, usage and disposal has given rise to new pathways by which metals can enter the hydrological cycle, in addition to increasing the magnitude of metal inputs through existing pathways. From a hydrological viewpoint, the two-fold distinction between point source contamination at the site of production and non-point contamination through atmospheric dispersal can be clearly identified. The implications of such dispersal processes on various parts of the fluvial system are shown schematically in Figure 3 and discussed in the next section.

METALS IN THE HYDROLOGICAL CYCLE

Of particular significance in quantifying the spatial and temporal trends in metal transport and in assessing the relative significance of the various heavy metal pathways (Figure 3) is the partitioning of metals between suspended sediment and water; the mechanisms by which heavy metals are retained in the particulate fraction and the post-depositional stability of sediment-associated metals within the river corridor (cf. Stumm and Morgan, 1981; Förstner and Wittmann, 1983; Förstner, 1983; Salomons and Förstner, 1984; Lunt et al., 1989; Fergusson, 1990; Horowitz, 1991). This section considers the role of sediment–water partitioning and transport and depositional processes affecting the distribution of heavy metals in the fluvial system.

Sediment–water partitioning

In river systems, trace metal concentrations in suspended sediments are greater than in the water column, although trace metal partitioning is a function of the characteristics of the metal ion, particle size, organic content and sediment concentration (Rygwelski, 1984). The relative significance of the dissolved fraction to the total metal load transported, in general, increases with increasing total metal concentrations, particularly in industrially contaminated river systems (Figure 4; Table II). Despite significant advances in the understanding of sorption/desorption processes and in measuring the partitioning of trace metals under laboratory conditions, Gardner and Gunn (1989) have emphasized the difficulty in applying models based on laboratory studies to natural systems.

Metal concentrations in the fine-grained sediments of fluvial substrates in river systems can be more than 100 000 times higher than the dissolved concentrations of the same metal (Horowitz, 1991). Not only are there significant differences in concentration, but even small concentrations associated with suspended sediment, generally, can make a major contribution to the total load of metals transported by river systems, particularly because suspended sediment concentrations usually increase with increased river discharge.

An understanding of the contribution made by the sediment-associated fraction to the total load requires knowledge of the effects of particle size, organic matter content, sediment surface area and the surface coatings on the suspended particles. Furthermore, the individual and combined effects of Eh and pH have a well documented impact on the partitioning of metals between sediments and the water column (Stumm and Morgan, 1981; Salomons and Förstner, 1984; Fergusson, 1990).

Of the fraction of heavy metals carried with suspended sediment, particle size is one of the most important physical controls (Figure 4). In general, there is a good negative correlation between sediment size and metal concentrations, although metals have been shown to accumulate on the surfaces of coarser materials, including sand, pebbles and even boulders (Robinson, 1982; 1983). High metal concentrations are more usually associated with fine sediments, particularly those in the silt and clay fractions, in both freshwater and marine sediments (Dossis and Warren, 1980; Förstner and Patchineelam, 1980; Morrison et al., 1984; Thoms, 1987a; 1987b; Burrus et al., 1990; Horowitz, 1991). In part, this is a surface area effect, because the surface area per unit mass of sediment decreases with an increase in particle diameter. Clay minerals are not spherical and the surface area of fluvial suspended sediments has been shown to be much larger than that calculated for a perfect sphere of equivalent diameter (Förstner and Wittmann, 1983). Despite the complication of grain shape, the effects of grain size on metal concentration can usually be determined and adjusted experimentally (Figure 4).

As the silt- and clay-sized fractions of river sediments comprise clay minerals, iron hydroxides, manganese oxides and organic matter, ion-exchange processes between the positively charged metals and negatively charged surfaces are likely to occur. The chemical partitioning of metals on or within sediments is often

---

Figure 4. Metal associations in river waters, fluvial suspended sediments and lake sediments. (A) Transport phases of Cr, Ni and Cu in the Amazon and Yukon rivers (Horowitz, 1991). (B) Relationship between metal concentration and the percentage of the sub-16 μm grain size fraction for the river Ems (deGroot et al., 1982; figure modified from Horowitz, 1991). (C) Concentrations of Zn by particle size for sediments from two sampling stations upstream (16·6 river miles) and downstream (0·5 river miles) of the Saddle River, near Lodi, New Jersey (Wilber and Hunter, 1979). (D) Variation in Pb and Zn concentrations with particle size for the sub-63 μm fraction (4ø) of (i) urban and (ii) downstream of urban fluvial substrates in the River Tame, UK. Modified from Thoms (1987a). (E) Particulate Pb and Cd fractions in sewer samples. All sites are in Paris except sample 7 (Bordeaux). Separated overflows: 1 and 2, Circular Highway; 3, Savigny Sewer; 4, La Moreé sewer; 5, Créteil lake SSO; 6, Orly Airport SSO; and 7, Béquigneaux SSO. Combined Sewers: 8 and 9, La Molette CSO; and 10, Clichy CSO (Flores-Rodriguez et al., 1993). (F) Heavy metals in lake Coeur d'Alene sediments (Idaho) based on sediment mineralogy of sample CDA 128: (i) concentration and (ii) percentage contribution. After Horowitz et al. (1993)

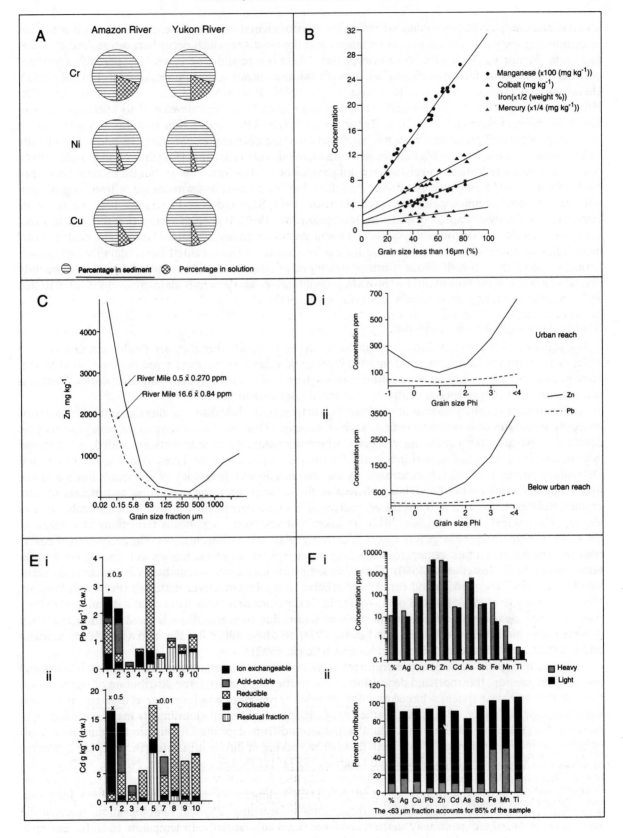

Figure 4

seen as important, as the key to understanding the environmental impact of trends in metal behaviour lies in quantifying the metal associations in sediments and the reactions which occur between sediments, water and biota (Figure 4). Despite this acknowledgement, there is a bewildering variety of approaches to metal partitioning based on factors such as grain size, chemical association and mineralogy (cf. Förstner, 1983; Håkanson and Jansson, 1983; Engstrom and Wright, 1984; Bengtsson and Enell, 1986; Horowitz, 1991; Horowitz et al., 1993). Most chemical extraction techniques have been developed to selectively remove metals in different chemical forms (e.g. Tessier et al., 1979). Other extraction techniques, notably those developed in palaeolimnological research, suggest that some chemical extractions can selectively identify autochthonous and allochthonous chemical components in lake sediments (Engstrom and Wright, 1984). No one extraction technique is likely to have unique selectivity (Horowitz, 1991), but the method developed by Tessier et al. (1979), or variants of this procedure, have been used in a wide range of hydrological, limnological and palaeolimnological studies (cf. Harrison, 1987; Samanidou and Fytianos, 1987; Foster et al., 1991a; 1991b; Horowitz et al., 1993; Flores-Rodriguez et al., 1993). It is often argued that those metals held in organic associations, or in weakly adsorbed form, are more readily available to biota and that the total metal concentrations in sediments provide little information on the potential bioavailability and, hence, toxicological effect of metal release. Further development, refinement and standardization of extraction procedures are clearly important for providing comparable heavy metals data from different national and research monitoring programmes (deGroot et al., 1982).

Transport and deposition in catchments

The route that metals take through a catchment depends on whether they are discharged directly into water courses, into the atmosphere or onto the land, or whether they are discharged in liquid, solid or gaseous phases. Figure 3 summarizes possible pathways into the fluvial system and also shows dominant sources and sinks of anthropogenically derived metal contamination.

Because metals are often associated with particulate transport, their dispersal through catchment systems is largely a function of hydraulic conditions which control: (1) sorting, according to differences in particle density and particle size; (2) mixing processes, where uncontaminated sediments are added o the fluvial system; and (3) storage and deposition on the floodplain, fluvial substrate, lakes, reservoirs and estuaries. Deposition on the floodplain is controlled by the magnitude and frequency of overbank flows and the hydraulic conditions on the floodplain, whereas in fluvial substrates, the deposition and ingress of fine-grained sediments into and through the armour layer is often associated with hydraulic controls, such as stream power (Graf, 1990; Hughes, 1992). In lakes and reservoirs, significant reductions in velocity at the point of inflow, and the large water storage capacity relative to inflow, results in the deposition of coarse sediments in deltas and fine sediments in the deep water regions; a process known as focusing. (Håkanson and Jansson, 1983; Foster et al., 1990). These fine sediments may contain high heavy metal concentrations simply as a function of sorting and enrichment relative to catchment source materials (Peart and Walling, 1982; Novotney et al., 1986). From point source inputs, particularly those associated with mineral extraction activities, downstream reductions in contamination due to metal dispersal and dilution are often observed and modelled (Wolfenden and Lewin, 1978; Bradley, 1982; Bradley and Cox, 1986; Macklin and Dowsett, 1989; Macklin, 1992; Macklin and Klimek, 1992).

Soils and floodplain sediments are transient stores of metal contamination. As they are intrinsically linked to the erosion, transport and deposition of fine particulate material, the distribution of heavy metals through the catchment system is linked to sediment delivery processes which operate at different spatial and temporal scales (Walling, 1983; 1989; Hadley et al., 1985). Contaminated sediments may be eroded from catchment topsoils as a result of agricultural activities and from deposited floodplain alluvium as a result of historical metal enrichment by overbank flows or reworking of fluvial substrates (Bradley, 1982; Bradley and Cox, 1986; Duijsings, 1986; Foster and Dearing, 1987a; 1987b; Leenaers, 1989a; Niriagu; 1993; Tipping et al., 1993).

Heavy metals incorporated into soils and floodplain sediments have been shown to have long residence times and are slowly depleted from these stores by leaching. Physical remobilization, through the erosion of deposited sediment, may shorten residence times substantially. In temperate soils, for example,

Salomons and Förstner (1984) calculated soil residence times for the easily leached elements Cd, Ca, Mg and Na of 75–380 years, for Hg of 500–1000 years and for As, Cu, Ni, Pb, Se and Zn of 1000–3000 years.

The depositional and post-depositional stability of heavy metals in lake, reservoir and estuarine sediments is considerably more complex due to the operation of physical, geochemical and biological controls. Physical controls on deposition include the particle size, settling speed and flocculation characteristics of the sediments (Lick, 1984; Bruk, 1985), whereas it was shown by Hart (1982), Sigg (1983; 1987) and Jannasch et al. (1988) that settling particles, especially biogenic organic material, play a dominant part in binding heavy metals and transferring them into lake sediments. From an analysis of lakes Constance and Zurich, Switzerland, Sigg (1983) suggested that a secondary metal scavenging cycle operates in deep water, which is controlled by the release and oxidation of Mn and Fe from the sediments to the overlying water.

Metal release from sediments at the sediment–water interface is largely controlled by Eh and pH. The significant effect of Eh on the partitioning of Fe and Mn between the interstitial waters and sediments has been demonstrated in several studies. For example, Tessenow and Baynes (1975) for the Feldsee, Germany showed that Mn was more mobile than Fe under reducing conditions, which resulted in the up-core displacement of Mn relative to Fe in the sediments. Van de Meent et al. (1985) were able to demonstrate significant associations between the organic component of sediments and heavy metals deposited in freshwater sedimentation basins of the River Rhine. In estuarine and marine systems, trace metal adsorption has also been shown to decrease with increasing salinity (Bourg, 1987).

Despite significant increases in laboratory experimentation on the kinetics of metal adsorption/desorption processes, the prediction of the scavenging, residence times and post-depositional stability of heavy metals in natural systems remains problematic (Jenne and Zachara, 1984; Davison, 1985; Norton, 1986; Honeyman and Santschi, 1988; Jannasch et al., 1988; Gardner and Gunn, 1989; Gobeil and Silverberg, 1989; Benoit and Hammond, 1990; Santschi et al., 1990; Varvas and Punning, 1993).

In large lake basins, considerable spatial variability exists in the heavy metal concentration of surface sediments (Håkansson and Jansson, 1983; Nriagu and Simmons, 1984; Horowitz et al., 1993), which partly relates to the point at which contamination enters the lake, and to processes operating within the lake. In-lake processes include currents and wave action, post-depositional bioturbation and possible metal release to interstitial water, which in turn will be controlled by changes in Eh and pH.

In small lakes and reservoirs, significant spatial variability in sedimentation rates has been observed from high density coring (Bloemendal et al., 1979; Dearing, 1983; Foster et al., 1985; 1986; Foster and Charlesworth, 1994). The variation in rates has a significant impact on the quantitative assessment of metal loadings to lake basins. Furthermore, coupled with often unknown rates of sorption and desorption in small eutrophic lakes, the trap efficiency of the reservoir (including atmospheric and catchment inputs) may have a significant impact on the amount of sediment retained. Conventional trap efficiency calculations are often based on the capacity/inflow ratio or on particle size specific settling velocities (Rausch and Heinemann, 1984; Foster et al., 1990). Comparison of particle size curves in sediments and soils in the catchment of the Old Mill Reservoir, Devon (Foster and Walling, 1994) has shown no depletion of the finer sediment fractions in the reservoir, which may be a result of particle aggregation and more rapid settling than would be predicted from the settling velocities of individual grains.

The processes controlling heavy metal scavenging, trap efficiency and post-depositional stability in lakes and reservoirs will significantly affect the preservation of the heavy metal record within the sediment column. These processes also will affect the distribution of radionuclides in the sediment column, which are often used to provide an absolute chronology and are considered more fully in the following section.

TRENDS IN HEAVY METALS IN RIVERS

Fluvial processes are the primary mechanisms responsible for the transport and distribution of heavy metals in drainage basins. Trends are evident in the concentration of heavy metals in solution and particulate phases, and in the total load transported at a range of spatial and temporal scales. However, little detailed information is available on heavy metal loads transported by rivers, although concentration

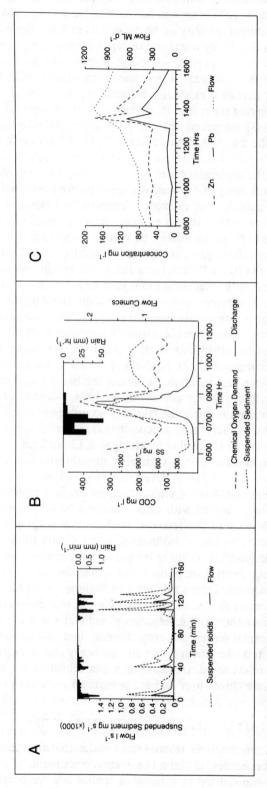

Figure 5. Storm period suspended sediment and metal behaviour. (A) Short-term trends in suspended sediment concentration from a 270 m² parking lot in the city of Lund. Modified from Spångberg and Niemczynowicz (1993). (B) Short-term trends in suspended sediment concentration in the 3·5 km² Graham Park drainage basin, north-west London. Modified from Ellis (1979). (C) Total Pb and Zn concentrations in the 262 km² drainage basin of the River Sowe at Stoneleigh, UK, in August 1992 (National Rivers Authority, Severn-Trent Region, unpublished data)

data are more readily available, and the following summary of trends reflects the imbalance between the availability of concentration and load data. The spatial variation in the trace metal concentration of stream sediments was represented conceptually by Dahlberg (1968) as a simple functional equation:

$$T = f(L, H, G, C, V, M, e) \tag{1}$$

where T is the trace metal concentration, L is the influence of lithology, H is the hydrological effect, G is the geological effect, C is the cultural (human) effect, V is the vegetation effect, M is the effect of mineralized zones and e is the error and non-explicit factors.

In fluvial sediment studies, the most significant problems relate to the difficulty in quantifying: firstly, the relative significance of anthropogenic and natural contributions to the total metal concentration, or load, and, secondly, the relative contribution of point and non-point (particularly atmospheric) anthropogenic contributions.

Small-scale studies of urban storm drainage systems have shown that total metal concentrations in a water sample generally increase with increasing discharge, largely as a function of the increased amount of sediment transporting metallic elements. In studies of urban storm drainage during storm events, a 'first flush' effect, associated with the removal of accumulated contaminants from roofs, road surfaces and storm drains is usually observed (Ellis and Harrop, 1984; Brinkmann, 1985; Ellis et al., 1986; Förster, 1993; Hahn, 1993; Xanthopolous and Hahn, 1993). Figure 5 shows the impact of the 'first flush' effect at three different scales. At the small scale, represented by the experimental plots of Spångberg and Niemczynowicz (1993, Figure 5A), sediment flushing occurs on the rising limb of the storm hydrograph and the amount of sediment mobilized decreases as the source is exhausted during a series of five consecutive storms on a 270 m^2 parking lot in the city of Lund. For a small urban catchment in north-west London, a similar effect was observed by Ellis (1979; Figure 5B) but at this scale, the sediment peak preceded the discharge peak by as much as 0·4 h. Figure 5C shows the short-term trend in total heavy metal concentrations in the 262 km^2 drainage basin of the River Sowe, a major headwater tributary of the Avon catchment in Midland England. Urban storm drainage is derived from the city of Coventry, which occupies ca. 30–40% of the total catchment. Both total Zn and Pb peak in concentration about 0·5 h before peak discharge with a secondary, and much lower, peak coincident with peak discharge. The research into the characteristics of metal transport in urban storm drainage by Wickliffe and Steele (1987) demonstrated that high metal concentrations were significantly associated with high fluvial suspended sediment concentrations in the particle size range of 0·3–0·4 μm.

Seasonal variations in heavy metal concentrations are also evident in many European rivers. Schleichert (1975) and Salomons and Eysink (1981) have shown that the concentrations of Cd and Cr in the suspended sediments of the Rhine at Koblenz and in the Netherlands, respectively, decline with increasing discharge. For many metals, the relationship between sediment-associated metal concentration and discharge is seasonal, with lower concentrations in sediments during spring runoff (Figure 6A). This trend was related to the management of flows in the regulated Neckar and Main rivers where, at low flows, metal-rich sediments were trapped by locks. The autumn/winter increase in sediment-associated metal concentration was attributed to the lack of flow regulation in these two river valleys allowing finer metal-rich sediments to be mobilized. Cortesao and Vale (1994) also note strong seasonality in total heavy metal concentrations in rivers, with contributions from mine drainage in the Sado River (Portugal), but suggest that in addition to the limits imposed by sediment supply at low flows, there may be a biological control on metal adsorption to suspended sediments. In part, therefore, trends in metal concentration at a single point in the drainage basin are a function of the spatial distribution of the polluting sources in the catchment and the regulation of these sources through time. Figure 6B, for example, shows the increase in Cd concentration through the lower Rhine between Cologne and Rotterdam. Upstream of the Ruhr valley, Cd concentrations are around 6–7 μg g^{-1}. Downstream of the Ruhr valley, Cd levels jump to over 30 μg g^{-1} and decline to between 15 and 20 μg g^{-1} at the German–Dutch border. The large increase in Cd from the Ruhr was associated with metal discharges from the Duisburg copper plant.

Establishing a relationship between metal concentrations and discharge is important for calculating the total load transported by using rating curves (Bradley and Lewin, 1984; Leenaers, 1989b). However, the relationship between dissolved or sediment-associated metal concentrations and discharge may be negative where significant dilution occurs either from less contaminated water sources or due to the incorporation of coarser and/or less metal-rich sediment (Bradley, 1982; Bradley and Lewin, 1982; Bird, 1987). Despite the decrease in concentration, total metal loads usually increase with increasing discharge because of the greater water and sediment mass transported. A recent study by Leenaers (1989b) of the polluted River Geul (The Netherlands) shows discharge–metal concentration relationships to be complex, and often statistically weak, even when the data are subdivided on the basis of river flow. The consequence of this is that load estimates based on rating curves may be subject to considerable error.

There are few long-term records of heavy metal concentrations in rivers, a notable exception being that of the sediment-associated metal concentrations in the rivers Scheldt, Meuse, Ems and Rhine in the Netherlands (Salomons and Eysink, 1981; Salomons and Förstner, 1984; Figure 6C). Metal concentrations increased steadily throughout the 20th century in these rivers. The higher sampling frequencies in the 1970s showed a rapid increase in metal concentrations at this time, with the first major decline being recorded in the post-1975 period. The decline has been attributed to increasing efforts to reduce the wastewater inputs (Figure 6C). Of the total amount of metals transported by the rivers Rhine and Meuse, Salomons *et al.* (1982) estimated that about 67% accumulated in the Netherlands, which acted as an important settling lagoon and reduced contaminant loadings to the North Sea. In a more detailed study by Malle and Müller (1982) on the Rhine at the German–Dutch border for the period 1971–1979, it was shown that the total Cr, Hg and Pb load associated with suspended sediments declined significantly. In contrast, the Cu, Pb and Zn loads associated with suspended sediments changed little, despite an overall reduction in the total loads of all metals (Figure 6D). Analysis of average metal concentrations in the River Aire, Yorkshire, UK and the River Tame in the West Midlands, UK shows declining concentrations in total Zn, Cu, Ni and Pb, although concentrations of Zn, Cu and Ni in the Tame for 1993 were higher than for 1974 in the River Aire (Figure 6E).

As metal concentrations have been measured on both filtered and unfiltered water samples for the River Aire for the 20 year period, an estimate of sediment-associated metals concentration has been made. The data show that, with decreasing metal concentrations, more Pb is sediment-associated, Cu and Zn show little change in the dissolved/particulate fraction and the dissolved Ni concentrations increase (Figure 6Fi). The relationship between total Pb concentration and the percentage sediment-associated Pb is significant at the 99·9% level (Figure 6Fii), and shows that a greater proportion of Pb is transported in association with suspended sediment in less polluted rivers.

Temporal trends in heavy metal concentrations have also been identified on the basis of repeat sampling of fluvial substrates (Buffa, 1976; Förstner and Müller, 1976; Förstner and Wittman, 1983). Figure 7 gives two examples of the Neckar valley, Germany between 1968 and 1970 and the River Sherbourne, UK between 1986 and 1993. In both examples, heavy metal concentrations in the urban and downstream reaches of these river systems have decreased substantially. In the former, elevated Cd levels were caused by a dye production plant which released an annual total of 10–20 t of Cd in dissolved or particulate form into the river. Improved water treatment decreased Cd discharges and resulted in a sharp decline in Cd concentrations in the pelitic sediments. The latter comparison shows marked improvements in sediment quality, largely as a result of the connection of several industrial effluent discharges, particularly from a Zn plating works, to the nearby sewage treatment plant. The decline in Pb concentrations, although less

Figure 6. Seasonal and long-term trends in metal behaviour. (A) Relationship between sediment-associated metal concentration and discharge for the River Rhine at Koblenz. Modified from Schleichert (1975). Spring/summer relationships are indicated by crosses, autumn/winter relationships are indicated by closed circles. (B) Cadmium concentrations in the suspended sediments of the lower River Rhine. Modified from Salomons and Förstner (1984). (C) Long-term trends in heavy metal concentrations in suspended sediments of the River Rhine. Modified from Salomons *et al.* (1982). (D) Temporal changes in dissolved and sediment-associated metals loadings in the River Rhine, 1971–1979. Modified from Malle and Müller (1982). (E) Long-term trends in total metal concentrations: (i) River Aire, Yorkshire, UK and (ii) River Tame, West Midlands, UK (National Rivers Authority, Yorkshire and Severn-Trent Regions, unpublished data). (F) Metals concentrations in the River Aire, UK: (i) temporal variations in sediment-associated metals concentrations and (ii) relationship between sediment-associated Pb concentration and total Pb concentration (National Rivers Authority, Yorkshire Region, unpublished data)

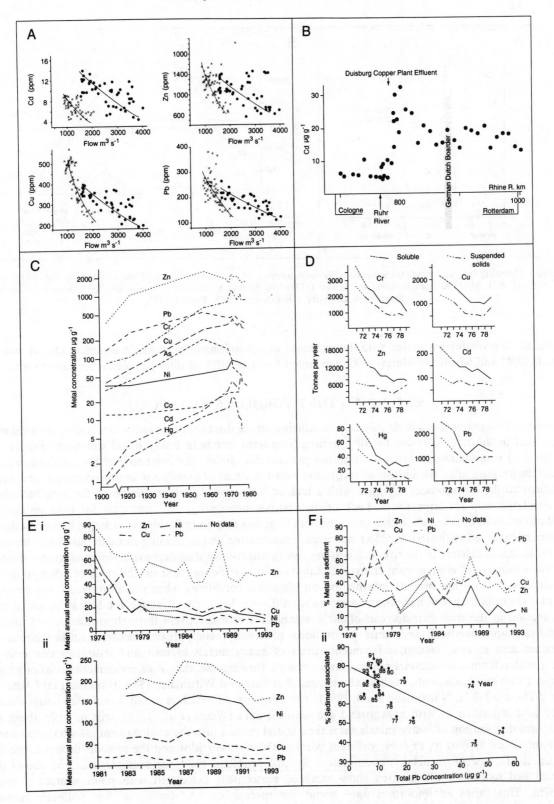

Figure 6

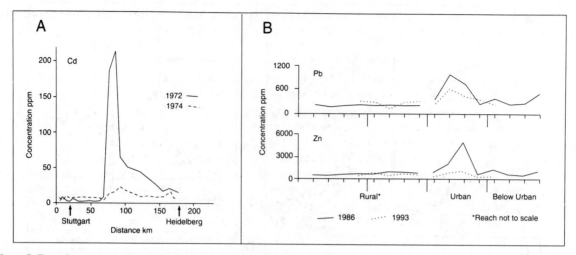

Figure 7. Detecting temporal trends from studies of fluvial substrates. (A) Cd content of fluvial bed sediments in the Neckar valley in 1972 and 1974. Modified from Förstner and Müller (1976). (B) Zn and Pb content of fine fluvial bed sediments in the River Sherbourne, Coventry, UK (Thoms, 1987b; Proffitt, 1993)

dramatic, is probably associated with the increased use of unleaded fuels, which were introduced into the UK in 1987 and which, by March 1993, accounted for over 50% of total petroleum consumption.

EXTENDING THE HYDROLOGICAL RECORD

There are two major limitations inherent in utilizing either direct water quality monitoring or sediments deposited in the river corridor for determining long-term trends in heavy metal transport. Firstly, the monitoring programmes reviewed here do not provide data before the beginning of the 20th century and hence begin long after the industrial revolution and the onset of significant anthropogenic excess metal loadings to the environment. Coupled with a lack of long-term data is the tendency for long records to be available only for large rivers. Lack of distributive monitoring, sufficient data for load estimation, and information on transfers within the catchment (e.g. loss or gain to floodplain storage) makes it difficult to determine whether the trends reflect increased concentration and/or loading through time, or a change in metal/sediment delivery to the river (cf. Walling, 1983). Short-term studies provide high quality information on the dynamics of contaminant transport, liquid–solid partitioning and on the rates of adsorption and desorption, but do not provide information on historical conditions when pollution levels and environmental controls may well have been considerably different. Secondly, although the history of metal contamination in the river corridor can often be established retrospectively through an analysis of alluvial sequences and overbank sediments as outlined previously, comparisons between anthropogenically enriched and natural background concentrations of heavy metals in soils and sediments can only be made with reference to 'uncontaminated' sites through, for example, the use of enrichment ratios or a comparison with stable elements, such as aluminium (Förstner and Wittmann, 1983; Håkanson and Jansson, 1983; Thoms, 1987a; Norton and Kahl, 1991). This comparison, which is usually conducted with subsoils, deep lake sediments or with background geochemical data (Webb et al., 1978), tells us little about the temporal distribution of heavy metals since their initial release into the environment; at what time metal contamination arrived in, or from, different parts of the river corridor and the post-disturbance recovery of the fluvial system to metal contamination. Specific problems exist, for example, with the use of fine fluvial bed sediments which often show localized enrichment associated with point sources (Thoms, 1987b). This raises an important issue about the mechanism and timing of fine sediment transfer between the active fluvial sediments and the fluvial substrate (Hughes, 1992; Foster et al., 1994).

Palaeolimnological studies of lake sediment chemistry often provide an alternative source of proxy hydrological data for a range of environments identified in Figure 3. The following sections consider the value of sediment-based studies for reconstructing trends in heavy metal contamination through time.

Reconstructing trends through time: the chronology

The value of the preserved historical heavy metal record requires consideration of the possible metal sources and their mode of transport, the initial distribution of fine sediments deposited at the lake bed, the post-depositional stability of metals within the sediment column and the accuracy of the derived chronology. The sources, modes of transport and post-depositional stability of heavy metals were considered previously. Spatial variability in sedimentation rates and, in consequence, heavy metal distributions at the lake or reservoir bed will affect the interpretation of trends observed in a single core and can produce large errors in the calculation of heavy metal budgets (Foster and Dearing, 1987a; Dearing and Foster, 1993; Foster and Charlesworth, 1994). The latter study has shown that in small (<6 ha) polluted lake basins in Midland England, the spatial variability in the heavy metal concentrations of surface sediments is considerably lower than the downcore variability. This finding suggests that long- term trends might be identified in the sediment column with some confidence because the trends are unlikely to reflect changing depositional patterns through time.

A key element in the reconstruction of trends from sediment analysis is the derivation of an accurate chronology. Although a range of radionuclides has been used for determining the age of deposited sediment, ^{210}Pb and ^{137}Cs have been the most widely used on sediments deposited over the last 100 years. ^{210}Pb is a radionuclide produced as part of the ^{238}U decay series, has a half-life of 22·26 years and reaches lakes and reservoirs with eroded catchment sediments (the supported ^{210}Pb) and via atmospheric fallout (the unsupported ^{210}Pb) (cf. Wise, 1980). A range of models has been developed to derive a chronology based on the activity of the unsupported component (cf. Oldfield and Appleby, 1984; Appleby *et al.*, 1986; 1988; 1990; 1991). ^{137}Cs is not a natural radionuclide, and initially appeared in fallout following thermonuclear tests in 1952 (Perkins and Thomas, 1980). More recently, significant quantities of ^{137}Cs also have been released from nuclear reactors (Walling *et al.*, 1986). ^{137}Cs has a half-life of 30 years and is transferred to lake sediments directly from atmospheric fallout and with eroded catchment topsoils which contain ^{137}Cs (Ritchie and McHenry, 1990; Walling and He, 1993). A chronology is usually obtained through the identification of the first measurable occurrence of ^{137}Cs in environmental samples in 1954 and a major peak in atmospheric fallout in 1964. Some workers have suggested that a third peak in atmospheric fallout might be identified in 1971, but the evidence for this is poor (Ritchie and McHenry, 1990).

The factors which are likely to influence the mobility and post-depositional stability of ^{210}Pb and ^{137}Cs have been identified previously. Early studies using ^{137}Cs, however, assumed that this radionuclide was strongly adsorbed to clay particles in catchment soils or suspended lake sediments, was subsequently deposited at the lake bed and suffered little post-depositional migration (Pennington *et al.*, 1973; Ritchie *et al.*, 1973). More recent studies have shown the possible effects of physical mixing by bioturbation to be potentially important in redistributing ^{137}Cs through the sediment column (Robbins *et al.*, 1979). In both acid and soft water lakes, independent studies have shown that ^{137}Cs mobility increases significantly and can result in either upward or downward migration relative to the initial point of deposition or to the age derived from ^{210}Pb analysis (Longmore *et al.*, 1983; Davis *et al.*, 1984). Pardue *et al.* (1989) have shown through laboratory studies that ^{137}Cs fixation and remobilization in lake sediments also is strongly controlled by redox potential and interstitial water chemistry. In part, the problems associated with the use of ^{137}Cs derive from its higher solubility relative to ^{210}Pb in the lake water column. For example, Appleby and Smith (1992) have shown that, for Galloway lakes, the radionuclide transfer function (i.e. the fraction sorbed by sediment in the lake water column) is ca. 0·85–0·9 for ^{210}Pb but only 0·35–0·5 for ^{137}Cs.

Research on ^{210}Pb and ^{137}Cs dating, briefly reviewed here, has raised important questions regarding the accuracy of the derived chronology and in general suggests that the ^{210}Pb dating technique is more robust due to both its lower solubility and higher post-depositional stability. Nevertheless, physical mixing by bioturbation will affect both radionuclides and any discrepancy between the chronology derived from both techniques is only likely to reveal the effects of chemical remobilization. Evidence of post-depositional radionuclide mobility within the sediment column has important implications, not only for the accuracy of the chronology, but also for the possible remobilization of heavy metals in the same core (Longmore *et al.*, 1983).

Identifying the atmospheric input

Reconstruction of the atmospheric contamination history in upland regions has often been made on the basis of metal contents of dated peats or of ice cores (Livett *et al.*, 1976; Wolff and Peel, 1985; Boutron, 1987; Hvatum *et al.*, 1987; Pacyna and Winchester, 1990). Also, the history of atmospheric Cd pollution has been reconstructed from a soil archive (Jones *et al.*, 1987). The general lack of peat and ice in the areas of highest contamination, particularly in lowland urbanized river basins, precludes the widespread application of such techniques for the reconstruction of long-term trends. Furthermore, although atmospheric heavy metal enrichment is found in urban soils (Thornton, 1990), a chronology cannot be directly established. However, small natural or man-made lake basins exist in many parts of the fluvial system, as demonstrated in Figure 3, and, with improved resolution on radiometric dating, offer a unified aquatic environment for identifying long-term trends in contamination history from a range of sources. From a hydrological viewpoint, establishing trends in atmospheric inputs to lake basins is vital to discriminate them from trends associated with catchment inputs.

Increased concentrations of trace metals in lake sediments have typically been used as an indication of trends in late 19th and 20th century air pollution (Müller and Barsch, 1980; Reuther *et al.*, 1981; Rippey *et al.*, 1981: Jones, 1984; Norton, 1986; Renburg, 1986; Smith, 1986; Foster and Dearing, 1987b; Lamberts *et al.*, 1987; Battarbee, 1988; Norton *et al.*, 1990; Foster *et al.*, 1991b; Williams, 1991). Figure 8A shows the anthropogenic excess metal profiles from Deep Lake in the Adirondack Mountains, New York and the Fish River Lake, northern Maine (Norton and Kahl, 1991). Background concentrations in both lakes are similar and concentrations begin to increase at about 1850 (Deep Lake) and 1860 (Fish Lake), and are consistent with recognizable air pollution in eastern North America caused by industrial development and the burning of coal. Maxima are reached between 1965 and 1975 and concentrations decline thereafter, probably as a result of the diminishing use of leaded fuels. Similar atmospheric pollution histories have been reconstructed at many upland sites in the UK (Battarbee, 1988; Figure 8B).

Preservation of the atmospheric Zn contamination history, particularly in upland lakes under highly acidified conditions, appears to be affected by the increased mobilization of Zn relative to Pb at low pH (compare Figures 8Bi and 8Bii), although the general differences in the Zn:Pb ratios were attributed to regional differences in pollution sources rather than differences in solubility (Battarbee, 1988). Figure 8C shows that the atmospheric contamination record is equally well preserved in lowland lakes and reservoirs in the UK, for which there is a clearly identifiable post-1850 increase in heavy metal concentrations in the sediments of Slapton Ley, a coastal barrier lake in Devon, UK. Spatial variations in atmospheric heavy metals transport have also been established from the analysis of lake sediments which are often enriched in close proximity to smelters (Allan, 1974), as has the early history of non-ferrous smelting in South Wales from an analysis of the metal content of sediment in Llangorse lake (Jones *et al.*, 1991; Figure 8D).

Evidence from lake sediments can be used to distinguish the historical switch from the earlier metal production/refinement activities of Figure 8D to consumptive and power-generating activities. The widespread dispersal of contaminants in the 20th century is due to the burning of fossil fuels and the increased Pb emissions from motor vehicles; a trend in metal contamination history identified in the Hudson–Rariton river basin by Tarr and Ayres (1990; Figure 2F). Furthermore, the analysis of hydrocarbons, mineral

Figure 8. Atmospheric influx to lake sediments. (A) Anthropogenic net accumulation rates of Pb based on dated cores in: (i) Deep Lake and (ii) Fish River Lake. Modified from Norton and Kahl (1991). (B) Atmospheric heavy metals, reconstructed pH and the carbonaceous particle content of bottom sediments from: (i) Loch Urr, Scotland and (ii) Llyn Hir, Wales. Modified from Battarbee (1988). (C) Concentrations of Pb and Zn in the bottom sediments of Slapton Ley, Devon, UK (dates derived from core correlation with a dated core published by Heathwaite, 1993). (D) Non-ferrous smelting in South Wales: (i) number of smelter works and (ii) Zn and Cu concentrations in bottom sediments of Llangorse Lake, Wales. Modified from Jones *et al.* (1991). (E) Concentrations of Pb, Zn, Cd and benzonapthothiophene (BNT) in bottom sediments of Lake Constance. Modified from Müller *et al.* (1977). (F) Concentration and loading of Zn in bottom sediments of Seeswood Pool, Warwickshire, UK. Modified from Foster and Dearing (1987b)

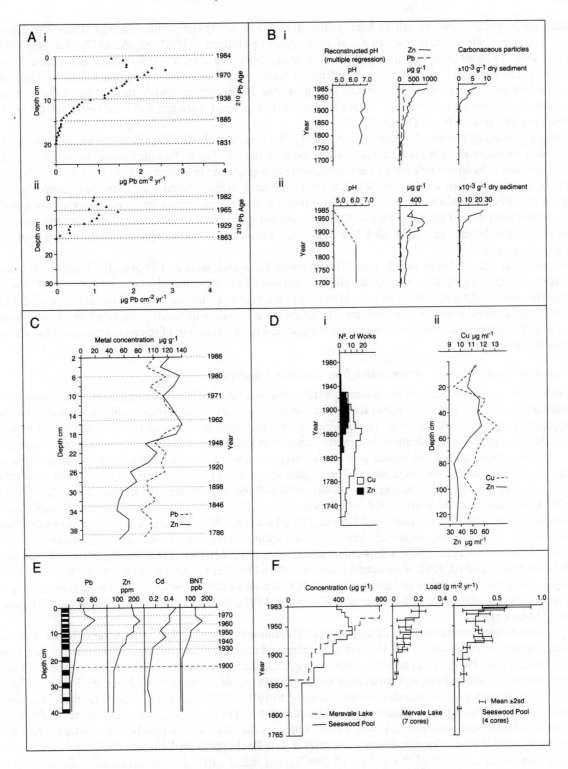

Figure 8

magnetic properties and fly ash in lake sediments provides comparative information which may allow discrimination between different forms of atmospheric pollution (Müller *et al.*, 1977; Oldfield, 1981; Hunt, 1986; Oldfield and Thompson, 1986; Wik *et al.*, 1986; McLean, 1991; Rose and Battarbee, 1991). The research by Müller *et al.* (1977) on sediments from Lake Constance, Switzerland, for example, has shown that the increase in Zn concentrations during the 1960s coincides with an increase in the concentration of benzonaphthothiophene, which is the product of coal burning, but is not emitted by petroleum-powered road vehicles (Figure 8E).

Interpretation of the metals record is complicated by variations in sediment delivery to the lake and by changing sediment accumulation rates at a single point in the lake basin. This may result, for example, from increased soil erosion or the effects of extreme events bringing coarser sediments into the lake. Smith *et al.* (1983), for example, in a study of Blagdon reservoir, have demonstrated that individual floods have a significant impact on both heavy metal concentrations and chemical speciation. It was shown from a sediment core analysis that total Pb, Zn and Cd concentrations decreased in the coarse flood deposits and that the exchangeable, carbonate, Fe and Mn associated forms of Zn were depleted at the expense of the residual fraction.

Comparison of the heavy metals record in Merevale Lake and Seeswood Pool in the English Midlands (Foster and Dearing, 1987b; Figure 8F) shows a decrease in metal concentrations in the post-1950 sediments of Seeswood Pool, compared with Merevale Lake, despite an increase in total metal influx. The dilution of metal concentrations was attributed to an increase in sediment yields dominated by channel bank erosion. The increase in total metal loading was suggested to be a function of re-eroded metal-contaminated topsoils reaching the lake.

Long-term trends in metal contamination from catchment sources

In addition to diffuse inputs from atmospheric pollution, lake sediments can also reveal the history of metal contamination related to mineral exploitation activities, urbanization and industrialization. Mineral exploitation has been shown to be responsible for the liberation of heavy metals into the fluvial system directly through the incorporation of contaminants in eroded mine tailings and drainage waters during the mining period and, as discussed previously, may become an important source of contamination in the later history of the drainage basin. The impact of metal exploitation has been recorded in several lake sediment studies (Maxfield *et al.*, 1974a; 1974b; Rabe and Bauer, 1977; Harding and Whitton, 1978; Elner and Happey-Wood, 1980; O'Sullivan *et al.*, 1987; 1989; Davison *et al.*, 1985; Coard, 1987; Horowitz *et al.*, 1988; 1993; Moore and Luoma, 1990; Dearing, 1992). The impact of urban storm drainage and effluent discharge on metal enrichment in lake sediments also has been established in several case studies (Hamilton-Taylor, 1979; Christensen and Chien, 1981; Michler, 1983; Smith, 1986; Johnson and Nicholls, 1988; Gaskell, 1992; Varvas and Punning, 1993). Detailed research on heavy metal contamination conducted in the Great Lakes of Canada, in the USA, Western Europe, the former Soviet Union and Japan in the 1970s, are referenced in Förstner and Wittmann (1983), Nriagu and Simmons (1984) and Salomons and Förstner (1984).

Figure 9A illustrates an example of the impact of tin mining in Cornwall, UK, on long-term trends in heavy metal release to the fluvial system through a geochemical analysis of sediments accumulating in Loe Pool (O'Sullivan *et al.*, 1989). A 3 m core was dated using ^{137}Cs and varve counting and the 3 m sequence covered the period from 1875 to 1979, with an average sedimentation rate of $2 \cdot 7 \, \mathrm{cm \, yr^{-1}}$. At 2 m in the 3 m core, a large peak in Sn concentration is recorded, which was dated to 1910–1915 by varve counting. This peak in concentration coincided with the operations of the Helston Valley Tin Company, which extracted Sn from the alluvial material deposited in the Lower Cober valley in the late 19th and early 20th century. Over this period of active mine working, sedimentation rates in Loe Pool approached $10 \, \mathrm{cm \, yr^{-1}}$.

Further direct evidence of the impact of non-ferrous metal extraction on lake sediments is given in Figure 9B. Here, 18th century Pb production in the catchment of Llyn Geirionydd produced a large peak in Pb concentrations in the lake sediments, which is supported by documented mining operations in the region. The late 20th century increase was correlated by Dearing (1992) with the extraction of Zn and Pb ores from the Pandora mine complex and the increase in concentration towards the surface was

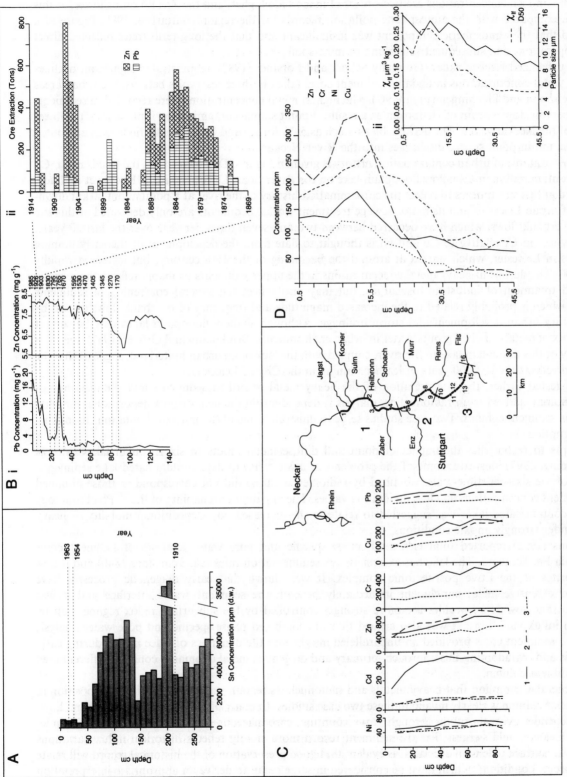

Figure 9. Catchment influx of heavy metals to lake sediments. (A) Long-term effect of tin mining on bottom sediments of Loe Pool, Cornwall. Modified from O'Sullivan et al. (1989). (B) History of Pb and Zn mining near Lake Geirionydd, Wales: (i) Pb and Zn concentrations in sediments and (ii) Pb and Zn extraction from the Pandora mine complex. Modified from Dearing (1992). (C) History of metal contamination in the bottom-sediments of three reservoirs in the Neckar valley. Modified from Schoer and Förstner (1980) (D) History of metal contamination and other characteristics from Leicester, UK as recorded in floodplain lake sediments of the River Soar (Quorn Lake): (i) metal concentrations and (ii) magnetic susceptibility and particle size distribution. Metal concentrations are expressed on a minerogenic basis. D_{50} is the median particle size of the lake sediments determined by laser granulometry

linked to the reworking of mine spoil in the late 20th century by fluvial processes and construction. The fact that the Pb and Zn concentrations are two orders of magnitude higher, and the Zn : Pb ratios lower, at this site when compared with the atmospheric pollution records for the region (Battarbee, 1988; Figure 8B), suggested that the atmospheric component was insignificant and that the long-term trend reflected direct mining operations and subsequent reworking of mine spoil.

In the urbanized Neckar valley, Germany Schoer and Förstner (1980) demonstrated significant declines in heavy metal concentrations in undated sediment cores taken in three reservoirs between the urban areas of Gundelsheim and Plochingen (Figure 9C). The highest metal concentrations were recorded in sediments of the reservoir downstream of Heilbronn as a result of point source contamination inputs along the lower reaches of the Neckar valley. The decline, by as much as 50% for Zn and Cd concentrations, was attributed to a reduction in point source discharges into the River Neckar.

Another example of urban contamination history is given in Figure 9D, which shows the trends in Cr, Cu and Ni contamination in a shallow floodplain lake on the River Soar, Leicestershire, UK. As yet undated, the record at this site appears to reflect the contamination history of the river at a point about 10 km downstream of urban Leicester and its major sewage treatment works. Sediments are only deposited at this site during overbank flows, which have occurred between two and seven times per year over the last 20 years. The early rise in Cr relative to Cu and Ni is thought to date from the development of the early tanning industries in Leicester, which started at around the beginning of the 19th century, but increased rapidly after 1850. The decline in heavy metal concentrations in the upper sediments probably reflects the recently improved treatment of industrial discharges, but may also reflect the general coarsening of sediments up-core, which is probably related to the increased magnitude and frequency of overbank flows caused by 20th century urban development and storm drainage. Although shallow floodplain lakes may be subject to an influx of metal-enriched groundwater in urban environments (Brinkmann and Grüger, 1987), the close proximity of this floodplain lake to the river, coupled with the lack of an urban hinterland, suggests that the Cr record reflects the last 200 years of leather tanning in the City of Leicester.

As suggested earlier, the interpretation of the heavy metal record in sediments and the accuracy of the chronology derived from radionuclide analysis is dependent on the lack of post-depositional mobility within the sediment column. Two case studies serve to illustrate some of the inherent limitations in making this assumption.

Attempts to reconstruct the impact of point and diffuse metal inputs to six Estonian lakes (Varvas and Punning, 1993) illustrates some of the problems in using ^{210}Pb to date contaminated lake sediments. At four of the six sites, time-scales obtained by radiometric dating did not correspond to those obtained from element concentrations and loss on ignition values. The apparent unreliability of the ^{210}Pb chronology in these contaminated lakes was suggested to relate to the increased post-depositional mobility of heavy metals under strongly anoxic conditions.

Williams (1992) reported total elemental, phase-specific and pore water analyses of sediment cores from Loch Ba, Scotland. The lake has extremely low sedimentation rates (ca. 3 cm since 1840) and drains granodiorites of the Etive post-tectonic complex. It was shown that early diagenetic processes have promoted extensive metal enrichment immediately beneath the sediment–water interface and it was suggested that down-core profiles are more strongly controlled by the subsurface redox regime than by historical influx variations. It was also argued that the combined phase-specific and pore water analysis applied to sediment cores provided an unparalleled insight into the dynamics of trace metals during early diagenesis and, by adopting this approach, primary and diagenetic metal anomalies could be differentiated in the sediment column.

The general assumption that heavy metals and radionuclides remain stable at the point of deposition in the sediment column is clearly invalid in these two case studies. In many of the studies discussed here, however, independent corroboration through varve counting, carbonaceous particle content and other independent means would suggest that the sediment record more closely reflects historical influx variations than the subsurface redox conditions. It is evident that good preservation of the historical record will relate to site-specific conditions, which must be considered in some detail to derive an appropriate interpretation of the trends obtained.

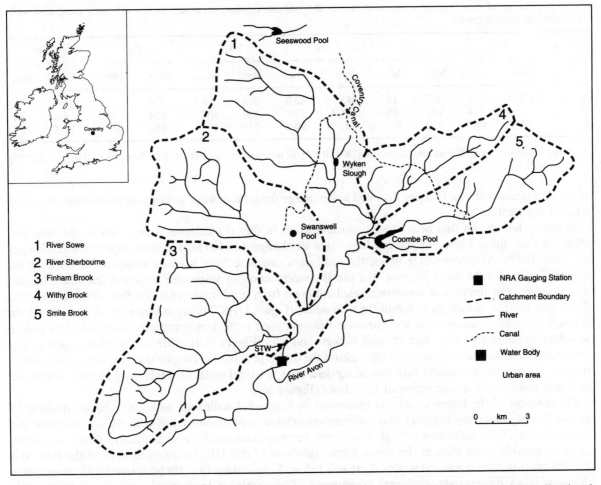

Figure 10. Catchments of the River Sowe around the urban area of Coventry showing locations of Swanswell Pool, Wyken Slough and Seeswood Pool. The main sewage treatment works (STW) is located upstream of the National Rivers Authority river gauging station

Before considering the implications of these case studies in more detail, the following section contains a brief review of our recent research into the dynamics of heavy metal transport in Midland England rivers and the reconstruction of trends based on an analysis of lake sediments.

HEAVY METALS IN THE URBANIZED RIVER SOWE, MIDLAND ENGLAND

The River Sowe rises to the north-west of the city of Coventry and flows for 30 km through a 262 km^2 catchment to its confluence with the River Avon. The city straddles the headwaters of the River Sowe, the 59·9 km^2 basin of the River Sherbourne and the 66·7 km^2 basin of the Finham Brook, which also includes the largely residential urban centre of Kenilworth. The predominantly rural catchments of Smite Brook (38·3 km^2) and Withy Brook (30·1 km^2) enter the upper Sowe catchment from the north-east, although most of the sediment from Smite Brook is trapped in Coombe Pool (Figure 10). Unlike many cities in the UK, Coventry only began to expand in the early 20th century, reaching a population of c. 300 000 by 1960 and remaining at this level to the present time. Major industries currently include motor vehicle assembly, aero-engine production, light engineering, metal plating and finishing. About 80% of the effluent from the city, including urban storm drainage and industrial discharge, is treated at the Finham sewage treatment works, which has an output of 120 Ml d^{-1} into the lower Sowe downstream

Table III. Range of heavy metal concentrations in the $<63\,\mu m$ fraction of urban fluvial subtrates compared with suspended sediments (ppm)

Source	Zn		Pb		Ni		Cr		Cu		Cd	
	Max.	Min.	Max.	Min.	Max.	Min.	Max.	Min.	Max.	Min.	Max.	Min.
A	1586	114	957	45·7	843	30·0	964	14·3	270	7·1	24·3	1·4
B	525	313	171	85·4	71·4	57·0	129	100	114	56·9	8·6	7·1
C	1482	1067	719	281	141	89·8	225	135	852	248	33·7	11·1

A, Urban fluvial substrates; B, high magnitude flow ($2400\,Ml\,d^{-1}$); C, intermediate flow ($400\,Ml\,d^{-1}$). Based on data in Proffitt (1993)

of its confluence with the Finham Brook. The discharge from the sewage works is approximately equal to natural dry weather flow.

A survey has shown that heavy metal concentrations in fluvial sediments transported by the river are three to four times higher at intermediate river discharges (c. $400\,Ml\,d^{-1}$) than high river discharges (c. $2400\,Ml\,d^{-1}$). At intermediate flows, the sediments originate from predominantly urban stormwater runoff (Table III). As flows increase, the metal concentration of suspended sediment decreases, largely due to larger contributions of uncontaminated sediment from rural catchments (Proffitt, 1993). In general, Pb is least soluble and Zn most soluble in river water (Figure 11A). Trends in heavy metal concentrations through the urban area are complex. For some elements, such as Pb, concentrations are significantly higher at urban contaminated sites than at rural background sites (Figure 11B), whereas for others, such as Zn, concentrations are relatively uniform throughout the drainage network. Despite decreases in metal concentrations in suspended sediment with increasing discharge, the total metal load increases because suspended sediment concentrations are higher at high flows (Figure 5C).

Comparison of the heavy metal concentrations in suspended sediments with the $<63\,\mu m$ fraction of urban fluvial substrates suggests that sediment-associated contaminants are unlikely to penetrate the armour layer of the sediments at high discharges, because concentrations in the active fluvial sediments are considerably lower than in the urban fluvial substrate (Table III). In urban reaches of the river, the Zn:Pb ratio in fine fluvial substrates decreases below 2, indicating that Pb associated with stormwater drainage is an increasingly significant component of contaminant loadings (Foster et al., 1994). This difference has been exacerbated by the reduction of point source Zn discharges to the lower Sherbourne (Figure 7B).

Within the upper Sowe drainage basin, two lakes have been studied in detail (Foster et al., 1991a; 1991b; Charlesworth and Foster, 1993; Charlesworth, 1995). These are the Swanswell Pool and Wyken Pool (Figure 10). Swanswell Pool is a small (0·73 ha) freshwater lake situated in the city centre and surrounded by a small (< 1 ha) park, which is effectively the drainage basin. No urban storm drainage enters the lake and, since it was isolated from its contributing catchment in 1850, water levels have been maintained from either local springs or municipal water supplies. The major contribution to the sediments of this lake, since 1850, have been atmospheric. In contrast, Wyken Pool is a small (2·25 ha) urban marginal lake, which receives water from two major tributaries in its $4·5\,km^2$ catchment. In addition to atmospheric fallout, heavy metals enter the main river system from industrial discharges, urban storm drainage and from landfill leachates. The lake basin was formed by mining subsidence in the mid-19th century and the lake is used to balance flow in the upper Sowe River.

Figure 11. Heavy metals in fluvial and reservoir bottom sediments in the Sowe catchments, West Midlands, UK. (A) Sediment-associated and dissolved metal concentrations in samples from the River Sowe, December 1992. (B) Downstream trends in the River Sherbourne subcatchment, December 1992: (i) Pb concentrations and (ii) Zn concentrations (dissolved and sediment associated). (C) Trends in total Zn concentrations in bottom sediments: (i) Swanswell Pool and (ii) Wyken Pool. After Charlesworth and Foster (1993). (D) Pb and Zn storage in the Wyken Pool catchment. After Charlesworth (1995). (E) Estimated anthropogenic excess loading from atmospheric and catchment sources for six lowland England lakes and reservoirs: (i) Zn and (ii) Pb

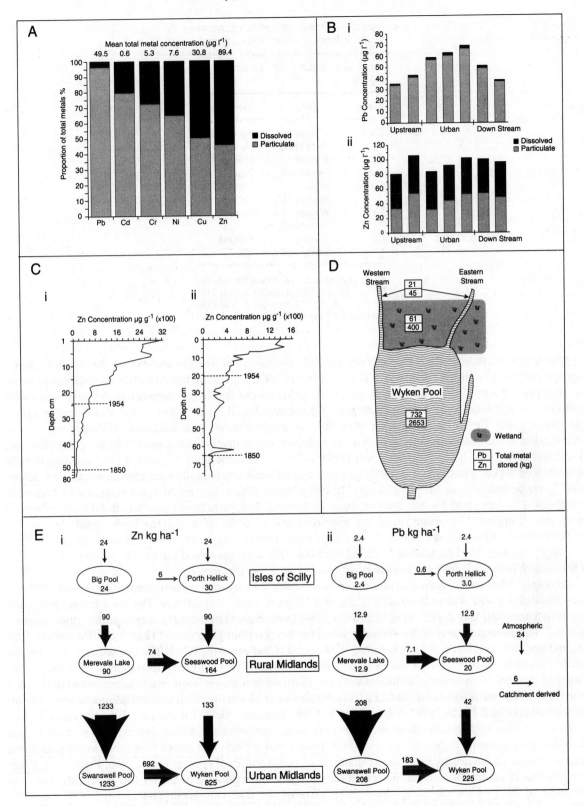

Figure 11

Table IV. Dominant forms of metals
in the sediments of the Wyken Slough
and Swanswell Pool. Based on the
extraction method of Tessier *et al.*
(1979)

Metal	Site	Metal form
Zn	Wyken	Fe/OM
	Swanswell	Fe/Ex
Pb	Wyken	OM/DT
	Swanswell	Fe/OM
Cu	Wyken	OM/DT
	Swanswell	Ex/Ca
Ni	Wyken	Fe/OM
	Swanswell	Fe/OM
Cd	Wyken	Fe/OM
	Swanswell	XX

Ex, Extractable; Ca, carbonate bound; Fe,
Fe/Mn bound; OM, organic bound; DT,
detrital; XX, no dominant form. Based on
data in Charlesworth (1995)

Heavy metal concentrations increase dramatically up-core in both lakes and mirror the trends in atmo-spheric pollution recorded elsewhere in the UK (Figure 11C). Maximum heavy metals concentrations occur in the upper 10 cm of the sediments in each lake, although the dominant chemical association varies with individual metal ions (Table IV) and with time (Charlesworth and Foster, 1993). Analysis of element ratios, particularly the Fe:Mn ratio, suggest that little post-depositional remobilization of heavy metals has occurred in these lake sediments and that the trends in concentration are more likely to reflect historical loadings to the lake basin (Charlesworth, 1995). The concentrations in Swanswell Pool are significantly higher than Wyken Pool, which is partly a function of an inner city location with high atmospheric loadings and lower sedimentation rates (see later). In comparison with heavy metal concentrations of lake sedi-ments directly impacted by non-ferrous metal extraction, concentrations of metals in lake and reservoir sediments dominated by urban runoff are approximately an order of magnitude lower (Table V).

Radiometric dating using ^{210}Pb and ^{137}Cs has proved problematic in Wyken and Swanswell pools. There is a high background of supported ^{210}Pb in both lakes which has prevented the establishment of an accurate chronology before the first occurrence of ^{137}Cs (Appleby, pers. comm.).

Average minerogenic accumulation rates (correcting for organic carbon content) since ca. 1850 in Swanswell Pool and Wyken Pool are 0·125 and 0·148 g cm^{-2} yr^{-1}, respectively. The pre-1954 accumulation rates in Swanswell and Wyken are 2·5 and 2·1 times lower than the post-1954 accumulation rates, respect-ively. Computation of equivalent sediment yields for Wyken Pool gives ca. 4·7 t km^{-2} yr^{-1} for the pre-1954 period and ca. 10·0 t km^{-2} yr^{-1} for the post-1954 period (Charlesworth, 1995). The low sediment yield in the early period is, in part, due to the existence of a wetland (Wyken Slough) upstream of the pool, which trapped significant amounts of sediment until ca. 1970, when a major flood relief channel was cut through the slough. The wetland has significant accumulations of sediment and high concentrations of heavy metals (Charlesworth and Foster, 1993; Charlesworth, 1995). Sediment yields for the post-1954 period at Wyken Pool are a little higher than those reconstructed from lake sediment accumulation rates in undisturbed forested environments in North Warwickshire (Foster *et al.*, 1985) and about four times lower than those reconstructed in the nearby cultivated catchment of Seeswood Pool (Foster *et al.*, 1986; Figure 10). From an analysis of sediment cores in Wyken Pool, the upstream wetland and fluvial bed sediments, the total amount of Pb and Zn stored in these three parts of the lake catchment system have been estimated (Figure 11D). About 3·2 t of Zn and 0·8 t of Pb have accumulated since the middle of the 19th century, of which over 80% is stored within the lake sediments. Of the total amount stored, an adjustment for

Table IV. Range of total heavy metal concentrations in the post-1850 sediments of selected lowland lakes and reservoirs unaffected by mining activity (ppm)

Location	Pb		Cu		Ni		Zn		Cd	
	Min.	Max.	Min.	Max.	Min.	Max.	Min.	Max.	Min.	Max.
Big Pool	tr	15	13	37	9	20	52	259	—	—
P. Hellick	tr	10	9	21	tr	22	37	93	—	—
Slapton	63·0	138·0	14·0	80·0	40·0	95·0	35·0	140·0	—	—
Merevale	37·7	77·7	64·0	114·3	61·8	143·5	52·5	785·7	—	—
Seeswood	18·4	61·7	33·3	80·0	41·9	62·8	62·8	124·0	—	—
Wyken	51·2	475·9	27·5	490·1	26·2	163·4	52·0	1000	3·2	29·0
Swanswell	66·0	311·5	25·5	292·2	75·0	165·3	60·0	1800	6·1	16·3
Holmer*	153·0	407·0	15·0	51·0	75·0	129·0	508	2045	10·0	21·0
Quorn	14·5	47·4	15·0	96·5	47·9	144·2	22·0	272·4	3·0	6·5
Neckar 1	127	216	157	300	47	68	470	941	14·5	37·4
Neckar 2	104	160	104	287	35	51	343	1096	3·5	9·0
Neckar 3	85	98	118	191	41	54	410	672	3·2	7·4

Big Pool: closed coastal lagoon, Isles of Scilly, UK (Foster et al., 1991b)
P. Hellick: coastal lagoon with small (1·6 km^2) cultivated drainage basin, Isles of Scilly, UK (Foster et al., 1991b)
Slapton: coastal lagoon, South Devon, UK
Merevale Lake: inland reservoir, Warwickshire, UK (Foster and Dearing, 1987a)
Seeswood Pool: inland reservoir, Warwickshire, UK (Foster and Dearing, 1987a)
Wyken Slough: urban balancing reservoir, Coventry, UK (Foster et al., 1991a)
Swanswell: urban closed basin, Coventry, UK (Foster et al., 1991a)
Holmer: urban balancing lake, Telford, UK (Gaskell, 1992)
Quorn: floodplain lake, River Soar, Leicestershire, UK (Carlyle, 1994)
Neckar 1–3: undated reservoir sediments in the lower Neckar Valley, Germany (Schoer and Förstner, 1980; Figure 9c)
* Constructed in 1969

natural background concentrations suggests that about 70% of both Zn and Pb are related to human disturbance over the last 150 years.

An attempt has been made to distinguish the atmospheric- from catchment-derived heavy metal inputs to these lakes over the last 150 years by comparing the heavy metal inventories with those established for nearby undisturbed soils and lake sediments where there is no catchment-derived influx (Foster et al., 1994). From these calculations, adjusting for background geochemical concentrations, the relative contribution of atmospheric- and catchment-derived inputs to the lake sediments can be estimated for the entire period of sedimentation since the first increase in heavy metal concentrations in the mid-19th century. Figure 11E summarizes these calculations for both the Wyken Pool and Swanswell Pool and provides comparable data for the rural Midlands and lakes in remote west coast locations in the UK (Isles of Scilly), where similar computations have been made. There are clear spatial trends evident in these data. Firstly, there is approximately an order of magnitude increase in atmospheric Zn and Pb fallout between remote rural locations and urban environments. Secondly, there is a strong gradient in atmospheric fallout between inner city sites and nearby rural sites, as evidenced from the data for Swanswell Pool, Wyken Pool, Seeswood Pool and Merevale Lake, which lie approximately on a south to north transect from the city centre of Coventry over a distance of only 18 km. Thirdly, there is evidence from both Porth Hellick and Seeswood Pool that eroded metal-enriched topsoils from the catchment can make a significant contribution towards the total influx to the lake (c. 20 and 40%, respectively, based on average Pb and Zn loadings).

Despite the inherent problems in interpreting the bottom sediment record discussed previously, heavy metal concentrations provide semi-quantitative estimates of temporal trends in metal delivery to these lakes as have been affected by atmospheric pollution and point source contamination within the catchment. The study has also shown that re-erosion of atmospheric pollutants may make a significant contribution to the total amount of heavy metals delivered to lakes.

SYNTHESIS AND CONCLUSIONS

Heavy metals enter the drainage basin from diffuse atmospheric sources and from point sources within the catchment (mine drainage, effluent discharge, urban storm runoff). An analysis of trends in metal production and product usage suggests that the latter now makes a greater contribution to the total flux of heavy metals to the environment compared with the late 19th and early 20th centuries. The transport of heavy metals through the fluvial system is, in part, a function of solubility and of chemical speciation controls in the suspended sediment. Experimental data also suggest that the solubility increases with increasing metal concentration in contaminated rivers. Even with a reduction in dissolved metal concentrations at high discharges, the total load increases because the water volume increases. With metals which are strongly adsorbed to sediments, the most significant controls on their distribution through the fluvial system relate to the hydraulic conditions in channels and on floodplains, to sediment delivery and storage within the catchment and to the post-depositional stability of metals within the sediment (either controlled by chemical solubility or the re-erosion of contaminated sediments).

Limited long-term data on heavy metal transport in European rivers show several trends. Firstly, there has been an increase in concentration throughout the 20th century until the early 1970s, after which the regulation of industrial and sewage effluent discharges has resulted in a decrease in metal concentration and loadings. The evidence from UK rivers has also shown that the sediment-associated concentration has increased for some metals and decreased for others over a period of generally declining total concentrations since the early 1970s.

Lake and reservoir bottom sediments have often been used to examine trends in heavy metals transport in drainage basins when long-term data from catchment monitoring programmes are not available. By the careful selection of suitable sites (Figure 3) and comparison with heavy metals data from other sources within the catchment, it may be possible to identify zones of metal enrichment within the river corridor, to quantify long-term trends in contaminant transport and to evaluate the relative significance of atmospheric and catchment contributions. Several case studies using dated lake and reservoir sediments have shown that it is possible to detect the temporal change from production to consumption-related sources of heavy metals in the environment and to independently assess the trends associated with atmospheric pollution, mineral extraction and urbanization.

Several limitations currently exist in using lake and reservoir bottom sediments to reconstruct the history of metal contamination. Firstly, the relative solubilities of different metals will give rise to a variable preservation of the total record within the sediment column. Secondly, it is not possible to assume constant ratios of dissolved to sediment-associated metal concentrations through time, as solubility often increases with increasing contamination. With increasing total metal concentrations, therefore, the preserved record is likely to underestimate the total metal load transported. Thirdly, where anoxic or low pH conditions are experienced in the overlying water or in the sediments, metals (including radionuclides used for dating sediments) may become mobile within the sediment column even if they are not lost in soluble form. This has led to difficulties in establishing reliable chronologies for lake and reservoir sediments as well as providing two possible explanations for observed up-core trends.

The fundamental question relating to an interpretation of the chemical record in bottom sediments, therefore, is whether the sediment record solely reflects the historical influx of metals to the basin or the pH and/or redox conditions which control post-depositional mobility within the sediment column. In reality, the sediment core record will undoubtedly reflect both processes in addition to the effects of physical mixing by bioturbation, which will not be readily separable by phase-specific or pore water chemical analysis.

Improvements in our understanding of the dynamics of metal transfers in fluvial and limnic systems will undoubtedly assist in providing the most appropriate interpretation of the sediment record, as will an analysis of both pore water and phase-specific metal and other elemental concentrations in the sediments. Furthermore, discrepancies between sediment chronologies based on ^{137}Cs and ^{210}Pb analysis may indicate the increasing significance of post-depositional mobility within the sediment column, a process which may be further investigated by comparison with other radionuclides, such as ^{241}Am (Appleby et al.,

1991). The use of independent chronological markers, such as pollen, diatoms, carbonaceous particles and varved sediments, may further reinforce the interpretation of the metals record and provide a more convincing argument that the trends in heavy metals clearly reflect historical loading to the lake basin.

Despite the practical difficulties in distinguishing the record of historical influx from diagenetic processes, an analysis of bottom sediments offers one of the few opportunities to assess trends in heavy metal transport through the fluvial system.

ACKNOWLEDGEMENTS

We are indebted to several individuals who have assisted in the collection and analysis of field samples and the preparation of the manuscript; including Kayzi Ambridge, Peter Appleby, Roger Cambray, James Carlyle, Heather Dalgleish, John Dearing, Rob Grew, David Keen, Liam Kelly, Tim Mighall, Phil Owens, Liz Turner and Des Walling. We are particularly grateful to Arthur Horowitz for making constructive comments on an earlier draft of this paper. The UK research has been funded and assisted by the Atomic Energy Research Establishment (Harwell), the British Geomorphological Research Group, Coventry City Council, Coventry University, the Duchy of Cornwall, the Isles of Scilly Environment Trust, the National Rivers Authority (Severn-Trent and Yorkshire Regions) and the Natural Environment Research Council. Special thanks go to Ruth Gaskell for providing the artwork in the figures.

REFERENCES

Allan, R. J. 1974. 'Metal contents of lake sediment cores from established mining areas. An interface of exploration and environmental geochemistry', *Geol. Surv. Can.*, **74–1/B**, 43–49.

Appleby, P. G., and Smith, J. T. 1992. 'The transport of radionuclides in lake-catchment systems', Presentation to the *Unesco workshop on the Hydrological Impact of Nuclear Power Plant Systems. Paris, 23–25 September 1992*.

Appleby, P. G., Nolan, P. J., Gifford, D. W., Godfrey, M. J., Oldfield, F., Anderson, N. J., and Battarbee, R. W. 1986. '^{210}Pb dating by low background gamma counting', *Hydrobiologia*, **143**, 21–27.

Appleby, P. G., Nolan, P. J., Oldfield, F., Richardson, N., and Higgitt, S. R. 1988. '^{210}Pb dating of lake sediments and ombrotrophic peats by gamma assay', *Sci. Total Environ.*, **69**, 157–177.

Appleby, P. G., Richardson, N., Nolan, P. J., and Oldfield, F. 1990. 'Radiometric dating of the United Kingdom SWAP sites', *Phil. Trans. R. Soc. London*, **B 327**, 233–238.

Appleby, P. G., Richardson, N., and Nolan, P. J. 1991. '^{241}Am dating of lake sediments', *Hydrobiologia*, **214**, 35–42.

Battarbee, R. W. 1988. *Lake acidification in the United Kingdom, 1800–1986*. Department of The Environment, HMSO, London.

Beckwith, P. R., Ellis, J. B., Revitt, D. M., and Oldfield, F. 1986. 'Heavy metal and magnetic relationships for urban source sediments', *Phys. Earth. Planet. Inter.*, **42**, 67–75.

Bengtsson, L., and Enell, M. 1986. 'Chemical analysis' in Berglund, B. (Ed.), *Handbook of Holocene Palaeoecology and Palaeohydrology*. Wiley, Chichester. pp. 423–451.

Benoit, G., and Hammond, H. F. 1990. '^{210}Po and ^{210}Pb remobilisation from lake sediments in relation to iron and manganese cycling', *Environ. Sci. Technol.*, **24**, 1224–1233.

Bird, S. C. 1987. 'The effect of hydrological factors on trace metal contamination in the River Tawe, South Wales', *Environ. Pollut.*, **45**, 87–124.

Bloemendal, J., Oldfield, F., and Thompson, R. 1979. 'Magnetic measurements used to assess sediment influx at Llyn Goddionduon', *Nature*, **280**, 50–53.

Bourg, A. C. M. 1987. 'Trace metal adsorption modelling and particle–water interactions in estuarine environments', *Continental Shelf Res.*, **7**, 1319–1332.

Boutron, C. F. 1987. Changes of lead, cadmium, zinc and copper concentrations in Antarctic ice during the last 30,000 years' in Lindberg, S. E., and Hutchinson, T. C. (Eds), *Proceedings of the International Conference on Heavy Metals in the Environment (New Orleans)*. CEP Consultants, Edinburgh. pp. 183–189.

Bradley, S. B. 1982. 'Sediment quality related to discharge in a mineralised region of Wales' in *Recent Developments in the Explanation and Prediction of Erosion and Sediment Yield (Proc. Exeter Symp.)*. IAHS Publ., **137**, 341–350.

Bradley, S. B., and Cox, J. J. 1986. 'Heavy metals in the Hamps and Manifold Valleys, North Staffordshire, UK: distribution in floodplain soils', *Sci. Total Environ.*, **50**, 103–128.

Bradley, S. B., and Lewin, J. 1982. 'Transport of heavy metals on suspended sediments under high flow conditions in a mineralised region of Wales', *Environ. Pollut. (Ser. B)*, **4**, 257–267.

Bradley, S. B., and Lewin, J. 1984. 'Flood effects on the transport of heavy metals', *Int. J. Environ. Stud.*, **22**, 225–230.

Brinkmann, W. L. F. 1985. 'Urban stormwater pollutants: sources and loadings', *GeoJ.*, **11**, 277–283.

Brinkmann, W. L. F., and Grüger, B. 1987. 'Hydrochemical status of Pond Enkheimer Ried, Rhine-Main Area, FR Germany', *GeoJ.*, **14**, 299–309.

Brown, H. S., Kasperson, R. E., and Raymond, S. S. 1990. 'Trace pollutants' in Turner, B. L., Clark, W. C., Kates, R. W., Richards, J. F., Mathews, J. T., and Meyer, W. B. (Eds), *The Earth as Transformed by Human Action*. Cambridge University Press with Clark University, Cambridge. pp. 437–454.

Bruk, S. (Ed.) 1985. *Methods of Computing Sedimentation in Lakes and Reservoirs.* Unesco, Paris.

Burrus, D., Thomas, R. L., Dominik, B., Vernet, J.-P., and Dominik, J. 1990. 'Characteristics of suspended sediment in the upper Rhone river, Switzerland, including the particulate forms of phosphorous', *Hydrol. Process.*, **4**, 85–98.

Buffa, L. 1976. 'Review of environmental control on mercury in Japan', *Econ. Technol. Rev. Rep.*, **EPS 3-WP- 76–7**, Canada Environmental Protection Service.

Carlyle, J. D. 1994. Reconstruction of regional heavy metal contamination in discontinuous sedimentary sequences: the case for floodplain lake studies', *Unpub. BSc Dissertation*, Coventry University.

Cawse, P. A. 1980. *Deposition of Trace Elements from the Atmosphere in the UK. Inorganic Pollution and Agriculture. Reference Book 326.* Ministry of Agriculture Fisheries and Food, HMSO, London. pp. 22–46.

Cawse, P. A. 1987. 'Trace and major elements in the atmosphere at rural locations in Great Britain, 1972–81' in Coughtry, P. J., Martin, M. H., and Unsworth, M. H. (Eds), *Pollutant Transport and Fate in Ecosystems.* Blackwell, Oxford. pp. 89–112.

Chambers, F. 1993. *Climate Change and Human Impact on the Landscape.* Chapman and Hall, London.

Charlesworth, S. M. 1995. The pollution history of two urban lakes in Coventry, UK', *Unpubl. PhD Thesis*, Coventry University.

Charlesworth, S. M., and Foster, I. D. L. 1993. 'Effects of urbanisation on lake sedimentation; the history of two lakes in Coventry, UK — preliminary results' in McManus, J., and Duck, R. W. (Eds), *Geomorphology and Sedimentology of Lakes and Reservoirs.* Wiley, Chichester. pp. 15–29.

Christensen, E. R., and Chien, N.-K. 1981. 'Fluxes of arsenic, lead, zinc and cadmium to Green Bay and Lake Michigan', *Environ. Sci. Technol.*, **15**, 553–558.

Coard, M. 1987. 'Palaeolimnological study of the history of Loe Pool, Helston and its catchment', *Unpubl. PhD Thesis*, Polytechnic South-West.

Cortesao, C., and Vale, C. 1994. 'The seasonal variability of metals in a river contaminated with mine drainage (Sado River, Portugal)' in *International Symposium on Particulate Matter in Rivers and Estuaries (Preprints) (Reinbek near Hamburg, 21–25 March, 1994).* Univ. Hamburg. pp. 13–20.

Dahlberg, E. C. 1968. 'Application of a selective simulation and sampling technique to the interpretation of stream sediment copper anomalies near south mountain, PA', *Econ. Geol.*, **63**, 409–417.

Davies, B. E. 1980. 'Trace element pollution' in Davies, B. (Ed.), *Applied Soil Trace Elements.* Wiley, Chichester. pp. 287–351.

Davis, R. B., Hess, C. T., Norton, S. A., Hanson, D. W., Hoagland, K. D., and Anderson, D. S. 1984. '[137]Cs and ^{210}Pb dating of sediments from soft-water lakes in New England (U.S.A.) and Scandinavia, a failure of ^{137}Cs dating', *Chem. Geol.*, **44**, 151–185.

Davison, W. 1985. Conceptual models for transport at a redox boundary' in Stumm, W. (Ed.), *Chemical Processes in Lakes.* Wiley, New York. pp. 31–53.

Davison, W., Hilton, J., Lishman, J. P., and Pennington, W. 1985. 'Contemporary lake transport processes determined from sedimentary records of copper mining activity', *Environ. Sci. Technol.*, **19**, 356–360.

Dearing, J. A. 1983. 'Changing patterns of sediment accumulation in a small lake in Scania, Southern Sweden', *Hydrobiologia*, 103, 59–64.

Dearing, J. A. 1992. 'Sediment yields and sources in a Welsh upland lake catchment during the past 800 years', *Earth Surf. Process. Landforms*, **17**, 1–22.

Dearing, J. A., and Foster, I. D. L. 1993. 'Lake sediments and geomorphological processes: some thoughts', in McManus, J., and Duck, R. W. (Eds), *Geomorphology and Sedimentology of Lakes and Reservoirs.* Wiley, Chichester. pp. 5–14.

deGroot, A., Zschuppe, K., and Salomons, W. 1982. 'Standardisation of methods of analysis for heavy metals in sediments', *Hydrobiologia*, **92**, 689–695.

Dossis, P., and Warren, L. J. 1980. 'Distribution of heavy metals between the minerals and organic debris in a contaminated marine sediment', in Baker, R.A. (Ed.), *Contaminant and Sediments.* Vol. 1. *Fate and Transport, Case Studies, Modeling, Toxicity.* Ann Arbor Science, Michigan. pp. 119–139.

Driver, N. E., and Lystrom, D. J. 1987. 'Estimation of urban storm runoff loads and volumes in the United States', *Proc. IVth Int. Conf. Urban Storm Drainage, Lausanne, Switzerland.* pp. 74–82. B. C. Yen (Ed). Published by Secretariat of the 22nd Congress Int. Assn. Hydraul. Res. Laboratoire d'Hydraulique Ecole Polytechnique, Fedende, Lausanne.

Duijsings, J. J. H. M. 1986. 'Seasonal variation in the sediment delivery ratio of a forested drainage basin in Luxembourg', in *Drainage Basin Sediment Delivery. (Proc. Albuquerque Symp.). IAHS Publ.*, **159**, 153–164.

EC 1991. Directive concerning urban waste water treatment (91/271/EEC)', *Off. J.*, **L135/40**, May, Council of the European Communities.

Ellis, J. B. 1979. 'The nature and sources of urban sediments and their relation to water quality: a case study from north-west London', in Hollis, G. E. (Ed.), *Man's Impact on the Hydrological Cycle in the United Kingdom.* GeoBooks, Norwich. pp. 199–216.

Ellis, J. B., and Harrop, D. O. 1984. 'Variations in solids loadings to roadside gully pots', *Sci. Total Environ.*, **33**, 203–212.

Ellis, J. B., and Revitt, D. M. 1982. 'Incidence of heavy metals in street surface sediments: solubility and grain size studies', *Wat. Air Soil Pollut.*, **17**, 87–100.

Ellis, J. B., Harrop, D. O., and Revitt, D. M. 1986. 'Hydrological control of pollutant removal from highway surfaces', *Wat. Res.*, **20**, 589–595.

Elner, J. K., and Happey-Wood, C. M. 1980. 'The history of two linked but contrasting Welsh lakes', *Brit. Phycol. J.*, **13**, 341–360.

Engstrom, D. R., and Wright, H. E. 1984. 'Chemical stratigraphy of lake sediments as a record of environmental change' in Hawarth, E. Y. and Lund, J. W. G. (Eds), *Lake Sediments and Environmental History.* Leicester University Press, Leicester. pp. 11–69.

Fergusson, J. E. 1990. *The Heavy Elements: Chemistry, Environmental Impact and Health Effects.* Pergamon Press, Oxford.

Flores-Rodriguez, J., Bussy, A.-L., and Thévenot, R. 1993. 'Toxic metals in urban runoff: physico-chemical mobility assessment using speciation schemes', in *Proc. VIth Int. Conf. Urban Storm Drainage, Niagara Falls, Ontario, Canada.* pp. 182–187 J. Marsalek and H. C. Torno (eds) Seapoint Publishing, British Columbia.

Förster, J. 1993. 'The influence of atmospheric conditions and storm characteristics on roof runoff pollution: studies with an experimental roof system' in *Proc. VIth Int. Conf. Urban Storm Drainage, Niagara Falls, Ontario, Canada.* pp. 411–416.

Förstner, U. 1983. 'Assessment of metal pollution in rivers and estuaries' in Thornton, I. (Ed.), *Applied Environmental Geochemistry*. Academic Press, London. pp. 395–423.

Förstner, U. 1986. 'Chemical forms and environmental effects of critical elements in solid waste materials: combustion residues' in Bernhard, M., Brinckmann, F. E., and Sadler, P. J. (Eds), *The Importance of Chemical Speciation in Environmental Processes*. Springer Verlag, Berlin. pp. 465–491.

Förstner, U., and Müller, G. 1976. 'Heavy metal pollution by monitoring river sediments', *Fortschr. Mineral.*, **53**, 271–288.

Förstner, U., and Patchineelam, S. R. 1980. 'Chemical associations of heavy metals in polluted sediments from the Lower Rhine River' in Kavanaugh, M. C., and Leckie, J. O. (Eds), *Particulates in Water; Characterisation, Fate, Effects and Removal. Adv. Chem. Ser.*, **189**, American Chemical Society, Washington. pp. 177–193.

Förstner, U., and Wittmann, G. T. W. 1983. *Metal Pollution in the Aquatic Environment*. Springer-Verlag, Berlin.

Foster, I. D. L., and Charlesworth, S. M. 1994. 'Variability in the physical, chemical and magnetic properties of reservoir sediments; some implications for environmental reconstruction and sediment source modelling' in *Variability in Stream Erosion and Sediment Transport. (Canberra Symp.). IAHS Publ.*, **224**, 153–160.

Foster, I. D. L., and Dearing, J. A. 1987a. 'Quantification of long term trends in atmospheric pollution and agricultural eutrophication: a lake–watershed approach' in *The Influence of Climate Change and Climatic Variability on the Hydrologic Regime and Water Resources (Proc. Vancouver Symp.). IAHS Publ.* **168**, 173–189.

Foster, I. D. L., and Dearing, J. A. 1987b. 'Lake-catchments and environmental chemistry: a comparative study of contemporary and historical catchment processes in Midland England', *GeoJ.*, **14**, 285–297.

Foster, I. D. L., and Walling, D. E. 1994. 'Using reservoir sediments to reconstruct changing sediment yields and sources in the catchment of the Old Mill reservoir, South Devon, UK', *Hydrol. Sci. J.*, **39**, 347–368.

Foster, I. D. L., Dearing, J. A., Simpson, A. D., Carter, A. D., and Appleby, P. G. 1985. 'Lake catchment based studies of erosion and denudation in the Merevale catchment, Warwickshire, UK', *Earth Surf. Process. Landforms*, **10**, 45–68.

Foster I. D. L., Dearing, J. A., and Appleby, P. G. 1986. 'Historical trends in catchment sediment yields: a case study in reconstruction from lake sediment records in Warwickshire, UK', *Hydrol. Sci. J.*, **31**, 427–443.

Foster, I. D. L., Dearing, J. A., Grew, R., and Orend, K. 1990. 'The sedimentary data base: an appraisal of lake and reservoir sediment based studies of sediment yield' in *Erosion, Transport and Deposition Processes (Proc. Jerusalem Workshop). IAHS Publ.*, **189**, 19–43.

Foster, I. D. L., Charlesworth, S. M., and Keen, D. H. 1991a. 'A comparative study of heavy metal contamination in four reservoirs in the English Midlands', *Hydrobiologia*, 214, 155–162.

Foster, I. D. L., Charlesworth, S. M., Dearing, J. A., Keen, D. H., and Dalgleish, H. 1991b 'Lake sediments: a surrogate measure of sediment-associated heavy metal transport in fluvial systems?' in *Sediment and Stream Water Quality in a Changing Environment; Trends and Explanation (Proc. Vienna Symp.). IAHS Publ.*, **203**, 321–328.

Foster, I. D. L., Charlesworth, S. M., and Proffitt, S. B. 1994. 'Sediment associated heavy metal distribution in urban fluvial and limnic systems; a case study of the River Sowe, UK' in *Int. Symp. Particulate Matter in Rivers and Estuaries (Preprints) (Reinbek near Hamburg, 21–25 March, 1994)*. University of Hamburg. pp. 91–100.

Gardner, M. J., and Gunn, A. M. 1989. 'The effect of natural ligands on trace metal partitioning', *Wat. Research Centre Rep. PRU 2183-M*, WRC, Medmenham, 17 pp.

Gaskell, R. 1992. 'Lake sediment properties from an urban catchment: South Telford, Shropshire', *Unpubl. BSc Dissertation*, Coventry University.

Gobeil, C., and Silverberg, N. 1989. 'Early diagenesis of lead in Laurentian Tough sediments', *Geochim. Cosmochim. Acta*, 53, 1889–1895.

Graf, W. L. 1990. 'Fluvial dynamics of ^{230}Th in the church rock event, Puerco River, New Mexico', *Ann. Assoc. Am. Geogr.*, **80**, 327–342.

Hadley, R. F., Lal, R., Onstad, C., Walling, D. E., and Yair, A. 1985. *Recent Developments in Erosion and Sediment Yield Studies. Unesco Tech. Documents in Hydrology*. Unesco, Paris. 127 pp.

Hahn, H. H. 1993. 'Pathways of anthropogenic pollutants within the urban drainage system' in *Proc. VIth Int. Conf. Urban Storm Drainage, Niagara Falls, Ontario, Canada. pp. 397–404.*

Håkanson, L., and Jansson, M. 1983. *Principles of Lake Sedimentology*. Springer-Verlag, Berlin.

Hamilton-Taylor, M. 1979. 'Enrichments of zinc, lead and copper in recent sediments of Windermere, England', *Environ. Sci. Technol.*, **13**, 693–697.

Harding, J. P. C., and Whitton, B. A. 1978. 'Zinc, cadmium and lead in water, sediments and submerged plants of the Derwent reservoir, Northern England', *Wat. Res.*, **12**, 307–316.

Harrison, R. M. 1987. 'Physico-chemical speciation and chemical transformations of toxic metals in the environment' in Coughtrey, P. J., Martin, M. H., and Unsworth, M. H. (Eds), *Pollutant Transport and Fate in Ecosystems*. Blackwell, Oxford. pp. 239–247.

Harrop, D. O., Ellis, J. B., and Revitt, D. M. 1983. 'Temporal loadings of sediment and heavy metals to roadside gully pots' in Perry, R., Müller, G., and Förstner, R. (Eds), *Heavy Metals in the Environment*. CEP Consultants, Edinburgh. pp. 876–879.

Hart, B. T. 1982. 'Uptake of trace metals by sediments and suspended particulates: a review', *Hydrobiologia*, **91**, 299–313.

Heathwaite, A. L. 1993. 'Lake sedimentation' in Burt, T. P. (Ed.), *A field Guide to the Geomorphology of the Slapton Region. Field Stud. Coun. Occ. Pub.*, **27**, 31–41.

Hindy, K. T., Farag, S. A., El-Taib, N. M., Rizk, H. F., and Ibrahim, J. M. 1987. 'Spectrographic study of heavy metals in an industrial area in Northern Cairo' in Lindberg, S. E., and Hutchinson, T. C. (Eds), *Int. Conf. Heavy Metals in the Environment, New Orleans, September 1987*. CEP Consultants, Edinburgh. pp. 134–136.

Honeyman, B. D., and Santschi, P. H. 1988. 'Metals in aquatic systems', *Environ. Sci. Technol.*, **8**, 862–871.

Horowitz, A. J. 1991. *A Primer on Trace Metal Sediment Chemistry*. 2nd edn. Lewis, Chelsea.

Horowitz, A. J., Elrick, K., and Callender, E. 1988. 'The effect of mining on the sediment-trace element geochemistry of cores from the Cheyenne River arm of Lake Oahe, South Dakota, USA', *Chem. Geol.*, **67**, 17–33.

Horowitz, A. J., Elrick, K. A., and Cook, R. B. 1993. 'Effect of mining and related activities on the sediment trace element geochemistry of Lake Coeur d'Alene, Idaho, USA. Part I: surface sediments', *Hydrol. Process.*, **7**, 403–424.

Hughes, N. 1992. 'Heavy mineral distribution in upland gravel-bed rivers', *Unpub. PhD Thesis*, Department of Geography, University of Loughborough.

Hunt, A. 1986. 'The application of mineral magnetic methods to atmospheric aerosol discrimination', *Phys. Earth Planet. Inter.*, **42**, 10–21.

Hutton, M., and Symon, C. 1987. 'Sources of cadmium discharge to the UK environment' in Coughtry, P. J., Martin, M. H., and Unsworth, M. H. (Eds) *Pollutant Transport and Fate in Ecosystems*. Blackwell, Oxford. pp. 223–237.

Hvatum, O. Ø., Steinnes, E., and Bølriken, B. 1987. 'Regional differences and temporal trends in heavy metal deposition from the atmosphere studied by analysis of ombrotrophic peat' in Lindberg, S. E., and Hutchinson, T. C. (Eds), *Heavy Metals in the Environment*. CEP Consultants, Edinburgh. pp. 201–203.

Imhoff, K. R., Koppe, P., and Dietz, F. 1981. Heavy metals in the Ruhr river and their budget in the catchment area', *Progr. Wat. Technol.*, **12**, 735–749.

Jannasch, H. W., Honeyman, B. D., Ballistieri, L. S., and Murray, J. W. 1988. 'Kinetics of trace element uptake by marine particles', *Geochim. Cosmochim. Acta*, **52**, 567–577.

Jarvis, P. J. 1983. *Metal Pollution — an Annotated Bibliography 1976–80*. GeoBooks, Norwich.

Jenne, E. A., and Zachara, J. M. 1984. 'Factors influencing the sorption of metals' in *Fate and Effects of Sediment-Bound Chemicals in Aquatic Systems*. Pergamon, New York. pp. 83–98.

Johnson, M. G., and Nicholls, K. H. 1988. 'Temporal and spatial trends in metal loads to sediments of lake Simcoe, Ontario', *Wat. Air Soil Pollut.*, **39**, 337–354.

Jones, K. C., Symon, C. J., and Johnston, A. E. 1987. 'Retrospective analysis of an archived soil collection II: cadmium', *Sci. Total Environ.*, **67**, 75–89.

Jones, R. 1984. 'Heavy metals in the sediments of Llangorse Lake, Wales, since Celtic–Roman times', *Verh. Int.. Verein. Limnol.*, **22**, 1377–1382.

Jones, R., Chambers, F. M., and Benson-Evans, K. 1991. 'Heavy metals (Cu and Zn) in recent sediments of Llangorse lake, Wales: non-ferrous smelting, Napoleon and the price of wheat — a palaeoecological study', *Hydrobiologia*, **214**, 149–154.

Lamberts, L., Thomas, M., and Petit, D. 1987. '210Pb dating, Pb isotopic ratios and heavy metals fallout in freshwater environments' in Lindberg, S. E., and Hutchinson, T. C. (Eds), *Heavy Metals in the Environment*. CEP Consultants, Edinburgh. pp. 41–43.

Leenaers, H. 1989a 'The dispersal of metal mining wastes in the catchment of the River Geul (Belgium–the Netherlands)', *Nederlandse Geogr. Stud.*, **102**, Geografische Institut, Rijksuni-versitat Utrecht, Amsterdam.

Leenaers, H. 1989b. 'The transport of heavy metals during flood events in the polluted river Geul (The Netherlands)',. *Hydrol. Process.*, **3**, 325–338.

Lick, W. 1984. 'The transport of sediments in aquatic systems' in Ward, C.H., and Walton, B. T. (Eds), *Fate and effects of Sediment-Bound Chemicals in Aquatic Systems*. Pergamon Press, New York. pp. 61–82.

Livett, E. A., Lee, J. A., and Tallis, J. H. 1976. 'Lead, zinc and copper analysis of British blanket peats', *J. Ecol.*, **67**, 865–891.

Longmore, M. E., O'Leary, B. M., and Rose, C. W. 1983. 'Caesium-137 profiles in the sediments of a partial-meromictic lake on Great Sandy Island (Fraser Island), Queensland, Australia', *Hydrobiologia*, **103**, 21–27.

Lunt, D. O., Gunn, A. M., Roddie, B. D., Rogers, H. R., Gardner, M. J., Dobbs, A. J., and Watts, C. D. 1989. 'Investigation of partitioning of contaminants between water and sediment', *Wat. Res. Centre Rep.*, **PRS 2262-M**, WRC, Medmenham, 84 pp.

Macklin, M. G. 1992. 'Metal pollution of soils and sediments: a geographical perspective' in Newson, M. D. (Ed.), *Managing the Human Impact on the Natural Environment: Patterns and Processes*. Belhaven, London. pp. 172–195.

Macklin, M. G., and Dowsett, R. B. 1989. 'The chemical and physical speciation of trace metals in fine grained overbank flood sediments in the Tyne Basin, north-east England', *Catena*, **16**, 135–151.

Macklin, M. G., and Klimek, K. 1992. 'Dispersal, storage and transformation of metal contaminated alluvium in the Upper Vistula basin, southwest Poland', *Appl. Geogr.*, **12**, 7–30.

Malle, K. G., and Müller, G. 1982. 'Metallgehalt und Schwebstoffgehalt im Rhein', *Zeit. Wass. Abwass. Forsch*, **15**, 11–15.

Marsalek, J. 1991. 'Pollutant loads in urban stormwater: review of methods for planning level estimates', *Wat. Resour. Bull.*, **27**, 283–291.

Martin, J. M., and Meybeck, M. 1979. 'Elemental mass balance of material carried by major world rivers', *Mar. Chem.*, **7**, 173–206.

Maxfield, D., Rodriguerz, J. M., Buettner, M., Davis, J., Forbes, L., Kovacs, R., Russel, W., Schultz, L., Smith, R., Stanton, J., and Wai, C. M. 1974a. 'Heavy metal pollution in the sediments of the Coeur d'Alene River Delta', *Environ. Pollut.*, **7**, 1–6.

Maxfield, D., Rodriguerz, J. M., Buettner, M., Davis, J., Forbes, L., Kovacs, R., Russel, W., Schultz, L., Smith, R., Stanton, J., and Wai, C. M. 1974b. 'Heavy metal content in the sediments of the southern part of the Coeur d'Alene Lake', *Environ. Pollut.*, **6**, 263–266.

McLean, D. 1991 'Magnetic spherules in recent lake sediments', *Hydrobiologia*, **214**, 91–97.

Mészáros, A., Friedland, A. J., Haszpra, L. Meszáros, E., Lásztity, A., and Horváth, Z. S. 1987. 'Lead and cadmium deposition rates and temporal patterns in central Hungary' in Lindberg, S. E., and Hutchinson, T. C. (Eds), *Int. Conf. Heavy Metals in the Environment, New Orleans, September 1987*. CEP Consultants, Edinburgh. pp. 44–48.

Meybeck, M., and Helmer, R. 1989. 'The quality of rivers from pristine stage to global pollution', *Palaeogeogr. Palaeoclimatol. Palaeoecol.*, **75**, 283–309.

Michler, G. 1983. 'Heavy metal content in sediments of lakes in southern Bavaria as a sign of long term environmental impact' in *Dissolved Loads of Rivers and Surface Water Quality/Quantity Relationships (Proc. Hamburg Symp.)*. *IAHS Publ.* **141**, 405–419.

Moore, J. N., and Luoma, S. N. 1990. 'Hazardous wastes from large scale metal extraction', *Environ. Sci. Technol.*, **24**, 1278–1285.

Morrison, G. M., Revitt, D. M., Ellis, J. B., Svensson, G., and Balmér, P. 1984. 'The physico-chemical speciation of zinc, cadmium lead and copper in urban stormwater' in *Proc. IIIrd Int. Conf. Urban Storm Drainage*, **3**, 989–1000. P. Balmer, P. A. Malmqvist and A. Sjoburg (Eds). Goteburg, Sweden. Chalmers University of Technology (publisher).

Müller, G., and Barsch, D. 1980. 'Anthropogenic lead accumulation in the sediments of a high arctic lake, Ooblayough Bay, N. Ellesmere Island, N.W.T. (Canada)', *Environ. Technol. Lett.*, **1**, 131–140.

Müller, G., Grimmer, G., and Böhnke, H. 1977. 'Sedimentary record of heavy metals and polycyclic aromatic hydrocarbons in a sediment core from Lake Constance', *Naturwissenschaften*, **64**, 427–431.

Norton, S. A. 1986. 'A review of the chemical record in lake sediment of energy related air pollution and its effect on lakes', *Wat. Air Soil Pollut.*, **30**, 331–345.

Norton, S. A., and Kahl, J. S. 1991. 'Progress in understanding the chemical stratigraphy of metals in lake sediments in relation to acid precipitation', *Hydrobiologia*, **214**, 77–84.

Norton, S. A., Dillon, P. J., Evans R. D., Mierle, G., and Kahl, J. S. 1990. 'The history of atmospheric pollution and deposition of Cd, Hg and Pb in North America' in Lindberg, S. E., Page, A. L., and Norton, S. A. (Eds), *Sources, Deposition and Canopy Interactions*, 3, Springer-Verlag, New York. pp. 73–102.

Novotney, V., Simsiman, G. V., and Chesters, G. 1986. 'Delivery of pollutants from non-point sources' in *Drainage Basin Sediment Delivery (Proc. Albuquerque Symp.)*. *IAHS Publ.*, **159**, 133–140.

Nriagu, J. O. 1979. 'Global inventory of natural and anthropogenic emissions of trace metals to the atmosphere', *Nature*, **279**, 409–411.

Nriagu, J. O. 1980. *Cadmium in the Environment*. Wiley, New York.

Nriagu, J. O. 1989. 'A global assessment of natural sources of atmospheric trace metals', *Nature*, **338**, 47–49.

Nriagu, J. O. 1990. 'Human influence on the global cycling of trace metals', *Palaeogeogr. Palaeoclimatol. Palaeoecol.*, , **82**, 113–120.

Nriagu, J. O. 1993. 'Legacy of mercury pollution', *Nature*, **363**, 589.

Nriagu, J. O., and Davidson, C. I. (Eds), 1986. *Toxic Metals in the Atmosphere*. *Wiley Ser. Adv. Environ. Sci. Technol.*, **17**. Wiley, New York.

Nriagu, J. O., and Pacyna, J. M. 1988. 'Quantitative assessment of worldwide contamination of air, water and soils by trace metals', *Nature*, **333**, 134–139.

Nriagu, J. O., and Simmons, M. S. (Eds), 1984. *Toxic Contaminants in the Great Lakes*. Wiley, New York.

Oldfield, F. 1981. 'History of particulate atmospheric pollution from magnetic measurements in dated Finnish peat profiles', *Ambio*, **10**, 185–188.

Oldfield, F., and Appleby, P. G. 1984. 'A combined radiometric and mineral magnetic approach to recent geochronology in lakes affected by catchment disturbance and sediment redistribution', *Chem. Geol.*, **44**, 67–83.

Oldfield, F., and Thompson, R. 1986. *Environmental Magnetism*. Allen and Unwin, London.

O'Sullivan, P. E., Coard, M. A., Cousen, S. M., and Pickering, D. A. 1987. 'Studies of the formation and deposition of annually laminated sediments in Loe Pool, Cornwall, UK', *Verh. Int. Verein. Limnol.*, **22**, 1383–1387.

O'Sullivan, P. E., Heathwaite, A. L., Farr, K. M., and Smith, J. P. 1989. 'Southwest England and the Shropshire–Cheshire Meres', *Guide to Excursion A, Vth Int. Symp. Palaeolimnology, Ambleside, Cumbria, UK (1–6 September 1989)*.

Pacyna, J. M., and Münch, J. 1987. 'Atmospheric emissions of As, Cd, Pb and Zn from industrial sources in Europe' in Lindberg, S. E., and Hutchinson, T. C. (Eds) *Int. Conf. Heavy Metals in the Environment, New Orleans, September 1987*. CEP Consultants, Edinburgh. pp. 20–25.

Pacyna, J. M., and Winchester, J. W. 1990. 'Contamination of the global environment as observed in the Arctic', *Palaeogeogr. Palaeoclimatol. Palaeoecol.*, **82**, 149–157.

Pardue, J. H., DeLaune, R. D., Patrick, W. H., and Whitcomb, J. H. 1989. 'Effect of redox potential on fixation of ^{137}Cs in lake sediment', *Health Phys.*, **57**, 781–789.

Peart, M., and Walling, D. E. 1982. Particle size characteristics of fluvial suspended sediment' in *Recent Developments in the Explanation and Prediction of Erosion and Sediment Yield (Proc. Exeter Symp.)*. *IAHS Publ.*, **137**, 397–407.

Pennington, W., Cambray, R. S., and Fisher, E. M. 1973. 'Observations on lake sediments using ^{137}Cs as a tracer', *Nature*, **242**, 324–326.

Perkins, R. W., and Thomas, C. W. 1980. 'Worldwide fallout' in Hansen, W. C. (Ed.), *Transuranic Elements in the Environment*. *USDOE/TIC-22800*, 53–82.

Proffitt, S. B. 1993. 'Heavy metal partitioning in the River Sowe catchment: a study of the extent, sources and processes of heavy metal contamination', *Unpub. BSc Dissertation*, Coventry University.

Quality of Urban Air Group 1993. *Urban Air Quality in the United Kingdom. First Report of the Quality of Urban Air Group*. Prepared for the Department of the Environment, HMSO, London.

Rabe, F. W., and Bauer, S. B. 1977. 'Heavy metals in lakes of the Coeur d'Alene River Valley, Idaho', *Northwest Sci.*, **51**, 183–197.

Rausch, D. L., and Heinemann, H. G. 1984. 'Measurement of reservoir sedimentation' in Hadley, R. F., and Walling, D. E. (Eds), *Erosion and Sediment Yield: Some Methods of Measurement and Modelling*. GeoBooks, Norwich. pp. 179–200.

Renburg, I. 1986. 'Concentration and annual accumulation values of heavy metals in lake sediments: their significance in studies of the history of heavy metal pollution', *Hydrobiologia*, **143**, 379–385.

Renfrew, C. and Bahn P. G. 1991. *Archaeology, Theories, Methods and Practice*. Thames and Hudson, London 543 pp.

Reuther, R., Wright, R. F., and Förstner, U. 1981. 'Distribution and chemical forms of heavy metals in sediment cores from two Norwegian Lakes affected by acid precipitation' in *Int. Conf. Heavy Metals in the Environment (Amsterdam)*. pp. 318–321. Ernst, W. H. O. (ed). CEP Consultants, Edinburgh.

Rippey, B., Murphy, R. J., and Kyle, S. W. 1981. 'Anthropogenically derived changes in the sedimentary flux of Mg, Cr, Ni, Cu, Zn, Hg, Pb and P in Lough Neagh, Northern Ireland', *Environ. Sci. Technol.*, **16**, 23–30.

Ritchie, J. C., and McHenry, J. R. 1990. 'Application of radioactive fallout cesium-137 for measuring soil erosion and sediment accumulation rates and patterns: a review', *J. Environ. Qual.*, **19**, 215–233.

Ritchie, J. C., McHenry, J. R., and Gill, A. C. 1973. 'Dating recent reservoir sediments', *Limnol. Oceanogr.*, **18**, 254–263.

Robbins, J. A., McCall, P. L., Fisher, J. B., and Krezoski, J. R. 1979. 'Effect of deposit feeders on migration of ^{137}Cs in lake sediments', *Earth Planet. Sci. Lett.*, **42**, 277–287.

Robinson, G. 1982. 'Trace metal adsorption potential of phases comprising black coatings on stream pebbles', *J. Geochem. Explor.*, **17**, 205–219.

Robinson, G. 1983. 'Heavy metal adsorption by ferromanganese coatings on stream alluvium: natural controls and implications for exploration', *Chem. Geol.*, **38**, 157–174.

Rose, N., and Battarbee, R. W. 1991. 'Fly-ash particles in lake sediments: extraction and characterisation', *Res. Pap.*, **40**, Palaeoecology Research Unit, Department of Geography, University College London.

Rygwelski, K. R. 1984. 'Partitioning of toxic trace metals between solid and liquid phases in the Great Lakes' in Nriagu, J. O., and Simmons, M. S. (Eds), *Toxic Contaminants in the Great Lakes*. Wiley, New York. pp. 321–333.

Salomons, W., and Eysink, W. 1981. 'Pathways of mud and particulate trace metals from rivers to the southern North Sea' in Nio, S. D., Schuenttenhelm, R. T. E and Weering, T. C. E. (Eds) *Holocene Marine Sedimentation in the North Sea Basin. Spec. Publ. Int. Assoc. Sedimentol.*, **5**, 429–450.

Salomons, W., and Förstner, U. 1984. *Metals in the Hydrocycle*. Springer-Verlag, Berlin.

Salomons, W., van Driel, W, Kerdijk, H., and Boxma, R. 1982. 'Help! Holland is plated by the Rhine (environmental problems associated with contaminated sediments)' in *Int. Symp. Effects of Waste Disposal on Groundwater (Proc. Exeter Symp.). IAHS Publ.*, **139**, 255–269.

Samanidou, V., and Fytianos, K. 1987. 'Transport mechanisms of heavy metals in rivers of Northern Greece' in Lindberg, S. E., and Hutchinson, T. C. (Eds), *Heavy Metals in the Environment*. CEP Consultants, Edinburgh. pp. 307–310.

Santschi, P., Höhener, H., Benoit, G., and Brink, M. 1990. 'Chemical processes at the sediment water interface. *Mar. Chem.*, **30**, 269–315.

Schleichert, U. 1975. 'Schwermetallgehalte der Schwebstoffe des Rheins bei Koblenz im Jahresablauf', *Dtsch Gewass. Mitt.*, **19**, 150–157.

Schoer, J., and Förstner, U. 1980. 'Die entwicklung der Schwermetallverschmutzung im mittleren Neckar', *Dtsch. Gewass. Mitt.*, **24**, 153–158.

Sigg, L. 1983. 'Metal transfer mechanisms in lakes; the role of settling particles' in Stumm, W. (Ed.), *Chemical Processes in Lakes*. Wiley, New York. pp. 283–310.

Sigg, L. 1987. 'Surface chemical aspects of the distribution and fate of metal ions in lakes' in Stumm, W. (Ed), *Aquatic Surface Chemistry; Chemical Processes at the Particle-Water Interface*. Wiley, New York. pp. 319–349

Smith, J. P. 1986. Mineral magnetic studies in two Shropshire–Cheshire Meres', *Unpubl. PhD Thesis*, Liverpool University.

Smith, J. P., Bradley, S. B., Macklin, M. G., and Cox, J. J. 1983. 'The influence of catastrophic floods on water quality as recorded in the sediments of Blagdon Lake, England' in *Dissolved Loads of Rivers and Surface Water Quality/Quantity Relationships (Proc. Hamburg Symp.). IAHS Publ.*, **141**, 421–430.

Smith, W. G. 1986. 'Heavy Metals in the New Zealand aquatic environment: a review', *Wat. Soil Misc. Publ.*, **100**, Ministry of Works and Development, Wellington.

Spångberg, Å, and Niemczynowicz, J. 1993. 'Measurements of pollution runoff from an asphalt surface' in *Proc. VIth Int. Conf. Urban Storm Drainage, Niagara Falls, Ontario, Canada*. pp. 423–428.

Stumm, W., and Morgan, J. J. 1981. *Aquatic Chemistry*. Wiley, New York.

Tarr, J. A., and Ayres, R. V. 1990. 'The Hudson Raritan River Basin' in Turner, B. L., Clark, W. C., Kates, R. W., Richards, J. F., Mathews, J. T., and Meyer, W. B.(Eds), *The Earth as Transformed by Human Action*. Cambridge University Press with Clark University, Cambridge. pp. 623–641.

Tessenow, U., and Baynes, Y. 1975 'Redox-dependent accumulation of Fe and Mn in a littoral sediment supporting *Isoetes lacustris*', *Naturwissenschaften*, **62**, 342.

Tessier, A., Campbell, P., and Bison, M. 1979. 'Sequential chemical extraction procedure for the speciation of particulate trace metals', *Anal. Chem.*, **51**, 844–851.

Thoms, M. C. 1987a. 'Channel bed sedimentation within the urbanised river Tame, United Kingdom', *Regul. Riv.*, **1**, 229–246.

Thoms, M. 1987b. 'Channel sedimentation within urban gravel bed rivers', *Unpubl. PhD Thesis*, Loughborough University of Technology.

Thornton, I. 1990 'Soil contamination in urban areas', *Palaeogeogr. Palaeoclimatol. Palaeoecol.* **82**, 121–140.

Tipping, E., Woof, C., and Clarke, K. 1993. 'Deposition and resuspension of fine particles in a riverine 'dead zone'', *Hydrol. Process*, **7**, 263–277.

Tylecote, R. F. 1987. *Early History of Metallurgy in Europe*. Longman, London.

Van de Meent, D., Leeuw, J. W., Schenck, P. A., and Salomons, W. 1985. 'Geochemistry of suspended particulate matter in two natural sedimentation basins of the river Rhine', *Wat. Res.*, **11**, 1333–1340.

Varvas, M. and Punning, J.-M. 1993. 'Use of the ^{210}Pb method in studies of the development and human impact history of some Estonian Lakes', *Holocene*, **3**, 34–44.

Walling, D. E. 1983. 'The sediment delivery problem', *J. Hydrol.*, **65**, 209–237.

Walling, D. E. 1989. 'The struggle against water erosion and a perspective on recent research' in Ivanov, K., and Pechinov, D. (Eds), *Water Erosion*. Unesco Tech. Documents in Hydrology. Unesco, Paris. pp. 39–60.

Walling, D. E., and He, Q. 1993. 'Towards improved interpretation of ^{137}Cs profiles in lake sediments' in McManus, J., and Duck, R. W. (Eds), *Geomorphology and Sedimentology of Lakes and Reservoirs*. Wiley, Chichester. pp. 31–53.

Walling, D. E., Bradley, S. B., and Wilkinson, C. J. 1986 'A caesium-137 budget approach to the investigation of sediment delivery from a small agricultural drainage basin in Devon, U.K' in *Drainage Basin Sediment Delivery (Proc. Albuquerque Symp.). IAHS Publ.*, **159**, 423–435.

Webb, J. S., Thornton, I., Howarth, R. J., Thompson, M., and Lowenstein, P. L. 1978. *The Wolfson Geochemical Atlas of England and Wales*. Oxford University Press,Oxford.

Wickliffe, D. S., and Steele, K. F. 1987. 'Time series heavy metal analyses of river water following rainstorm events' in Lindberg, S. E., and Hutchinson, T. C. (Eds) *Heavy Metals in the Environment*. CEP Consultants, Edinburgh. pp. 311–313.

Wik, M., Renburg, I., and Darley, J. 1986. 'Sedimentary records of carbonaceous particles from fossil fuel combustion', *Hydrobiologia*. **143**, 387–394.

Wilber, W. G., and Hunter, J. V. 1979. 'The impact of urbanization on the distribution of heavy metals in bottom sediments of the Saddle River', *Wat. Resour. Bull.*, **15**, 790–800.

Williams, T. M. 1991. 'A sedimentary record of the deposition of heavy metals and magnetic oxides in the Loch Dee basin, Galloway, Scotland, since c. AD 1500', *Holocene*, **1**, 142–150.

Williams, T. M. 1992. 'Diagenetic metal profiles in recent sediments of a Scottish freshwater loch', *Environ. Geol. Wat. Sci.*, **20**, 117–123.

Wise, S. M. 1980. 'Caesium-137 and lead-210: 'a review of techniques and some applications in geomorphology' in Cullingford, R. A., Davidson, D. A., and Lewin, J. (Eds), *Timescales in Geomorphology*. Wiley, London. pp. 109–127.

Wolfenden, P. J., and Lewin, J. 1978. 'Distribution of metal pollution in active stream sediments', *Catena*, **5**, 67–78.

Wolff, E. W., and Peel, D. A. 1985. 'The record of global pollution in polar snow and ice', *Nature*, **313**, 535–540.

Wright, P. 1992. 'The impact of the EC Urban Waste Water Treatment Directive', *J. Inst. Wat. Environ. Manag.*, 6, 675–681.

Xanthopolous, C., and Hahn, H. H. 1993. 'Anthropogenic wash-off from street surfaces' in *Proc. VIth Int. Conf. Urban Storm Drainage, Niagara Falls, Ontario, Canada*. pp. 417–422. J. Marsalek and H. C. Torno (Eds). Seapoint Publishing, British Columbia.

6

TRENDS IN NUTRIENTS

A. LOUISE HEATHWAITE

Department of Geography, University of Sheffield, Sheffield, UK

PENNY J. JOHNES

Department of Geography, University of Reading, Reading, UK

AND

NORMAN E. PETERS

US Geological Survey, 3039 Amwiler Rd., Atlanta, GA 30360-2824, USA

ABSTRACT

The roles of nitrogen (N) and phosphorus (P) as key nutrients determining the trophic status of water bodies are examined, and evidence reviewed for trends in concentrations of N and P species which occur in freshwaters, primarily in northern temperate environments. Data are reported for water bodies undergoing eutrophication and acidification, especially water bodies receiving increased nitrogen inputs through the atmospheric deposition of nitrogen oxides (NO_x). Nutrient loading on groundwaters and surface freshwaters is assessed with respect to causes and rates of change, relative rates of change for N and P, and implications of change for the future management of lakes, rivers and groundwaters. In particular, the nature and emphasis of studies for N species and P fractions in lakes versus rivers and groundwaters are contrasted. This review paper primarily focuses on results from North America and Europe, particularly for the UK where a wide range of data sets exists. Few nutrient loading data have been published on water bodies in less developed countries; however, some of the available data are presented to provide a global perspective. In general, N and P concentrations have increased dramatically (>20 times background concentrations) in many areas and causes vary considerably, ranging from urbanization to changes in agricultural practices.

INTRODUCTION

One important aspect of water quality assessment involves the trophic status of waters (see Rast and Thornton, Eutrophication research and control, this issue). Two key chemical elements affecting the trophic status are nitrogen (N) and phosphorus (P). It is important to establish the relative importance of N and P in the eutrophication process and thereby to evaluate the availability and quality of long-term records to establish trends and rates of change.

A long-term perspective is important because water quality trends cannot be easily separated from environmental policy trends. Environmental policy takes time to establish; therefore, it takes time for the implementation to impact the environment. For example, European legislation on water quality has undergone major developments since the 1980s, including new directives on urban wastewater treatment and nitrate (NO_3^-) in surface waters. The need for such developments stemmed from concern over river water quality and the pollution of coastal seas, such as the North Sea. In the UK, water quality was the focus of new policy initiatives in the late 1980s, partly as a result of the privatization of the water industry in 1989, but also through evidence that river water quality, which had been improving from the 1960s, was showing no sign of further improvement. Water authorities blamed this on a lack of investment in sewage treatment and pollution stemming from intensive farming practices. At the same time that nutrient pollution was improving through wastewater treatment, the use of inorganic fertilizers increased, resulting from a decrease in the price of fertilizer and an increase in availability. The combination resulted in no appreciable change and potential future degradation with respect to the nutrient content of the receiving waters. Marked degradation was still occurring in subsurface waters, which eventually discharged groundwaters with high nutrient concentrations to surface waters.

The objective of this paper is to review reported long-term trends in the nutrient content (concentrations N species and P fractions) of surface and groundwaters. Although a global perspective is intended, publications are sparse for all but the temperate zone. Consequently, most trend analyses are from Europe and North America. However, these trends emphasize the magnitude of change that probably has occurred and is occurring world-wide.

NUTRIENT FORMS AND CYCLING IN FRESHWATERS

Nutrient forms

Historical research has focused on inorganic forms of N and P, including nitrate (NO_3^-), total oxidizable N ($NO_3^- + NO_2^-$), ammonium (NH_4^+) and soluble reactive phosphate (PO_4^{3-}) in receiving waters. Thus studies of N loss usually focus on NO_3^- (Wild and Cameron, 1980; Powlson et al., 1986; McGill and Myers, 1987), despite evidence for N losses as organic and NH_4^+, particularly from grassland (Ryden et al., 1984; Heathwaite et al., 1990; Johnes and Burt, 1991; Heathwaite and Johnes, in press). Similarly, particulate P is commonly measured in studies of P transfer from land to stream, although losses may also occur in the soluble organic and inorganic forms. Emphasis on transfers of particular species disregards fluxes that may occur between varying nutrient fractions during transport and as a result of in-stream, in-river or in-lake transformations. Within the aquatic environment, the predominant forms of N and P are largely dependent on oxygen availability and the pH of the immediate environment (Brady, 1984; O'Neill, 1985).

Catchment characteristics are also important in controlling nutrient species transport and transformation. In a relatively acidic environment, P is likely to be held strongly in soils through metal complex formation, or adsorption onto clay particles. Nitrogen, however, will predominate as NO_3^- and thus be susceptible to leaching. In strongly alkaline environments, P will form insoluble calcium complexes. Nitrogen will exist as NH_4^+, which is bioavailable, but NH_4^+ is usually bound to soil particles through cation exchange, which reduces the risks of leaching loss. The release of N and P in soluble form from organic compounds is largely determined by the characteristics of the decomposer environment, which, in turn, is controlled by oxygen availability. For example, aerobic conditions created by ploughing enable ammonification and subsequent nitrification that results in NO_3^- release. For P held in aquatic sediments, the degree of anoxia influences mobilization through association with redox elements such as iron (Fe).

To evaluate trends in surface water quality, emphasis should be on the total nutrient fraction, not only the inorganic component. Although the inorganic fraction is often analysed routinely, the dissolved organic fraction is also important (Johnes and Burt, 1991; Heathwaite, 1993a). Dissolved organic N and P in freshwaters are generated from nutrients contained in the organic matter of plant and organic biomass residues. Dissolved inorganic N and P are released through mineralization (N and P) and nitrification (N). The soluble fraction is generally thought to be bioavailable, although nutrients associated with particulate material have *potential* bioavailability. For example, about one-third of P associated with suspended sediment is thought to be biologically available (Sonzongni et al., 1982; Ryding and Rast, 1989; Thomas et al., 1991). For P, the bioavailable fraction is generally adsorbed onto hydrated non-crystalline oxides, especially Fe. The fraction of particulate P bound to organic matter or incorporated in minerals, e.g. apatite, is usually assumed to be unavailable for biological uptake.

N cycling

Nitrogen exists in several different forms, the majority of which are soluble and highly mobile. It may

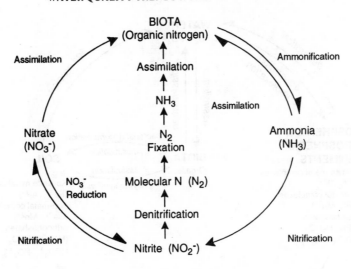

Figure 1. Nitrogen cycle

form an oxide (frequently NO_3^- and NO_2^-) or a reduced species, including the hydrides, ammonia (NH_3) and ammonium (NH_4^+), and a wide range of organic compounds derived mainly from the amine group (NH_2). Transfer processes between the various N species are determined by redox reactions primarily involving biota (see Figure 1).

In most rivers, lakes and groundwaters, NO_3^- dominates the total N pool for most of a year with the exception of the summer period (Stewart *et al.*, 1982; Johnes and Burt, 1991; Heathwaite, 1993b). Nitrate loading is usually highest in winter and spring resulting from soil–water recharge in autumn and winter, and N mineralization increases that occur when soil drying is following by re-wetting. In snow-covered forested watersheds, the highest surface water NO_3^- concentrations and loadings occur during moderate to high flow caused by snowmelt in winter and spring. Solutes stored in snowpacks, including NH_4^+ and NO_3^-, are differentially released during melting (Johannessen and Henriksen, 1978; Peters and Driscoll, 1987; Tranter *et al.*, 1992) and nitrification in unfrozen soils can contribute significant amounts of NO_3^- (Peters and Driscoll, 1987; Rascher *et al.*, 1987; Peters and Leavesley, 1995). Low summer NO_3^- loads correspond with low water throughflow and high biological uptake within a catchment or drainage network.

Denitrification and leaching cause most N loss from a catchment. Leaching losses are particularly important in arable soils where most soil N is present as NO_3^-. In forest and grassland soils, inorganic N concentrations are generally lower (Royal Society, 1983) and a larger soil NH_4^+ fraction may exist, but most N is in organic complexes associated with biological material. Ammonium losses tend to be associated with surface runoff and erosion rather than subsurface flow. Nitrogen losses resulting from denitrification may occur both within the terrestrial system, e.g. riparian zones, and within the aquatic system, e.g. anoxic waters or bottom sediments. Denitrification rates are controlled by several factors including temperature and the availability of organic matter. For instream losses, the ratio of stream to sediment NO_3^- (Hill, 1988) and the physical characteristics of a stream are important, such as the flow regime and water residence time (Terry and Nelson, 1975).

P cycling

Phosphorus transformations generally operate over a longer time-scale than those of N, occur in fewer steps, and are strongly linked to the cycling of sediments (see Figure 2). Mechanisms of P transfer are relatively simple, involving assimilation and dissimilation by biota, immobilization through complexation and chelation, mobilization through the hydrolysis of bivalent and trivalent cations, and adsorption to sediments. Phosphorus does not undergo redox reactions within its cycle. Geochemical fluxes of P are

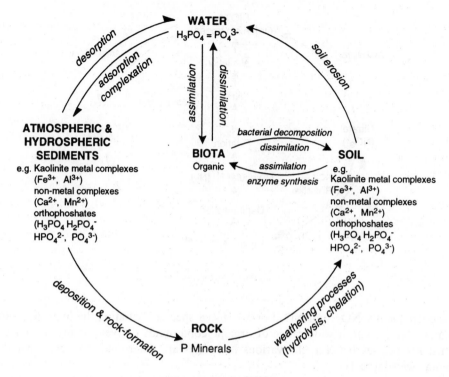

Figure 2. Phosphorus cycle

dependent on suspended sediment transport in water bodies and dust emissions to the atmosphere. The rate and magnitude of both fluxes are small. Newman (1995) suggests that phosphorus in the atmosphere from fine soil particles and the burning of plant material, coal and oil varies from 0·7 to 1·7 kg P ha^{-1} yr^{-1}, with input rates showing marked spatial variations. Newman (1995) also suggests that P released through weathering processes may produce yields from 0·05 to 1·0 kg P ha^{-1} yr^{-1}, with rates reaching a high of 5 kg P ha^{-1} yr^{-1} in some locations. In the UK, several meres of the Shropshire glacial plain are considered to be naturally eutrophic as a result of weathering of deposits of P-rich minerals, such as apatite, which occurs in localized deposits in their catchments (see Moss *et al.*, 1995). However, increases in the concentrations of P in the aquatic environment are largely attributable to increasing inputs of condensed inorganic phosphates from fertilizer sources, and dissolved P from sewage, livestock wastes and detergent sources (Withers, 1994). These additions to the P cycle exist as orthophosphate, commonly referred to as soluble reactive phosphorus (PO_4^{3-}) and P pentoxide compounds (P_2O_5), and supplement the soluble polyphosphate pool created by enzyme synthesis in plants and animals, and through release from P minerals.

CHANGES IN THE NUTRIENT SUPPLY TO FRESHWATERS

The *scale* of nutrient delivery, both spatial and temporal, from a variety of point and non-point sources and in many different forms affects the magnitude of change in water quality. Small-scale inputs, such as sewage and other wastewater discharges, are relatively easy to detect and quantify because their impact is great and highly localized. Large-scale inputs, such as atmospheric pollution, are both difficult to quantify and implicate as causes of change in water quality. Furthermore, the scale of impact tends to 'dilute' the magnitude of effect so that subtle changes in water quality (trends) are often recorded. Some of the different scales and sources of change in the nutrient concentrations are discussed in the following.

Local to regional scale

Delivery of N and P to aquatic environments can occur from point or non-point (diffuse) sources. Point sources have a fixed point of discharge, but the composition and rate of discharge can vary markedly from regular intervals and rates through time (e.g. consented discharges from industrial enterprises) or sporadically as land storage is exceeded (e.g. septic tanks). Non-point sources follow a wide range of routes to aquatic environments and depend on the hydrological balance of overland flow, throughflow and base flow.

Point sources. The importance of point courses in the aquatic environment depend on the relative proportion of urban land and industrial operations in a catchment, and the efficiency of pre-discharge treatment processes. Domestic (sewage) and industrial (effluent) sources of N and P are well documented (e.g. Alexander and Stevens, 1976; Smith, 1977; Wilkinson and Greene, 1982; Royal Society, 1983). Industrial sources include steel production (NH_4^+), petroleum production and refining (N and P are released during refining and in association with sediment discharges), pulp and paper, organic and inorganic chemicals, plastics and fertilizer production. All these manufacturing processes discharge high concentrations of N and P effluents directly to aquatic environments. In England and Wales, about 80% by weight of all domestic and industrial wastes are liquid effluents and are disposed of mainly to rivers. Sewage effluent may form a very high percentage of dry weather river flow. Its sources may be point or diffuse, depending on the method of disposal. In most developed areas of the world sewage is treated and discharged to water courses through effluent pipelines; the remainder is applied to land as sewage sludge or discharged to septic tanks (see National Water Council, 1977, Canter *et al.*, 1987). Septic tanks and landfill sites contain concentrated sources of both N and P, particularly if they are poorly maintained. The link between these sources and the drainage network is primarily through groundwater discharge to surface waters.

Non-point sources. Nitrogen, and to a lesser extent P, reach surface waters mainly from diffuse sources. Consequently, causality is difficult to prove, as is unravelling the interaction of a wide range of human-influenced variables such as land use, fertilizer application, soil type and the hydrological pathways linking the land to a stream (Casey and Clarke, 1979; Dermine and Lamberts, 1987). Increasing the intensity of crop production, or the extent of other land-use change in many areas, is reflected in long-term increases in the nutrient concentration of many rivers and lakes (Royal Society, 1983; Foster *et al.*, 1985; Roberts and Marsh, 1987; Meybeck *et al.*, 1989; Tsirkunov *et al.*, 1992; Casey *et al.*, 1993; Heathwaite, 1993b, Johnes and Burt, 1993; Johnes, in press). This has been clearly demonstrated by Tsirkunov *et al.* (1992) in their work on long-term trends in nitrate concentrations in Latvian rivers (see Figure 3).

Landuse influences the magnitude and timing of nutrient input to surface waters. For example, cultivation alters the soil structure and the rate of microbial breakdown of organic matter, releasing a potentially leachable N source (Skjemstad *et al.*, 1988; McEwen *et al.*, 1989). Arable cultivation has been frequently cited as a source of N enrichment (Addiscott *et al.*, 1986; Jenkinson *et al.*, 1987; Reinhorn and Avnimelech, 1974). The risk of NO_3^- leaching is particularly high after the harvest when plant uptake is zero, but N release through mineralization continues. Grassland also releases N when ploughed, with magnitudes around $4 \, t \, ha^{-1}$ for permanent grass (e.g. Whitmore *et al.*, in press) and 0.1–$0.2 \, t \, N \, ha^{-1}$ for temporary grass depending on the length of ley (Darby *et al.*, 1988). Where crop residues are incorporated, the pool of mobile N compounds is increased and loss through leaching may occur when crop demand is absent (see Schreiber and McDowell, 1985; Powlson *et al.*, 1987).

Less intensive land-use changes, where vegetation may be changed but the land surface is not physically altered, also release nutrients when disturbed. For example, nutrients are released following clear-felling of forests, regardless of the type of vegetation (Feller and Kimmins, 1984; Hornbeck *et al.*, 1986; Tiedemann *et al.*, 1988; Hornung *et al.*, 1990; Ahtiainen, 1992; Neal *et al.*, 1992; Reynolds *et al.*, 1995). However, forests in early to mid-stages of regrowth have a high nutrient demand and result in subsequent decreases in nutrient concentrations and loads. Similarly, fires can have a dramatic effect on water quality, such as observed in heathland catchments (Belillas and Roda, 1993) or forests (Wright, 1976; Tiedemann *et al.*, 1978; Richter *et al.*, 1984; Riggan *et al.*, 1994).

Even in the absence of a land-use change, nutrient transport and transformations can be affected by a combination of internal and external factors. For example, as a forest approaches a steady state, nutrient requirements decrease relative to the early growth period (Bormann and Likens, 1979). Consequently, increasing inputs of atmospheric N from increased emissions from fossil fuel combustion combined with

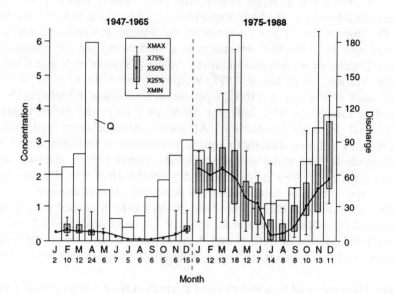

Figure 3. Mean long-term hydrographs ($m^3 s^{-1}$) and box and whisker plots of NO_3^- concentrations ($mg l^{-1}$) for the Venta River, Latvia (Tsirkunov *et al.*, 1992)

the decreasing nutrient requirement of old growth forests can result in N leaching (Aber, 1992), even though major land-use change had not occurred.

Most research on nutrient export from agricultural land has focused on inorganic fertilizer losses (e.g. Burns and Greenwood, 1982; Foster *et al.*, 1982). Although inorganic fertilizers are applied more widely than organic manure and slurries, the latter are nevertheless used extensively, especially in dairy and upland areas. For the UK, about 33% of the annual application of fertilizer N is lost through leaching (Foster *et al.*, 1982; Burns and Greenwood, 1982). Other losses occur through denitrification and volatilization (Frissel and Van Veen, 1982; Jenkinson, 1982; Stewart and Rosswall, 1982). Organic fertilizers contain appreciable quantities of nutrients, especially N, in soluble forms (MAFF, 1991). Livestock are particularly important where a lot of slurry is produced, e.g. dairy and pig farming. An intensive animal rearing unit may have a potential N load as high as $1600 kg N ha^{-1} yr^{-1}$ (Loehr, 1974). Livestock wastes and sewage sludge applied to agricultural land are discharged to water bodies both as point and non-point source discharge (Johnes, in press). In England and Wales between 1987 and 1989, slurry, silage effluent and dirty water (sewage derived) accounted for 28, 25 and 19% respectively, of the serious pollution incidents recorded (MAFF, 1991). Several routes for N transfer exist including subsurface NO_3^- leaching, surface runoff of NH_4^+-rich excreta from farmyards or grazed land, direct input of excreta at watering sites and leakage of silage effluent. Phosphorus influx, associated with eroded soil material (Olness *et al.*, 1975; Johnson *et al.*, 1976), may be supplemented by subsurface soluble P transport in macropore flow. Few workers have quantified subsurface pathways for P transport (Sharpley and Syers, 1976; Sharpley *et al.*, 1976; Duffy *et al.*, 1978), although research by Brookes *et al.* (in press), Dils and Heathwaite (in press), Heathwaite (in press), Heckrath *et al.* (in press); Sharply (in press) and Withers (1994) are addressing this imbalance.

Regional to global scale: atmospheric pollution

Nitrogen gas (N_2), together with NO_x fixed by lightning or derived from the emission of N from biological systems (NO_x and NH_3), constitute the atmospheric phase of N (Galbally and Roy, 1978; Johansson,

1984). Atmospheric P is primarily bound to dust particles and is found in phosphate minerals in soils and sediments. P release through weathering is slow and human activities have increased the rate of P cycling.

Documentation of N inputs from atmospheric sources is widely available owing to the role of nitrous oxides (NO_x) in atmospheric change (e.g. Rodgers, 1978; Hill et al., 1980; Galloway and Likens, 1981; Hahn and Crutzen, 1982; Soderlund et al., 1985; Malanchuk, and Nilsson, 1989; Mason 1990; Stoddard, 1994). Considerable interest has been generated in assessing the critical concentrations of N in the atmosphere and critical loads of N to ecosystems as N inputs can cause damage to ecosystems and forestry (Malanchuk and Nilsson 1989). The N species content of precipitation is spatially and temporally variable (e.g. Soderlund et al., 1985). Dry deposition of N species can be larger than wet deposition (Hanson and Lindberg, 1991; Geigert et al., 1994; Lovett, 1994). The Surface Waters Acidification Programme (SWAP) detected significant regional variations in both wet and dry deposition of N in the UK (Mason, 1990). Wet and dry atmospheric deposition are important sources of NO_3^- and NH_4^+ (Dollard et al., 1987). High concentrations of NO_3^- and NH_4^+ occur near large emission sources and in high rainfall areas. In the UK, some regions, such as the south-east, receive high N loading, of the order of 35–40 kg ha^{-1} yr^{-1} (Goulding, 1990), which is similar to the maximum deposition in other areas of Europe and southern Scandinavia (Malanchuk and Nilsson, 1989, Lovblad et al., 1992) and the north-eastern USA and eastern Canada (Venkatram et al., 1991). This is equivalent to 20% of the average annual application of N fertilizers to arable crops. Increases in the concentration of N in surface waters might be anticipated where such increases in N inputs are balanced by increased N losses. On a larger scale, Soderlund et al. (1985) present data from the European Air Chemistry Network dating from 1955 and show a build up of NO_x from anthropogenic sources in the atmosphere over this time period. Emissions of NO_x are expected to increase in the future, primarily from vehicular emissions.

Atmospheric P sources relative to N are not well documented, but the characteristics of the wet and dry deposition of P have been evaluated (Brezonik, 1976; Graham and Duce, 1979; Peters and Reese, 1995). Research has primarily focused on the atmospheric deposition of P because it was considered a potentially important source of nutrient to lakes (Murphy and Doskey, 1977; Delumyea and Petel, 1978) and oceans (Graham and Duce, 1982; Falkowska and Korzeniewski, 1988). Reactive P concentrations, PO_4^{3-}, are routinely measured in wet deposition monitoring networks and are typically below the detection limit. Reckhow and Simpson (1980) summarize wet atmospheric P deposition, although the atmosphere currently constitutes a relatively insignificant part of the P cycle. Studies on the nature of atmospheric P inputs suggest that the dry deposition may be the most important atmospheric source (Hendry et al., 1981; Lewis et al., 1985; Cole et al., 1990; Peters and Reese, 1995), although measuring dry P deposition may be extremely problematic (Peters and Reese, 1995).

TRENDS IN FRESHWATER N AND P CONCENTRATIONS

On a global scale, less than 10% of rivers may be classified as pristine (WHO/UNEP, 1987). For N, pristine rivers should have, on average, 0·015 mg NH_4-N l^{-1}, 0·001 mg NO_2-N l^{-1} and 0·1 mg NO_3-N l^{-1} (Meybeck, 1982). Currently, more than 10% of European rivers have NO_3^- concentrations ranging from 9 to 25 mg NO_3-N l^{-1}, which is about three orders of magnitude higher than background concentrations.

To evaluate the extent to which water quality deterioration has occurred, trends in water quality must be examined. However, the variable quality of environmental data is a major problem. Meybeck et al. (1989), in a survey of trends in water quality at regional (European) and global scales, highlight the problems of comparing data from several sources because of variability in sampling, handling and processing techniques, and analytical quality. These complications suggest that such data should only be used as an indicator of change rather than as an absolute record of water quality.

In the UK, NO_3^- data have been available since the adoption of the WHO recommended limit (10 mg N l^{-1}) on NO_3^- in drinking water in the early 1970s (e.g. Tomlinson, 1970). Although several studies have examined nutrient concentrations in rivers (e.g. Rodda and Jones, 1983; Betton et al., 1991), few provide analyses of long-term trends (José, 1989; Betton et al., 1991; Heathwaite and Burt, 1991; Johnes and Burt, 1991; 1993). Furthermore, most published material focuses on NO_3^-, partly because of the ease with which NO_3^- is transferred from terrestrial to aquatic environments. Also, NO_3^- is the focus for health impacts that have caused a gradual increase in the mandatory control of NO_3^- in drinking water and drinking water sources (Burt *et al.*, 1993). There are very few data on total N and P, or their dissolved and particulate fractions (both organic and inorganic) in the aquatic environment. This means that where increases in the concentration of, e.g. NO_3^-, in many UK rivers are claimed (Edwards and Thornes, 1973; Casey and Clarke, 1979; Burt *et al.*, 1988) it is not always possible to determine whether such changes are the result of an *absolute* increase in the concentration of NO_3^- in the river or a *relative* change in the proportions of the various N forms present. Furthermore, because most nutrients exhibit seasonal variations, the time of sampling must be taken into account in evaluating trends in nutrient concentrations. For N and P, two separate studies have found that dissolved inorganic N and P showed greater seasonal variation than organic forms (Lewis and Saunders, 1989; Johnes and Burt, 1991).

N and P in rivers

Global water quality. Global freshwater quality has been reviewed by Meybeck *et al.* (1989). General trends in the content of several national surface waters are presented in Table I. Although the global median in surface waters (excluding Europe) is $0.25 \, mg \, N \, l^{-1}$, the European median concentration is $4.5 \, mg \, N \, l^{-1}$. The generally high NO_3^- concentrations in Europe are attributed to high anthropogenic loading of N to surface waters in western Europe.

North America. Publications of historical water quality data are not abundant for North America. Furthermore, the most comprehensive water quality data collection activities began in the 1960s and 1970s.

Table I. General trends in surface water nitrate concentrations (Meybeck *et al.*, 1989)

Country	River	1970	1975	1980	1985
Belgium	Meuse, Heer/Agimont	1·800*	7·600	2·180	3·120
	Meuse, Lanaye	3·900	9·400	2·520	2·790
	Escaut, Doel	3·000*	7·350	4·170	3·910
Canada	St Lawrence	0·193	0·230	0·160	0·210†
	Mackenzie	0·084	0·111	0·110	0·090‡
	Fraser	0·049	0·300	0·060	0·120‡
	Nelson	0·040	0·400	0·060	0·090‡
France	Loire, Nantes	1·581*	1·445	1·987	
	Garonne, Bordeaux	1·152*	0·925	1·829	
Federal Republic of Germany	Rhine, Bimmen Lake	1·820	3·020	3·590	4·200
Italy	Po	0·945	1·350	1·630	3·280
	Tevere		1·500	1·370	
Japan	Ishikari	0·380		0·530§	
	Yodo			0·780§	
The Netherlands	Meuse, Keizersv	3·070	3·690	3·770	4·280
	Meuse, Eijsden	2·450	2·510	2·780	2·920
	Ijssel, Kampden	2·780	3·450	4·270	4·330*
Sweden	Dalaiven	0·120	0·107	0·136	0·106
USA	Delaware, Trenton		0·880†	1·080†	
	Mississippi, St Franc	0·380	1·040†	1·300†	1·230*†

* 1971 data.
† 1984 data.
‡ 1983 data.
§ 1979 data.
Adapted from OECD (1985; 1987) by Meybeck *et al.* (1989).

Shorter term (six year) nutrient trends have been assessed by Smith *et al.* (1982), Smith and Alexander (1985) and Smith *et al.* (1987), which also were reviewed by Meybeck *et al.* (1989), for 298 USA rivers based on the US Geological Survey's National Stream Quality Accounting Network (NASQAN). Increasing trends in NO_3^- outnumbered decreasing trends in 116 to 27 at a total of 383 stations during 1974–1981 (Smith *et al.*, 1987). Most increases were in the east, were strongly associated with agricultural activity, and were attributed, in part, to a 38% increase in fertilizer usage during 1974–1981. At 303 NASQAN stations, five- to eight-year trends were evaluated for flow-adjusted P concentrations and only 45 showed significant uptrends and 40 showed significant downtrends (Smith *et al.*, 1982). Increases in P concentrations were correlated with increases in suspended solids and the relation is attributed to adsorption of P from fertilizer application and with a delayed transport of the sediment compared with NO_3^- (Smith and Alexander, 1985). Although this analysis was only for six years, longer term changes in N and P in freshwaters would be anticipated given relatively large long-term changes in fertilizer use. In the USA, N from fertilizer use increased 20 times from 1945 to 1993 and P use tripled (Pucket, 1995).

Long-term NO_3^- trends have been evaluated for some specific rivers including the Passiac River (including NH_4^+ concentrations during a 50-year period, Cirello *et al.*, 1979), and the Illinois, Mississippi and Ohio Rivers since 1900 (Ackerman *et al.*, 1970), and for several long-term monitoring sites on lakes and streams in the northeastern USA (Peters *et al.*, 1982; Driscoll and Van Dreason, 1993; Stoddard, 1994). The results of these studies are similar to those for the short-term studies described earlier, particularly in the eastern states by agricultural land uses and in the Pacific coastal areas. The only rivers to show a decrease in NO_3^- concentrations were those draining grassland catchments, highlighting the importance of land use on water quality. Arable land and ploughed grassland in particular appear to generate high concentrations of NO_3^- in surface waters.

For streams of the Adirondack and Catskill Mountains, New York (Driscoll and Van Dreason, 1993; Murdoch and Stoddard, 1993), NO_3^- concentrations vary seasonally, with high concentrations primarily occurring during winter and spring snowmelt. A significant increase in NO_3^- concentrations of $1 \cdot 5$–$2 \, \mu equiv \, l^{-1} \, yr^{-1}$ from 1983 to 1989 was attributed to increasingly higher concentrations during the snowmelt period.

Historical data, with start dates ranging from the 1910s to 1940s, for several streams in the Catskill Mountains, New York indicate that NO_3^- concentrations have increased significantly, with the rate of increase being highest since 1970 (Stoddard, 1991; Mueller *et al.*, 1995). Although the changes in NO_3^- concentrations are significant, concentrations are typically low, less than $1 \cdot 0 \, mg \, N \, l^{-1}$. Many of the basins are forested and the recent increase is attributed to changes in the capacity of the watersheds to retain N and recent increases in atmospheric N deposition (Stoddard, 1991; Aber, 1992).

In 1991, the US Geological Survey began implementing the National Water Quality Assessment Program (NAWQA) to describe the status of and trends in the quality of US groundwater and surface water resources and to link the assessment of status and trends with an understanding of the natural and human factors that affect the quality of water (Gilliom *et al.*, 1995). NAWQA is designed to evaluate the water quality of 60 major hydrological basins (Figure 4), covering 50% of the area of the USA and 60–70% of the national water use. Twenty NAWQA basins will be monitored intensively for three years, preceded by a one-year retrospective analysis and succeeded by one year of report preparation plus low-level monitoring and an additional four years of low-level monitoring to complete a nine-year cycle. Continuous intensive data collection at 20 basins concurrently is accomplished by staggering start dates for the basins.

Some important results have already emerged from the NAWQA programme. Results for a retrospective analysis of the trends in P loads of the Chattahoochee River above and below Atlanta, Georgia, show: (1) an increase from 25 to $200 \, US \, t \, yr^{-1}$ during 1980–1993 above Atlanta; and (2) an increase from 900 in 1981 to $1800 \, US \, t \, yr^{-1}$ in 1984 followed by a decrease to $575 \, US \, t \, yr^{-1}$ in 1993. The higher discharges below Atlanta are attributed primarily to wastewater discharge. Restricted use of P detergents and upgrades to wastewater treatment facilities accounted for 50% of the observed decrease in P load from 1988 to 1993 (Wangsness *et al.*, 1994; Mueller *et al.*, 1995). Although N concentrations and loads typically have increased, improved wastewater treatment has caused a shift in speciation. In the Trinity River, Texas,

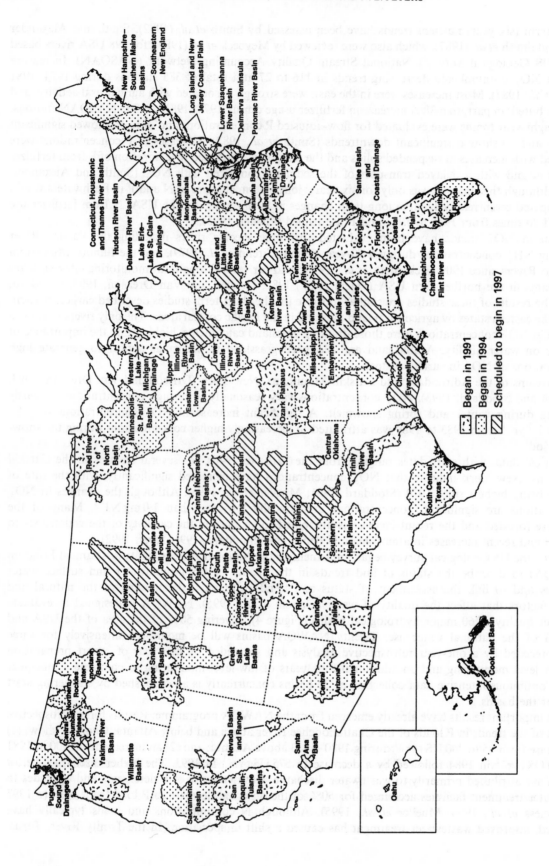

Figure 4. Study units of the US Geological Survey's National Water Quality Assessment Program (Gilliom *et al.*, 1995)

NH_4^+ concentrations have decreased and NO_3^- concentrations increased from 1974 to 1991 as a result of adding an oxidation step to convert NH_4^+ to NO_3^- during wastewater processing. Also, trends in N species of the White River, Indiana show the same pattern following the use of advanced wastewater treatment (Crawford and Wangsness, 1993). Advanced wastewater treatment has increased the dissolved oxygen in rivers and reduced the number of fish kills.

The importance of agriculture is noted in the retrospective analysis for the South Platte NAWQA, which has its headwaters primarily in the Rocky Mountains, Colorado (Dennehy et al., 1994). NO_3^- concentrations were highest in agricultural area streams and other nutrients were highest in either agricultural area or mixed urban and agricultural area streams; median NO_3^- concentrations in agricultural area streams were $3.2\,mg\,Nl^{-1}$ (Dennehy et al., 1994). Implementation of best management practices at a site in the Snake River NAWQA (Idaho and western Wyoming) caused the total P concentration to decrease at two stations and is attributed to a reduction in sediment reaching the stream (Clark, 1994). Also, for the Upper Snake River, NO_3^- concentrations increased downstream as the percentage of agricultural land increased, but temporal trends for stations where data were available (1980–1989) show no overall trend in nutrient concentrations (Clark, 1994). Also, in the Yakima River Basin, nutrient concentrations increase dramatically from pristine uplands to the urban and agricultural lowlands and more than 50% of the surface water sites had increasing trends for N species during 1974–1981, with a median increase of $0.06\,mg\,NO_3\text{-}Nl^{-1}\,yr^{-1}$ for the mainstem and canals, and $0.09\,mg\,NO_3\text{-}Nl^{-1}\,yr^{-1}$ for tributaries (Rinella et al., 1992). The rate of increase in NH_4^+ was an order of magnitude less than for NO_3^-, but organic N, for which data were available at one downstream site, increased by 10% per year.

In contrast with the increases generally noted in streamwater NO_3^- for the USA, no trends in NO_3^- concentrations of surface waters from south-eastern Canada were evident from 1983 to 1992, whereas concurrent increases were generally noted for NO_3^- in precipitation (Clair and Ehrman, 1995). Most of the Canadian sites are forested and are N limiting, particularly in the boreal forest.

Europe. Although a general increase in the concentration of nutrients, particularly NO_3^-, in European rivers is evident, marked seasonal and regional variations are recorded. In general, wet, mountainous regions (e.g. Scandinavia, Ryding and Forsberg, 1979; the Alps, Meybeck et al., 1989) have lower nutrient concentrations and rates of nutrient increases in surface waters than low-lying regions (e.g. The Netherlands, van der Weijden and Middelburg, 1989). In the forested Strengbach catchment of the Vosges Mountains, France (Probst et al., 1995), no trend in streamwater NO_3^- concentrations was discernible from 1985 to 1992, but a marked seasonal pattern was observed which was similar to that of the north-eastern USA. This contrast points to the significance of agricultural inputs in intensively farmed lowlands. Also, comparisons of data among different sized surface waters is problematic in that large rivers tend to mask the input from tributaries and groundwater sources through dilution, but can be sensitive to point source discharges.

Trends in the concentration of NO_3^- in European rivers have been analysed for the River Elbe in Czechoslovakia (Paces, 1982), for the River Maas (Roberts and Marsh, 1987) and the River Rhine (van der Weijden and Middelburg, 1989) in The Netherlands, for rivers in Finland (Kauppi, 1984) and surface waters in France (Henin, 1986). All data show a contemporaneous increase in NO_3^- loading over the past 20 years. For the River Elbe, Paces (1982) identified a 30-fold increase in ammonia and a 20-fold increase in NO_3^- concentrations for the period 1877–1976. Both NO_3^- and NH_4^+ concentrations and loads, in general, have increased three-fold in the River Rhine at Lobith, The Netherlands, since the 1930s; however, NH_4^+ concentrations and loads decreased from 1975 to 1985 while NO_3^- increased (van der Weijden and Middelburg, 1989). The subtle differences in behaviour of these N species is attributed to a combination of increased source of NO_3^- from manure and fertilizer and decreased source of NH_4^+ due to increased effectiveness of wastewater treatment (van der Weijden and Middelburg, 1989).

The P and N concentrations of inflows and a drinking-water mesotropic reservoir were compared with those of an oligotrophic reservoir for a 16-year period in southern part of eastern Germany (Horn et al., 1994). Gradual increases were observed in both N and P species in the more densely populated catchment of the mesotropic reservoir and were associated with the intensification of agriculture and high domestic sewage releases from settlements having no wastewater treatment compared with the relatively pristine drainage of the oligotrophic reservoir. The higher population density (79 versus 1·45 people km^{-2}) and higher percentage of agricultural lands (61 versus 16%) resulted in concentrations 20–30 times more for PO_4^{3-}, 10 times more for total P and two times more for NO_3^- in the mesotrophic reservoir than the oligotrophic reservoir. Concentrations of P species in the inflow did not change appreciably over the 16 years for the oligotrophic reservoir, whereas changes of up to 50% were noted for the mesotropic reservoir. However, the change in NO_3^- concentrations were comparable, averaging about $10 \, mg \, l^{-1}$. Changes in either the N or P species concentrations were higher in the mesotrophic reservoir than in the oligotrophic reservoir, but were less than in the inflows.

United Kingdom. The longest series of water quality data in the UK is available for the River Thames at Walton and the River Lee at Chingford. Data for both and PO_4^{3-} concentrations commencing in 1928 were presented in the Digest of Environmental Statistics (Department of the Environment, 1978). Long-term NO_3^- trends for the Thames have been reproduced by several workers (e.g. Onstad and Blake, 1980; Roberts and Marsh, 1987; Royal Society, 1983). Temporal variations in N species concentrations are not as well known for other UK rivers (Tomlinson, 1970; Nicholson, 1979; Marsh, 1980; Rodda and Jones, 1983; Royal Society 1983; Roberts and Marsh, 1987; José, 1989). Much of the research focus has been on the examination of trends in the concentration of NO_3^- in rivers (Burfield, 1977; Department of the Environment, 1978; Greene, 1978; Nicholson, 1979; Marsh, 1980; Slack and Williams, 1985; Robert and Marsh, 1987). Long-term NO_3^- trends for the River Stour at Langham, the River Tees at Broken Scar and the River Great Ouse at Bedford are illustrated together with the Thames record in Figure 5 (cf Roberts and Marsh, 1987). The pattern of NO_3^- concentration in UK rivers suggests that in southern and eastern England, particularly in East Anglia, NO_3^- concentrations have increased rapidly to significantly high concentrations. In a review of 25 British rivers (STACWQ, 1983), NO_3^- concentrations increased more from 1981 to 1984 than for any previous period. José (1989) reported a long-term NO_3^- concentration increase in the River Trent equivalent to an annual NO_3^- increase of $455 \, t \, N \, yr^{-1}$ ($0·43 \, kg \, ha^{-1}$). Meanwhile, an NH_4^+ concentration decrease was recorded that appears to be directly related the reduction in the number of effluents discharging into the Trent.

The pattern of increasing NO_3^- concentration through time is not necessarily consistent across the UK (Burt *et al.*, 1988; Casey, 1976; Casey and Clarke, 1979; Johnes and Burt, 1993; Webb and Walling, 1985). Various reasons are given to account for the recorded trends including climatic variation and agricultural intensification. Some rivers do not show increases in NO_3^- concentration; for example, 12 Welsh rivers reviewed by Brooker and Johnson (1984) during 1969–1979. Catchment land-use change is an important cause of water-quality change (Johnes, in press). Welsh river catchments have relatively little arable land and a low human population compared with most English rivers studied.

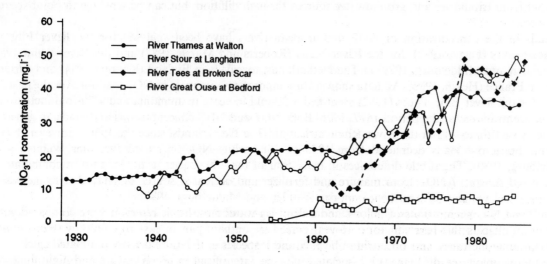

Figure 5. Long-term trends in mean annual NO_3^- concentrations for four British rivers, 1928–1985 (Roberts and Marsh, 1987)

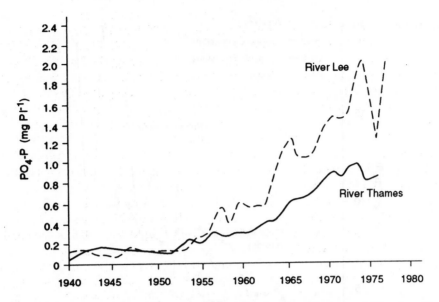

Figure 6. Long-term trends in mean annual PO_4^{3-} concentrations for the River Thames and River Lee, England, 1940–1980 (DOE, 1978)

Available data for long-term PO_4^{3-} concentrations in UK rivers are limited and, where available, are often presented only as general statements of trends rather than as absolute data. A generally upward P-trend showing periods of rapid increase in the P concentration of rivers is shown in Figure 6. For example, the PO_4^{3-} concentration in the River Frome, south-west England increased 21% from 1965 to 1972 (Casey and Clarke, 1979). Also for south-west England, Heathwaite and Johnes (in press) found that trends in PO_4^{3-} concentrations in the Slapton catchment drainage network paralleled the NO_3^- increases. Although much of the PO_4^{3-} concentration trend in many rivers may be explained by rapid post-war urbanization and overloading of sewage works in urban catchments, these examples were largely for agricultural catchments where P-loading increases have contributed to P concentrations increases in surface waters (Johnes and Heathwaite, in press).

N and P in groundwater

Global groundwater quality. Global trends in groundwater NO_3^- concentrations have been reviewed by Meybeck *et al.* (1989). In general, less data are available for temporal changes in the nutrient content of groundwater than for surface waters. However, analyses are available for some less developed countries (Lewis *et al.*, 1980), New Zealand (Burden, 1982), the USA (Gormly and Spalding, 1979; Robertson, 1979; Canter *et al.*, 1987; Fairchild, 1987), France (Royal Society, 1983; Roberts and Marsh, 1987), Hungary (Alfoldi and Homonnay, 1986) and Germany (Darimont *et al.*, 1985).

The general trend for groundwaters mimics that of surface waters, with a widespread NO_3^- concentration increase in boreholes and springs throughout the industrialized world, and particularly in areas of arable cultivated land. For example, in western Germany where the drinking water supplies come primarily from boreholes, 5–10% of boreholes exceed the EC limit, and average NO_3^- concentrations are increasing by $1–2\,mg\,N\,l^{-1}\,yr^{-1}$ in intensively cultivated areas (Saull, 1990). Agriculture is not the only major source of nutrient enrichment. Aquifer pollution from human sewage disposal is clearly a significant problem. For example, Canter *et al.* (1987) note the importance of septic-tank drainage as a source of aquifer contamination in the USA. Similar water quality deterioration is reported by Lewis *et al.* (1980) in less developed countries, particularly where domestic water supplies are drawn from shallow aquifers that also receive inputs from septic tanks and pit latrines. Typical concentrations of NO_3^- may reach $30–100\,mg\,N\,l^{-1}$ in central India, may exceed $113\,mg\,N\,l^{-1}$ in Botswana, and may exceed the WHO limit for NO_3^- in drinking water for more than 50% of wells in Nigeria.

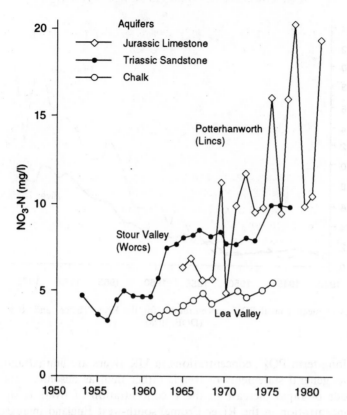

Figure 7. Long-term trends in NO_3^- concentrations for public supply boreholes in three British aquifers (after Wilkinson and Greene, 1982; Smith-Carrington *et al.*, 1983)

North America. A review of the occurrence of NO_3^- in groundwater in the conterminous USA indicates that concentrations decrease with depth (Spaulding and Exner, 1993; Hamilton and Helsel, 1995; Mueller *et al.*, 1995). The highest concentrations occurred in wells within 30 m of the land surface (Mueller *et al.*, 1995). Anomalously high NO_3^- concentrations also were associated with substandard well construction and improper siting of wells (Spalding and Exner, 1993). However, high NO_3^- concentrations displayed some regional patterns and, although associated primarily with agriculture, were probably caused by source availability, physical controls on transport and biogeochemical controls on transformation of N (Spalding and Exner, 1993). In some states, more than 20% of the observation wells had NO_3^- concentrations exceeding the WHO limit. According to the analysis of groundwater from five NAWQA studies by Hamilton and Helsel (1995), NO_3^- concentrations in 12–46% of the wells sampled in agricultural areas exceeded the $10\,\text{mg}\,l^{-1}$ maximum contaminant level set by the US Environmental Protection Agency. The highest NO_3^- concentrations were observed in heavily irrigated areas or areas underlain by well-drained surficial materials (Hamilton and Helsel, 1995).

United Kingdom. Groundwater quality trends in the UK concur with global patterns of increasing NO_3^- concentrations in groundwaters since the 1940s (e.g. Foster 1986; Foster and Young, 1981; Oakes *et al.*, 1981; Royal Society, 1983; Young *et al.*, 1983; Young, 1986). In the UK, 30% of public water supplies come from groundwater sources, with the Chalk and Triassic Sandstone aquifers providing 50 and 30% of the supply, respectively (Wyndle-Taylor, 1984). Figure 7 illustrates the trends for key aquifers in the Chalk, Jurassic Limestone and Triassic Sandstone. For the Jurassic Limestone in Lincolnshire, NO_3^- concentrations in groundwater equal and occasionally exceed the WHO limit. These high NO_3^- concentrations have been attributed to rapid movement of water through the highly fissured limestone (Wilkinson and Greene, 1982; Smith-Carrington *et al.*, 1983). This contrasts with the slower movement of water in the Chalk (Greene, 1980; Oakes *et al.*, 1981).

Contamination of groundwater supply with NO_3^- means that in 1983–1984, 125 groundwater sources supplying 1·8 million people exceeded the WHO limit (Lean, 1990). In contrast, 90 groundwater supplies exceeded the WHO limit in 1980 and 60 supplies in 1960 (DOE, 1986).

The NO_3^- load is thought to be primarily derived from fertilizer application (Foster et al., 1982), strongly associated with land use (Young et al., 1976; Young and Hall, 1977; Smith-Carrington et al., 1983; Cameron and Wild, 1984; Parker et al., 1990).

Legislation has attempted to reduce the N loading to groundwaters. In areas of Jurassic Limestone, such as the Cotswolds where NO_3^- concentrations are high, Nitrate Sensitive Areas (DOE, 1988a) have been designated. This scheme, introduced under the Water Act 1989, is an attempt to limit surface applications of NO_3^- that may eventually permeate through to the regional aquifer (Johnes and Burt, 1993).

Long-term studies on a Chalk borehole at Bridget's Farm, Winchester have been reviewed by Greene (1980), Hall et al. (1976) and Young and Gray (1978). Results indicate a time lag of approximately 40 years before peak NO_3^-, observed in boreholes, reach the aquifer. Similar trends are observed in a Triassic sandstone aquifer (Oakes et al., 1981; DOE, 1988b).

N and P in lakes

Unlike research into the quality of surface waters, limnological studies have traditionally considered both P and N species (Vollenweider, 1968; 1975; Dillon and Rigler, 1974; Schindler et al., 1974; Vollenweider and Kerekes, 1981). Their results have demonstrated the importance of the rates of interchange and flux between different forms of N and P in freshwaters (e.g. Ryding and Forsberg, 1979; Stevens and Stewart, 1982).

Once nutrients enter a lake, they may be recycled many times among sediments, aquatic plants and water column. On a longer time-scale, lake sediments generally form the ultimate nutrient sink, although this depends on the lake flushing rate and the nature of catchment inputs. For N, biological fixation can form a potentially significant diffuse source, particularly for eutrophic lakes. For P, the lake sediments are commonly the main nutrient store with re-release largely determined by redox conditions. This is particularly true for shallow lakes where the trophic status is strongly influenced by the internal release of P from lake sediments under anoxic conditions, or through the re-suspension of sediments during turbulent periods (see Moss, 1991).

In temperate lake ecosystems, nutrient concentrations commonly show winter maxima independent of lake trophic status. This winter maxima arise because surface and subsurface catchment inputs are high and microbial activity is low. Once nutrients have entered the lake, the way in which they are utilized through the spring to autumn growing season is dependent on the trophic status of the water body. In eutrophic, and especially shallow lakes, the nutrient concentration usually drops rapidly from winter to spring as the biotic demand exceeds supply once temperatures rise. In oligotrophic lakes, the relative decrease in nutrient concentration is less than that in eutrophic lake systems because biotic demand in the summer period is lower.

There have been remarkably few studies of long-term changes in the nutrient chemistry of lakes around the world, despite a wealth of research on the ecological manifestations of lake eutrophication (see Rast and Thornton, this issue). Research on lakes that have undergone nutrient enrichment has largely concentrated on shorter term seasonal and diurnal variations in nutrient chemistry, with most studies focusing on P rather than N. This is due to the assumption that biological productivity in most lakes is limited by the availability of P rather than N (e.g. Allcock and Buchanan, 1994). Although this is so in most lakes, research suggests that in the intensively used catchments of England and Wales, many of the more productive lakes receive proportionally high loads of P from urban and agricultural point sources, resulting in N limitation (Johnes et al., 1994a; 1994b).

Lough Neagh. The evidence presented in the few published studies on long-term nutrient trends in lakes suggests that the loading of both N and P has increased markedly in both European and North American lakes over the past 20 years. The mean annual concentration at NO_3^- in six rivers draining into Lough Neagh, Northern Ireland, increased from 1·41 mg N l^{-1} in 1969–1970 to 2·41 mg N l^{-1} in 1978–1979, with an increase in the maximum concentration of NO_3^- observed in each water year from 2·9 mg N l^{-1} to

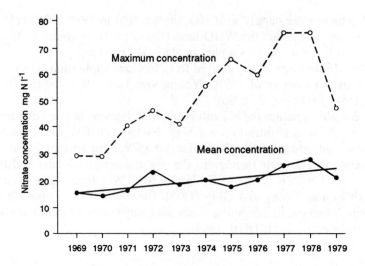

Figure 8. Mean and maximum NO_3^- concentrations for six rivers discharging into Lough Neagh, Northern Ireland, 1969–1979 (Smith *et al.*, 1982)

$7·5 \, \text{mg} \, \text{N} \, \text{l}^{-1}$ during the same time period (Smith *et al.*, 1982) (Figure 8). Smith *et al.* (1982) attribute these trends to changes in the intensity of grassland management and fertilizer application rates in the catchment. Foy *et al.* (1982) present comparable data for the P load delivered to Lough Neagh by the same six rivers during 1971–1979 (Table II). The PO_4^{3-} load remained relatively constant over this period at $210 \pm 20 \, \text{t} \, \text{P} \, \text{yr}^{-1}$ delivered to the lough. The relative stability of the P trend may reflect a parallel stability in the number of people living in the catchment, and the rates of P fertilizer applications to crops and grass in the catchment during 1971–1979.

Loch Leven. Harriman and Pugh (1994) reviewed the water chemistry of Scottish lochs stating that the long-term nutrient trends of most Scottish lochs are not available, despite eutrophication being identified as a major concern. The exception to this trend is Loch Leven in east-central Scotland, where the ecology and water chemistry have been studied in detail for the past 25 years (e.g. Holden and Caines, 1974; Bailey-Watts and Kirika, 1987; Bailey-Watts *et al.*, 1990). Variations in the annual maximum NO_3^- concentration observed in Loch Leven are described for 1968–1985 by Bailey-Watts *et al.* (1990). The annual maximum NO_3^- concentration has increased from $1·3 \, \text{mg} \, \text{N} \, \text{l}^{-1}$ in 1968–1969 to $3·5 \, \text{mg} \, \text{N} \, \text{l}^{-1}$ in 1985–1986 (Figure 9).

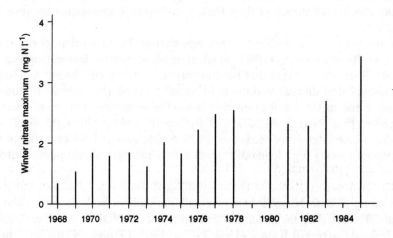

Figure 9. Variations in the annual maximum NO_3^- concentration of Loch Leven, Scotland, 1968–1985 (Bailey-Watts *et al.*, 1990)

Table II. Chemical losses (kg ha^{-1}) during 1971–1979 for six major rivers entering Lough Neagh, Northern Ireland (Foy *et al.*, 1982)

River	1971	1972	1973	1974	1975	1976	1977	1978	1979
Nitrate-N									
Six Mile Water	10·9	19·9	13·5	15·3	9·8	18·0	20·0	27·4	16·9
Main	12·0	19·0	14·5	15·8	10·4	16·5	24·1	32·5	23·5
Upper Bann	4·8	14·5	10·2	12·6	6·5	16·8	15·7	20·8	13·6
Blackwater	4·9	9·7	6·9	7·6	4·0	8·3	11·7	13·0	9·4
Moyola	7·1	13·1	10·0	13·9	7·9	13·0	16·7	20·3	15·0
Ballinderry	7·0	15·0	11·5	13·1	7·8	15·4	22·3	21·1	18·3
Kjeldahl-N									
Six Mile Water	6·8	6·6	5·2	8·1	4·3	6·1	5·8	5·8	6·7
Main	5·5	8·7	7·7	3·7	4·6	6·7	7·6	8·2	8·7
Upper Bann	4·5	7·6	10·7	3·6	4·1	9·2	6·4	6·4	9·4
Blackwater	3·5	7·0	6·0	3·7	3·1	6·6	6·3	6·3	6·7
Moyola	4·5	9·3	4·5	3·4	4·2	6·9	7·2	7·3	7·6
Ballinderry	3·8	7·7	6·8	3·7	8·8	8·4	9·5	7·7	9·1
SRP									
Six Mile Water	0·716	0·763	0·823	0·847	0·546	0·624	0·541	0·650	0·620
Main	0·704	0·646	0·729	0·612	0·422	0·522	0·624	0·655	0·707
Upper Bann	0·534	0·487	0·654	0·687	0·381	0·518	0·478	0·691	0·786
Blackwater	0·422	0·403	0·473	0·447	0·230	0·426	0·471	0·572	0·555
Moyola	0·368	0·492	0·446	0·422	0·264	0·405	0·412	0·503	0·588
Ballinderry	0·330	0·463	0·385	0·486	0·301	0·505	0·526	0·553	0·598
SOP									
Six Mile Water					0·100	0·183	0·154	0·120	0·138
Main					0·141	0·198	0·209	0·135	0·225
Upper Bann					0·151	0·183	0·171	0·165	0·215
Blackwater					0·137	0·140	0·097	0·097	0·145
Moyola					0·185	0·198	0·131	0·114	0·166
Ballinderry					0·135	0·170	0·140	0·097	0·164
PP									
Six Mile Water					0·151	0·229	0·236	0·439	0·268
Main					0·079	0·269	0·308	0·404	0·342
Upper Bann					0·129	0·271	0·324	0·400	0·416
Blackwater					0·088	0·338	0·314	0·341	0·362
Moyola					0·081	0·196	0·241	0·332	0·287
Ballinderry					0·187	0·328	0·382	0·497	0·482

SRP, Soluble reactive phosphorus; SOP, soluble organic phosphorus; and PP, particulate phosphorus.

Bailey-Watts *et al.* (1990) note that no ecological effect of this trend has been noted in the loch. However, although P concentrations have not been regularly monitored, periodic phases of monitoring indicate that concentrations of the total soluble P fraction in 1982–1983 are similar to those recorded in 1968–1969. They suggest that the P loading on Loch Leven has changed little over the past 15 years. An earlier study by Holden and Caines (1974) presents data on the total P concentrations observed in streams draining into Loch Leven in the period 1963–1971, and also concluded that the annual mean total P concentrations in each stream have remained relatively constant. From 1967–1968 to 1971–1972, although the total P load input to Loch Leven decreased from 151 to 130 kg, respectively, the total N load input increased from 4651 to 8381 kg (Holden and Caines, 1974). What is of interest ecologically is that, as with Lough Neagh, the elemental ratio of N : P delivered to Loch Leven has changed markedly from 1967–1968 to 1971–1972. In both lakes this may have changed the type of algae dominating the phytoplankton community. Smith *et al.* (1982) suggest that the result of increasing N availability as the N loading increased in Lough Neagh was to cause the disappearance of N-fixing algae.

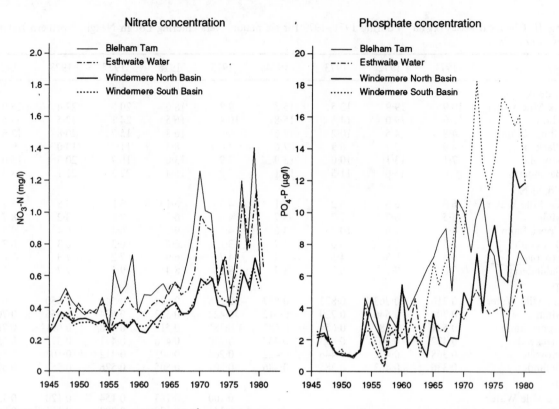

Figure 10. Long-term trends in NO_3^- and PO_4^{3-} concentrations for four lakes in the English Lake District (based on data from Sutcliffe *et al.*, 1982)

English Lake District. Several valuable reviews of nutrient trends in lakes of the English Lake District will be presented with emphasis on three lakes (Esthwaite Water, Blelham Tarn and Windermere) for which records of NO_3^- are available from 1945 to 1995 (Lund, 1969; 1972a; 1972b; 1978; Sutcliffe *et al.*, 1982; Heaney *et al.*, 1986; Talling and Heaney, 1988). Sutcliffe *et al.* (1982) reviewed data collected by the Freshwater Biological Association (FBA) on long-term and seasonal changes in the chemical composition of precipitation and surface waters in the Lake District. Within this review, NO_3^- and PO_4^{3-} concentrations of weekly samples collected from Blelham Tarn, Esthwaite Water and the northern and southern basins of Windermere are presented (Figure 10). The greatest changes in nutrient concentrations occurred after 1970 (Sutcliffe *et al.*, 1982). Sutcliffe *et al.* (1982) suggest that NO_3^- concentration increases in Blelham Tarn and Esthwaite Water are caused by increased use of N fertilizers in their catchments during this period. Increases in NO_3^- concentrations in both the northern and southern basins of Windermere are much more gradual, attributable to the greater proportion of upland rough grazing and bare rock in the Windermere catchment. In contrast, the catchments of Blelham Tarn and Esthwaite Water both lie in the lowland regions of the Lake District where agricultural production is, of necessity, intensive. Concentrations of PO_4^{3-} also increased from 1945 to 1980 in each lake (Sutcliffe *et al.*, 1982). However, these workers stress that the periods of change and the factors causing change differ among the four basins. An increase in PO_4^{3-} concentration in Blelham Tarn in the early 1960s was related to agricultural intensification (Sutcliffe *et al.*, 1982) and to the increased use of P-rich detergents which were discharged to the tarn via sewage effluent (Lund, 1969; 1972a; 1978). In Esthwaite Water, PO_4^{3-} concentration increases were primarily attributed to increased use of P-rich detergents (Lund, 1972b). Sutcliffe *et al.* (1982) also review the P data for Windermere and show little change from 1945 to 1980 in the northern basin, which has a largely upland catchment and a relatively low population density compared with the southern basin. The southern basin receives discharge from the northern basin and runoff from the surrounding lower lying land where both

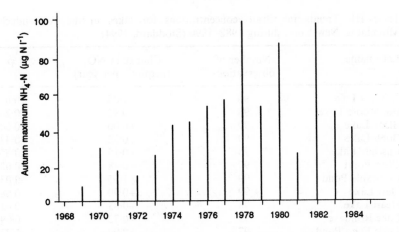

Figure 11. Long-term trends in the seasonal maximum weekly concentrations of NH_4^+ in the upper 7 m layer of the southern basin of Windermere (Talling and Heaney, 1988)

agriculture and human settlements are more extensive. A PO_4^{3-} concentration increase occurred in the southern basin of Windermere in the 1960s and was attributable to increased discharge of P-rich sewage effluent to the southern basin at this time, and increased use of P-rich detergents and the number of people living in and visiting the urban settlements surrounding the southern basin (Sutcliffe et al., 1982).

Talling and Heaney (1988) provide a more recent review of the FBA nutrient data for these lakes (see Figure 10), including an analysis of NH_4^+ concentrations in these four basins (see Figure 11 for southern basin of Windermere). Talling and Heaney (1988) note that concentrations of NH_4^+ have also increased in all four lake basins since 1940. The NH_4^+ concentrations for Esthwaite Water are reviewed in more detail by Heaney et al. (1986).

Analysis of land use, livestock, fertilizer applications and human population in these four basins by Johnes et al. (1994a; 1994b) supports these earlier conclusions. The amount of land used for agricultural production, the numbers of livestock and people, and rates of fertilizer applications of N and P to crops and grass have increased in the catchments of each lake basins. Agriculture has intensified most markedly in the catchments of Esthwaite Water and Blelham Tarn, and intensified least in the catchment of the northern basin of Windermere. Population changes are most marked around the town of Hawkshead in the Esthwaite Water catchment, and the towns of Windermere and Bowness lying along the shore of the southern basin of Lake Windermere.

The ecological implications of such nutrient enrichment are reflected in observed increases in algal biomass and increases in occurrence of blue–green algae, most notably of Esthwaite Water and Blelham Tarn but also in the southern basin of Windermere (Talling and Heaney, 1988). Contrasts in the ecological response to nutrient enrichment between the four lake basins may also reflect contrasts in basin morphometry. Esthwaite Water and Blelham Tarn are relatively shallow basins and therefore likely to receive significant internal P loading from the lake sediments. In contrast, Windermere is a deep, narrow lake where internal loading is unlikely to contribute significantly to the total nutrient loading on the water body.

North America. Although the increases in P concentrations in Lake Washington and several of the Great Lakes, together with associated ecological manifestations of eutrophication, have been widely reported (e.g. Edmonson, 1985), changes in N concentrations of lakes in the USA are less widely understood. Stoddard (1994) reviews research on long-term trends in the deposition of N on forested catchments in the USA, and the impact of such deposition on the N chemistry on lakes draining these catchments. Most of these lakes are in areas that have base-poor bedrock and are therefore susceptible to acidification. The sites studied have been monitored since 1982 by the US Environmental Protection Agency as part of its Long-Term Monitoring project (LTM) (Newell et al., 1987). Stoddard and Kellog

Table III. Trends in nitrate concentrations for lakes in the Adirondack Mountains, New York, during 1982–1990 (Stoddard, 1994)

Lake name	Number of observations	Change in NO_3 (μequiv./l per year)	p
Arbutus Lake	90	+1·06	<0·01
Big Moose Lake	98	+0·92	0·20
Black Lake	99	+0·00	0·60
Bubb Lake	99	+0·32	0·43
Cascade Lake	98	+0·15	0·65
Clear Pond	97	+0·48	0·02
Constable Pond	99	+1·65	0·03
Dart Lake	99	+0·90	0·08
Heart Lake	98	+0·83	0·01
Lake Rondaxe	99	+0·72	0·09
Little Echo Pond	97	+0·00	0·23
Moss Lake	99	+0·42	0·30
Otter Pond	94	+1·33	0·02
Squash Pond	93	+1·80	0·03
West Pond	99	+0·41	0·07
Windfall Lake	99	+0·46	0·40

(1993) evaluated NO_3^- concentrations in 24 Vermont lakes monitored during 1981–1989 in the LTM project and conclude that none of the sites shows an increase in NO_3^- concentrations over this period. However, NO_3^- concentrations in nine of 16 ponds with forested catchments in the Adirondack Mountains, New York increased markedly during 1982–1990, with a maximum rate of change of $1·80 \, \mu$equiv. $1^{-1} \, yr^{-1}$ in Squash Pond (Table III). For Constable Pond (Driscoll and Van Dreason, 1992), the NO_3^- concentration increased $1·65 \, \mu$equiv. $1^{-1} \, yr^{-1}$ (Figure 12). Driscoll and Van Dreason (1992), as summarized by Stoddard (1994), note that the baseflow NO_3^- concentrations in the Adirondack Mountain lakes remain relatively constant over this time period; the increase in mean annual NO_3^- concentrations is associated with spring snowmelt. A comparable decrease in acid neutralizing capacity (ANC), as shown in Figure 12, suggests that these trends are attributable to increased N deposition from atmospheric sources, rather than to catchment land use or population density as is the case for increasing P and N concentrations observed in the Great Lakes and Lake Washington. For lakes in Vermont, seasonality in NO_3^- concentrations is evident, but no trends were detected during 1983–1990 (Stoddard and Kellogg, 1993).

The trends in nutrient chemistry described for lakes in the USA show clear similarities to those described for UK lakes. For all lakes with increasing H^+ concentrations, H^+ increases could be attributed to anthropogenic impacts, either through increases in atmospheric N deposition derived from emission sources or the intensification of agricultural production in the basin. Upward trends in P concentrations are most commonly attributed to increased human population density, with associated increases in the use of P-rich detergents, and the temporal/spatial changes in quantity of sewage effluent discharged to each lake. Where downward trends in P concentrations were observed as in Lake Washington (Edmonson, 1985), these were associated with improved sewage treatment before discharge, and a move away from the use of P-rich detergents. However, in agricultural catchments where P concentrations have increased, the trend may also be attributable to the intensification of agricultural production and a switch from grassland to arable cultivation in the catchment leading to increased soil erosion and associated P export to adjacent aquatic environments. Ultimately, results from these studies support those presented for rivers and groundwaters, where most water bodies have experienced an increase in nutrient loading from their catchments. Where longer records exist such as those for the River Thames, Windermere, Esthwaite Water and Blelham Tarn, although there are clearly N and P concentrations increases in the water bodies from 1930–1940s onwards, the most marked changes have occurred during the past 25 years. What is perhaps less clear from these trends is the ecological impact of such increases

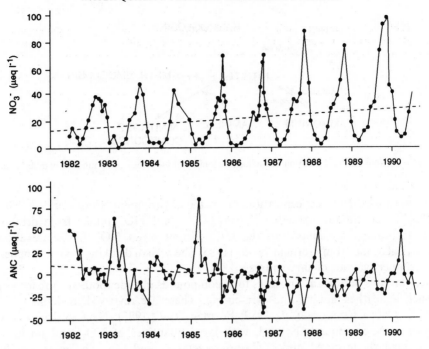

Figure 12. Temporal variations in NO$_3^-$ concentrations and acid neutralising capacity of Constable Pond, in the Adirondack Mountains, New York, 1982–1990 (Driscoll and Van Dreason, 1993)

in nutrient loading, as there is some debate over the factors controlling ecological structure and function in aquatic environments.

NUTRIENTS AND HEALTH

Health issues are largely related to concern over NO$_3^-$ concentrations in drinking water owing to the link with methaemoglobinaemia and stomach cancer (Burt *et al.*, 1993). However, opinion is divided over human health concerns and NO$_3^-$ concentrations (Shuval and Gruener, 1977; Jollans, 1983), particularly because the incidence of methaemoglobinaemia is low and the link between N and stomach cancer is not well understood (Kornberg, 1979; Beresford, 1985; Forman *et al.*, 1985).

Routes of exposure

Water treatment processes throughout the world do not remove or modify the NO$_3^-$ concentrations in source waters to any marked extent, although NO$_3^-$ stripping through resin columns is a widely available treatment technique for doing so. Thus drinking-water supplies frequently contain comparable NO$_3^-$ concentrations to the source waters. In areas where the source waters exhibit a rising trend in NO$_3^-$ concentration this is thus becoming an increasingly dominant proportion of the total dietary intake of N in humans. Conversely, NO$_2^-$ in the source waters is oxidized to NO$_3^-$ during water treatment, particularly if chlorination takes place (Ministry of Supply and Services, Quebec, 1980). Thus drinking water is not a key route of exposure to large quantities of NO$_2^-$.

The International Standing Committee on Water Quality and Treatment reported in 1974 that the majority of the world's population would be exposed to less than 5 mg NO$_3$-Nl^{-1} in drinking water supplies, but small populations could be exposed to at least 100 mg NO$_3$-Nl^{-1} (World Health Organisation, 1984). They calculated the typical individual daily exposure as 20 mg NO$_3$-N. However, in the 16 years since that report, NO$_3^-$ concentrations in surface waters throughout the world have increased significantly, in many instances to twice the 1974 level. Thus drinking water has become one of the key

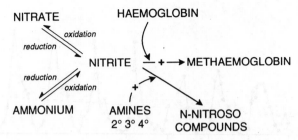

Figure 13. Reactions for the formation of methaemoglobin and N-nitroso compounds (Chamben *et al.*, 1983)

routes of exposure to high NO_3^- concentrations from the environment. Fraser *et al.* (1980) stated that in areas where NO_3^- concentrations exceeded $22 \cdot 6 \, mg \, N \, l^{-1}$, daily NO_3^- intake from drinking water could exceed 70% of the total daily intake. In the UK, current mean NO_3^- levels are about 33% of the threshold, and calculations from Forman *et al.* (1985) show that drinking water supplies 25% of daily intake of NO_3^- from exogenous dietary sources.

Other significant routes of human exposures to N compounds include food, air, tobacco smoke and some occupations, such as leather-tanning and tyre-curing. These are reviewed in detail by several workers (e.g. Fraser *et al.*, 1980; Chambon *et al.*, 1983; Forman *et al.*, 1985). Next to drinking water, the most important source of N compounds is food, with large quantities of NO_2^- and nitrite (NO_3^-) used as preservatives, particularly in cured meats, cheese, vegetables and fish. The most significant source of exposure to N-nitroso compounds come from smoking, with these compounds present in both tobacco and smoke as nitroconicotine.

Metabolism

Human NO_3^- metabolism has been studied by several workers with respect to the health impacts of increased NO_3^- intake (e.g. Fine *et al.*, 1977; Fraser *et al.*, 1980; Fraser and Chilvers, 1981; Beresford, 1985; Forman *et al.*, 1985). Results of these studies have been significant in determining the impacts of elevated NO_3^- concentration on human health. However, most NO_3^- metabolism studies on uptake, transport and excretion processes have been conducted on laboratory animals, the results of which are considered for extrapolation to human metabolism (National Research Council, 1977; World Health Organisation, 1984). Additional research is currently being conducted in this area.

Research on NO_3^- metabolism has indicated that NO_3^- is absorbed in the upper gastrointestinal tract and is ultimately concentrated in the saliva (National Research Council, 1977; World Health Organisation, 1984). Concern over elevated NO_3^- concentrations is not directly linked to the NO_3^- ion, but to the fact that bacteria exist in the mouth, stomach and bladder which can readily reduce the NO_3^- to NO_2^-, which then takes part in two reactions thought to have a direct impact on human health. The reactions are: (1) NO_2^--induced methaemoglobinaemia and (2) nitrozation reactions that produce N-nitroso compounds, which are known to be potent carcinogens in animals (Chambon *et al.*, 1983; Forman *et al.*, 1985). The reaction routes are summarized in Figure 13. The impacts of each reaction are discussed below in more detail.

Health effects: a review of the evidence

The health effects in humans of elevated NO_3^- concentrations have been reviewed by several workers (e.g. World Health Organisation, 1984), and in a report by the European Chemistry Industry Ecology and Toxicology Centre (ECETOC, 1988). These effects are now briefly reviewed.

Methaemoglobinaemia. Oxygen is transported in the human body by oxidation of haemoglobin in the blood to form oxyhaemoglobin. This is possible through the presence of the bivalent ferrous ion (Fe^{2+}) in the haemoglobin molecule. However, if the Fe becomes oxidized to ferric iron (Fe^{3++}), then oxyhaemoglobin liberates oxygen to form methaemoglobin, a compound incapable of fixing oxygen. When sufficiently high concentrations exist in blood, then a condition is induced known as methaemoglobinaemia. Nitrate plays a part in this process through its ability to oxidize the Fe^{2+} to Fe^{3+}, resulting in the formation of methaemoglobin. Reduction of this process to produce haemoglobin is usually catalysed by the enzyme methaemoglobin reductase. In young children, the production of this enzyme is inhibited by elevated NO_2^- (Wellburn, 1990).

Normally between 1 and 2% of the body's haemoglobin exists as methaemoglobin (World Health Organisation, 1984). If the amount of methaemoglobin increases to 5% of the total haemoglobin count, then although this will have no significant effect in adults, signs of cyanosis may appear in infants (Chambon et al., 1983). Below 5%, although the clinical manifestations of infantile methaemoglobinaemia may not be apparent, reduced oxygen availability in the body is considered undesirable (World Health Organisation, 1984). Above 10%, adults experience a lack of muscular oxygen and exhibit signs of cyanosis, particularly in the body extremities (e.g. fingertips and feet), and infants experience clinical effects of methaemoglobinaemia, or the 'blue-baby syndrome', with the skin taking on a blue–grey colour owing to the low oxygen content of the blood. At and above 30%, anoxia occurs (World Health Organisation, 1984) and at 50%, anoxia is irreversible, leading to death (Chambon et al., 1983).

The particular sensitivity of infants (below 6 months old) to this condition has been attributed to several factors: (1) a greater intake of water in relation to body weight, especially when bottle-fed; (2) the absence of two enzymes present in adults that catalyse the reaction converting methaemoglobin to haemoglobin (International Standing Committee on Water Quality and Treatment, 1974); and (3) the presence of NO_3^- reducing bacteria in both the upper intestinal tract and the stomach, owing to the neutral pH of the infant stomach and greater susceptibility of fetal haemoglobin to oxidation (World Health Organisation, 1984). These factors explain why methaemoglobinaemia has not been observed in adults. Although elevated concentrations have been recorded in some pregnant women, these data have not been confirmed (ECETOC, 1988).

Impacts of high aqueous NO_3^- concentrations on the incidence of infant methamoglobinaemia has been well documented (e.g. US Environmental Protection Agency, 1977; Ministry of Supply and Services, Quebec, 1980; ECETOC, 1988). However, uncertainty exists regarding the threshold NO_3^- concentration that, when exceeded, will induce the condition (International Standing Committee on Water Quality and Treatment, 1974; Fraser et al., 1980). In addition, only 10 cases have been confirmed in the UK since the first recognized case in 1951, including one fatality, with 2000 cases reported for Europe and USA before 1977, having a 7–8% mortality rate (Burfield, 1977). For the UK, although this disorder is not currently considered a widespread problem (e.g. Fraser and Chilvers, 1981; Forman et al., 1985), pregnant women and infants are supplied with bottled water with low NO_3^- concentrations where groundwater supplies have now reached or exceeded the World Health Organisation $22 \cdot 6 \, \text{mg} \, NO_3^- \text{-N} \, l^{-1}$ limit, such as in East Anglia (World Health Organisation, 1985).

Human cancer. N-nitroso compounds are known to be potent carcinogens in animals and are readily formed in the human body through the reaction of NO_2^- with secondary, tertiary and quaternary N groups in an acid environment (Fraser et al., 1980; Chambon et al., 1983; Forman et al., 1985). Reactions readily occur in the stomach, using salivary and dietary NO_2^- sources, but can also occur in an achlorhydic stomach (acidified), using additional NO_2^- derived from bacterial reduction (Fraser et al., 1980). N-nitroso compounds also can be formed in the colon or an infected bladder, and are potential agents of cancer in these areas, although this link is yet to be proved (Burfield, 1977; Fraser et al., 1980).

Several epidemiological studies have been conducted throughout the world on the incidence of cancer in relation to elevated NO_3^- intake. These are fully reviewed by Fraser et al. (1980) and the evidence for the UK is assessed by Forman et al. (1985) and Pocock (1985). Thus Fraser and co-workers conclude that although a large body of research is currently available on NO_3^- intake and gastric cancer incidence, this is by no means conclusive, and all conclude that further research is needed to evaluate the interrelations of environmental parameters, NO_3^- concentrations in different localities, and incidences of gastric cancer, and to detail the behaviour of NO_3^- in under different conditions. Forman et al. (1985) also observe that although evidence for a health impact from elevated NO_3^- concentrations in drinking water in the UK is poor, high NO_3^- concentrations in drinking water may take some time to occur owing to the 30–40 year time lag in percolation to groundwater in certain areas. Results have been widely misused either to prove or disprove the case for a casual link between elevated NO_3^- concentrations and the incidence of gastric cancer. Until additional research has been conducted to distinguish other operative environmental parameters from NO_3^- effects, and to detail the role of NO_3^- in complex interactions, accurate conclusions cannot be drawn.

Drawing on the research, World Health Organisation (1970) concluded that results of some studies were sufficient to merit setting a limit on NO_3^- concentrations in drinking water of $22 \cdot 6 \, mg \, NO_3\text{-}N \, l^{-1}$ (or $100 \, mg \, NO_3 \, l^{-1}$), and suggesting that it was desirable to keep concentrations below $11 \cdot 3 \, mg \, NO_3\text{-}N \, l^{-1}$. The World Health Organisation NO_3^- limits have been reviewed with respect to results of current research and a new and lower limit for NO_3^- concentrations of $10 \, mg \, NO_3\text{-}N \, l^{-1}$ in drinking water has been recommended (World Health Organisation, 1984). This new limit is currently being considered to be adopted by the European Community.

FUTURE CONSIDERATIONS

This review of causes and manifestations of trends in nutrient concentrations indicates important issues for consideration when devising management strategies for waters enriched with nutrients. Before implementing expensive management plans, efforts should be made to evaluate the long-term trends in nutrient concentrations to determine the origins and nature of any observed trends. To do this effectively, monitoring programmes for both rivers and lakes need to include both N and P and to determine concentrations of all N species and P species rather than the typical inorganic and most readily bio-available forms. The inclusion of all species in a comprehensive monitoring programme will be necessary to more fully elucidate: (1) the role of N and P in the eutrophication of waters; (2) the relative contributions of atmospheric N deposition, agricultural and domestic sources in a catchment; or (3) the relative contributions of P from point and non-point sources in waters undergoing enrichment. Only by providing long-term comprehensive data of N and P concentrations can we hope to understand the causes and effects of nutrient enrichment and devise cost-effective and sustainable management strategies for nutrient-rich waters.

CONCLUSIONS

Trends in nutrient concentrations in freshwaters, and the potential causes of these trends, have been examined for a range of water bodies across Europe and North America. Information on N trends, particularly NO_3^- concentrations, are widely available for surface waters and for groundwater systems and tend to demonstrate a general trend of increase, most notably in the past three decades. Where longer records exist, such as those for some British rivers (e.g. Figure 5), trends can be constructed since the 1930–1940s. This largely reflects the intensification and expansion of agricultural production, and increases in human population density, which were not reflected in improvements in wastewater treatment facilities (WWTF) until recently. Before this period, the human population density was generally lower and agricultural production was largely confined to areas with gently sloping land, underlain by soils with a high nutrient retention capacity. Also, fertilizer use was dominated by animal manure, rather than inorganic fertilizer, which became popular in the post-war agriculture industry.

Far fewer records exist for P in river systems in either Europe or North America, which may be linked to the widely held assumption that P is so strongly bound within the soil matrix and in such short supply in natural environments (relative to N) that it is never exported to aquatic environments. This assumption ignores the movement of soil particles from land to stream by soil erosion processes that provide an active pathway for the transport of sediment-associated P (particulate P) to adjacent aquatic environments. Recent evidence demonstrates that P may be transported to streams as PO_4^{3-} along preferred hydrological pathways during storm events, when the soil has a high moisture content and becomes supersaturated with P. Supersaturation with P typically results from the over-application of P-rich fertilizers, causing a reduction in the net P retention capacity of the soil. Examination of the few available long-term records of P concentrations for rivers supports this mechanism, demonstrating trends of increase in P concentrations parallel to those observed for N during the past 50 years (see Table II and Figure 6).

Trends in N and P concentrations in rivers in lowland regions of the UK are considerably higher than those reported for lowlands in North America. This probably reflects the higher population density, greater stocking densities for all categories of livestock on grazing land and in stock-rearing units, and the greater intensity with which land is used for cultivation of grass and arable crops in the UK than in North America. Where data are available for less developed countries, N-concentration trends in rivers tend to be lower again than in North America.

For lakes, more data are available for P than N concentrations. This perhaps reflects the widely held view among limnologists that the trophic status of lakes is limited by the availability of P rather than N. Recent studies, however, indicate that more lakes are becoming N limited than were held to be so, either through P enrichment from domestic sewage sources or through P availability from natural sources in their catchments. Lake data are available for total P and PO_4^{3-} concentrations and generally indicate P loading increases to lakes during the past 50 years. Increased P loading is most strongly correlated with increases in human population density within catchments and, in particular, to point source discharge of P-rich effluent to lake waters. Recent trends in the domestic use of P-free detergents and the introduction of P stripping at major WWTF are occasionally reflected in P concentrations observed in some lakes, as reported for lakes in North America. More recently, modelling studies and field based manipulation experiments have clearly demonstrated the importance of non-point source P loading on lakes from their catchments as a causal factor in the trends of increasing P concentrations, particularly in intensive agricultural regions of Europe. Consequently, introduction of expensive P stripping at WWTF and the implementation of sewage diversion schemes around some lakes, to address the wastewater contributions of P, have not led to expected reductions in lake P concentrations. Internal P release from lake sediments may be an important process which may continue, perhaps for decades, after the implementation of such management strategies, especially in shallow lakes. However, this may also be linked to continued external P loading on the lake delivered via non-point source pathways in intensively farmed catchments.

Trends in N concentrations in lakes, where available, mirror those for N in rivers in lowland regions (see Figures 9, 10 and 11), and typically reflect non-point source N loading from agricultural catchments. These trends of increasing N concentrations are repeated in studies of several lakes with forested or moorland catchments, which are undergoing acidification and in this case reflect increased rates of N deposition in the form of NO_x from atmospheric sources (see Figure 11). Atmospheric N deposition also can cause nutrient enrichment of lowland lakes, although the effect is masked by extremely high rates of nutrient export from land-based agricultural sources, and is therefore proportionally less significant than in predominantly forested upland regions.

REFERENCES

Aber, J. D. 1992. 'Nitrogen cycling and nitrogen saturation in temperate forest ecosystems', *Trends Ecol. Evol.*, **7**, 220–224.
Ackermann, W. C., Harmeson, R. H., and Sinclair, R. A. 1970. 'Some long-term trends in water quality of rivers and lakes', *EOS, Trans. Am. Geophys. Union*, **561**, 516–522.
Addiscott, T. M., Heys, P. J., and Whitmore, A. P. 1986. 'Application of simple leaching models in heterogeneous soils', *Geoderma*, **38**, 185–194.
Ahtiainen, M. 1992. 'The effects of forest clear-cutting and scarification on the water quality of small brooks, *Hydrobiologia*, **243**, 465–473.
Alexander, G. C. and Stevens, R. J. 1976. 'Per capita phosphorus loading from domestic sewage', *Wat. Res.*, **10**, 757–764.
Alfoldi, L. and Hommonay, A. 1986. 'Groundwater pollution in Hungary with special regard to nitrate', in Solbe, J. F. de L. G. (Ed.), *Effects of Land Use on Freshwaters*. Ellis Horwood, Chichester. pp. 238–250.
Allcock, R. and Buchanan, D. 1994. 'Agriculture and fish farming', in Maitland, P. S., Boon, P. J., and McLusky, D. S. (Eds), The Fresh Waters of Scotland. Wiley, Chichester. pp. 365–384.
Bailey-Watts, A. E., and Kirika, A. 1987. 'A re-assessment of the phosphorus inputs to Loch Level (Kinross, Scotland): rationale and an overview of results in instantaneous loadings with special reference to run-off', *Trans. R. Soc. Edinburgh Earth Sci.*, **78**, 351–367.
Bailey-Watts, A. E., Kirika, A., May, L., and Jones, D. H. 1990. 'Changes in phytoplankton over various time scales in a shallow, eutrophic lake: the Loch Leven experience with special reference to the influence of flushing rate', *Freshwater Biol.*, **23**, 85–111.
Beresford, S. A. A. 1985. 'Is nitrate in the drinking water associated with the risk of cancer in the urban UK?' *Int. J. Epidemiol.*, **14**, 57–63.

Belillas, C. M. and Roda, F. 1993. 'The effects of fire on water quality, dissolved nutrient losses and the export of particulate matter from dry heathland catchments', *J. Hydrol.*, **150**, 1–17.

Betton, C., Webb, B. W., and Walling, D. E. 1991. 'Recent trends in NO_3-N concentration and loads in British Rivers', in Peters, N. E. and Walling, D. E. (Eds) Sediment and Stream Water Quality in a Changing Environment: Trends and Explanation. *IAHS Publ.*, **203**, 169–180.

Bormann, F. H. and Likens, G. E. 1979. *Pattern and Processes in a Forested Ecosystem.* Springer-Verlag, New York, 253 pp.

Brady, N. C. 1984. *The Nature of Properties of Soils.* 9th edn. Collier Macmillan, New York, 639 pp.

Brezonik, P. L. 1976. 'Nutrients and other biologically active substances in atmospheric precipitation', *J. Great Lakes Res.*, **2**, 166–186.

Brooker, M. P. and Johnson, P. C. 1984. 'The behaviour of phosphate, nitrate, chloride and hardness in twelve Welsh rivers', *Wat. Res.*, **18**, 1155–1164.

Brookes, P. C., Heckrath, G., De Smet, J., Hoffman, G. and Vanderleen, J. 'Losses of phosphorus in drainage water' in Tunney, H., Brookes, P. and Caton, O. (Eds). *Proc. Int. Conf. Phosphorus Loss to Water From Agriculture, September 1995*. CABI, Oxford.

Burden, R. J. 1982. 'Elevated levels of nitrates in groundwater beneath intensively grazed pastureland in New Zealand', in *Mem. Int. Assoc. Hydrogeol. XVI (3) Proc. Prague Congress.* pp. 197–206, IAHS, Paris.

Burfield, I. 1977. 'Public health aspects of nitrates in Essex water supplies', *Public Health Eng.*, **5**(5), 116–124.

Burns, I. G. and Greenwood, D. J. 1982. 'Estimation of the year to year variations in nitrate leaching in different soils and regions of England and Wales', *Agric. Environ.*, **7**, 34–45.

Burt, T. P., Arkell, B. P., Trudgill, S. T., and Walling, D. E. 1988. 'Stream nitrate levels in a small catchment in south west England over a period of 15 years, 1970–1985', *Hydrol. Process.*, **2**, 267–284.

Burt, T. P., Heathwaite, A. L., and Trudgill, S. T. (Eds) 1993. *Nitrate: Process, Patterns and Management.* Wiley, Chichester.

Cameron, K. C. and Wild, A. 1984. 'Potential aquifer pollution from nitrate leaching following the plowing of temporary grassland', *J. Environ. Qual.*, **13**, 274–278.

Canter, L. W., Knox, R. C., and Fairchild, D. M. (Eds) 1987. *Ground Water Quality Protection.* Lewis, Boca Rabon.

Casey, H. 1976. 'Variation in chemical composition of the River Frome, England, from 1965–1972', *Freshwater Biol.*, **5**, 507–514.

Casey, H. and Clarke, R. T. 1979. 'Statistical analysis of the nitrate concentrations from the River Frome (Dorset) for the period 1865–1976', *Freshwater Biol.*, **9**, 91–97.

Casey, H., Clarke, R. T., and Smith, S. M. 1993. 'Increases in nitrate concentration in the River Frome (Dorset) catchment related to changes in land use, fertilizer applications and sewage input', *Chem. Ecol.*, **8**, 232–242.

Chambon, P., Coin, L., and Vial, J. 1983. 'Risks to human health from certain nitrogen compounds usually or possibly present in drinking waters; nitrates, nitrites and N-nitroso compounds', *La Technique de L'Eau de L'Assainissement*, **438/439**, 33–28.

Cirello, J., Rapaport, R. A., Strom, P. F., Matulewich, V. A., Morris, M. L., Goetz, S., and Finstein, M. S. 1979. 'The question of nitrification in the Passiac River, New Jersey: an analysis of historical data and experimental investigation', *Wat. Res.*, **13**, 525–537.

Clair, T. A. and Ehrman, J. M. 1995. 'Acid precipitation-related chemical trends in 18 rivers of Atlantic Canada — 1983 to 1992', *Environ. Monit. Assess.*, **35**, 165–179.

Clark, G. M. 1994. 'Assessment of selected constituents in surface water of the Upper Snake River Basin, Idaho and western Wyoming, water years 1975–89, *US Geol. Surv. Wat.-Resour. Invest. Rep.*, **93-4229**, 49 pp.

Cole, J. J., Caraco, N. F., and Likens, G. E. 1990. 'Short-range atmospheric transport: a significant source of phosphorus to an oligotropic lake', *Limnol. Oceanogr.*, **35**, 1230–1237.

Crawford, C. G. and Wangsness, D. J. 1993. 'Effects of advanced treatment of municipal wastewater on the White River near Indianapolis, Indiana: trends in water quality, 1978–86', *US Geol. Surv. Wat. Supply Pap.*, **2393**, 23 pp.

Darby, R. J., Hewitt, M. V., Penny, A., Johnson, A. E., and McEwen, J. 1988. The effects of increasing length of ley on the growth and yield of winter wheat', *Rothamsted Rep.*, **1987 Part I**, 101–102.

Darimont, T., Lahl, U., and Zeschmar, B. 1985. 'Agriculture and the protection of groundwater — catalogue of measurements for the reducing of nitrate levels, *Watterwirtschaft*, **75**, 106–110.

Delumyea, R G. and Petel, R. L. 1978. 'Wet and dry deposition of phosphorus into Lake Huron', *Wat. Air Soil Pollut.*, **10**, 187–198.

DOE (Department of Environment) 1978. *Digest of Environmental Statistics.* HMSO, London.

DOE 1986. 'Nitrate in water', *Department of Environment Pollution Pap.*, **26**, HMSO, London.

DOE 1988a. *The Nitrate Issue.* HMSO, London.

DOE 1988b. *The Hatton Catchment Nitrate Study: a Report of a Joint Investigation on the Control of Nitrate in Water Supplies.* Department of Environment, Severn-Trent Water, MAFF, London.

Dennehy, K. F., Like, D. W., McMahon, P. B., Tate, C. M., and Heiny, J. S. 1994. 'Water-quality assessment of the South Platte River Basin, Colorado, Nebraska and Wyoming — analysis of available nutrient, suspended-sediment and pesticide data, water years 1980–1992' *US Geol. Surv. Wat.-Resour. Invest. Rep.*, **94-4095**, 145 pp.

Dermine, B. and Lamberts, L. 1987. 'Nitrate nitrogen in the Belgian course of the Meuse River — fate of the concentrations and origins of the inputs', *J. Hydrol.*, **93**, 91–99.

Dillon, P. J. and Rigler, F. H. 1974. 'The chlorophyll–phosphorus relationship in lakes', *Limnol. Oceanogr.*, **19**, 767–773.

Dils, R. M. and Heathwaite, A. L. 1996. 'Phosphorus transport in agricultural runoff', in Brooks, S. and Anderson, M. (Eds), *Advances in Hillslope Processes.* Wiley, Chichester.

Dollard, G. J., Atkins, D. H. F., Davies, T. J., and Healy, C. 1987. 'Concentrations and dry deposition velocities of nitric acid', *Nature*, **326**, 481–483.

Dowdell, R. J. and Webster, C. P. 1980. 'A lysimeter study using ^{15}N on the uptake of fertiliser nitrogen in perennial ryegrass swards and losses by leaching', *J. Soil Sci.*, **31**, 65–75.

Driscoll, C. T. and Van Dreason, R. 1993. 'Seasonal and long-term temporal patterns in the chemistry of Adirondack lakes', *Wat. Air Soil Pollut.*, **67**, 319–344.

Duffy, P. D., Schreiber, J. D., McClurkin, D. C., and McDowell, L. L. 1978. 'Aqueous- and sediment-phase phosphorus yields from five southern pine watersheds', *J. Environ. Qual.*, **7**, 45–50.

ECETOC 1988. 'Nitrate and drinking water', *European Chemistry Industry Ecology and Toxicology Centre (ECETOC) Tech. Rep.*, **27**.

Edmonson, W. T. 1985. 'Recovery of Lake Washington from eutrophication', *Proc. Int. Congress on Lake Pollution and Recovery, European Water Pollution Control Association, Rome, 15–18 April*. pp. 228–234.

Edwards, A. M. C. and Thornes, J. B. 1973. 'Annual cycle of river water quality; a time series approach', *Wat. Resour. Res.*, **9**, 1286–1295.

Fairchild, D. M. 1987. *Groundwater Quality and Agricultural Practices*. Lewis, Boca Rabon.

Falkowska, L. and Korzeniewski, K. 1988. 'Deposition of airborne nitrogen and phosphorus on the coastal zone and coastal lakes of southern Baltic', *Pol. Arch. Huydrobiol.*, **35**, 141–154.

Feller, M. C. and Kimmins, J. P. 1984. 'Effects of clearcutting and slash burning on streamwater chemistry and watershed nutrient budgets in southwestern British Columbia', *Water. Resour. Res.*, **20**, 29–40.

Fine, D. H., Rounbehler, D. P., Fan, T., and Ross, R. 1977. 'Human exposure to N-nitroso compounds in the environment', in Hiatt, H., Watson, J. and Winsten, J. (Eds) *Origins of Human Cancer*. Cold Spring Harbor Laboratory. pp. 293–307.

Forman, D., Al-Dabbagh, S., and Doll, R. 1985. 'Nitrates, nitrites and gastric cancer in Great Britain', *Nature*, **313**, 620–625.

Foster, S. S. D. 1986. 'The groundwater nitrate problem: a summary of research on the impact of agricultural practices on groundwater quality between 1976–1985'. *BGS, Wallingford Ser. Hydrogeol. Rep.*, **86/2**.

Foster, S. S. D. and Young, C. P. 1981. 'Effects of agricultural land-use on groundwater quality with special reference to nitrate' in *Royal Society, a Survey of British Hydrogeology*. Royal Society London. pp. 47–60.

Foster, S. S. D., Cripps, A. C., and Smith-Carington, A. 1982. 'Nitrate leaching to groundwater', *Phil. Trans. R. Soc. London.*, **B296**, 477–489.

Foster, S. S. D., Geake, A. K., Lawrence, A. R., and Parker, J. N. 1985. 'Diffuse groundwater pollution: lessons of the British experience' in *Mem. 18th Congr. Int. Assoc. Hydrogeol., Cambridge*. pp. 168–177.

Foy, R. H., Smith, R. V., Stevens, R. J., and Stewart, D. A. 1982. 'Identification of factors affecting nitrogen and phosphorus loadings to Lough Neagh', *J. Environ. Manage.*, **15**, 109–129.

Fraser, P. and Chilvers, C. 1981. 'Health aspects of nitrate in drinking water', *Sci. Total Environ.*, **18**, 103–116.

Fraser, P., Chilvers, C., Beral, V., and Hill, M. J. 1980. 'Nitrate and human cancer: a review of evidence', *Int. J. Epidemiol.*, **9**, 3–11.

Frissel, M. J. and Van Veen, J. A. 1982. 'A review of models for investigating the behaviour of nitrogen in soils', *Phil. Trans. R. Soc. London*, **B296**, 341–349.

Galbally, I. E. and Roy, C. R. 1978. 'Loss of fixed nitrogen from soils by nitric oxide exhalation', *Nature*, **275**, 734–735.

Galloway, J. N. and Likens, G. E. 1981. 'Acid precipitation: the importance of nitric acids', *Atmos. Environ.*, **15**, 1081–1085.

Geigert, M. A., Nikolaidis, N. P., Miller, D. R., and Heitert, J. 1994. 'Deposition rates for sulfur and nitrogen to a hardwood forest in northern Connecticut, U.S.A.', *Atmos. Environ.*, **28**, 1689–1697.

Gibson, C. E. 1976. 'The problem in Northern Ireland' in *Inst. Public Health Engin. Symp. Eutrophication of Lakes and Reservoirs*. pp. A1–A7.

Gilliom, R. J., Alley, W. M., and Gurtz, M. E. 1995. 'Design of the National Water-Quality Assessment Program: occurrence and distribution of water-quality conditions', *US Geol. Surv. Circ.*, **1112**, 33 pp.

Gormly, J. R. and Spalding, R. F. 1979. 'Sources and concentrations of nitrate-nitrogen in ground water of the Central Platte region, Nebraska', *Groundwater*, **17**, 291–301.

Goulding, K. W. T. 1990. 'Nitrogen deposition to arable land from the atmosphere', *Soil Use and Management*, **6**, 61–63.

Graham, W. F. and Duce, R. A. 1979. 'Atmospheric pathways of the phosphorus cycle', *Geochim. Cosmochim. Acta*, **43**, 1195–1208.

Graham, W. F. and Duce, R. A. 1982. 'The atmospheric transport of phosphorus to the western North Atlantic', *Atmos. Environ.*, **16**, 1088–1097.

Greene, L. A. 1978. 'Nitrates in water supply abstractions in the Anglian Region: current trends and remedies under investigation', *Wat. Pollut. Control*, **77**, 478–491.

Greene, L. A. 1980. 'Nitrate in groundwater in the Anglian region', *Proc. Inst. Civil Eng.*, **69**(2), 73–86.

Hahn, J. and Crutzen, P. J. 1982, 'The role of fixed nitrogen in atmospheric photochemistry', *Phil. Trans. R. Soc. London*, **B296**, 521–541.

Hall, E. S., Oakes, D. B., and Young, C. P. 1976. 'Nitrate in groundwater — studies on the chalk near Winchester, Hampshire,' *Wat. Res. Center Tech. Rep.*, **TR31**.

Hamilton, P. A. and Helsel, D. R. 1995. 'Effects of agriculture on ground-water quality in five regions of the United States', *Groundwater*, **33**, 217–226.

Hanson, P. J. and Lindberg, S. E. 1991. 'Dry deposition of reactive nitrogen compounds: a review of leaf, canopy and non-foliar measurements', *Atmos. Environ.*, **25A**, 1615–1634.

Harriman, R. and Pugh, K. B. 1994. 'Water chemistry', in Maitland, P. S., Boon, P. J., and McLusky, D. S. (Eds), *The Fresh Waters of Scotland*. Wiley, Chichester, pp. 89–112.

Heaney, S. I., Smyly, W. J. P., and Talling, J. F. 1986. 'Interactions of physical, chemical and biological processes in depth and time within a productive English lake during summer stratification', *Int. Revue Ges. Hydrobiol.*, **71**, 441–494.

Heathwaite, A. L. 1993a. 'The impact of agriculture on dissolved nitrogen and phosphorus cycling in temperate ecosystems', *Chem. Ecol.*, **8**, 217–231.

Heathwaite, A. L. 1993b. 'Nitrogen cycling in surface water and lakes' in Burt, T. P., Heathwaite, A. L., and Trudgill, S. T. (Eds), *Nitrate: Processes, Patterns and Management*. Wiley, Chichester, 99–140.

Heathwaite, A. L. 'Sources and pathways of phosphorus loss from agriculture' in Tunney, H., Brookes, P. and Caton, O. (Eds), *Proc. Int. Conf. Phosphorus Loss to Water From Agriculture, September 1995*. CABI, Oxford.

Heathwaite, A. L. and Burt, T. P. 1991. 'Predicting the effect of land use on stream water quality' in Peters, N. E. and Walling, D. E. (Eds), *Sediment and Stream Water Quality in a Changing Environment: Trends and Explanation. IAHS Publ.*, **203**, 209–218.

Heathwaite, A. L., Burt, T. P., and Trudgill, S. T. 1991. 'The effect of agricultural land use on nitrogen, phosphorus and suspended sediment delivery to streams in a small catchment in south west England', in Thornes, J. B. (Ed.), *Vegetation and Erosion*. Wiley, Chichester. pp. 161–179.

Heathwaite, A. L. and Johnes, P. J. 1996. 'The contribution of nitrogen species and phosphorus fractions to stream water quality in agricultural catchments', *Hydrol. Process.*, 10 in press.

Heckrath, G., Brooks, P. C., Poulton, P. R., and Goulding, K. W. T. 'Phosphorus losses in drainage waters from arable clay loam soil' in Tunney, H., Brookes, P. and Caton, O. (Eds), *Proc. Int. Conf. Phosphorus Loss to Water from Agriculture*. September 1995, CABI, Oxford.

Hendry, C. D., Brezonik, P. L., and Edgerton, E. S. 1981. 'Atmospheric deposition of nitrogen and phosphorus in Florida' in Eisenreich, S. J. (Ed.), *Atmospheric Pollutants in Natural Waters*. Ann Arbor Science, Ann Arbor. pp. 199–215.

Henin, S. 1986. 'Water quality — the French problem' in Solbe, J. F. de L. G. (Ed.), *Effects of Land Use on Fresh Waters*. Ellis Horwood, Chichester. pp. 210–220.

Hill, A. R. 1988. 'Factors influencing nitrate depletion in a rural stream', *Hydrobiologia*, **160**, 111–122.

Hill, R. D., Rinker, R. G., and Wilson, H. D. 1980. 'Atmospheric nitrogen fixation by lightning', *J. Atmos. Sci.*, **37**, 179–192.

Holden, A. V. and Caines, L. A. 1974. 'Nutrient chemistry of Loch Leven, Kinross', in *Royal Society of Edinburgh, The Loch Leven IBP Project., Proc. Roy. Soc. Edinburgh*, **B74**, 101–122.

Horn, W., Horn, H., and Paul, L. 1994. 'Long-term trends in the nutrient input and in-lake concentrations of a drinking water reservoir in a densely populated catchment area (Erzebirge, Germany)', *Int. Rev. Ges. Hydrobiol*, **79**, 213–227.

Hornbeck, J. W., Martin, C. W., Pierce, R. S., Bormann, F. H., Likens, G. E., and Eaton, J. S. 1986. 'Clearcutting northern hardwoods: effects on hydrologic and nutrient ion budgets', *Forest Sci.*, **32**, 667–686.

Hornung, M., Roda, F., and Langan, S. J. 1990. 'A review of small catchment studies in Western Europe producing hydrochemical budgets', *Air Pollut. Res. Rep.*, **28**. Commission of the European Communities, Brussels. 186 pp.

International Standing Committee on Water Quality and Treatment 1974. 'Nitrates and water supplies report', *Aqua*, **1**, 5–24.

Jenkinson, D. S. 1982. 'An introduction to the global nitrogen cycle', *Soil Use Manage.*, **6**, 56–61.

Jenkinson, D. S., Hart, P. B., Rayner, J. S., and Parry, L. C. 1987. 'Modelling the turnover of organic matter in long-term experiments at Rothamsted', *INTECOL Bull.*, **15**, 1–16.

Johannessen, M., and Henrikson, A. 1978. 'Chemistry of snow meltwater: changes in concentration during melting', *Wat. Resour. Res.*, **14**, 615–619.

Johansson, C. 1984. 'Field measurements of emissions of nitric oxide from fertilized and unfertilized forest soils in Sweden', *J. Atmos. Chem.*, **1**, 429–442.

Johnes, P. J. 1990. 'An investigation of the effects of land use upon water quality in the Windrush catchment', *DPhil. Thesis*, University of Oxford.

Johnes, P. J. 1996. 'Evaluation and management of the impact of land use change on the nitrogen and phosphorus load delivered to surface waters: the export coefficient modelling approach', *J. Hydrol.*, in press.

Johnes, P. J. and Burt, T. P. 1991. 'Water quality trends in the Windrush catchment: nitrogen speciation and sediment interactions' in Peters, N. E. and Walling, D. E. (Eds), *Sediment and Stream Water Quality in a Changing Environment: Trends and Explanation. IAHS Publ.*, **203**, 349–357.

Johnes, P. J. and Burt, T. P. 1993. 'Nitrate in surface waters' in Burt, T. P., Heathwaite, A. L. and Trudgill, S. T. (Eds), *Nitrate: Processes, Patterns and Control*. Wiley, Chichester. pp. 269–320.

Johnes, P. J. and Heathwaite, A. L. 'Modelling the impact of land use change on water quality in agricultural catchments', *Hydrol. Process.*, in press.

Johnes, P. J., Moss, B., and Phillips, G. L. 1994a. 'Lakes — classification and monitoring. A strategy for the classification of lakes', *NRA Res. Devel. Note*, **253**. NRA, Bristol.

Johnes, P. J., Moss, B., and Phillips, G. L. 1994b. Lakes — classification and monitoring. A strategy for the classification of lakes', *NRA Res. Devel. Note*, **286/6/A**. NRA, Bristol.

Johnson, A. H., Bouldin, D. R., Gayette, E. A., and Hedges, A. M. 1976. 'Phosphorus loss by stream transport from a rural watershed: quantities, processes and sources', *J. Environ. Qual.*, **5**, 148–157.

Jollans, J. L. 1983. *Fertiliser in UK Farming*, Centre for Agricultural Strategy, Reading.

José, P. 1989. 'Long-term nitrate trends in the River Trent and four major tributaries', *Regul. Riv.*, **4**, 43–57.

Kauppi, L. 1984. '1: Contribution of agricultural loading to the deterioration of surface waters in Finland: 2: Nitrate in runoff and river waters in Finland in the 1960s and 1970s', *Public Wat. Res. Inst. & Nat. Board Waters, Finland*, **57**, 24–40.

Kornberg, H. 1979. *Pollution and Agriculture. Royal Commission on Environmental Pollution 7th Report*. HMSO, London.

Lean, G. 1990. 'Ministers weaken nitrate pollution control', *The Observer*, 21 January.

Lewis, W. J., Foster, S. S. D. and Drabsar, B. S. 1980. 'The risk of groundwater pollution by on-site sanitation in developing countries: a literature review', *Int. Refer. Centre Waste Disposal, Rep.*, **01/82**.

Lewis, W. M. and Saunders, J. F. 1989. 'Concentration and transport of dissolved and suspended substances in the Orinoco River', *Biogeochemistry*, **7**, 203–240.

Lewis, W. M. J., Grant, M. C., and Hamilton, S. L. 1985. 'Evidence that filterable phosphorus is a significant atmospheric link in the phosphorus cycle', *Oikos*, **45**, 428–432.

Loehr, R. C. 1974. 'Characteristics and comparative magnitude of non-point sources', *J. Wat. Pollut. Control Fed.*, **46**, 1849–1872.

Lovblad, G., Amann, M., Andersen, B., Hovmand, M., Joffre, S., and Pedersen, U. 1992. 'Deposition of sulfur and nitrogen in the Nordic Countries — present and future', *Ambio*, **21**, 339–347.

Lovett, G. M. 1994. 'Atmospheric deposition of nutrients and pollutants in North America: an ecological perspective', *Ecol. Appl.*, **4**, 629–650.

Lund, J. W. G. 1969. 'Phytoplankton' in Rohlich, G. A. (Ed.), *Eutrophication: Causes, Consequences and Correctives*. National Academy of Sciences, Washington. pp. 306–330.

Lund, J. W. G. 1972a. 'Eutrophication', *Proc. R. Soc. London*, **B180**, 371–382.

Lund, J. W. G. 1972b. 'Changes in the biomass of blue-green and other algae in an English lake from 1945–69' in Desikachary, T. V. (Ed.), *Taxonomy and Biology of Blue-Green Algae, Madras Symp. 1970*. pp. 305–327.

Lund, J. W. G. 1978. 'Changes in the phytoplankton of an English lake, 1945–1977', *Hydrobiol. J.*, **14**, 6–21.

Malanchuk, J. L. and Nilsson, J. 1989. 'The role of nitrogen in the acidification of soils and surface waters', *Miljørapport*, Nordic Council of Ministers, Copenhagen.

Marsh, T. J. 1980. 'Towards a nitrate balance for England and Wales', *Water Services*, Oct, 601–606.

Mason, B. (Eds) 1990. *The Surface Waters Acidification Programme*. Cambridge University Press, Cambridge.

McEwen, J., Darby, R. J., Hewitt, M. V., and Yeoman, D. P. 1989. 'Effects of field beans, fallow, lupins, oats, oilseed rape, peas, ryegrass, sunflowers and wheat on nitrogen residues in the soil on the growth of a subsequent wheat crop', *J. Agric. Sci.*, **115**, 209–219.

McGill, W. B. and Myers, R. K. J. 1987. 'Controls of dynamics of soil and fertiliser nitrogen', in Follett, R. F. *et al.* (Eds), *Soil Fertility and Organic Matter as Critical Components of Production Systems*. Soil Science Society of America and American Society Agronony, Madison, 2247.

Meybeck, M. 1982. 'Carbon, nitrogen and phosphorus transport by world rivers', *Am. J. Sci.*, **282**, 401–450.

Meybeck, M., Chapman, D., and Helmer, R. 1989. *Global Freshwater Quality: a First Assessment*. Global Environmental Monitoring System/UNEP/WHO.

Ministry of Supply and Services, Quebec. 1980. 'Guidelines for Canadian drinking water quality 1978; in WHO (1984) *Guidelines for Drinking Water Quality. Vol. 2. Health Criteria and Other Supporting Information*, **130**. WHO, Geneva.

MAFF (Ministry of Agriculture, Fisheries and Food and Welsh Office Agricultural Department) 1991. *Code of Good Agricultural Practice for the Protection of Water*. HMSO, London.

Moss, B. 1991. 'The role of nutrients in determining the structure of lake ecosystems and implications for the restoration of submerged plant communities to lakes which have lost them' in *Danish Research Programme on Nitrogen, Phosphorus and Organic Matter (NPO)*. Danish Ministry of the Environment, Copenhagen. pp. 75–85.

Moss, B., McGowan, S., and Carvalho, L. 1994. 'Determination of phytoplankton crops by top-down and bottom-up mechanism in a group of English lakes, the West Midland Meres', *Limnol. Oceanogr.*, **39**, 1020–1029.

Mueller, D. K., Hamilton, P. A., Helsel, D. R., Hitt, K. J., and Ruddy, B. C. 1995. 'Nutrients in ground water and surface water of the United States — an analysis of data through 1992', *US Geol. Surv. Wat.-Resour. Invest. Rep.*, **95-4031**.

Murdoch, P. S. and Stoddard, J. L. 1993. 'Chemical characteristics and temporal trends in eight streams of the Catskill Mountains, New York', *Wat., Air Soil Pollut.*, **67**, 367–395.

Murphy, T. J. and Doskey, P.V. 1977. 'Inputs of phosphorus from precipitation to Lake Michigan', *J. Great Lakes Res.*, **2**, 60–70.

National Research Council. 1977. 'Drinking water quality and health, Washington DC, National Academy of Sciences' in WHO (1984) *Guidelines for Drinking Water Quality. Vol. 2. Health Criteria and Other Supporting Documentation*, **130**. WHO, Geneva.

National Water Council. 1977. *Report of the Working Party on the Disposal of Sewage Sludge to Land*. Department of the Environment, HMSO, London.

Neal, C., Fisher, R., Smith, C. J., Hill, S., and Neal, M. 1992. 'Effects of tree harvesting on stream-water quality at an acidic and acid-sensitive spruce forested area: Plynlimon, Mid-Wales', *J. Hydrol.*, **135**, 305–319.

Newell, A. D., Powers, C. F., and Christie, S. J. 1987. 'Analysis of data from long-term monitoring of lakes', US Environmental Protection Agency Rep., **EPA600/4-87/014**. US-EPA, Corvallis, Oregon.

Newman, E. I. 1995. 'Phosphorus inputs to terrestrial ecosystems', *J. Ecol.*, **83**, 713–726.

Nicholson, N. J. 1979. 'A review of the nitrate problem', *Chem. Ind.*, **1979**, 189–195.

Oakes, D. B., Young, C. P., and Foster, S. S. D. 1981. 'The effects of farming practices on groundwater quality in the UK', *Sci. Total Environ.*, **21**, 17–30.

Olness, A., Smith, S. J., Rhoades, E. D., and Menzel, R. G. 1975. 'Nutrient and sediment discharge for watersheds in Oklahoma', *J. Environ. Qual.*, **4**, 331–336.

O'Neill, P. 1985. *Environmental Chemistry*. George, Allen and Unwin, London.

Onstad, C. A. and Blake, J. 1980. 'Thames Basin nitrate and agricultural relations' in *Proc. Symp. Watershed Management*. American Society of Civil Engineers. 961–973.

Paces, T. 1982. 'Long-term changes in concentration and fluxes of nitrogen species in the Elbe River basin' in *Impact of Agriculture Activities on Groundwater. Int. Symp. IAH Mem., Prague*, **16**(2), 299–315.

Parker, J. M., Young, C. P., and Chilton, P. J. 1990. 'Rural and agricultural pollution' in *Survey of British Hydrogeology*. Royal Society, London.

Peters, N. E. and Driscoll, C. T. 1987. 'Sources of acidity during snowmelt of 1984 and 1985 at a forested site in the west central Adirondack Mountains, New York' in Swanson, R. H., Bernier, P. Y., and Woodard, P. D. (Eds), *Proceedings of an International Symposium on Forest Hydrology and Watershed Management. IAHS Publ.*, **167**, 99–108.

Peters, N. E. and Leavesley, G. H. 1995. 'Biotic and abiotic processes controlling water chemistry during snowmelt at Rabbit Ears Pass, Rocky Mountains, Colorado', *Wat. Air Soil Pollut.*, **79**, 171–190.

Peters, N. E. and Reese, R. S. 1995. 'Variations of weekly atmospheric deposition for multiple collectors at a site on the shore of Lake Okeechobee, Florida', *Atmos. Environ.*, **29**, 179–187.

Peters, N. E., Schroeder, R. A., and Troutman, D. E. 1982. 'Temporal trends in the acidity of precipitation and surface waters of New York from 1965–1978', *US Geol. Surv. Wat.-Supply Pap.*, **2188**.

Pocok, S. J. 1985. 'Nitrates and gastric cancer', *Human Toxicol.*, **4**, 471–474.

Powlson, D. S., Pruden, G., Johnson, A. E., and Jenkinson, D. S. 1986. 'Recovery of ^{15}N labelled fertiliser applied in autumn to winter wheat at four sites in eastern England', *J. Agric. Sci.*, **107**, 611–620.

Powlson, D. S., Brookes, P. C., and Christensen, B. T. 1987. 'Measurement of soil microbial biomass provides an early indication of changes in total soil organic matter due to straw incorporation', *Soil Biol. Biochem.*, **19**, 154–164.

Probst, A., Fritz, B., and Viville, D. 1995. 'Mid-term trends in acid precipitation, streamwater chemistry and element budgets in the Strengbach catchment (Vosges Mountains, France)', *Wat., Air and Soil Pollut.*, **79**, 39–59.

Pucket, L. J. 1995. 'Identifying the major sources of nutrient water pollution', *Environ. Sci. Technol.*, **29**, 408–414.

Rascher, C. M., Driscoll, C. T., and Peters, N. E. 1987. 'Concentration and flux of solutes from snow and forest floor during snowmelt in the west-central Adirondack region of New York' in Rudd, J. W. M. (Ed.), *Acidification of the Moose River System in the Adirondack Mountains of New York State. Biogeochemistry*, **3**, 209–224.

Reckhow, K. H. and Simpson, J. J. 1980. 'A procedure using modelling and error analysis for prediction of lake phosphorus concentration and land use information', *Can. J. Fish. Aquat. Sci.*, **37**, 1439–1448.

Reinhorn, T. and Avnimelech, Y. 1974. 'Nitrogen release associated with the decrease in soil organic matter in newly cultivated soils', *J. Environ. Qual.*, **3**, 118–121.

Reynolds, B., Stevens, P. A., Hughes, S., Parkinson, J. A., and Weatherley, N. S. 1995. 'Stream chemistry impacts of conifer harvesting in Welsh catchments', *Wat. Air Soil Pullut.*, **79**, 147–170.

Richter, D. D., Ralston, C. W., and Harms, W. R. 1984. 'Prescribed fire: effects on water quality and forest nutrient cycling', *Science*, **215**, 661–663.

Riggan, P. J., Lockwood, R. N., Jacks, P. M., Colver, C. G., Weirich, F., Debano, L. F., and Brass, J. A. 1994. 'Effects of fire severity on nitrate mobilization in watersheds subject to chronic atmospheric deposition', *Environ. Sci. Technol.*, **28**, 369–375.

Rinella, J. F., McKenzie, S. Q., and Fuhrer, G. J. 1992. 'Surface-water-quality assessment of the Yakima River basin, Washington: analysis of available water-quality data through 1985 water year', *US Geol. Surv. Open-File Rep.*, **91-453**.

Roberts, G. and Marsh, T. 1987. 'The effects of agricultural practices on the nitrate concentrations in the surface water domestic supply sources of Western Europe' in *Water for the Future: Hydrology in Perspective. IAHS Publ.*, **164**, 365–380.

Robertson, F. N. 1979. 'Evaluation of nitrate in the ground water in the Delaware Coastal Plain', *Groundwater*, **17**, 328–337.

Rodda, J. C. and Jones, G. N. 1983. 'Preliminary estimates of loads carried by rivers to estuaries and coastal waters around Great Britain derived from the harmonised monitoring scheme', *J. Inst. Wat. Engrs. Scient.*, **37**, 529–539.

Rodgers, G. A. 1978. 'Dry deposition of atmospheric ammonia at Rothamsted in 1976 and 1977', *J. Agric. Sci. Cambridge*, **90**, 537–542.

Royal Society. 1983. *The Nitrogen Cycle of the United Kingdom: a Study Group Report*. Royal Society, London.

Ryden, S.-O., Ball, P. R., and Garwood, E. A. 1984. 'Nitrate leaching from grassland', *Nature*, **311**, 50–54.

Ryding, S.-O. and Forsberg, C. 1979. 'Nitrogen, phosphorus and organic matter in running waters: studies from six drainage basins', *Vatten*, **1**, 46–58.

Ryding, S.-O. and Rast, W. (Eds) 1989. *The Control of Eutrophication in Lakes and Reservoirs. Man and the Biosphere Ser.*, Vol. 1. Parthenon, Paris, France.

Saull, M. 1990. 'Nitrates in soil and water', *New Scientist*, 15 September.

Schindler, D. W., Welch, H. E., Kalff, J., Brunskill, G. J., and Kritsch, N. 1974. 'Physical and chemical limnology of Char Lake, Cornwallis Island (75° N lat.),' *J. Fish. Res. Board Can.*, **31**, 585–607.

Schreiber, A. N. and McDowell, L. L. 1985. 'Leaching of nitrogen, phosphorus and organic carbon from wheat straw residues', *J. Environ. Qual.*, **14**, 251–260.

Sharpley, A. N. 1996. 'Identification of critical source areas for phosphorus export from agricultural catchments' in Brooks, S. and Anderson, M. (Eds), *Advances in Hillslope Processes*. Wiley, Chichester. (September 1996).

Sharpley, A. N. and Syers, J. K. 1976. 'Phosphorus transport in surface runoff as influenced by fertiliser and grazing cattle', *N. Z. J. Sci.*, **19**, 277–282.

Sharpley, A. N., Syers, J. K., and O'Connor, P. W. 1976. 'Phosphorus inputs into a stream draining an agricultural watershed. I: sampling, *Water, Air Soil Pollut.*, **6**, 39–52.

Shuval, H. and Gruener, N. 1977. 'Infant methaemoglobinaemia and other health effects of nitrates in drinking water' in *Progr. Water Tech.* **8**(415), 183–193.

Skjemstad, J. O., Vallis, I., and Myers, R. K. J. 1988. 'Decomposition of soil organic nitrogen' in Wilson, J. R. (Ed.), *Advances in Nitrogen Cycling in Agricultural Ecosystems*. CABI, Wallingford. pp. 134–144.

Slack, J. G. and Williams, D. N. 1985. 'Long term trends in Essex river water nitrates and hardness', *Aqua*, **2**, 77–78.

Smith, R. A. and Alexander R. B. 1985. 'Trends in concentrations of dissolved solids, suspended sediments, phosphorus and inorganic nitrogen at U.S. Geological Survey National Streams Quality Accounting Network Stations' in *National Water Summary 1984 — Hydrologic Perspectives. US Geol. Surv. Wat. Supply Pap.*, **2275**, 66–73.

Smith, R. A., Alexander, R. B., and Wolman, M. G. 1987. 'Water-quality trends in the Nation's rivers', *Science*, **235**, 1607–1615.

Smith, R. A., Hirsch, R. M., and Slack, J. R. 1982. 'A study of trends in total phosphorus measurements at NASQAN stations', *US Geol. Surv. Wat-Supply Pap.*, **2190**.

Smith, R. V. 1977. 'Domestic and agricultural contributions to the inputs of phosphorus and nitrogen to Lough Neagh', *Wat. Res.*, **11**, 453–459.

Smith, R. V., Stevens, R. J., Foy, R. H., and Gibson, C. E. 1982. 'Upward trend in nitrate concentrations in rivers discharging into Lough Neagh for the period 1969–1979', *Wat. Res.*, **16**, 183–188.

Smith-Carrington, A. K., Bridge, L. R., Robertson, A. S., and Foster, S. S. D. 1983. 'The nitrate pollution problem in groundwater supplies from Jurassic limestones in central Lincolnshire', *Inst. Geol. Sci. Rep.*, **83/3**, HMSO, London.

Soderlund, R., Granat, L., and Rodhe, H. 1985. 'Nitrate in precipitation — a presentation of data from the European Air Chemistry Network', *Univ. Stockholm Rep.*, **CM-69**.

Sonzongni, W. C., Chapra, S. C., Armstrong, D. E., and Logan, T. J. 1982. 'Bioavailability of phosphorus inputs to lakes', *J. Environ. Qual.*, **11**, 555–563.

Spalding, R. F. and Exner, M. E. 1993. 'Occurrence of nitrate in groundwater — a review', *J. Environ. Qual.*, **22**, 392–402.

STACWQ. 1983. '4th Biennial report of the Standing Advisory Committee on Water Quality', *Department of Environment/National Water Council, Standing Technical Committee Rep.*, **37**, HMSO, London.

Stevens, R. J. and Stewart, B. M. 1982. 'Concentration, fractionation and characterisation of soluble organic phosphorus in river water entering Lough Neagh', *Wat. Res.*, **16**, 1507–1519.

Stewart, W. D. P. and Rosswall, T. 1982. 'The nitrogen cycle', *Phil. Trans. R. Soc. London*, **B296**, 299–576.

Stewart, W. D. P., Preston, T., Peterson, H. G., and Christofi, N. 1982. 'Nitrogen cycling in eutrophic freshwaters', *Phil. Trans. R. Soc. London*, **B296**, 491–509.

Stoddard, J. L. 1991. 'Trends in Catskill stream water quality: evidence from historical data', *Wat. Resour. Res.*, **27**, 2855–2864.

Stoddard, J. L. 1994. 'Long-term changes in watershed retention of nitrogen' in Baker, L. A. (Ed.), *Environmental Chemistry of Lakes and Reservoirs*. Advances in Chemistry Series, **237**, American Chemical Society, Washington, 223–284.

Stoddard, J. L. and Kellogg, J. H. 1993. 'Trends and patterns in lake acidification in the state of Vermont — evidence from the long-term monitoring project', *Water Air Soil Pollut.*, **67**, 301–317.

Sutcliffe, D. W., Carrick, T. R., Heron, J., Rigg, E., Talling, J. F., Woof, C., and Lund, J. W. G. 1982. 'Long term and seasonal changes in the chemical composition of precipitation and surface waters of lakes and tarns in the English Lake District', *Freshwater Biol.*, **12**, 451–506.

Talling, J. F. and Heaney, S. I. 1988. 'Long term changes in some English (Cumbrian) lakes subjected to increased nutrient inputs' in Round, F. E. (Ed.), *Algae and the Aquatic Environment*. Biopress, Bristol. pp. 1–29.

Terry, R. E. and Nelson, D. W. 1975. 'Factors influencing nitrate transformations in sediments', *J. Environ. Qual.*, **4**, 549–554.

Thomas, R. L., Santiago, S., Gandais, V., Zhang, L., and Vernet, J. 1991. 'Forms of particulate phosphorus and the sediment carbon/nitrogen ratio as indicators of phosphorus origins in aquatic systems', *Can. J. Wat. Pollut. Res.*, **26**, 433–451.

Tiedemann, A. R., Helvey, J. D., and Anderson, T. D. 1978. 'Stream chemistry and watershed nutrient economy following wildfire and fertilization in eastern Washington', *J. Environ. Qual.*, **7**, 581–588.

Tiedemann, A. R., Quigley, T. M., and Anderson, T. D. 1988. 'Effects of timber harvest on stream chemistry and dissolved nutrient losses in northeast Oregon', *Forest Sci.*, **34**, 344–358.

Tomlinson, T. E. 1970. 'Trends in nitrate concentration in English rivers in relation to fertiliser use', *Water Treat. Exam.*, **19**, 277–293.

Tranter, M., Tsiouris, S., Davies, T. D., and Jones, H. G. 1992. 'A laboratory investigation of the leaching of solute from snowpack by rainfall', *Hydrol. Process.*, **6**, 169–178.

Tsirkunov, V. V., Nikanorov, A. M., Laznik, M. M., and Dongwei, A. 1992. 'Analysis of long-term and seasonal river water quality changes in Latvia', *Wat. Res.*, **26**, 1203–1216.

US Environmental Protection Agency. 1977. 'Health effects of nitrate in water', *US Environ. Protection Agency Rep.*, **EPA-600/1-77-030**, US-EPA, Washington.

van der Weijden, C. H. and Middelburg, J. J. 1989. 'Hydrogeochemistry of the River Rhine: long-term and seasonal variability, elemental budgets, base levels and pollution', *Wat. Res.*, **23**, 1247–1266.

Venkatram, A., McNaughton, D., Karmchandani, P. K., Shannon, J., Fernau, M., and Sisterson, D. L. 1991. 'NAPAP Report 8' in P. M. Irving (Ed.) *Relationships Between Atmospheric Emissions and Deposition/Air Quality, Acidic Deposition: State of Science and Technology*. Vol. 1. US Government Printing Office, Washington, pages 8.1–8.110.

Vollenweider, R. A. 1968. 'Scientific fundamentals of the eutrophication of lakes and flowing waters, with particular reference to nitrogen and phosphorus as factors in eutrophication', *Tech. Rep.*, **DAS/CSI/68.27**, Environmental Directorate, OECD, Paris, 154 pp.

Vollenweider, R. A. 1975. 'Input–output models with special reference to phosphorus loading concept in limnology', *Schweiz. Zeit. Hydrol.*, **37**, 53–84.

Vollenweider, R. A. and Kerekes, J. J. 1981. 'Background and summary results of the OECD cooperative programme on eutrophication. Appendix 1' in *The OECD Cooperative Programme on Eutrophication Canadian Contribution* (compiled by L. L. Janus and R. A. Vollenweider). *Environment Canada Sci. Ser.*, **131**.

Wangsness, D. J., Frick, E. A., Buell, G. R., and DeVivo, J. C. 1994. 'Effect of the restricted use of phosphate detergent and upgraded wastewater-treatment facilities on water quality in the Chattahoochee River near Atlanta, Georgia', *US Geol. Surv. Open-File Rep.*, **94–99**.

Webb, B. W. and Walling, D. E. 1985. 'Nitrate behaviour in streamflow from a grassland catchment in Devon, UK', *Wat. Res.*, **19**, 1005–1016.

Wellburn, A. R. 1990. 'Environment today: the nitrate issue, a reply', *Geogr. Rev.*, **2**(6), 27.

Whitmore, A. P., Bradbury, N. J., and Johnson, P. A. 'The potential contribution of ploughed grassland to nitrate leaching', *Agric. Ecosyst. Environ.*, in press.

WHO/UNEP. 1987. *Global Pollution and Health. Results of Health Related Environmental Monitoring*. World Health Organisation, Geneva and United Nations Environment Programme, Nairobi. Yale Press, London. 24 pp.

Wild, A. and Cameron, K. C. 1980. 'Soil nitrogen and nitrate leaching' in Tinker, P. B. (Ed.), *Soils and Agriculture*. Blackwell, Oxford.

Wilkinson, W. B. and Greene, L. A. 1982. 'The water industry and the nitrogen cycle', *Phil. Trans. R. Soc. London*, **B296**, 459–475.

Withers, P. J. A. 1994. 'The significance of agriculture as a source of phosphorus pollution to inland and coastal waters in the UK', Unpublished Report to MAFF, London, 90 pp.

World Health Organisation. 1970. *European Standards for Drinking Water*. 2nd edn. WHO, Geneva.

World Health Organisation. 1984. *Guidelines for Drinking Water Quality*. Vol. 2. *Health Criteria and Other Supporting Information*. WHO, Geneva.

World Health Organisation. 1985. *Health Hazards from Nitrates in Drinking Water*. WHO, Copenhagen.

Wright, R. F. 1976. 'The impact of forest fire on the nutrient influxes to small lakes in northeastern Minnesota', *Ecology*, **57**, 649–663.

Wyndle-Taylor, E. 1984. 'Nitrates in water supplies; Report to the International Standing Committee on Water Quality and Treatment', *Aqua*, **1**, 5–25.

Young, C. P. 1986. 'Nitrate in groundwater and the effects of ploughing on release on nitrate' in Solbe, J. F. de L. G. (Ed.), *Effects of Land Use on Fresh Waters*. Ellis Horwood, Chichester. pp. 221–237.

Young, C. P. and Gray, E. M. 1978. 'Nitrate in groundwater'. Water Research Centre Tech. Rep., **TR69**.

Young, C. P. and Hall, E. S. 1977. 'Investigations into factors affecting the nitrate content of groundwater' in *Groundwater Quality, Measurement, Prediction and Protection*. Water Research Centre, Medmenham, UK.

Young, C. P., Hall, E. S., and Oakes, D. B. 1976. 'Nitrate in groundwater — studies on the Chalk near Winchester, Hampshire', *Water Research Centre Tech. Rep.*, **TR31**.

Young, C. P., Oakes, D. B., and Wilkinson, W. B. 1983. 'The impact of agricultural practices on the nitrate content of groundwater in the principal United Kingdom aquifers' in Golubev, G. (Ed.), *Environmental Management of Agricultural Watersheds. IIASA Collab. Proc.*, **CP-83-S1**, *Int. Inst. Applied Systems Analysis*, 165–197.

7

TRENDS IN EUTROPHICATION RESEARCH AND CONTROL

WALTER RAST

Freshwater Programme, United Nations Environment Programme, PO Box 30552, Nairobi, Kenya

AND

JEFFREY A. THORNTON

Environmental Planning Division, Southeastern Wisconsin Regional Planning Commission, PO Box 1607, Waukesha, Wisconsin 53187–1607, USA

ABSTRACT

Eutrophication is the natural ageing process of lakes. It is characterized by a geologically slow shift from in-lake biological production driven by allochthonous (external to the water body) loading of nutrients, to production driven by autochthonous (in-lake) processes. This shift typically is accompanied by changes in species and biotic community composition, as an aquatic ecosystem is ultimately transformed into a terrestrial biome. However, this typically slow process can be greatly accelerated by human intervention in the natural biogeochemical cycling of nutrients within a watershed; the resulting cultural eutrophication can create conditions inimical to the continued use of the water body for human-driven economic purposes. Excessive algal and rooted plant growth, degraded water quality, extensive deoxygenation of the bottom water layers and increased fish biomass accompanied by decreased harvest quality, are some features of this process.

Following the Second World War, concern with cultural eutrophication achieved an intensity that spurred a significant research effort, culminating in the identification of phosphorus as the single most significant, and controllable, element involved in driving the eutrophication process. During the late 1960s and throughout the 1970s, much effort was devoted to reducing phosphorus in wastewater effluents, primarily in the developed countries of the temperate zone. These efforts generally resulted in the control of eutrophication in these countries, albeit with varying degrees of success. The present effort in the temperate zone, comprising mostly developed nations, has now shifted to the control of diffuse sources of a broader spectrum of contaminants that impact human water use.

In the developing countries of the inter-tropical zone, however, rapidly expanding populations, a growing industrial economy and extensive urbanization have only recently reached an intensity at which cultural eutrophication can no longer be ignored. Further, initial attempts at applying temperate zone control measures in this region have been largely unsuccessful. Modification of the temperate zone eutrophication paradigm will be needed, especially to address the differing climatic and hydrological conditions, if cultural eutrophication is to be contained in this region, where eutrophication-related diseases continue to be a primary cause of human distress.

INTRODUCTION

If the term 'eutrophication' is mentioned in most temperate zone countries, many listeners will conjure up a notion of 'weed'-infested, 'scum'-encrusted water bodies that look, smell and feel 'bad'; waters that certainly do not meet the US Clean Water Act objective of being fishable and swimmable. In contrast, if the term is mentioned elsewhere, particularly in developing countries, blank looks may be encountered from fish-farmers who work hard to enhance the biological productivity of waters for the specific purpose of generating a high yield of protein-rich fish or cash-rich prawns for the export market. At the same time, however, you also may encounter suitable looks of horror from hydroelectric plant and water treatment plant operators, even in developing countries. Accordingly, the purpose of this paper is to examine the concept of eutrophication in some detail and to attempt to reconcile the divergent viewpoints surrounding this descriptive and often emotive term. We will draw extensively on personal knowledge and experiences, as they reflect the concerns and emphases of the scientific and technical literature on the subject of eutrophication over the past 50 years.

Eutrophication defined

Eutrophication — the enrichment of water bodies with plant nutrients, typically nitrogen and phosphorus, and the subsequent effects on water quality and biological structure and function — is a process, rather than a state. It represents the aging process of lakes, whereby external or allochthonous sources of nutrients and organic matter of terrestrial origin accumulate in a lake basin, gradually decreasing the depth of the water body, and increasing autochthonous production, to the point that the lake begins to take on a marsh-like character and, ultimately, a terrestrial character (Figure 1). Under natural conditions (i.e. without human interferences), this process typically takes place over geological time. However, human influences in a drainage basin can greatly accelerate this enrichment process, rapidly diminishing the utility of a water body, sometimes within only decades. This latter process, termed cultural eutrophication, can be distinguished from natural eutrophication in this way. The former is a consequence of natural lake ageing, whereas the latter is a symptom of human-induced imbalances in the biogeochemical cycling of nutritive elements, such as nitrogen and phosphorus (a discussion of the role of carbon in the eutrophication process is presented in a following section).

In further refining this definition, the Organisation for Economic Cooperation and Development (OECD, 1982) described the process of eutrophication as 'the nutrient enrichment of waters which results

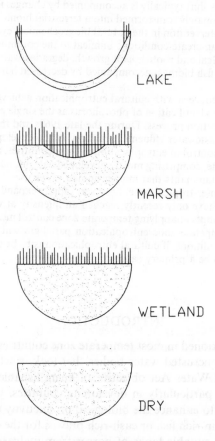

Figure 1. Eutrophication process, showing the progression from aquatic environment to terrestrial environment (After Wetzel, 1975a)

in the stimulation of an array of symptomatic changes, among which increased production of algae and macrophytes, deterioration of water quality and other symptomatic changes, are found to be undesirable and interfere with water uses.' As discussed further in the following, this definition guided the OECD international study of eutrophication (OECD, 1982), as well as introducing the concept of impairment or loss of beneficial human uses of freshwater resources as a principle defining feature of eutrophication (see Gregor and Rast, 1981).

Historical overview

Weber (see Hutchinson, 1969) was the first to use the terms 'eutrophic' and 'oligotrophic' to describe the general nutrient conditions of soils in German bogs. Naumann (1919) subsequently introduced these terms into the aquatic sciences in the 20th century. He used the term 'eutrophy' to distinguish lakes containing much planktonic algae from those containing little algae ('oligotrophy'). Naumann further observed that the algal-rich lakes were associated with fertile and populous lowland regions. The algal-poor lakes tended to occur in areas dominated by primary rocks. He drew the link between human activities in the lowland areas and the productive state of the lakes and identified nitrogen, phosphorus and calcium as the primary determinants of algal productivity.

Concurrently, Thienemann (1925) developed other biological indicators of enrichment to which he applied Naumann's terminology. Thienemann noted that alpine and sub-alpine lakes were typically unproductive, deep and well oxygenated, even when stratified. In contrast, eutrophic lakes were more productive, less deep and exhibited reduced hypolimnetic oxygen concentrations during summer stratification. Midge larvae (e.g. *Tanytarsus*) were common in the former, whereas more robust species (e.g. *Chironomus*) tended to dominate in the latter, mainly because of their ability to tolerate the typically lower oxygen concentrations.

These initial moves towards the classification of lakes on the basis of observed natural features led to a period of single lake studies which refined and divided this type of classification to the point where '[t]he terminology and dichotomies of classification that developed, while instructive, were excessive . . . [to the point of] requiring taxonomic keys for lake types (proposed by Zafar, 1959)' (Wetzel, 1975a:638–638; see also National Academy of Sciences, 1969, for a further discussion of the derivation of the concept of trophic state). Although such a spectrum of classification, wherein eutrophy was simply one stage of a larger spectrum, might have been instructive, its existence highlights a further aspect of trophic state classification — the fact that the system is based on a continuum of states from very nutrient-poor, ultra-oligotrophic systems to very nutrient-rich, hyper-eutrophic waters. This fact highlights yet another truth about trophic state classification — namely that lakes can exhibit several states during their lifetimes; not only moving sedately from a nutrient-poor state to a nutrient-rich state, but also moving from a nutrient-rich state to a less-rich state to a more-rich state. This fact has been demonstrated in newly formed, man-made lakes (reservoirs), as documented in Lake Kariba by Balon and Coche (1974). Similar evidence has been presented for natural lakes of glacial origin (Hutchinson, 1973). Nevertheless, both natural lakes and reservoirs generally reach a type of equilibrium state following these initial oscillations, which reflect the allochthonous inputs from the weathering of soils and rock within the watershed. It is the largely human-induced disruption of this process that leads to cultural eutrophication, as mentioned earlier.

During this developmental period, roughly the half century from 1900 to 1950, much of the necessary refinement of analytical methodologies and many of the data collections for developing both the conceptual and operational definitions of eutrophy were completed. This foundation allowed the identification and quantification of the major causal mechanisms governing aquatic productivity and eutrophication during the next quarter century, at least for the temperate zone of the world (Sawyer, 1966; Rodhe, 1969; Schindler, 1971a; 1971b; Hutchinson, 1973; Schindler and Fee, 1974; Vallentyne, 1974). It was during the later years of this period that many of the key empirical relationships governing eutrophication responses were developed (Sakamoto, 1966; Vollenweider, 1968; 1975; 1976; Dillon and Rigler, 1974; 1975; Dillon and Kirchner, 1975; Larsen and Mercier, 1976; Rast and Lee, 1978; Lee *et al.*, 1978; OECD, 1982; Rast *et al.*, 1983). Indeed, short of any further dramatic revelations within this decade, it appears that the fundamental principles inherent in these relationships will govern the assessment and control of eutrophication into the next century.

Thus, by 1975, the process of eutrophication had been characterized as a biogeochemical process generally controlled by the rate of inorganic nutrient input to a lake or reservoir from its surrounding drainage basin. Oligotrophic water bodies were typically found to have a small external nutrient load or input and, hence, low levels of planktonic production. These low rates of production, when combined with a morphology favouring a high hypolimnion to epilimnion ratio, resulted in low rates of decomposition and little internal nutrient cycling. Microbial populations were also low. As external nutrient inputs increased, the rates of photosynthetic production increased; the internal nutrient regeneration also assumed greater importance. Increased planktonic productivity reduces the depth of light penetration to the point where self-shading by the sheer density of the algal mass itself ultimately limits further phytoplankton growth. As planktonic biomass approaches such levels, production exceeds even the high rate of decomposition characteristic of eutrophic lakes. This excess organic matter sediments into the hypolimnion, engendering anaerobiasis and promoting further internal nutrient release (both biochemical and geochemical) from the lake sediments. The degree to which these hypolimnetic nutrients contribute to continued algal growth in the surface waters depends on both diffusive transport across the metalimnion/thermocline and the rate at which the hypolimnion is broken down by the convective processes associated with overturn.

Considerable debate occurred during the late 1960s and early 1970s concerning the nutrient(s) responsible for driving the eutrophication process. The rationale was that if the specific nutrient(s) responsible for fuelling the excessive algal and aquatic plant growth characteristic of cultural eutrophication could be identified, then we could attempt to control (limit) the magnitude of the eutrophication problem by reducing the input of the nutrient(s) to a lake or reservoir. The debate centred primarily on nitrogen, phosphorus and carbon. Some attention also was given to inorganic micronutrients (e.g. iron, molybdenum, sulphate, silicon) and organic micronutrients (e.g. vitamins). However, these latter substances generally were found to be of more importance in marine systems than freshwater systems (Riley and Chester, 1971; Provasoli and Carlucci, 1974; Rast and Lee, 1978).

For freshwater systems, of the three macronutrients, both nitrogen and carbon were generally found to be abundantly available as a result of atmospheric inputs. Nitrogen inputs were further enhanced by the fixation of molecular nitrogen by heterocystous, blue–green algae (cyanobacteria). This conclusion was dramatically reinforced by the whole-lake experiments of Schindler and co-workers in the early 1970s (cf. Schindler and Fee, 1974). Exceptions to this generalization do exist, an example being soft water lakes with extremely low carbonate concentrations or extremely hard water lakes (Allen, 1972; James and Lee, 1974). Nevertheless, phosphorus, and to a lesser extent, nitrogen, were ultimately implicated as the principle nutrient(s) of concern (Vollenweider, 1968; Wetzel, 1975b; Rast and Lee, 1978; Lee et al., 1980; OECD, 1982).

Ryding and Rast (1989) summarize many of the water quality parameters, and critical boundary values, used to describe the trophic condition of lakes and reservoirs. However, as discussed further in the following, the identification of the 'limiting nutrient' was an important consideration in developing appropriate eutrophication control measures, primarily because the ultimate effectiveness of such measures depends largely on the 'controllability' by humans of the nutrient(s) driving the eutrophication process.

ASSESSMENT AND CONTROL OF EUTROPHICATION IN THE TEMPERATE ZONE

Having defined the concept of eutrophication and identified its primary causative factors, the next logical step was the quantification of the magnitude of the problem. To this end, several major national and international surveys of eutrophication were undertaken during the 1970s. These included the National Eutrophication Survey (NES) of approximately 800 lakes and reservoirs in the USA by the US Environmental Protection Agency (US EPA, 1975), the bi-national study of non-point source pollution of the Laurentian Great Lakes conducted by the Pollution from Land Use Activities Reference Group (PLUARG) of the US–Canada International Joint Commission (IJC), and the international Cooperative Programme on Eutrophication conducted by the Organisation for Economic Cooperation and Development (OECD,

1982). Examples of several other national studies are those reported from Spain (Margalef Lopez *et al.*, 1976), the USA (Rast and Lee, 1978), South Africa (Walmsley and Butty, 1980) and Canada (Janus and Vollenweider, 1981). These studies helped to identify commonalities between systems and to define the limits of eutrophy on a global basis. (Conversely, the differences between water bodies and the eutrophication process in various climatic zones are only now being examined; Thornton and Rast (1993).) These studies also showed that the cultural eutrophication of lakes and reservoirs was sufficiently widespread and of such magnitude as to be considered a serious global problem. Accordingly, the various studies identified above provided an impetus for a substantial research effort directed at the control of eutrophication in lakes and reservoirs (e.g. Rast and Kerekes, 1981; US EPA, 1981; 1984; Ryding and Rast, 1989). Such studies also resulted in the promulgation and/or amendment of several important pieces of water pollution control legislation that encompassed the link between water quality and water availability (Holland and Rast, 1985; Goldfarb, 1988; Rast and Holland, 1988; Davidson and Delogu, 1989; Schlickman *et al.*, 1991).

The international eutrophication study conducted by OECD (1982) merits specific mention in regard to the historical chronology of eutrophication assessment and control. This study involved approximately 100 water bodies around the world, including (1) Nordic lakes, (2) alpine lakes, (3) shallow lakes and reservoirs and (4) a range of lakes and reservoirs in North America (OECD, 1982). A major outcome of this systema-

Table I. Selection of OECD models forecasting eutrophication-related response parameters (after OECD, 1982)

OECD project	Derived relationship	n	r
I. Annual mean total phosphorus concentration (μg l^{-1})			
Combined OECD study	$1{\cdot}55\,P^{0{\cdot}82}$	87	0·93
Shallow lakes and reservoirs	$1{\cdot}02\,P^{0{\cdot}88}$	24	0·95
Alpine lakes	$1{\cdot}58\,P^{0{\cdot}83}$	18	0·93
Nordic lakes	$1{\cdot}12\,P^{0{\cdot}92}$	14	0·86
USA lakes	$1{\cdot}95\,P^{0{\cdot}79}$	31	0·95
II. Annual mean chlorophyll concentration (μg l^{-1})			
Combined OECD study	$0{\cdot}37\,P^{0{\cdot}79}$	67	0·88
Shallow lakes and reservoirs	$0{\cdot}54\,P^{0{\cdot}72}$	22	0·87
Alpine lakes	$0{\cdot}47\,P^{0{\cdot}78}$	12	0·94
Nordic lakes	$0{\cdot}13\,P^{1{\cdot}03}$	13	0·82
USA lakes	$0{\cdot}39\,P^{0{\cdot}79}$	20	0·89
III. Annual maximum chlorophyll concentration (μg l^{-1})			
Combined OECD study	$0{\cdot}74\,P^{0{\cdot}89}$	45	0·89
Shallow lakes and reservoirs	$0{\cdot}77\,P^{0{\cdot}86}$	21	0·88
Alpine lakes	$0{\cdot}83\,P^{0{\cdot}92}$	11	0·96
Nordic lakes	$0{\cdot}47\,P^{1{\cdot}00}$	13	0·77
USA lakes	—	—	—
IV. Annual mean secchi disc transparency (μg l^{-1})			
Combined OECD study	$14{\cdot}7\,P^{-0{\cdot}39}$	67	−0·69
Shallow lakes and reservoirs	$8{\cdot}5\,P^{-0{\cdot}26}$	26	−0·55
Alpine lakes	$15{\cdot}3\,P^{-0{\cdot}30}$	18	−0·74
Nordic lakes	—	—	—
USA lakes	$20{\cdot}3\,P^{-0{\cdot}52}$	22	−0·82
V. Areal hypolimnetic oxygen depletion (g O$_2$ m^{-2} d^{-1})			
Combined OECD study	$\pm 0{\cdot}1\,P^{0{\cdot}55}$	—	—
Shallow lakes and reservoirs	—	—	—
Alpine lakes	—	—	—
Nordic lakes	$0{\cdot}085\,P^{0{\cdot}47}$	—	—
USA lakes	$0{\cdot}115\,P^{0{\cdot}67}$	—	—

$P = [(L_P/q_s)\,(1 + T_w^{-1/2})^{-1}]$ = the flushing corrected average annual phosphorus inflow concentration (OECD, 1982).

tic, co-ordinated international study was the development of quantitative, empirical relationships between phosphorus loads (cause), and a number of eutrophication-related, water quality response parameters (effect). As summarized by Ryding and Rast (1989), these latter parameters included in-lake phosphorus and chlorophyll concentrations, primary productivity, hypolimnetic oxygen depletion, Secchi disc transparency and fish yield. A selection of these models is summarized in Table I, and the specific relationship between phosphorus load and in-lake phosphorus concentration is illustrated in Figure 2. A subsequent, similar but independent, international study was carried out by the Pan American Health Organization (PAHO/CEPIS) and is described in the following.

Thus the 1970s and, to a lesser extent the 1980s, were the period in which the nations of the temperate zone faced and defeated the spectre of cultural eutrophication. Although this 'headline' may be dramatic and overstated, the identification of phosphorus as the key element for limiting algal growth (hence the magnitude of the eutrophication problem) provided water resource managers with a means of attempting to control the symptoms of eutrophication, as well as improving the reduced utility of water bodies affected by it. As previously noted, phosphate (the naturally occurring chemical form of phosphorus in aquatic systems) readily lends itself to management efforts. It is an highly reactive, easily measured, water-soluble element — although its analytical forms do not always bear an exact relationship to its chemical species. It also is an essential nutrient for plant growth and has a finite reservoir in the natural environment (it does not occur in the atmosphere in gaseous form under ambient conditions). Finally, compared with nitrogen and carbon, it exhibits a relatively simple biogeochemical cycle, even though its physical chemistry — especially relative to iron and manganese — can be fairly complex (Stumm and Morgan, 1970).

Eutrophication control measures, at least in developed countries, have focused primarily on the reduction of the external phosphate load to water bodies. As noted by Ryding and Rast (1989), this is generally thought to be the most effective, long-term measure for attempting to control cultural eutrophication. This is typically achieved by removing phosphate from wastewaters at municipal wastewater treatment plants, via its precipitation from the wastewater before its release in effluents (usually termed tertiary treatment). In extreme cases, phosphorus removal can also be applied directly to a lake or reservoir by applying aluminium or iron salts, or in some instances calcium salts, directly to the water column (although trivalent cations are generally more effective in removing phosphorus from the water column). Numerous case studies from throughout the world illustrating phosphorus removal from wastewater effluents and/or water bodies are given in Dunst et al. (1974) and Ryding and Rast (1989).

The above-noted point source control, based on an advanced level of wastewater treatment (tertiary treatment), is typically used via an activated sludge treatment process. It is based on microbially mediated

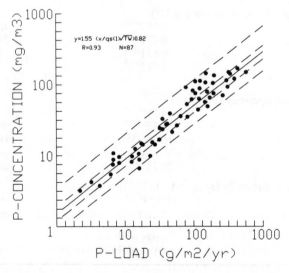

Figure 2. OECD–Vollenweider phosphorus loading–trophic state model (after OECD, 1982)

phosphate fixation and denitrification, in alternating aerobic and anaerobic chambers, to trap phosphorus and drive off nitrogen from sewage waters. This process can result in an overall reduction in phosphorus concentrations from about 7–10 to about 1 mg l^{-1}. In fact, 1 mg l^{-1} P is generally accepted as the most feasible economically achievable level of phosphorus reduction based on conventional technologies (Goldfarb, 1988).

In some instances, municipal wastewater treatment facilities have used chemical precipitation techniques, using aluminium or iron salts or polyelectrolyte solutions, to achieve a nearly complete removal of phosphorus from the water. The use of such extreme (and usually very costly) measures is in situations where conventional treatment will not achieve a sufficient reduction in phosphorus concentrations to effect a diminution of the trophic state of a water body. These alternatives were considered in North America by the IJC with respect to the Laurentian Great Lakes (Rast et al., 1979; IJC, 1980; Vollenweider et al., 1980; Rast, 1981) and used for Wahnbach Reservoir in Germany (Bernhardt, 1983).

Determination of the 'acceptable' nutrient load to a lake or reservoir is increasingly being accomplished through the determination of total maximum daily loads, using a process known as waste-load allocation. In this process, all the possible nutrient sources (point and non-point) in a drainage basin are identified and quantified. This is either by direct measurement or by prescribed estimation techniques. One example of the latter is regionally relevant unit area loads (sometimes termed nutrient export coefficients; see Rast and Lee, 1983). This information is then compared with the maximum permissible nutrient load that will still allow achievement of the 'desired' trophic state in a downstream lake or reservoir. The term 'desired' is based on the trophic condition which allows a specific human-designed water use(s) to be achieved. If the estimated nutrient load exceeds the level conducive to a desired water use(s), the required reduction in the nutrient load to achieve the desired tropic state can be calculated. The previously identified models of the international OECD eutrophication study (1982) have proved useful in such efforts. This load can then be apportioned to all the nutrient sources within the drainage basin, using a predetermined formula — ranging from equal reduction of all sources within a drainage basin to a weighted allocation based on technologies, type of source (industry, municipality, farm), or distance from the affected water body. As a general observation, point sources (e.g. municipal wastewater treatment plants) are considered more important than non-point sources as nutrient reduction targets, especially as an initial control target, primarily because of the relative ease with which the former can be controlled. It is noted, however, that some degree of non-point source control may prove more cost-effective beyond a certain advanced level of point source control (e.g. see PLUARG, 1978; Rast et al., 1990).

The control of nutrients from non-point sources in a drainage basin has lagged behind the control of nutrients from point sources, both in terms of available technologies and the legal requirements for implementation. Available technologies for achieving control of urban non-point source nutrients, metals and sediments have been reviewed by Stahre and Urbonas (1990). These technologies consist primarily of detention, retention and infiltration techniques. They are based on a system of containment and release that parallels traditional wastewater or flood control engineering; it relies primarily on the stilling of stormwaters to allow sedimentation of particulates and adsorbed contaminants. The subsequently produced supernatant can then either be discharged downstream with a reduced sediment and contaminant load, transferred to a wastewater treatment facility for final 'polishing', or allowed to percolate into the substratum. In non-urban areas, similar approaches can be used to control agricultural runoff and construction erosion. Examples include swales, ditches, various cultivation techniques (such as conservation tillage or contour farming) and temporary diking (using hay bales, plastic sheeting, geofabrics or similar materials) commonly used to reduce the rate of runoff and the intensity of overland flow (Rast, 1981; Novotny and Chesters, 1981; Novotny and Olem, 1994). In the USA, it is noted that, legally, point source controls have been required under the Federal Water Pollution Control Act — now called the Clean Water Act — since 1972. In contrast, non-point controls have been mandated only since 1987, when the Clean Water Act was extensively overhauled and expanded (Davidson and Delogu, 1989).

In-lake control of eutrophication, via the application of alum or other multivalent cation salts directly to a water body, has been undertaken much less frequently than the control of point sources. Such applications are costly, logistically difficult and usually only partially effective, compared with reducing the external nutrient load to a water-body. Examples of the application of this method are given in Dunst *et al.* (1974). In fact, this control method is usually attempted only when point and/or non-point source control measures applied in a drainage basin have failed to achieve the desired reduction in trophic state, or in situations in which the nutrient content in the bottom sediments of a lake or reservoir has accumulated to the extent that internal loading from the sediments is anticipated. In some water bodies, in fact, the internally generated phosphorus load can equal or even exceed the external nutrient load (cf. Ryding, 1981; 1985). Such extraordinary internal nutrient loads usually can be moderated by alum treatments. However, the removal of the contaminated sediments, via dredging, is the more common alternative. This is especially the case for shallow lakes, in which alum addition to the water column cannot generally establish a cohesive flocculent 'blanket' at the sediment–water interface to induce nutrient sedimentation, due to wind-induced turbulent mixing (Ryding, 1982).

To date, in the temperate zone nations of North America and Europe, the widespread implementation of point source pollution control measures, and judicious use of in-lake and non-point source control measures, has largely removed the issue of eutrophication from the mainstream of hydrological concern (e.g. National Research Council, 1991). In fact, attention in this geographical/climatic region has now shifted from controlling nutrient enrichment and its symptoms to the control of more invidious and diffuse substances, such as acidic compounds, salts and toxic organic chemicals (e.g. pesticides, herbicides) (Thornton, 1991). It must be emphasized, however, that this change in focus reflects more a heightened concern with the latter types of water quality degradation than with the actual elimination of cultural eutrophication as a widespread and continuing water quality problem in temperate zone countries. The reality, therefore, is that cultural eutrophication persists to the present and will continue to be a significant source of water quality degradation, with its associated impacts on human water uses; it remains a widespread, pervasive water quality problem on a global scale.

ASSESSMENT AND CONTROL OF EUTROPHICATION IN THE INTER-TROPICS

Ryding (1992), in his seminal compendium of environmental management philosophy, observes that '[t]he geographical distribution of the various water-related problems throughout the world indicates some important differences with regard to the character of the environmental stress and the extent of preventive efforts.' He elucidates these differences as (1) a shift towards control of 'widespread contamination by a large variety of compounds' in the developed countries (as noted above), (2) an emphasis on the control of intense localized/regionalized pollution in countries with economies in transition (i.e. the legacy of long-term environmental neglect) and (3) the control of incipient, localized pollution by a limited number of well-defined pollutants in the developing countries, reminiscent of the situation in the developed countries 25 years earlier. Thus, in this discussion (as well as elsewhere; e.g. Thornton and Rast, 1993), we can draw a distinction between eutrophication assessment and control in the temperate zone on the one hand, and elsewhere in the world on the other, where these concerns are still very much in vogue — although the wider concerns of toxic chemicals, acidic substances and salts (particularly salination) also intrude (Thornton, 1991; Thornton *et al.*, 1991).

That there are real differences between the characteristics and behaviour of water bodies in the inter-tropic climatic zone versus those in the temperate zone (the climatic region in which the fundamental basis of classical limnology, as well as the concept of eutrophication, was derived) was initially suggested by Williams (1988). Tacit acknowledgment of this fact has driven eutrophication research and limnological research, at least in southern Africa, for many years (Davies and Walmsley, 1985; Thornton and Rast, 1987). However, only recently have these differences been quantified (e.g. see Thornton, 1986; 1987; Salas and Martino, 1991; Thornton and Rast, 1989, 1993).

With respect to eutrophication, Ryding and Rast (1989:39), note that these differences generally are more a matter of degree than of substance. For example, Davies and Walmsley (1985) note four specific characteristics of the southern hemisphere, largely warm-water, lakes. These characteristics are: (1) their temperature regime, which predisposes warm-water lakes to stable stratification (and hence hypolimnetic deoxygenation, given their generally higher rate of microbial production); (2) their hydrological regime, which is dependent on seasonal rainfall events which themselves exhibit significant inter- and intra-annual variations; (3) their seasonality and its effect on the production of aquatic biomass; and (4) their turbidity. With regard to the latter, Thornton and Rast (1993) ascribe the tendency towards higher turbidity to the larger watershed:lake surface area ratio (largely a function of the greater number of man-made lakes or reservoirs in this portion of the globe) and the higher erodibility of the generally organic-poor soils.

These same features contribute to a generally higher nutrient concentration representing the boundary between oligotrophic, mesotrophic and eutrophic conditions for warm-water lakes and reservoirs (although moderated by typically shorter water residence times). For example, as noted by Thornton and Rast (1987; 1993) and Salas and Martino (1991), the phosphorus concentration at the mesotrophic–eutrophic boundary for warm-water lakes is between 0·06 and 0·07 mg l^{-1} P. This value is in contrast with the OECD boundary value of 0·035 mg l^{-1} for water bodies in the temperate zone (cf. Gregor and Rast, 1982; Ryding and Rast, 1989). Thornton and Rast (1993) also identified other subtleties in the assessment of eutrophication within semi-arid, man-made lakes. One example is a shift in algal classes symptomatic of eutrophic conditions in semi-arid, tropical (blue–green algae — cyanophytes and cyanobacteria — tend to dominate) and semi-arid, mid-latitude (chrysophytes and diatoms tend to dominate) lakes. Another example is the inverse relationship between largely inorganic turbidity and increasing trophic state (resulting from the flocculation of particulates under the saline conditions that typically occur at higher trophic states in these areas). It is noted that Salas and Martino (1991) were unable to resolve these features in their study of lakes in the humid tropics at the time their paper went to press, given their more diverse and less abundant data set.

Notwithstanding, the work of Salas and Martino (1991) in the Regional Programme for the Development of Simplified Methodologies for the Evaluation of Eutrophication in Warm-Water Tropical Lakes, like that in the OECD (1982) study, deserves special mention. This study was undertaken under the auspices of the Pan American Health Organization, and covered 42 lakes and reservoirs throughout South and Central America. Because of its focus on warm-water lakes and reservoirs in Central and South America, this study is probably of greater relevance to many water bodies in the developing countries than the OECD study. Nevertheless, both methodologies have proved to be useful, accurate and quantitative management tools for assessing in-lake, eutrophication-related water quality and the probable responses to alternative eutrophication control strategies for lakes and reservoirs (based on reducing external phosphorus loads; Rast et al., 1983; 1990; Ryding and Rast, 1989).

In contrast with the indicators of eutrophication, it should be noted that fundamental differences do tend to exist between the possibilities for control of eutrophication in the inter-tropics (the climatic/geographical location of many developing countries) versus the temperate zone. In fact, control measures typically used in the temperate zone should also work in the inter-tropic regions, although with some differing efficiencies. A more fundamental factor in the selection of control measures for the latter region, however, relates to the resource base of the latter. As a general observation, the developed countries focus on technology-based control measures, such as municipal wastewater treatment plants, which typically require a substantial financial and technical resource base. Many developing nations lack this resource base. They do, however, often have a relatively large, cheap labour force. As a result, the developed nations in the temperate zone generally utilize more 'high-tech' eutrophication control measures (e.g. treatment plants), whereas developing nations may focus on measures that are less expensive and/or that can be enhanced by abundant, cheap labour (e.g. harvesting and other in-lake measures).

RECENT ADVANCES IN EUTROPHICATION RESEARCH AND CONTROL

Two principal lines of investigation are presently being followed in the field of eutrophication assessment and control. To some extent, these two lines are parallel and equally relevant to the determination of the existence of eutrophic conditions. The first is a continuation of the research initiated by Thornton and Rast (1989) and Salas and Martino (1991) and relates directly to the existence of climatic and regional differences between lakes. Recognizing the existence of inter-regional variability, and building on the work of Omernik (1987), numerous workers have reported refinements to the indicators commonly used to define the existence of eutrophic conditions within the USA, examples being Heiskary *et al.* (1987), Reckhow (1988) and Jones and Knowlton (1993). Cooke *et al.* (1993) presented a useful review of these studies and their implications for the assessment of eutrophication in the north temperate zone. The implications generally centre on the ability of management interventions to 'return' a lake to a 'non-eutrophic' state, in circumstances where the determination of 'non-eutrophic' is based on an inappropriate criterion.

The second major research initiative relates to public perceptions of eutrophication. Canfield and Hoyer (1989:18) lament the confusion in the public mind between the concepts of eutrophication and pollution. However, although the concepts have scientifically distinct differences, they share some similarities in that they both impair the use of water bodies, reducing their utility for human exploitation, whether for domestic consumption, recreation or environmental purposes. For this reason, it is important to consider public perceptions of water pollution in general, and eutrophication in particular. To this end, a number of recent studies in both the temperate zone and inter-tropical zone have explored this issue (Thornton *et al.*, 1989; Heiskary *et al.*, 1992; Quick and Johansson, 1992), and several have attempted to quantify the relationship between public perceptions and water quality (Thornton and McMillan, 1989; Heiskary *et al.*, 1992). Both types of studies have built on the foundation laid in the early days of eutrophication research in North America (e.g. David, 1971; Kooyoomijian and Clesceri, 1974).

Indeed, even though the notion of trophic state, and terms such as 'oligotrophic', 'mesotrophic' and 'eutrophic', are commonly used to describe the condition of lakes and reservoirs throughout the world, there is no universal acceptance of the scientific or technical basis for applying such terms. In fact, when used descriptively, these terms are largely subjective in nature. For example, an in-lake chlorophyll concentration of 10 μg l^{-1} would probably be considered indicative of impaired water quality in the transparent lake waters of central Wisconsin (USA; e.g. the lakes used as the basis for much of the pioneering work in the field of classical limnology — cf. Birge and Juday, 1911). This chlorophyll level would manifest itself as a green colour in the water and be viewed in a negative manner by most individuals observing the water body. In contrast, this same chlorophyll concentration in a turbid reservoir in south-west Texas (USA) would probably not even be noticed by most individuals viewing the water body. The in-lake chlorophyll concentration is identical in both instances; what differs is the interpretation of the degree to which beneficial uses are being impaired — the degree of impairment being perceived to be greater in Wisconsin because of the customary human uses of these waters compared with the customary uses of the semi-arid Texas reservoirs. This observation, quantified in southern Africa by Thornton and co-workers (Thornton and McMillan, 1989; Thornton *et al.*, 1989; Quick and Johansson, 1992), provides an interesting confirmation of the work of Thornton and Rast (1993) and Salas and Martino (1991).

The OECD (1982), in recognition of such differences in perception and in addition to the previously mentioned empirical eutrophication models, developed a probabilistic framework for assigning a trophic state classification to specific water bodies. This framework was based on the range of interpretations, by a number of professional limnological scientists from a variety of geographical and climatic settings, of a given value of an eutrophication-related, water quality parameter (as in the chlorophyll example). The graphic form of this probabilistic classification scheme is shown in Figure 3A. In this scheme, a given chlorophyll concentration has a maximum probability of being interpreted as a characteristic of a specific trophic state, but recognizes the probability that others might interpret the same concentration as indicative of a different trophic state. A similar scheme has been developed by the Pan American Health Organization for warm-water lakes (Figure 3B; Salas and Martino, 1991).

As noted, both lines of research are parallel to some extent. The user-based studies have proved particularly useful in determining public use objectives for both individual water bodies and for water bodies within a particular region, and in providing guidance to public works officials directing eutrophication management interventions. Examples include the studies of Thornton *et al.* (1989) and Quick and Johansson (1992), as well as Wiseman and Sowman (1992), all of which have had direct application in establishing restoration goals in several southern African water bodies. These studies have also proved to be valuable educational and informational tools, enabling greater public understanding of, and participation in, water body restoration and monitoring efforts in the USA (Heiskary, 1989), as well as leading to greater responsiveness by public officials throughout North America to public demands in setting of water quality standards and guidelines (Heiskary *et al.*, 1992).

A further major research line seems to have stemmed from these two main efforts. This line of research includes the refinement — or extension — of: (1) models (e.g. Davic and DeShon, 1989; Carlson, 1991; Salas and Martino, 1991; Meyer and Rossouw, 1992); (2) control measures (e.g., Thanh and Biswas, 1990; Carroll and Hendrix, 1992; Haarhoff *et al.*, 1992; Harding and Quick, 1992; Grigg, 1993; Harding, 1993; Wilson and O'Sullivan, 1993); and (3) alternative eutrophication indicators — or sets of indicators (e.g. Welch, 1989; Chapman, 1992; Bazzanti *et al.*, 1993), to better meet local/regional water quality management needs. Contributing to this effort is the recent publication and/or re-issue of several useful guidance manuals (e.g. Ryding and Rast, 1989; 1992; 1994; US EPA 1990a; 1990b; 1993; Jorgensen and Vollenweider, 1989; Jorgensen and Loffler, 1990; Hashimoto and Barrett, 1991; Matsui *et al.*, 1991; Jorgensen, 1993; Cooke *et al.*, 1993) and regional syntheses (e.g. Dejoux, 1988; Allanson *et al.*, 1990; Bloom *et al.*, 1991; Dudgeon and Lam, 1994).

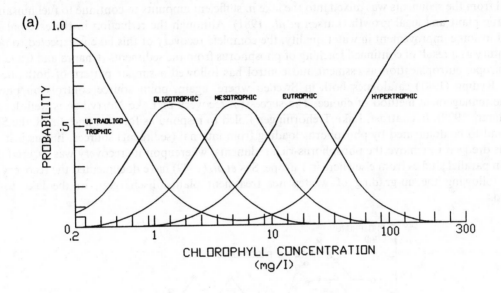

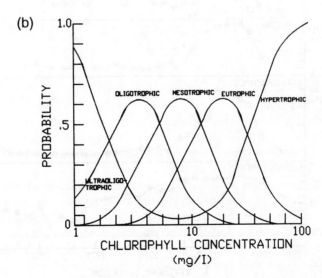

Figure 3. (A) OECD probabilistic trophic state classification scheme (after OECD, 1982). (B) PAHO/CEPIS probabilistic classification scheme (after Salas and Martino, 1991)

SOME EXAMPLES

Numerous examples of the assessment and control of eutrophication exist, although most are overwhelmingly from lakes situated in the northern hemisphere (Chapman, 1992). Perhaps most well-known (and well-documented) is the case of Lake Washington (USA) (Figure 4). This lake rapidly eutrophied in the post-Second World War years as a result of the discharge of wastewater from several neighbouring municipalities (Edmondson *et al.*, 1956). Subsequent installation of wastewater treatment plants and the diversion of wastewaters away from the lake resulted in nearly complete recovery of the water body to its former (natural) state (Edmondson and Lehman, 1981). In contrast, Lake Shagawa (USA) is probably the antithesis of successful eutrophication management. This lake also received municipal wastewater, which was subsequently treated to tertiary standards using a calcium carbonate flocculation technique (Larsen *et al.*, 1979). Unlike Lake Washington, however, Shagawa Lake failed to respond as predicted and remained eutrophic. Further investigation determined that the lake was subject to extremely intense internal loading from anoxic hypolimnetic sediments. Owing to its shallow nature, much of the nutrient released from the sediments was mixed into the lake in sufficient amounts to continue to fuel nuisance levels of aquatic plant and algal growth (Larsen *et al.*, 1981). Although the reduction in the external load has resulted in some improvement in water quality, the complete recovery of this lake is expected to take close to a century as a result of continued leaching of phosphorus from the sediments (Chapra and Canale, 1991).

In Europe, eutrophication assessment and control has followed a similar pattern of both success and failure. Ryding (1981) catalogues both in Sweden where, again, point source controls on phosphorus were the management method of choice. The successful recovery of Lake Norrviken is well-documented (cf. Ahlgren, 1978). In contrast, Lake Trehorningen failed to respond as forecast and, like Lake Shagawa, was found to be dominated by phosphorus loading from internal (sediment) sources. In this instance, the lake was dredged to remove the phosphorus-rich sediments, whereupon its recovery was assured (Ryding, 1982). In parallel studies from elsewhere in Europe, Sas *et al.* (1989) have documented the recovery of Lake Zurich following the up-grading of wastewater treatment plants discharging to the lake to tertiary standards.

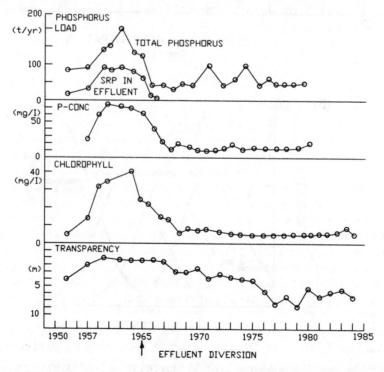

Figure 4. Effects of wastewater inflows and diversion on Lake Washington, USA (after Ryding and Rast, 1989)

Elsewhere, few documented cases of eutrophication assessment and control have been widely published. Many countries are still at the reconnaissance stage, undertaking country-wide assessments of the frequency and severity of eutrophication problems; for example, South Africa (Walmsley and Butty, 1980), Zimbabwe (Thornton, 1980), Brazil (Tundisi and Matsumura-Tundisi, 1986), China (Xiangcan *et al.*, 1990) and Indonesia (Bloom *et al.*, 1991). Many of these countries are also conducting eutrophication assessments of, and preparing management plans for, a variety of single lakes, such as Laguna de Bay (Philippines) and Saguling Reservoir (Indonesia) (Jorgensen and Vollenweider, 1989). In terms of eutrophication control, Lake McIlwaine (now Lake Chivero, Zimbabwe; Thornton, 1982) and Hartbeespoort Dam (South Africa; NIWR, 1985; Chutter, 1989) provide two of the very few documented cases. Both lakes suffered from cultural eutrophication as a result of urban growth and development in the post-Second World War years. Tertiary wastewater treatment became mandatory in Zimbabwe in 1976 and in South Africa in 1985, and advanced wastewater treatment methods were implemented shortly thereafter. Tertiary treatment of wastewater in Harare, Zimbabwe, was initially provided by land application of secondary effluent and, more recently, by an activated sludge process; in Johannesburg, South Africa, treatment was by means of the activated sludge process throughout. Lake Chivero responded well, recovering to a meso-trophic state by 1979, before continued non-point source contamination caused the lake to re-eutrophy (as forecast) in the late 1980s (Thornton, 1982; B.E. Marshall and E. Khaka, pers. comm.) (Figure 5). Hartbeespoort Dam also responded as predicted (NIWR, 1985), but remains eutrophic as its principal water source is treated sewage effluent (Chutter, 1989). Nevertheless, application of advanced wastewater treatment technologies to the latter reservoir did cause a shift in the lake's N:P ratio, which resulted in a (temporary?) change in the dominant alga away from the nuisance cyanophyte, *Microcystis aeruginosa*, to a mixed population of less undesirable chlorophytes.

Finally, in an attempt to synthesize the available information on the state of the world's lake environments, the Lake Biwa Research Institute and International Lake Environment Committee is compiling an encyclopaedic collection of data on individual lakes from around the world. This on-going project provides an useful and comprehensive source of information on eutrophication assessment, research and control from all six continents (Asia, Africa, Europe, North America, Oceania and South America) and numerous individual countries and water bodies. A brief perusal of the four published volumes (to date) reveals the seriousness, and currency, of the problems associated with cultural eutrophication on the global scale (ILEC, 1987; 1989; 1990; 1994).

SYNTHESIS AND CONCLUSIONS

The enrichment of natural waters (lakes, reservoirs, rivers) with plant nutrients can be many things to many people. Generally, however, and at some point, the benefits of enhanced fish and biotic production are outweighed by the impairment of other uses of a water body. Although the exact point at which this loss of utility occurs differs between global ecoregions, the majority of water users are likely to agree on the fact of impairment if not its precise nature; to some, it will be represented as an 'explosive' growth of floating plants, whereas to others it will be identified by taste and odour problems, while to yet others it will be characterized by a 'coarsening' of the fish population, with the more desirable (higher value) sport fish being replaced by less desirable species. Thus it may seem to some that we have failed in our mission to reconcile these divergent definitions of eutrophication — that is, there appears to be an increasing re-fragmentation of thought as the apparent global consensus achieved with the OECD Cooperative Programme on Eutrophication is massaged and manipulated into regionally relevant criteria (e.g. see Salas and Martino, 1991). Indeed, by introducing the concept of impairment of human water uses and public perceptions, there would appear to be a danger that eutrophication assessment and control could drift completely from its scientific and technical roots.

On the other hand, of course, the perception to be gained from the scientific and technical literature — at least the literature published in the temperate zone — might be that the 'problem of eutrophication', so prevalent in the 1960s, has been solved forever. As previously noted, this is definitely *not* the case, as is borne out by the inclusion of eutrophication-related concerns in Agenda 21 (UN, 1993), and the fact that

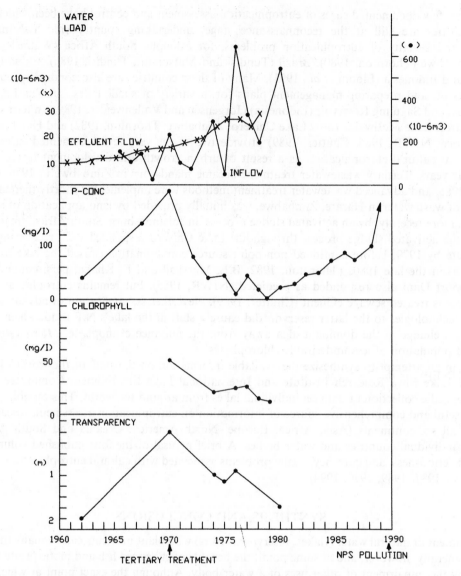

Figure 5. Effects of wastewater inflows, tertiary treatment and non-point source (NPS) pollution on Lake Chivero, Zimbabwe (after Thornton, 1982, with additional unpublished data from E. Khosa, pers. comm.)

eutrophication-associated disease continues to be the single most significant cause of child mortality in the world, especially in developing countries (e.g. see Black, 1987:180; Ryding, 1992:204).

The problems associated with cultural eutrophication will continue to require monitoring and management for the foreseeable future, and for as long as mankind chooses to dwell in densely populated urban centres and to engage in large-scale, agro-industrial activities to meet its demands for goods and food (Ryding, 1992). Indeed, as is presently the case in the developed nations, it can be anticipated that a shift in the developing world, and in nations with economies in transition, from controlling point source nutrients (e.g. wastewater) to the control of numerous widely distributed contaminants (including toxics, acidic substances and metals) will also occur. Further, it can also be anticipated that the greatest challenge to adequately managing cultural eutrophication will be economic in nature, and that the countries most in need of adequate water resources management (including both water supply and wastewater treatment) will be those least able to afford it (Black, 1987). Thus there is a pressing need for low-cost, appropriate technologies for such situation. This is not meant to be second-rate, stop-gap measures (Thornton *et al.*, 1991), but rather a type that will allow nations to augment and improve their infrastructure and processes as their economic situation improves. It must also be recognized that, rather than simply applying existing technologies developed in the temperate zone to inter-tropical settings, the development of appropriate technologies for the developing world will require the transfer of knowledge, concepts and information, particularly to encourage the creation of a resident body of professionals, and to fully develop a suite of regionally defined indices of, and control measures for, eutrophication (Thornton *et al.*, 1991; Thornton and Rast, 1993).

Finally, in assessing the broader context of eutrophication research and control, it is important that consideration be given to the potential long-term impacts of climate change on water resources. Not only will cultural factors continue to impugn global water resources, but also changes in the periodicity of precipitation (e.g. changes in drought frequencies, rainfall volumes, evaporative losses) and warmer surface temperatures (hence, warmer water temperatures) may challenge the very paradigms on which eutrophication management is founded (e.g. see Abu-Zeid and Biswas, 1992; Herrmann, 1992). This could have potentially severe consequences for continued management of (especially) temperate zone waterbodies (Harmancioglu and Alpaslan, 1992). In this regard, the investment of effort in developing appropriate eutrophication control measures for the inter-tropical zone could, in the future, result also in large dividends to the developed nations of the temperate zone.

ACKNOWLEDGEMENTS

The discussions presented herein are the personal, professional views of the authors, and should not be interpreted as representing the official views or policies of either the United Nations Environment Programme (UNEP), or the Southeastern Wisconsin Regional Planning Commission (SEWRPC).

REFERENCES

Abu-Zeid, M. A. and Biswas, A. K. 1992. *Climatic Fluctuations and Water Management.* Butterworth-Heinemann, Oxford. 356 pp.

Ahlgren, I. 1978. 'Response of Lake Norrviken to reduced nutrient loading', *Verh. Int. Verein. Limnol.*, **20**, 846–850.

Allanson, B R., Hart, R. C., O'Keeffe, J. H., and Robarts, R. D. 1990. *Inland Waters of Southern Africa: an Ecological Perspective, Monogr. Biol.*, **64**. Kluwer Academic, The Hague. 458 pp.

Allen, H. L. 1972. 'Phytoplankton photosynthesis, micro-nutrient interactions, and inorganic carbon availability in a soft-water Vermont lake.' In Likens, G. E. (Ed.), *Nutrients and Eutrophication: the Limiting-Nutrient Controversy, Spec. Symp. 1*, Am. Soc. Limnol. Oceanogr. 63–83.

Balon, E. K. and Coche, A. G. 1974. *Lake Kariba: a Man-made Tropical Ecosystem in Central Africa, Monogr. Biol.*, **24**. Junk, The Hague.

Bazzanti, M., Seminara, M., and Tamorri, C. 1993. 'Eutrophication in a deep lake: depth distribution of profundal benthic communities as an indicator of environmental stress', *Verh. Int. Verein. Limnol.*, **25**, 784–789.

Bernhardt, H. 1983. 'Input control of nutrients by chemical and biological methods', *Wat. Supply*, **1**, 187–206.

Birge, E. A. and Juday, C. 1911. 'The inland waters of Wisconsin. The dissolved gases of the water and their biological significance', *Bull. Wis. Geol. Nat. Hist. Surv.*, **22**, Sci. Ser. 7, 259 pp.

Black, M. 1987. *The Children and the Nations: Growing Up Together in the Postwar World.* Macmillan, Melbourne. 339 pp.

Bloom, H., Dodson, J. J., Soetarmi Tjitrosomo, S., Umaly, R. C. and Sukimin, S. 1991. *Inland Aquatic Environmental Stress Monitoring, Biotrop Spec. Publ.*, **43**. Southeast Asian Regional Center for Tropical Biology, Bogor. 262 pp.

Canfield, D. E. and Hoyer, M. V. 1989. 'Managing lake eutrophication: the need for careful lake classification and assessment', *Enhancing States' Lake Management Programs*, **1989**, 17–26.

Carlson, R. E. 1991. 'Expanding the trophic state concept to identify non-nutrient limited lakes and reservoirs', *Enhancing States' Lake Management Programs*, **1991**, 59–72.

Carroll, M. S. and Hendrix, W. G. 1992. 'Federally protected rivers: the need for effective local involvement', *J. Am. Plan. Assoc.*, **58**, 346–352.

Chapman, D. 1992. *Water Quality Assessments: a Guide to the Use of Biota, Sediments and Water in Environmental Monitoring.* Chapman and Hall, London. 585 pp.

Chapra, S. C. and Canale, R. P. 1991. 'Long-term phenomenological model of phosphorus and oxygen for stratified lakes', *Wat. Res.*, **25**, 707–715.

Chutter, F. M. 1989. 'Evaluation of the impact of the 1 mg/l phosphate-P standard on the water quality and trophic state of Hartbeespoort Dam', *WRC Rep.*, **181/1/89**. Water Research Commission, Pretoria. 69 pp.

Cooke, G. D., Welch, E. B., Peterson, S. A., and Newroth, P. R. 1993. *Restoration and Management of Lakes and Reservoirs*, 2nd edn, Lewis, Boca Raton. 548 pp.

Davic, R. D. and DeShon, J. E. 1989. 'The Ohio Lake condition index: a new multiparameter approach to lake classification', *Lake Reserv. Manage.*, **5**, 1–8.

David, E. L. 1971. 'Public perceptions of water quality', *Wat. Resour. Res.*, **7**, 453–457.

Davidson, J. H. and Delogu, O. E. 1989. *Federal Environmental Regulation. Vol. 1*. Butterworth Legal, Salem. pp. 2-1–2-38B.

Davies, B. R. and Walmsley, R. D. 1985. 'Perspectives in Southern Hemisphere limnology', *Dev. Hydrobiol.*, **28**, Junk, The Hague. 263 pp.

Dejoux, C. 1988. *La Pollution des Eaux Continentales Africaines: Experience Acquise Situation Actuelle et Perspectives*. Editions de l'ORSTOM Collection Travaux et Documents, **213**. Institut Francais de Recherche Scientifique pour le Developpement en Cooperation, Paris. 513 pp.

Dillon, P.J. and Kirchner, W.B. 1975. 'The effects of geology and land use on the export of phosphorus from watersheds', *Wat. Res.*, **9**, 135–148.

Dillon, P. J. and Rigler, F. H. 1974. 'The chlorophyll–phosphorus relationship in lakes', *Limnol. Oceanogr.*, **19**, 767–773.

Dillon, P. J. and Rigler, F. H. 1975. 'A simple method for predicting the capacity of a lake for development based on lake trophic status', *J. Fish. Res. Board. Can.*, **32**, 1519–1531.

Dudgeon, D. and Lam, P. K. S. 1994. 'Inland waters of tropical Asia and Australia: Conservation and management', *Mitt. Int. Verein. Limnol.*, **24**, 386 pp.

Dunst, R. C., Born, S. M., Uttormark, P. D., Smith, S. A., Nichols, S. A., Peterson, J. O., Knauer, D. R., Serns, S. L., Winter, D. R., and Wirth, T. L. 1974. *Survey of Lake Rehabilitation Techniques and Experiences. Wisconsin Department of Natural Resour. Tech. Bull.* **75**, 179 pp.

Edmondson, W. T. and Lehman, J. R. 1981. 'The effect of changes in the nutrient income on the condition of Lake Washington', *Limnol. Oceanogr.*, **26**, 1–29.

Edmondson, W. T., Anderson, G. C., and Peterson, D. R. 1956. 'Artificial eutrophication of Lake Washington', *Limnol. Oceanogr.*, **1**, 47–53.

Goldfarb, W. 1988. *Water Law*. 2nd edn, Lewis, Boca Raton. pp. 190–192.

Gregor, D. J. and Rast, W. 1981. 'Benefits and problems of eutrophication control' in US EPA, *Restoration of Lakes and Inland Waters*, **EPA 440/5–81–010**, 166–171.

Gregor, D. J. and Rast, W. 1982. 'Simple trophic state classification of the Canadian nearshore waters of the Great Lakes', *Wat. Resour. Bull.*, **18**, 565–573.

Grigg, N. S. 1993. 'New paradigm for coordination in water industry', *J. Wat. Res. Plan. Mgmt.*, **119**, 572–587.

Harding, W. R. 1993. 'Faecal coliform densities and water quality criteria in three coastal recreational lakes in the SW Cape, South Africa', *Water SA*, **19**, 235–246.

Harding, W. R. and Quick, A. J. R. 1992. 'Management options for shallow hypertrophic lakes, with particular reference to Zeekoevlei, Cape Town', *S. Afr. J. Aquat. Sci.*, **18**, 3–19.

Haarhoff, J., Langenegger, O., and van der Merwe, P.J. 1992. 'Practical aspects of water treatment plant design for a hypertrophic impoundment', *Water SA*, **18**, 27–36.

Harmancioglu, N. B. and Alpaslan, N. 1992. 'Risk factors in water quality assessment' in Herrmann, R. (Ed.), *Managing Water Resources During Global Change. Amer. Wat. Resour. Assoc. Tech. Publ. Ser.* TPS-92-4, Bethesda, pp. 299–308.

Hashimoto, M. and Barrett, B. F. D. 1991. *Guidelines of Lake Management. Vol. 2. Socio-economic Aspects of Lake Reservoir Management*. International Lake Environment Committee Federation, Otsu. 229 pp.

Heiskary, S. A. 1989. 'Lake assessment program: a cooperative lake study program', *Lake Reserv. Manage.*, **5**, 85–94.

Heiskary, S. A., Wilson, C. B., and Larsen, D. P. 1987. 'Analysis of regional patterns in lake water quality: using ecoregions for lake management in Minnesota', *Lake Reserv. Manage.*, **3**, 337–344.

Heiskary, S. A., Bryant, N., Butkus, S., Dennis, J., Duda, A. M., Larsen, D. P., Raschke, R., Ratcliffe, S. and Smeltzer, E. 1992. *Developing Eutrophication Standards for Lakes and Reservoirs*. North American Lake Management Society, Alachua. 51 pp.

Herrmann, R. 1992. *Managing Water Resources During Global Change, Am. Wat. Resour. Assoc. Tech. Publ. Ser.*, **TPS-92-4**, 860 pp.

Holland, M. M. and Rast, W. 1985. 'Lake management: the influence of the scientist in policy formulation and decision-making' in *Lake and Reservoir Management: Practical Applications. Proc. 4th Int. Symp. Lake and Watershed Management*. North American Lake Management Society, Washington. pp. 49–52.

Hutchinson, G. E. 1969. 'Eutrophication, past and present' in National Academy of Sciences, *Eutrophication: Causes, Consequences, Correctives*. National Academy Press, Washington. pp. 197–209.

Hutchinson, G. E. 1973. 'Eutrophication. the scientific background of a contemporary practical problem', *Am. Sci.*, **61**, 269–279.

International Joint Commission, Phosphorus Management Task Force (IJC) 1980. *Phosphorus Management for the Great Lakes*, IJC, Windsor. 125 pp.

International Lake Environment Committee (ILEC) 1987. Lake Biwa Research Institute and International Lake Environment Committee, *Data Book of World Lake Environments — a Survey of the State of World Lakes*. Vol. I. *1988*. ILEC and United Nations Environment Programme, Otsu.

International Lake Environment Committee (ILEC) 1989. Lake Biwa Research Institute and International Lake Environment Committee, *Data Book of World Lake Environments — a Survey of the State of World Lakes*. Vol. II. *1989*. ILEC and United Nations Environment Programme, Otsu.

International Lake Environment Committee (ILEC) 1990. Lake Biwa Research Institute and International Lake Environment Committee, *Data Book of World Lake Environments — a Survey of the State of World Lakes*. Vol. III. *1990*. ILEC and United Nations Environment Programme, Otsu.

International Lake Environment Committee (ILEC) 1994. Lake Biwa Research Institute and International Lake Environment Committee; *Data Book of World Lake Environments — a Survey of the State of World Lakes*. Vol. IV. *1994*. ILEC and United Nations Environment Programme, Otsu.

James, D. W. and Lee, G. F. 1974. 'A model of inorganic carbon limitation in natural waters', *Wat. Soil Air Pollut.*, **3**, 315–320.

Janus, L. L. and Vollenweider, R. A. 1981. *The OECD Cooperative Programme on Eutrophication — Canadian Contribution. Sci. Ser. Rep.*, **131**, Canada Centre for Inland Waters, Burlington.

Jones, J. R. and Knowlton, M. K. 1993. 'Limnology of Missouri reservoirs: an analysis of regional patterns', *Lake Reserv. Manage.*, **8,** 13–16.

Jorgensen, S. E. 1993. *Guidelines of Lake Management. Vol. 5. Management of Lake Acidification*, International Lake Environment Committee Federation, Otsu. 195 pp.

Jorgensen, S. E. and Loffler, H. 1990. *Guidelines of Lake Management. Vol. 3. Lake Shore Management*. International Lake Environment Committee Federation, Otsu. 174 pp.

Jorgensen, S. E. and Vollenweider, R. A. 1989. *Guidelines of Lake Management*. Vol. 1. *Principles of Lake Management*, International Lake Environment Committee Federation, Otsu. 199 pp.

Kooyoomijian, K. H. and Clesceri, N. L. 1974. 'Perceptions of water quality by select respondent groupings in inland-water-based recreational environments', *Wat. Resour. Res.*, **10,** 728–744.

Larsen, D. P. and Mercier, H. T. 1976. 'Phosphorus retention capacity of lakes', *J. Fish. Res. Board. Can.*, **33,** 1742–1750.

Larsen, D. P., Van Sickle, J., Malueg, K. W. and Smith, P. D. 1979. 'The effect of wastewater phosphorus removal on Shagawa Lake, Minnesota: phosphorus supplies, lake phosphorus and chlorophyll *a*', *Wat. Res.*, **13,** 1259–1272.

Larsen, D. P., Schultz, D. W., and Malueg, K. W. 1981. 'Summer internal phosphorus supplies in Shagawa Lake, Minnesota', *Limnol. Oceanogr.*, **26,** 740–753.

Lee, G. F., Rast, W., and Jones, R. A. 1978. 'Eutrophication of waterbodies: insights into an age-old problem', *Environ. Sci. Technol.*, **12,** 900–908.

Lee, G. F., Jones, R. A., and Rast, W. 1980. 'Availability of phosphorus to phytoplankton and its implications for phosphorus management strategies' in Loehr, R.C., Martin, C. and Rast, W. (Eds), *Phosphorus Management Strategies for Lakes*. Interscience, Ann Arbor. pp. 259–308.

Margalef Lopez, R., Planas Mont, D., Armengol Bachero, J., Vidal Celma, A., Prat Fornells, N., Guiset Serra, A., Toja Santillana, J., and Estrada Miyares, M. 1976. *Limnologia de Los Embalses Espanoles*. Ministerio de Obras Publicas, Madrid. 452 pp.

Matsui, S., Barrett, B. F. D. and Banerjee, J. 1991. *Guidelines of Lake Management*. Vol. 4. *Toxic Substances Management in Lakes and Reservoirs*. International Lake Environment Committee Foundation, Otsu. 170 pp.

Meyer, D. H. and Rossouw, J. N. 1992. 'Development of the reservoir eutrophication model (REM) for South African reservoirs', *Water SA*, **18,** 155–164.

National Academy of Sciences 1969. *Eutrophication: Causes, Consequences, Correctives*. National Academy Press, Washington. 661 pp.

National Research Council 1991. *Opportunities in the Hydrologic Sciences*. National Academy Press, Washington. 348 pp.

Naumann, E. 1919. 'Nagra synpunkter angaende limnoplankton', *Svensk Bot. Tidskr.*, **13,** 129–163.

National Institute for Water Research (NIWR) 1985. *The Limnology of Hartbeespoort Dam, South African National Scientific Programmes Rep.*, **110.** Council for Scientific and Industrial Research, Pretoria. 269 pp.

Novotny, V. and Chesters, G. 1981. *Handbook of Nonpoint Pollution*. Van Nostrand Reinhold, New York. 555 pp.

Novotny, V. and Olem, H. 1994. *Water Quality: Prevention, Identification, and Management of Diffuse Pollution*. Van Nostrand Reinhold, New York. 1054 pp.

Organisation for Economic Cooperation and Development (OECD) 1982. *Eutrophication of Waters: Monitoring, Assessment, and Control*. OECD, Paris. 154 pp.

Omernik, J. M. 1987. 'Ecoregions of the coterminous United States', *Ann. Assoc. Am. Geogr.*, **77,** 118–125.

Pollution from Land Use Activities Reference Group (PLUARG) 1978. *Environmental Management Strategy for the Great Lakes Ecosystem, Final Report*. International Joint Commission, Windsor. 115 pp.

Provasoli, L. and Carlucci, A. F. 1974. 'Vitamins and growth regulators' in Stewart, W. D. P. (Ed.), *Algal Physiology and Biochemistry*, Univ. California Press, Berkeley. pp. 741–787.

Quick, A. J. R. and Johansson, A. R. 1992. 'User assessment survey of a shallow freshwater lake, Zeekoevlei, Cape Town, with particular emphasis on water quality', *Water SA*, **18,** 247–254.

Rast, W. 1981. 'Quantification of the input of pollutants from land use activities in the Great Lakes Basin' in Steenvoorden, J. H. A. M. and Rast, W. (Eds), *Impact of Non-point Sources on Water Quality in Watersheds and Lakes: Field Measurements and the Use of Models*, Workshop Proceedings. United Nations Education, Scientific and Cultural Organisation Man And the Biosphere Programme, Amsterdam. pp. 98–141.

Rast, W. and Holland, M. M. 1988. 'Eutrophication of lakes and reservoirs: a framework for making management decisions', *Ambio*, **17,** 2–12.

Rast, W. and Kerekes, J. J. 1981. *International Workshop on the Control of Eutrophication. Proceedings*. United Nations Education, Scientific and Cultural Organisation, International Institute for Applied Systems Analysis, and Organisation for Economic Cooperation and Development, Laxenburg. 107 pp.

Rast, W. and Lee, G. F. 1978. *Summary Analysis of the North American (US Portion) OECD Eutrophication Project: Nutrient Loading—Lake Response Relationships and Trophic State Indices*, **EPA 600/3–78–008,** 454 pp.

Rast, W. and Lee, G. F. 1983. 'Evaluation of nutrient loading estimates for lakes', *J. Environ. Engrg. Div., ASCE*, **109,** 502–517.

Rast, W., Lee, G. F., and Jones, R. A. 1979. 'Use of OECD eutrophication modeling approach for assessing Great Lakes water quality' in Lee, G. F. and Jones, R. A. (Eds), *Water Quality Characteristics of the U.S. Waters of Lake Ontario During the IFYGL and Modeling Contaminant Load–Water Response Relationships in the Nearshore Waters of the Great Lakes, Tech. Rep.* National Oceanic and Atmospheric Administration, Great Lakes Environmental Research Laboratory, Ann Arbor. Ch 6.

Rast, W., Lee, G. F., and Jones, R. A. 1983. 'Predictive capability of U.S. OECD phosphorus loading—lake response models', *J. Wat. Pollut. Control Fed.*, **55,** 990–1003.

Rast, W., Holland, M. M., and Ryding, S.-O. 1990. 'Eutrophication of lakes and reservoirs: a management framework', *MAB Digest 1*. Unesco, Paris. 84 pp.

Reckhow, K. H. 1988. 'Empirical models for trophic state in Southeastern U.S. lakes and reservoirs', *Wat. Resour. Bull.*, **24,** 723–734.

Riley, J. P. and Chester, R. 1971. *Introduction to Marine Chemistry*. Academic Press, London. 465 pp.

Rodhe, W. 1969. 'Crystallization of eutrophication concepts in northern Europe' in National Academy of Sciences, *Eutrophication: Causes, Consequences, Correctives*. National Academy Press, Washington. 50–64.

Ryding, S.-O. 1981. 'Reversibility of man-induced eutrophication: experiences of a lake recovery in Sweden', *Int. Rev. ges. Hydrobiol.*, **66**, 449–503.

Ryding, S.-O. 1982. 'Lake Trehorningen Restoration Project: changes in water quality after sediment dredging', *Hydrobiology*, **92**, 549–558.

Ryding, S.-O. 1985. 'Chemical and microbiological processes as regulators of the exchange of substances between sediment and water in shallow, eutrophic lakes', *Int. Rev. ges. Hydrobiol.*, **70**, 657–702.

Ryding, S.-O. 1992. *Environmental Management Handbook. The Holistic Approach — from Problems to Strategies.* Lewis, Boca Raton. 777 pp.

Ryding, S.-O. and Rast, W. 1989. *The Control of Eutrophication of Lakes and Reservoirs. Man and the Biosphere Ser. 1*, Parthenon, Carnforth. 314 pp.

Ryding, S.-O. and Rast, W. 1992. *El Control de la Eutrofizacion en Lagos y Pantanos.* Ediciones Piramide, Madrid. 375 pp.

Ryding, S.-O. and Rast, W. 1994. *Le Controle de l'Eutrophisation des Lacs et des Reservoirs, Sciences de l'Environnement 9.* Masson, Paris. 294 pp.

Sakamoto, M. 1966. 'Primary production by the phytoplankton community in some Japanese lakes and its dependence on lake depth', *Arch. Hydrobiol.*, **62**, 1–28.

Salas, H. J. and Martino, P. 1991. 'A simplified phosphorus trophic state model for warm-water tropical lakes', *Wat. Res.*, **25**, 341–350.

Sas, H., Ahlgren, I., Bernhardt, H., Bostrom, B., Clasen, J., Forsberg, C., Imboden, D., Kamp-Nielson, L., Mur, L., de Oude, N., Reynolds, C., Schreurs, H., Seip, K., Sommer, U., and Vermij, S. 1989. *Lake Restoration by Reduction of Nutrient Loading: Expectations, Experiences, Extrapolation.* Academia-Verlag, Richarz.

Sawyer, C. N. 1966. 'Basic concept of eutrophication', *J. Wat. Pollut. Control Fed.*, **38**, 737–744.

Schindler, D. W. 1971a. 'A hypothesis to explain differences and similarities among lakes in the Experimental Lakes Area', *J. Fish. Res. Board. Can.*, **28**, 295–301.

Schindler, D. W. 1971b. 'Carbon, nitrogen and phosphorus, and the eutrophication of freshwater lakes', *J. Phycol.*, **7**, 321–329.

Schindler, D. W. and Fee, E. J. 1974. 'Experimental lakes area: whole-lake experiments in eutrophication', *J. Fish. Res. Board. Can.*, **31**, 937–953.

Schlickman, J. A., McMahon, T. M., and Van Riel, N. 1991. *International Environmental Law and Regulation.* Butterworth Legal Publishers, Salem. 555 pp.

Stahre, P. and Urbonas, B. 1990. *Stormwater Detention for Drainage, Water Quality and CSO Management.* Prentice Hall, Englewood Cliffs. 338 pp.

Stumm, W. and Morgan, J. J. 1970. *Aquatic Chemistry: an Introduction Emphasizing Chemical Equilibria in Natural Waters.* Wiley-Interscience, New York. 583 pp.

Thanh, N. C. and Biswas, A. K. 1990. *Environmentally-Sound Water Management.* Oxford University Press, Delhi. 276 pp.

Thienemann, A. 1925. 'Die Binnengewasser Mitteleuropas. Eine limnologische Einfuhrung', *Die Binnengewasser*, **1**, 1–255.

Thornton, J. A. 1980. *A Review of Limnology in Zimbabwe: 1959–1979. National Water Quality Surv. Rep.*, **1**, 86 pp.

Thornton, J. A. 1982. *Lake McIlwaine: the Eutrophication and Recovery of a Tropical African Lake, Monogr. Biol.*, **49**, Junk, The Hague, 251 pp.

Thornton, J. A. 1986. 'Nutrients in African lake ecosystems: do we know all?', *J. Limnol. Soc. S. Afr.*, **12**, 6–21.

Thornton, J. A. 1987. 'Aspects of eutrophication management in tropical/sub-tropical regions', *J. Limnol. Soc. S. Afr.*, **13**, 25–43.

Thornton, J. A. 1991. 'People, politics, pollution and public perceptions: an international look at emerging environmental issues in the 90s' in *Environmental Threats to Wisconsin's Water Resources: Concerns for the '90s, Abstr., 15th Annu. Meeting of the Wisconsin Section, Am. Wat. Resour. Assoc.*, 32.

Thornton, J. A. 1993. 'Perceptions of public waters: water quality and water use in Wisconsin' in van Valey, T., Crull, S. and Walker, L. (Eds), *The Small City and Regional Community.* Vol. 10. Stevens Point Foundation Press, Stevens Point. pp. 469–478.

Thornton, J. A. and McMillan, P. H. 1989. 'Reconciling public opinion and water quality criteria in South Africa', *Water SA*, **15**, 221–226.

Thornton, J. A. and Rast, W. 1987. 'Application of eutrophication modelling techniques to man-made lakes in semi-arid southern Africa' in Nix, S. J. and Black, P. E. (Eds), *Proceedings of a Symposium on Monitoring, Modeling, and Mediating Water Quality, Am. Wat. Resour. Assoc. Tech. Publ. Ser.*, **TPS-87-2**, 547–558.

Thornton, J. A. and Rast, W. 1989. 'Preliminary observations on nutrient enrichment of semi-arid man-made lakes in the northern and southern hemispheres', *Lake Reserv. Manage.*, **5**, 59–66.

Thornton, J. A. and Rast, W. 1993. 'A test of hypotheses relating to the comparative limnology and assessment of eutrophication in semi-arid man-made lakes' in Straskraba, M. L., Tundisi, J. and Duncan, A. (Eds), *Comparative Reservoir and Water Quality Management.* Kluwer Academic, The Hague. 219 pp.

Thornton, J. A., McMillan, P. H., and Romanovsky, P. 1989. 'Perceptions of water pollution in South Africa: case studies from two water bodies (Hartbeespoort Dam and Zandvlei)', *S. Afr. J. Psychol.*, **19**, 197–204.

Thornton, J. A., Rast, W., and Ryding, S.-O. 1991. 'The role of socio-economic determinants in lake management in developing countries', *Lake Reserv. Manage.*, **7**, 115–120.

Tundisi, J. G. and Matsumura-Tundisi, T. 1986. 'Eutrophication processes and trophic state for 23 reservoirs in S. Paulo State, Southern Brasil' in *Fourth Brasil/Japan Symposium on Science and Technology, Suppl. Vol.* Academy of Sciences, Sao Paulo. 26 pp.

United Nations (UN) 1993. *Report of the United Nations Conference on Environment and Development, Rio de Janeiro, 3–14 June 1992.* Vol. I. *Resolutions Adopted by the Conference, Document,* **A/CONF.151/26/Rev. 1**, (Vol. I). United Nations, New York. pp. 275–314.

United States Environmental Protection Agency (US EPA) 1975. *Lake Eutrophication: Results from the National Eutrophication Survey.* US EPA, Corvallis.

United States Environmental Protection Agency (US EPA) 1981. *Restoration of Lakes and Inland Waters*, **EPA 440/5-81-010**, 552 pp.

United States Environmental Protection Agency (US EPA) 1984. *Lake and Reservoir Management*, **EPA 440/5-84-001**, 604 pp.

United States Environmental Protection Agency (US EPA) 1990a. *The Lake and Reservoir Restoration Guidance Manual.* 2nd edn, **EPA 440/4–90–006,** 326 pp.

United States Environmental Protection Agency (US EPA) 1990b. *Monitoring Lake and Reservoir Restoration: Technical Supplement to the Lake and Reservoir Restoration Guidance Manual,* **EPA 440/4–90–007,** 109 pp.

United States Environmental Protection Agency (US EPA) 1993. *Fish and Fisheries Management in Lakes and Reservoirs: Technical Supplement to the Lake and Reservoir Restoration Guidance Manual,* **EPA 841/R-93–002,** 109 pp.

Vallentyne, J. R. 1974. *The Algal Bowl — Lakes and Man. Misc. Spec. Publ.,* **22.** Environment Canada, Ottawa. 185 pp.

Vollenweider, R. A. 1968. *Scientific Fundamentals of the Eutrophication of Lakes and Flowing Waters, with Particular Reference to Nitrogen and Phosphorus as Factors in Eutrophication, Tech. Rep.,* **DAS/CSI/68.27.** OECD, Paris. 154 pp.

Vollenweider, R. A. 1975. 'Input–output models, with special reference to the phosphorus loading concept in limnology', *Schweiz Z. Hydrol.,* **37,** 53–84.

Vollenweider, R. A. 1976. 'Advances in defining critical loading levels for phosphorus in lake eutrophication', *Mem. Ist. Ital. Idrobiol.,* **33,** 53–83.

Vollenweider, R. A., Rast, W., and Kerekes, J. J. 1980. 'The phosphorus loading concept and Great Lakes eutrophication' in Loehr, R. C., Martin, C. and Rast, W. (Eds), *Phosphorus Management Strategies for Lakes.* Interscience, Ann Arbor. pp. 207–234.

Walmsley, R. D. and Butty, M. 1980. *Limnology of Some Selected South African Impoundments.* Water Research Commission and National Institute for Water Research, Pretoria. 229 pp.

Welch, E. B. 1989. 'Alternative criteria for defining lake quality for recreation', *Enhancing States' Lake Management Programs,* **1989,** 7–16.

Wetzel, R. G. 1975a. *Limnology.* Saunders, Philadelphia. 743 pp.

Wetzel, R. G. 1975b. 'General Secretary's Report — 19th Congress of the Societas Internationalis Limnologiae', *Verh. Int. Verein. Limnol.,* **19,** 3232–3292.

Williams, W. D. 1988. 'Limnological imbalances: an antipodean viewpoint', *Freshwater Biol.,* **20,** 407–420.

Wilson, H. M. and O'Sullivan, P. E. 1993. 'The control of eutrophication of small shallow lakes in the United Kingdom: the legal framework', *Verh. Int. Verein. Limnol.,* **25,** 461–464.

Wiseman, K. A. and Sowman, M. R. 1992. 'An evaluation of the potential for restoring degraded estuaries in South Africa', *Water SA,* **18,** 13–20.

Xiangcan, J., Hongliang, L., Qingying, T., Zongshe, Z., and Xuan, Z. 1990. *Eutrophication of Lakes in China.* Chinese Research Academy of Environmental Sciences, Beijing.

8

PATHOGENIC AGENTS IN FRESHWATER RESOURCES

EDWIN E. GELDREICH

Drinking Water Research Division, National Risk Management Research Laboratory, US Environmental Protection Agency, Cincinnati, OH 45268, USA

ABSTRACT

Numerous pathogenic agents have been found in freshwaters used as sources for water supplies, recreational bathing and irrigation. These agents include bacterial pathogens, enteric viruses, several protozoans and parasitic worms more common to tropical waters. Although infected humans are a major source of pathogens, farm animals (cattle, sheep, pigs), animal pets (dogs, cats) and wildlife serve as significant reservoirs and should not be ignored. The range of infected individuals within a given warm-blooded animal group (humans included) may range from 1 to 25%. Survival times for pathogens in the water environment may range from a few days to as much as a year (*Ascaris, Taenia* eggs), with infective dose levels varying from one viable cell for several primary pathogenic agents to many thousands of cells for a given opportunistic pathogen.

As pathogen detection in water is complex and not readily incorporated into routine monitoring, a surrogate is necessary. In general, indicators of faecal contamination provide a positive correlation with intestinal pathogen occurrences only when appropriate sample volumes are examined by sensitive methodology.

Pathways by which pathogens reach susceptible water users include ingestion of contaminated water, body contact with polluted recreational waters and consumption of salad crops irrigated by polluted freshwaters. Major contributors to the spread of various water-borne pathogens are sewage, polluted surface waters and stormwater runoff. All of these contributions are intensified during periods of major floods. Several water-borne case histories are cited as examples of breakdowns in public health protection related to water supply, recreational waters and the consumption of contaminated salad crops. In the long term, water resource management must focus on pollution prevention from point sources of waste discharges and the spread of pathogens in watershed stormwater runoff.

INTRODUCTION

Pristine water resources, both surface and groundwaters, are becoming more scarce because of global increases in population and the active intervention of humans in the environment. In the process, an increasingly adverse impact on the earth's water resources has developed in terms of the available reserves, quality, and natural self-purification capacity. Once it has been determined that a given water resource can satisfy the maximum daily demand for water supply in the community, the next important consideration is water quality and the degree of treatment necessary to satisfy a specific use. The latter determines the limits for chemical contaminants and microbial agents and the level of acceptable health risk to animals, especially to humans. The entry of pathogenic agents into the water resource intended for drinking, recreation, livestock feed, irrigation and aquaculture can become a serious human health risk or economic disaster in agriculture. How this adverse situation can develop in freshwater systems and the evidence for pathogens in freshwater resources are the themes that will be examined in this paper.

CHARACTERIZING FRESHWATER QUALITY

In remote areas where the human and farm animal populations are sparse, it is possible to find little

Table I. Pathogen pathways from faeces via wastewater to contaminated drinking water*

Pathogen agents	Quantity excreted by infected individual/ g-faeces	Maximum survival in water (days)	Infective dose†	Contaminated water supply outbreaks		
				USA‡	Sweden§	UK¶
Bacterial						
Toxigenic *E. coli*	10^8	90	10^2-10^9	5	1	—
Salmonella	10^6	60–90	10^{6-7}	37	2	9
Shigella	10^6	30	10^2	52	2	4
Campylobacter	10^7	7	10^6	5	5	5
Vibrio	10^6	30	10^8	1	—	—
Yersinia enterocolitica	10^5	90	10^9	1	—	—
Aeronomas	—	90	10^8	—	1	—
Viruses						
Enterovirus	10^7	90	1–72	—	—	—
Heptatitis A	10^6	5–27	—	51	1	—
Rotavirus	10^6	5–27	—	1	1	—
Norwalk	—	5–27	—	16	—	8
Parasites						
Entamoeba	10^7	25	10–100	3	—	1
Giardia	10^5	25	1–10	84	1	1
Cryptosporidium‖	—	—	5–10	2	—	2
Balantidium coli	—	20	25–100	—	—	—
Ascaris	10^3	365	2–5	—	—	—
Taenia	10^3	270	1	—	—	—
Unknown agents	?	?	?	266	19	—

* Data selected from Geldreich (1978), Kowal (1982) and Prost (1987)
† Infectious dose to provoke clinical symptoms in 50% of individuals tested
‡ Water-borne outbreaks in USA during 1961–1983; Craun (1985)
§ Water-borne outbreaks in Sweden during 1975–1984: Andersson and Stenstrom (1986)
¶ Water-borne outbreaks in England and Wales during 1937–1986: Galbraith *et al.* (1987)
‖ *Cryptosporidium* outbreaks: D'Antonio *et al.* (1985) and Smith (1992)

evidence of faecal contamination of the water resource. except low residuals from occasional wildlife inhabiting the immediate vicinity. Beaver and muskrats on the river banks, and significant concentrations of deer, elk and other game animals in forest reserves, can add faecal contamination and detectable pathogenic agents. As these waters travel down the watershed, agricultural and industrial activities increase and the river becomes laden with a variety of domestic and industrial wastes. Faecal discharges entrapped on soil will also be moved into the drainage basin by the flushing action of storm events.

In a similar fashion, high elevation lakes in remote portions of the watershed will contain high quality water (unless inhabited by flocks of aquatic birds), whereas those in the lower part of the watershed are fed with surface drainage from metropolitan areas and intense agricultural activity. For large lakes such as the Great Lakes, long holding times and vast volumes of water serve to buffer these magnificent water resources from the impact of stormwater runoff. Unfortunately, pollution plumes around wastewater discharges may be temporarily spread to new areas by wind-driven currents. At greatest risk are the numerous small lakes that become surrounded by residential developments that ultimately result in sporadic drainage from septic systems and runoff of fertilizers from neighbourhood lawns.

The microbial quality of groundwater is often superior to that of surface waters because of an effective barrier of soil on top of the impervious rock strata that cap the aquifer. As a consequence, quality is uniformly excellent with little influence from climatic changes and stormwater migrations through shallow depths of soil. The exception is rock strata composed of limestone, which is very porous and often results in sink holes and caverns through which surface water passes without the effective entrapment of microorganisms. In other situations, excessive surface application of minimally treated wastewaters may inundate

Table II. Major infectious agents found in contaminated drinking waters world-wide*

Bacteria	Viruses	Protozoa
Campylobacter jejuni	Adenovirus (31 types)	*Balantidium coli*
Enteropathogenic *E. coli*	Enteroviruses (71 types)	*Entamoeba histolytica*
Salmonella (1700 spp.)	Hepatitis A	*Giardia lamblia*
Shigella (4 spp.)	Norwalk agent	*Cryptosporidium*
Vibrio cholerae	Reovirus	
Yersinia enterocolitica	Rotavirus	
Helminths	Cox sackie virus	
Ancylostoma duodenale		
Ascaris lumbricoides		
Dracunculus medinensis		
Echinococcus granulosis		
Necator americanus		
Strongyloides stercoralis		
Taenia (spp.)		
Trichuris trichiura		

*Data from Geldreich (1990)

the natural soil barrier (Kowal, 1982). Land application of animal wastes from feedlot operations, water plant sludges or garbage wastes which are improperly located can also contribute significant pathogen releases in leachates which reach the groundwater or emerge to drain into surface waters. For this reason, the treatment of manure by composting, drying and storage before land application would have a significant effect on reducing pathogen releases to water. Once the aquifer becomes contaminated, the restoration of water purity is very slow, even with an intervention such as pumping the water to a treatment site and then returning it to the aquifer.

Much of the groundwater contamination is found in shallow wells, less than 100 ft deep. In these wells, the source water is influenced by surface water runoff that percolates through the soil. As there is no protective bedrock perched on top to seal off surface contaminants, the water quality is erratic.

WATER-BORNE PATHOGENIC AGENTS

Numerous pathogenic agents have been isolated from ambient waters used for water supply, recreational bathing and the irrigation of garden salad crops (Geldreich, 1972a; Rosenberg *et al.*, 1976; Craun, 1988; Cordano and Virgilio, 1990; Rose, 1990). The serious nature of these pathogen occurrences in wastewater and the consequences to water supply source protection and treatment failure can be seen in Table I. The list of water-borne agents in both temperate and tropical regions of the world will increase as new methodologies evolve to detect the more elusive organisms that cause gastroenteritis or other human illnesses. Perhaps *Helicobacter pylori*, which is associated with stomach ulcers and water ingestion, should be included in this expanding list of recognized water-borne pathogenic agents (Klein *et al.*, 1991; West, *et al.*, 1992).

All of these micro-organisms can be classified within four broad groups: bacteria, viruses, protozoans and helminths (Table II). The listing of water-borne agents could be further expanded to include exposure to pathogenic leptospires, parasitic protozoa such as *Naegleria fowleri* and parasitic worms that include *Schistosoma mansoni*, *S. japonicum* and *Diphyllobothrium latum*. Exposure to these aquatic pathogens occurs in recreational waters and irrigation water farming. Aerosol exposure to water supply used in the air cooling of buildings or through contaminated shower attachment devices can lead to respiratory illness caused by fungal spores and *Legionella* (Geldreich, 1972a). Beyond this grouping there are additional pathogens which are at this time 'non-culturable', largely because of their fastidious requirements for unique nutrients. In nature these substances are found in the metabolic wastes of other organisms in the aquatic flora. *Legionella* can be cited as an example of an organism that is ubiquitous in the environment,

Table III. Isolation of salmonellae from muds of various quality waters compared with isolation from water

Source	Faecal coliforms per 100 ml water	Total examinations	Isolation of *Salmonella*	
			Positive occurrence	Percentage
Mud	1–200	21	4	19·1
	201–2000	12	6	50·5
	Over 2000	15	12	80·0
Fresh water	1–200	29	8	27·6
	200–2000	27	19	70·3
	Over 2000	54	53	98·1
Estuarine water	1–200	258	33	12·8
	201–2000	91	40	44·0
	Over 2000	75	45	60·0

Data from Geldreich (1970) and Van Donsel and Geldreich (1971)

yet could not be cultured in the laboratory until a few years ago. *Giardia* and *Cryptosporidium* are examples of organisms that can be identified under the microscope, but cannot be successfully cultivated to prove their viability.

Opportunistic water-borne pathogens deserve considerable attention. Unlike primary pathogens that can infect all ages, these organisms are particularly invasive to susceptible individuals (the elderly, newborn infants, patients with AIDS and irrigation farm workers). In contrast with the gastroenteritis pathogens, many of these organisms cause infections of the respiratory organs, skin, ear, eye and urinary tract. Exposure to opportunistic pathogens in the outdoor environment may occur from the aerosol inhalation of spraying sewage effluents and from body contact with contaminated recreational or irrigation water.

PATHOGEN/SURROGATE RELATIONSHIP

Pathogen detection in water is complex and not readily incorporated into routine monitoring. As these organisms are present in relatively low densities, the search for pathogens begins with 1-l samples for bacterial agents, whereas samples of 100 l or more are necessary for the detection of viruses and even larger volumes for protozoans. The problem is further complicated by the limitation of methodology: hepatitis A cannot be cultured; many agents of gastroenteritis remain unknown; and the recovery of *Giardia* cysts and *Cryptosporidium* oocysts ranges from 4·5 to 14·2% and 5·8 to 11·6% respectively (Clancy *et al.*, 1993). These difficulties have led to a reliance on surrogate indicators for the presence of pathogens in aquatic environments.

As many of these agents are intestinal pathogens, shed in faecal material from infected individuals, the most logical alternative to multiple pathogen analysis is an indicator of proved correlation with faecal contamination in water. Various indicators of faecal contamination in water have been proposed: heterotrophic plate counts at 28 and 35°C; total coliform; faecal coliform; *E. coli*; faecal streptococcus; enterococcus; acid-fast bacteria; sulphite-reducing clostridia; *Bifidobacteria*; *Bacteroides*; and bacterial phages. None of these has proved to be the perfect answer because of differences in the survivability of various pathogens and the possibility that some bacterial indicators multiply in environmental waters. In many instances, the presence of surrogates suggests the occurrence of pathogenic organisms; however, the absence of indicators may not ensure the absence of some pathogen. Part of this conflict in interpretation is caused by the selection of small (100 ml) sample volumes used for indicator detection versus large sample analysis (100 l or more) for virus and protozoan cysts. In essence, water quality indicators should be compared with pathogens on a more realistic scale by increasing the sample volume size examined for surrogate occurrence.

Table IV. Percentage of warm-blooded animals excreting pathogens world-wide

Pathogen agent	Percentage of individual excretors*						
	Human	Cattle	Sheep	Pig	Dog	Cat	Wildlife†
Bacteria							
Salmonella	1–3·9	13	3·7–15	7–22	1–5	—	1–21·3
Shigella	0·33–2·4	—	—	—	—	—	—
Leptospira	<1–3·0	2·3	—	2·5	26·6	—	5·9–33
Campylobacter	3·0	—	—	—	—	—	1–10
Pathogenic E. coli	1·2–15·5	3·5	2·0	1·5–9	—	—	—
Vibrio cholera	1·9–9·0	—	—	—	—	—	—
V. cholera El Tor (Asia)	9·5–25·0	—	—	—	—	—	—
Versinia	—	—	—	1–10	—	—	1–10
Parasites							
Entamoeba histolytica	10–17	—	—	—	—	—	—
Cryptosporidium	0·6–4·3	16–100	40	95	80	87	1–68
Giardia	7·4	10	33	—	13	2·5	6
Viruses							
Enteroviruses	0·88	—	—	—	—	—	—

* Data revised from Geldreich (1989)
† Birds, beaver, deer, coyote, mice, etc
(—) Occurence unknown.

Of the indicator systems currently in use, the faecal coliform group, does satisfy many of the requirements: positive correlation with faecal contamination; simplicity of test performance; and minimal cost to a monitoring programme (Geldreich, 1966). As faecal coliforms have an excellent correlation with faecal contamination, it would be logical to project a positive correlation between water-borne pathogens and increasing densities of faecal coliforms in the same body of water. Such a record of correlation between the frequency of *Salmonella* occurrence and the range of faecal coliform density has been found in both fresh and estuarine waters (Slanetz *et al.*, 1965; 1968; Clemente and Christensen, 1967; Geldreich, 1970, Brezenski, 1971). These findings (Table III) demonstrate that in freshwater, *Salmonella* detection increased from 27 to 85% as the faecal coliforms increased from 200 to over 1000 organisms per 100 ml. In fact, *Salmonella* could be isolated from nearly all samples when the faecal coliform concentration exceeded 2000 organisms per 100 ml. Thus the chances of detecting *Salmonella* increased in direct relationship to the amount of faecal contamination present. Furthermore, a sharp increase occurred in the *Salmonella* detected in the bottom deposits of various lakes, rivers and creeks when the faecal coliform densities of the overlying waters were greater than 200 organisms per 100 ml. In these waters, *Salmonella* were recovered from two-thirds of the mud samples examined (Van Donsel and Geldreich, 1971).

Attempts to correlate virus and protozoan occurrences with faecal coliforms or other surrogates have been few, largely because of the cost of virus and protozoan analyses. The complexity of the analyses and the special training needed in virology and parasitology have also had an impact on the availability of such field data on ambient waters. Another confounding factor has been the disparity among sample volumes used to make these comparisons. Levels of enteric viruses in raw sewage and receiving waters are expected to be 1000 virus units per liter or less (Akin and Hoff, 1978), whereas *Giardia* cyst densities are often found in the one to ten cysts per 100 liter range. Thus, when 'relatively clean' natural waters are reported to have no faecal coliforms in a 100 ml portion of the sample, yet the water does contain virus, *Giardia* or *Cryptosporidium*, the problem is due to mis-matched sample volumes. As a consequence, much of the field data on faecal coliform and pathogen occurrences is limited to polluted waters where the differences in test volumes was not a factor in finding the surrogate present. Field studies attempting to correlate pathogen occurrence with surrogate organisms must increase the standard volume of water used for coliform analyses (Bitton *et al.*, 1986).

Monitoring faecal bacterial surrogates cannot be expected to give meaningful correlations with those pathogens that cause respiratory infections or infections of the skin, ear, eye, nose and throat. These opportunistic pathogens are important agents associated with the inhalation of contaminated water aerosols or through body contact sports (swimming, water-skiing). Recent epidemiological evidence suggests that staphylococcus detection in recreational waters is a more realistic indicator of swimming-associated skin infections (Seyfried et al., 1985). A suitable surrogate indicator of respiratory risk from water-borne pathogens is less clearly defined.

ORIGINS OF WATER-BORNE PATHOGENS

Varying inputs of faecal wastes from all warm-blooded animals (humans, domestic pets, farm animals and wildlife) reach the water and soil environments directly or indirectly through seepage from sanitary landfills or poorly processed sewage effluents from municipal treatment facilities and malfunctioning septic tanks. The frequency of pathogen shedders in the human population, domestic animals and wildlife has been estimated to range from less than 1 to 25% of a given animal species (Table IV). Although pathogenic bacteria, protozoans and parasitic worms may be found in a wide range of warm-blooded animal hosts, human viral pathogens are shed only by infected people. Several other significant pathogens such as *Legionella* and *Naegleria* do not have an animal host reservoir, but appear to be ubiquitous in the environment (free-living organisms) and comprise both non-pathogenic and pathogenic strains (Chang, 1970; Thornsberry et al., 1984). It is not clear why some strains in the environment are predisposed to be non-pathogenic, whereas others of the same genus, or even of the same species but different in serotype reaction, are pathogenic. These latter strains can quickly become opportunistic, invading the human body under conditions of stress, weakened immune system or through the general physical degeneration associated with advancing age.

The degree and frequency of pathogen exposure is exacerbated by expanding human populations worldwide. Crowding and the unregulated development of satellite communities places undue pressures on sanitation infrastructural barriers (water supply and distribution, sewage collection and treatment, solid waste disposal). Add to these problems the mobility of people on an international scale and it becomes apparent that disease can quickly reach epidemic proportions.

The 1991 cholera outbreak in South and Central America indicates the role that poor water quality plays in this public health problem. The pathogenic agent, *Vibrio cholera* 01, biotype El Tor, was first detected in Indonesia during 1934. Initially, this pathogen remained a regional pandemic occurrence in south-east Asia until the 1960s, whereupon it appeared in several countries in the Near East. By 1970 it was detected in outbreaks in Russia and South Korea. Until January 1991, the Americas had not been exposed to this pathogenic strain. Suddenly, the first cases of El Tor cholera were identified in the port city of Lima, Peru. Within a few days the disease had spread north to coastal cities in Peru and then quickly moved into neighbouring countries in South America. Within weeks, cholera was reported in almost all the countries of Central America.

The initial exposures were thought to be the result of eating contaminated raw fish. However, the pathogenic agent spread rapidly due to poor sanitation practices, appearing in contaminated groundwater and surface waters used for water supply. Inadequate treatment of public water supplies then hastened the escalation of the illness. The epidemic quickly spread to neighbouring countries in South America as a direct result of infected people travelling to previously unaffected areas. No quarantine was in effect at the borders to screen out ill people nor to inspect for the presence of raw fish products in personal luggage. As a consequence, isolated cases of cholera in the USA were traced to the consumption of raw fish brought back by travellers from outbreak areas of South America.

The rapid spread of disease is not as great a probability in developed nations because of public health policy protecting water resources and the use of multiple treatment barriers against pathogens in the aquatic environment. The multiple barrier concept includes: (1) the collection and treatment of municipal sewage and food processing wastes; (2) the enhancement of natural self-purification in rivers and lakes by the management of point source effluents and releases of stormwater runoff; (3) the appropriate treatment

Table V. Microbial densities in municipal raw sewage from two cities in South Africa

Organism or microbial group	Average count per 100 ml	
	Worcester sewage	Pietermaritzburg sewage
Aerobic plate count (37°C; 48 h)	1 110 000 000	1 370 000 000
Total coliforms	10 000 000	—
E. coli, type 1	930 000	1 470 000
Faecal streptococci	2 080 000	—
C. perfringens	89 000	—
Staphylococci (coagulase positive)	41 400	28 100
Ps. aeruginosa	800 000	400 000
Salmonella	31	32
Acid-fast bacteria	410	530
Ascaris ova	16	12
Taenia ova	2	9
Trichuris ova	2	1
Enteroviruses and reoviruses (TC1D$_{50}$)	2890	9500

*TC1D, Tissue culture infective dose
Source: Grabow and Nupen (1972)

of raw waters used to produce water supply; and (4) the continuous protection of drinking water in distribution systems.

Cattle feedlots and poultry operations result in an unusual concentration of farm animals and their faecal wastes in a confined space. In feedlot operations, the density of beef cattle per square mile may approach 10 000 animals. Under such restrictions, the removal of faecal wastes is a major disposal operation. The closeness of these animals in confined feeding operations invites the spread of disease in a healthy herd or poultry flock. Some animal pathogens such as *Salmonella* and *E. coli* 0157 : H7 are also human pathogens. Unless careful application of treated animal manures is practised, faecal material in stormwater runoff from feedlots and poultry farms becomes a major source of contamination in rural watersheds, polluting streams and lakes in its drainage path. Recycling untreated farm animal wastes by application to fields may become an intensive contributor of pathogens to streams unless contour cultivation is carried out to reduce runoff.

Wildlife refuges are also a significant source of faecal contamination, often on a seasonal basis, as many of these animals migrate seasonally in their search for food. The largest threat of wildlife pathogens is from those warm-blooded animals such as beaver, deer, coyote and gulls that are permanent residents of a watershed. These animals and others serve as reservoirs for *Giardia*, *Cryptosporidium*, *Salmonella*, *Campylobacter* and *Yersinia* among the water-borne bacterial and protozoan pathogens.

Wildlife is also attracted to protected watershed areas where human activities are more restricted (Walter and Bottman, 1967). Protected nearshore water environments are often the location of large beaver colonies, including individual animals infected with *Giardia*. Infected coyotes, muskrats and voles are other wild animals that may be involved in the shedding of *Giardia* cysts and other pathogens into the aquatic environment. Terrestrial birds and waterfowl can be sources of bacteria pathogens. The songbird population includes individuals that may be infected with *Salmonella*. Gulls are scavengers that frequent open garbage dumps, eat contaminated food wastes and contribute *Salmonella* in their faecal droppings to coastal lakes (Alter, 1954; Fennel *et al.*, 1974). In one instance, gulls were the contributors of *Salmonella* to an untreated surface supply in an Alaskan community, causing several cases of salmonellosis (Anon., 1954). Wildlife are also believed to be the source of *Campylobacter* that contaminated untreated or poorly treated streams and reservoirs of low turbidity in Vermont (Vogt *et al.*, 1982) and British Columbia (Health and Welfare, Canada, 1981). Vacationers to national parks in Wyoming became ill after drinking water from mountain streams, resulting in a 25% increase statewide in *Campylobacter* enteritis (Taylor *et al.*, 1983). Therefore, watershed management policies should include an annual programme to assess the potential pathogen threat from carriers in the wildlife population. This information can be invaluable to water supply treatment operations and to the laboratory strategy for monitoring water quality in distribution.

PATHOGEN PATHWAYS

In the aquatic environment, pathogen passage to susceptible water users is generally through the ingestion of contaminated water, body contact in recreational waters, or by the consumption of shellfish raised in contaminated harvesting beds and salad crops irrigated with polluted waters. Major contributors to the spread of various water-borne pathogens are sewage, polluted surface waters and stormwater runoff.

Although sewage collection systems have decreased the public health risk in urban centres, this practice only transports the collected wastes to some destination where treatment is supposed to be applied before release into a water course. Raw sewage discharges have often been shown to contain a variety of pathogens. The density and variety of human pathogens released is related to the population served by the sewage collection system, by seasonal patterns for certain diseases and by the extent of community infections at a given time. Some indication of the relative occurrences of various pathogens in raw sewage is given in Table V for two cities in South Africa (Grabow and Nupen, 1972). As the methodology differs for the wide spectrum of pathogens that might be present, these findings in sewage represent only a portion of the health threat that was identified. If socio-economic factors and epidemic eruptions are overlaid on these expected occurrences of pathogenic agents in sewage, the exposure risk to people, farm animals and wildlife will be seen as being unacceptable.

It has been hypothesized that a sewage collection network of 50–100 homes is the minimum size required before there is a chance of the successful detection of *Salmonella bacilli* (Callagan and Brodie, 1969). *Salmonella* strains were regularly found in the sewage system of a residential area of 4000 people (Harvey *et al.*, 1969). In another study, 32 *Salmonella* serotypes were found in sewage effluent samples and downstream sites on the Oker River in Germany (Popp, 1974).

In major river systems receiving discharges of meat processing wastes, raw sewage and effluents from ineffective sewage treatment plants, the densities of *Salmonella* species may be substantial. It has been calculated that the Rhine and Meuse rivers carried approximately 50×10^6 and 7×10^6 *Salmonella* bacilli per second, respectively (Kampelmacher and Van Noorle Jansen, 1973). The Missouri River represents another example of a pollution conduit, transporting a faecal pollution load from raw sewage, effluents from treatment plants of differing efficiencies, runoff from cattle feed lots and waste discharges from meat and poultry processing plants. As a consequence, it is not surprising that various *Salmonella* serotypes and viruses have been detected at the public water supply treatment plant intakes at Omaha, St Joseph and Kansas City. In this 250 mile stretch of the river, faecal coliforms ranged from 1950 to 8300 organisms per 100 ml during late autumn 1968, indicating that the pollution loading was beyond the capacity for stream self-purification.

The documentation of gross pollution of surface water quality in developing countries is sparse, but nevertheless represents an even greater threat because the treatment of domestic and food processing wastes is often marginal or non-existent. In India, for example, the high population densities along river valleys has led to considerable pollution problems. Of India's 3119 cities, only 209 have partial treatment of sewage and just eight cities utilize secondary treatment schemes (WRI, 1986). In this situation, it is not surprising that for a 48 km stretch of the Yamuna river (above New Delhi) there may be at least 7500 coliforms per 100 ml. After an estimated discharge of 200×10^6 l of raw sewage from New Delhi, the coliform density in the Yamuna River suddenly escalates to 24×10^6 coliforms per 100 ml. Along with such a dramatic rise in faecal waste discharges there is also the certainty of continuous releases of a variety of pathogens shed by infected individuals living in this densely populated area.

Stormwater and regional floods are often the major cause of transient deteriorations in water resources. The impact of stormwater runoff on water quality relates to all land uses over the drainage basin. Rural runoff of stormwater can be a very significant contributor of *Giardia, Cryptosporidium, Campylobacter*

Table VI. Summary of microbiological data from Detroit and Ann Arbor overflows

Month	Analysis	Separate system — Ann Arbour			Combined system — Detroit		
		Density per 100 ml*	Per cent faecal coliform	FC/FS ratio	Density per 100 ml*	Per cent faecal coliform	FC/FS ratio
April	Total coliform	340 000	2·9		2 400 000	37·1	
	Faecal coliform	10 000			890 000		
	Faecal streptococci	20 000		0·5	—		—
May	Total coliform	510 000	10·0		4 400 000	34·1	
	Faecal coliform	51 000			1 500 000		
	Faecal streptococci	200 000		0·26	320 000		4·7
June	Total coliform	4 000 000	2·0		12 000 000	22·4	
	Faecal coliform	78 000			2 700 000		
	Faecal streptococci	120 000		0·65	740 000		3·7
July	Total coliform	4 000 000	3·0		37 000 000	20·5	
	Faecal coliform	120 000			7 600 000		
	Faecal streptococci	390 000		0·31	350 000		21·7
August	Total coliform	1 700 000	20·6		26 000 000	16·9	
	Faecal coliform	350 000			4 400 000		
	Faecal streptococci	310 000		1·1	530 000		8·3

Data of 1964 adapted from Burm and Vaughan (1966)
* Geometric means per 100 ml

and *Yersinia* from wildlife and farm animals living on the watershed. Heavy loads of faecal pollution are common to stormwater runoff from cattle and poultry feedlots. One study equated feedlot operations runoff near Hereford, Texas to the discharge of raw sewage for a city of approximately 10 000 people (Geldreich, 1972b). The prevalent pathogens shed by infected cattle are *Salmonella*, *E. coli* 0157 : H7 and *Cryptosporidium*, all of which are also pathogenic to humans.

Urban stormwater runoff can be a major factor in the fluctuating quality of surface waters. For those cities that collect both stormwater and domestic wastes in the same pipe system, the problem of treatment capacity is critical. Sudden large inflows of stormwater from major storms may overwhelm the treatment capacity, resulting in the need to by-pass some portion of the mixed waste directly to receiving waters. In recent years, Chicago and several other large cities have constructed huge underground storage tunnels to hold untreated combined wastes for a few days until the excess can be treated. Separate collection systems for storm water and domestic wastes are also a common alternative (Table VI), but the storm drainage is often discharged to the receiving waters untreated. The most concentrated pathogens in urban stormwater around the Baltimore area were found to be *Pseudomonas aeruginosa* and *Staphylococcus aureus* at levels of 10^3 to 10^5 per ml and from 10^1 to 10^3 per ml, respectively (Oliveri et al, 1977). *Salmonella* and enteroviruses were often isolated, but at much lower densities, ranging from 10^1 to 10^4 per 10 liters of urban runoff. These pathogens originated in sewage that was diluted with rainwater in storm sewer drains and through the combined sewer overflows.

The Midwest flood of 1993 is an example of a natural disaster that had a tremendous impact on water quality. Among the public health concerns cited were: manure and dead animals from cattle and poultry feedlots; fertilizers; herbicides; pesticides; and other chemical industrial products. Among the more bizarre concerns in this flood was the movement of coffins and body parts out of a cemetery. A total of 707 coffins were washed out of an old rural cemetery near Hardin, Missouri, and carried as far as 14 miles. During the peak of the flood, 18 wastewater plants in the Missouri river basin were subject to flooding. In Janesville, Wisconsin the city sewage treatment plant became inundated and spilled more than one million gallons of raw sewage into the Rock River. Floodwaters overflowed the water works in Des Moines, Iowa, St Joseph, Missouri and several smaller water utilities along the Des Moines and Missouri rivers (McMullen, 1994). Throughout the flooded areas of Missouri, Kansas, Illinois and Iowa, there were 250 contaminated public water supplies (Reid, 1994). Sand filters became clogged with a thick layer of fine silt and the distribution pipe networks had to be flushed of sediments and high dosages of chlorine applied to decontaminate the distribution system.

Thousands of private water supplies were inundated by flooding and had to be vigorously pumped to flush out contaminants, then chlorinated to restore water quality. In a nine state water quality survey (Missouri, Minnesota, Nebraska, North Dakota, South Dakota, Illinois, Iowa, Kansas and Wisconsin), carried out after the flood period, several facts emerged. Private wells that were less than 100 ft deep were slow to recover from contamination because of soil saturation. These are the wells that are most often under the influence of surface water penetration because of poor soil barrier conditions. Deep wells which were properly protected by casing and well head protection were quick to be restored to excellent quality, with little permanent damage to the aquifer. Apparently, hydrostatic pressure from the aquifer was an effective barrier to flood water penetration. The Iowa Hygienic Laboratory surveillance for water-borne outbreaks observed a decrease in *Salmonella, Shigella, Campylobacter* and *Cryptosporidium* episodes during the flood period, probably because recreational activities were shut down. Three cases of Leptosporosis did occur in the state and many have been the result of wading in flood waters. After the flood waters crested, the dilution effect of the floodwaters and the relatively quiescent state of the slowly receding flood led to the sedimentation of fine silt and associated micro-organisms. Only the heroic efforts of public health officials involved in emergency operations were able to prevent water-borne outbreaks by providing a safe water supply and curbside portable toilets to a sanitation-conscious public.

Flooding sewage and stormwater contaminate both surface waters and groundwater systems, and many private wells may become unsafe because the soil barrier is breached. Over the years, a variety of water-borne disease outbreaks have been attributed to contaminated aquifers or poorly protected well sites that contained pathogenic bacteria, virus, protozoan and worms. Specific bacterial pathogens that have been isolated from well waters include enteropathogenic *E. coli, Vibrio cholera, Shigella flexneri, S. sonnei, Salmonella typhimuruim, Yersinia enterocolitica* and *Campylobacter* (Greenberg and Jongerth, 1966; Schroeder *et al.*, 1968; Lassen, 1972; Center for Disease Control, 1973; 1974; 1980; Evison and James, 1973; Lindel and Quinn, 1973; Woodward *et al.*, 1974; Dragas and Tradnik, 1975; Highsmith *et al.*, 1977; Schieman, 1978; Mentzing, 1981). Poliovirus and enterovirus have been isolated from a well water used to supply water to restaurant patrons (Vander Velde and Mack, 1973).

Concentrated reservoirs of pathogens may also contaminate the aquatic environment from alternative methods of waste disposal. Land applications of minimally treated wastewaters can contaminate ground-waters in areas where there is rapid infiltration of soils (high percolation rates) and surface waters may also become polluted through excessive overland spraying in the drainage basin (Kowal, 1982). Under these conditions, the natural self-purification processes (desiccation, acid soil contact, sunlight exposure, soil organism competition, antagonism and predation) become either inoperative or ineffective. Improperly located or poorly engineered sanitary landfills for animal wastes from feedlot operations, water plant sludges and refuse wastes can also contribute significant pathogen releases to ground and surface waters (Geldreich, 1978). Urban solid wastes (refuse) contain not only food discards, trash, plastics, cloth, paper products, soil, rock and ash residues, but also faecal material. This faecal material is derived from disposable baby diapers, animal pet wastes in litter materials and the droppings of rodents and birds (gulls, etc.) foraging for food in exposed waste dumps. Dredging of river channels has also been noted to recirculate viable pathogens from the sludge banks that accumulate around sewage outfalls and boat harbors (Grimes, 1980).

PATHOGEN PERSISTENCE IN AMBIENT WATERS

Upon discharge into environmental waters, the persistence of pathogens is a variable determined by many factors. For example, *Salmonella* was detected with regularity in surface water up to 250 m downstream from a wastewater treatment plant, but never at sample sites 1·5–4 km upstream (Kampelmacher and Van Noorle Jansen, 1976). *Salmonella* transported by stormwater through a wastewater drain at the University of Wisconsin experimental farm were isolated regularly at a swimming beach 800 m downstream

(Claudon *et al.*, 1971). Excessive biochemical oxygen demand (BOD) or total organic carbon (TOC) in a poor quality waste water effluent and low stream temperatures can also depress stream self-purification processes. For example, *Salmonella* were isolated in the Red River of the North (North Dakota, Minnesota), 22 miles downstream of sewage discharges from Fargo, North Dakota and Moorhead, Minnesota during September (US Department of Health, Education and Welfare, 1965). By November, *Salmonella* strains were found 62 miles downstream of these two sites. In January, with the beginning of the sugar beet processing season, wastes reaching the stream under the cover of ice brought high levels of bacterial nutrients. *Salmonella* were then detected 73 miles downstream — four days flow time from the nearest point source discharges of warm-blooded animal pollution.

Sedimentation is one of the natural self-purification factors that can rapidly remove pathogens from the overlying water into bottom deposits. This action takes place in quiet segments of a stream or during long retention periods in lakes and impoundments. Over a one year period, approximately 90% of *Salmonella* species obtained from the North Oconee river (Georgia) were recovered in bottom sediments. The adsorption of the organisms to sand, clay and sediment particles resulted in their concentration in the bottom deposits (Hendricks, 1971). A maximum density of 11 *Salmonella* organisms per 100 ml was detected in canal water, whereas canal sediments yielded *Salmonella* densities up to 150 organisms per 100 g (Andre *et al.*, 1967).

WATER-BORNE PATHOGEN INVASIONS: CASE HISTORY EXAMPLES

How serious are the exposures to pathogenic agents in the aquatic environment? Several case histories involving exposure to contaminated water supply, recreational bathing in sewage-laden river water and irrigation of salad crops with grossly polluted surface waters illustrate the concern:

Water supply risks

As everyone drinks water, most reported outbreaks have involved the contamination of water supply. Many of these outbreaks go unreported in developing countries, thus much of the documentation describes case histories in North America and Western Europe. Three case histories will be noted to illustrate the consequences of barrier failures, either as a result of a change in treatment or the loss of the protective integrity of the distribution network.

Cryptosporidium outbreak. The City of Milwaukee experienced a water-borne outbreak starting in April 1993 that was caused by *Cryptosporidium*. Approximately 403 000 people developed diarrhoea with flu-like symptoms of nausea and stomach cramps. The outbreak was preceded by a series of weather events, including seasonal lake turnover in the nearshore areas, heavy rainfall that exceeded the storm sewer capacity and springtime increases in agricultural runoff. All of these events resulted in an increase in flow from small streams draining into Lake Michigan and a drift of the pollution plume in the direction of the raw water intake for one of the two treatment plants.

Coincidental to the weather events, the conventional water treatment process was being modified to lower pH to reduce lead released from lead water pipes on service lines. To do this, the traditional alum coagulant was replaced by polyaluminium hydroxychloride (PAC) directly in the process basin. The combined changes in raw water quality and chemical treatment brought about sudden irregularities in process water turbidities, although still within the limits specified by Federal Regulations. With these 'spikes' in turbidity came a filter breakthrough of the *Cryptosporidium* oocysts into the filter effluent. It is conceivable that recycling the filter backwash water may have exacerbated the release of more oocysts flowing through the process basins. As post-chlorination is not effective for *Cryptosporidium* inactivation, the pathogenic agent passed into the distribution system to the consumer's tap.

E. coli 0157 : H7 outbreak. An outbreak of hemorrhagic *E. coli* serotype 0157 : H7 occurred in the small farm community of Cabool, Missouri (population 2090) during the period from 15 December 1989 to 20 January 1990 (Geldreich *et al.*, 1992), resulting in four deaths, 32 admissions to hospital and a total of 243 cases of diarrhoea. This pathogenic organism was found in the faeces of some infected individuals and the initial investigation sought to locate a common source of contaminated raw beef or tainted milk, foods that had been implicated as the source of this agent in several published studies (Martin *et al.*, 1986; Borczyk *et al.*, 1987). When the initial investigation proved negative, attention was focused on the water supply. Based on a household survey conducted by the Centers for Disease Control, it was concluded that persons living inside the city (using municipal water) were 18·2 times more likely to develop bloody diarrhoea than persons outside the city using private well water supplies. At this point a 'boil water' order was issued and the number of new cases rapidly declined.

Investigation revealed that the Cabool water supply was untreated groundwater supplied by four municipal wells drilled between 1000 and 1300 f into an aquifer capped by limestone formation and sink holes. Although the type of soil barrier was a prime suspect, monitoring data revealed no coliforms in any sample tested during the period from 9 November 1989 to 11 January 1990. As a further check on water quality in the aquifer, records at the local dairy on their private wells drilled 1000 ft into the same aquifer revealed that the water supply met the total coliform standard. These findings suggested that the source water quality was adequately protected. Conclusion: groundwater was not a factor in the water-borne outbreak.

Attention then focused on the distribution system. Record low temperatures at the start of the outbreak caused ice blockage in 43 service meters, and two major distribution line breaks occurred near the centre of the community. Some meter boxes were reported to be partially submerged by surface water drainage during replacement, which could have introduced contamination. Based on customer recollections, localized reductions in water pressure from the two line breaks created opportunities for back-siphonage and a pathway for contaminant infiltration during several hours that elapsed before breaks were repaired. There was no line disinfection following the repairs or the meter replacements.

A possible source of pathogen contamination was thought to be the deteriorating sewage collection system that provided opportunities for stormwater infiltration during periods of heavy rainfall. On-site inspections of sewer manholes revealed evidence of overflow problems. Various paper products associated with sewage littered the area around a few manhole covers and several other entry structures had erosion gullies around their peripheries, indicating that overflow conditions had occurred. Clearly, there were infrastructural problems with the sewage collection system which provided entry of stormwater into broken pipes. This situation also served as a vehicle for pathogen passage via leaking sewer pipes into the untreated water supply distribution pipe network whenever there were main repairs or service meter replacements.

Verifying the existence of this surface contamination pathway took two directions: a search for residual pockets of contaminated water at the ends of the distribution system and the use of a distribution system hydraulic model to show patterns of contaminant flow from the line breaks towards the location of cases of illness. Because the invitation to investigate this outbreak was not received until four weeks after the outbreak had subsided, the chances of isolating the pathogenic agent were remote. However, an effort was made to sample a segment of the waters from the ends of the system that might represent old water from the period of the outbreak. These samples contained coliform densities of 55, 68 and 95 organisms per 100 ml. Verification and identification of these coliforms revealed the presence of *Escherichia hermanii*, a possible faecal organism. Although *E. hermanii* is not known to cause gastroenteritis, its presence is significant because this organism closely resembles *E. coli* 0157:H7 in its biochemical profile and serological reactions and has been found in raw milk, ground beef and faeces (Lior and Borczyk, 1987). Some of the coliform isolates were tetracycline resistant, a characteristic shared with the outbreak strain, *E. coli* 0157:H7, suggesting that these organisms may have originated from a common source of contamination. The pathogenic *E. coli* 0157:H7 was not detected in any samples.

A hydraulic model was used first to plot the pattern of water movement in the pipe network under average cold weather demand. Based on the computer-generated patterns of water flow, it became evident that the public water sources (Wells #5 and #6) or a possible dairy interconnection were not in the contamination pathway. A more likely scenario was a disturbance in the system close to most of the outbreak cases, such as the cluster of meter replacements and two line breaks. Using the computer program to simulate flow conditions during the water line breaks, then inserting a hypothetical contamination dose of the pathogenic agent (assuming no die-off of the organism during a four-hour passage in the pipe network), it was possible to overlay 85% of all illness case locations on the path followed by the pathogen from both line break locations. Meter replacements were not a major contributor, but could have accounted for several early cases before the line breaks.

Vibrio cholera *outbreak*. The international spread of cholera in Latin America during 1991, which is still ongoing, is the result of the lack of environmental control of raw sewage discharges, inadequate treatment of water supplies and deterioration of distribution system infrastructure. *V. cholera*, like other intestinal pathogens, is transmitted via the ingestion of contaminated food or water supply. The initial source of the El Tor strain of cholera was thought to have been from an infected merchant seaman on board a Chinese cargo ship in the harbour at Callao, Peru. From there the pathogen quickly spread to different population centres through raw sewage; it then infected thousands of people who either consumed raw seafood processed in contaminated water or drank water that was not adequately protected from faecal wastes.

The public water supply in Lima, Peru is obtained from a surface water source and various wells throughout the city. The quality of the river water at the water plant intake is poor because of the discharge of untreated sewage from upstream farm communities and excessive soil erosion over the watershed during major storms. Although there are 100 point source discharges within 20 miles of the intake, 30 of these discharges are responsible for about 90% of the river pollution. During March 1991 raw water turbidities at the water plant intake ranged from 400 to 60 000 nephelometric units (NTU); 20% of the turbidity values were over 10 000 NTU. Microbial quality was also poor and faecal coliform densities ranged from 50 000 to 130 000 organisms per 100 ml. Not surprising was the detection of the cholera agent in water collected at the intake. Considering the gross nature of this source water, it is remarkable that the water plant was able to produce a water quality during the flash flooding of March 1991 that met international standards of less than one coliform per 100 ml. *Vibrio cholerae* was not detected in plant effluent entering the distribution system. Not only were the redundancies in water treatment barriers effective for microbial control, but the water turbidity was reduced to approximately 0·5 NTU during the same period.

Unfortunately, the quality of water produced at the plant deteriorated during distribution. In a study of distribution water quality in the Lima system, the Pan American Health Organization collected 183 water samples and promptly tested them for faecal coliform bacteria. Fecal coliforms were found in 18·7% of all samples and increased in occurrence with distance from the treatment plant. During this two week survey, none of the special samples collected from the distribution system contained cholera organisms, although two wells on the system were releasing *V. cholerae*.

There are a variety of factors that could have contributed to the deterioration of water quality. The daily loss of line pressure, infiltration of contamination by back-siphonage from pipe breaks and increased chlorine demand in sediment-laden pipes were significant. An attempt was made to apply 0·5 mg/l free chlorine at different areas in the system, but free chlorine residuals were often undetectable in many areas because of contaminant incursions in broken pipes during breaks in water pressure. This situation created many opportunities for reversals of flow and suction that brought contaminated surface water and sewage overflows into the water supply pipe network.

Recreational water risks

There are only a few published reports of water-borne disease outbreaks associated with swimmers in polluted freshwaters, but the implications should not be ignored.

Shigella *outbreak*. A water-borne outbreak of shigellosis occurred among swimmers who used an 8 km stretch of the Mississippi River near Dubuque, Iowa (Rosenberg *et al.*, 1976). In this example, 31 of 45 cases of shigellosis were traced to this recreational activity. Bathers had been swimming in water where the mean faecal coliform density was 17 500 organisms per 100 ml. A sanitary survey of this section of the river indicated that at the time of the outbreak, the upstream municipal sewage treatment plant was producing a poor quality effluent. With a survival time greater than two days, *Shigella* discharged from this source could have easily reached the swimming area (Dolivo-Dobrovolskiy and Rossovkaya, 1956). It was noted that the attack rate for male swimmers under 20 years of age was more than twice that for male swimmers of 20 years and older, and that the illness was significantly associated with male bathers who ingested river water while swimming. One month after authorities banned swimming in this area, *Shigella sonnei* (with the same antibiogram, colicin type and phage type as the isolates from six swimmers) were isolated from a sample of

Table VII. *Salmonella* serotypes found in water and vegetables*

Salmonella serotype	Surface waters	Vegetables Santiago
Paratyphi B	130	46
Typhimurium	36	9
Agona	32	9
Derby	14	4
Kingston	3	—
Bredeney	2	1
Schwarzengrund	—	1
Oranienburg	12	1
Infantis	11	1
Bareilly	9	—
Montevideo	4	2
Thompson	4	2
Livingstone	2	—
Lomita	1	—
Mbandaka	1	1
Newport	2	2
Manhattan	1	1
Hadar	—	5
Albany	—	1
Typhi	17	—
Panama	12	11
Enteritidis	7	—
Anatum	56	15
Newington (Anatum 15)	1	—
Give	1	1
London	1	—
Senftenberg	11	5
Worthington	3	2
Cerro	2	—
Minnesota	1	3
	376	123

*Table modified from Cordano and Virgilio (1990).

river water containing 400 faecal coliforms per 100 ml. However, no shigellosis developed among those bathers who, during this later period, ignored the ban and swam in the waters of gradually improving bacterial quality. Apparently, the *Shigella* density had declined below the infective dose level, which is estimated to be 10–100 organisms.

Irrigation waters

Agriculture in the semi-arid and arid regions of the world depends on any available waters in the area for irrigation. Unfortunately, these waters are most often small streams with disproportionately large amounts of pollution from raw sewage, feedlot drainage, food processing waste discharges and return waters from upstream agricultural use. These inputs add varying numbers of pathogenic organisms to irrigation waters, and such organisms are ultimately in contact with field crops. *Salmonella*, *Shigella*, entero-pathogenic *E. coli*, amoebic cysts, *Ascaris* ova and enteroviruses have been detected on salad crops and berries grown in soil irrigated with sewage effluents (Geldreich and Bordner, 1971). The true extent of disease resulting from the consumption of these uncooked foods is unknown, but a number of epidemiological reports have shown that transmission of enteric diseases does occur.

Salmonella outbreak. Chile is a developing country whose economy includes a large contribution of agricultural produce (fruits, berries, salad crops) to the markets of North America during their winter

Table VIII. Water quality monitoring data excerpts*

Water source	Country	Water temperature (°C)	pH range	Suspended solids (mg l⁻¹)	BOD maximum (mg l⁻¹)	Faecal coliforms (no./100 ml)
Baseline stations (1982–1984)						
Rivers						
Temperate waters (minimum temperature <10°C)						
Glama	Norway	0·0–17·0	6·6–7·7	1–90	5 (TOC)	5–635
Cark Suyer	Turkey	5·0–25·0	7·0–8·6	1–39	3	240
Kiso	Japan	2·0–22·0	6·6–7·5		2	110–2600
Mapocho	Chile	4·5–18·8	6·5–8·2	4–648	2	2–790
Tropical waters (temperature above 10°C)						
Tapti	India	19·5–31·5	7·0–8·6		5	7–150
Godavari (Polavaron)	India	25·0–32·0	6·5–8·7		5	<1–34
Chaliyar	India	25·5–33·9	6·1–8·8		4	21–1300
Cauveri	India	23·0–30·0	7·4–8·9		1	<1–1800
Waikato	New Zealand	10·4–20·2	7·7–8·3		2	<1–11
Impact stations (1982–1984)						
Rivers						
Temperate waters (minimum temperature <10°C)						
Exe	UK	4·9–21·0	7·0–8·9	1–62	5	70–22000
Rhine	Netherlands	1·2–24·6	6·9–8·3	3–177	15 (TOC)	<1–540000
Maas	Netherlands	3·9–25·5	6·7–8·3	1–310	36	80–80000
Safami	Japan	4·5–24·8	7·0–7·4	3–119	4	130–4900
Shimano	Japan	0·5–24·5	6·7–7·5	9–989	4	10–5000
Yodo	Japan	4·1–28·4	7·1–7·9	13–65	4	9300–230000
Sefid	Iran	0·0–29·5	7·7–8·3	36–31700		170–16000
Tropical waters (temperature above 10°C)						
Sabaramati	India	23·0–37·0	7·0–8·8		284	20000–9900000
Narmada	India	19·0–32·0	7·0–8·8		11	<2–900000
Subernakekla	India	18·0–40·0	6·8–8·4		8	990–46000
Conchos	Mexico	15·0–36·0	6·0–8·3		10	200–100000
Balas	Mexico	28·0–30·0	6·7–8·4	10–11900	3	9–240000
Blanco	Mexico	19·0–27·0	6·0–8·5	4–995	40	4–110000
San Pedro	Ecuador	12·0–20·0	6·6–8·3	28–1250		2–100000
Ravi	Pakistan	11·0–26·0	7·1–7·7	220–360	3	363–19000
Brahmaputra	Bangladesh	20·0–30·0	6·6–7·9	21–69		300–7000
Chao Phyra	Thailand	24·0–32·0	5·5–8·0	25–280	3	50–35000
Prasak	Thailand	23·0–32·5	6·5–9·8	4–72	3	49–11000

* Source: GEMS/WATER Database, Canada Centre for Inland Waters, Burlington, Ontario
TOC = Total organic carbon
All data derived from 1–3 year monitoring results

season. Much of the agricultural activity relies on irrigation water to sustain the growth of crops. Unfortunately, rivers in Chile are heavily polluted with sewage, so it is not surprising that *Salmonella* is often detected in surface waters and is a serious public health problem among the people (Cordano and Virgilio, 1990). Although poorly treated water supplies cause some of these infections, it is believed that *Salmonella* transported on salad crops is a major contributor to the endemic nature of this pathogen. For example, 51·5% of 33 samples examined from the Canal Las Mercedes contained *Salmonella*. The same seven serotypes were found to be present throughout the 60 km course of the canal, which irrigates an extensive area of cultivated land. Vegetables examined from this agricultural area contained most of the serotypes present in the water and may be a major factor in the epidemic spread of certain serotypes among infected people in the country (Table VII). In an earlier study, 7·1% of 113 heads of lettuce cultivated on the farms around Santiago were contaminated with four of the *Salmonella* serotypes (*S. paratyphi B*, *S. newport*, *S. anatum* and *S. bredeney*) linked to river water and infected individuals in the community (Lobos *et al.*, 1976).

GLOBAL STATUS OF FRESHWATER RESOURCES

An opportunity to assess the quality of freshwater resources world-wide has been made possible through the Global Environmental Monitoring System (GEMS) sponsored by the World Health Organization and the United Nations Environmental Programme. Data from monitoring laboratories in different countries is being entered into a database located at the Canada Centre for Inland Waters. Although there may be some problems with data precision generated by laboratories using different test procedures in various countries, the general trends illustrate why water-borne pathogens are a continuing public health threat everywhere in the world. Such information could also identify specific limitations on the ambient water resource capacity to adsorb pollutants, and identify critical water pollution problems that need priority attention because of their impact on all water users in the area.

The mass of data being collected may be analysed in several ways. One approach (Table VIII) is to first identify baseline stations — sites chosen to establish the minimum level of pollution in waters that do not receive any major exposure to human waste discharges. These are divided into two categories: temperate waters (minimum temperatures reaching <10°C) and tropical waters (minimum temperatures above 10°C). The other grouping of monitoring data is from sites of known impact — areas of intense human activity in both temperate and tropical areas. Although none of these data collections include information on the occurrence and density of specific water-borne pathogens, there is obvious concern with the magnitude of faecal contamination that must also be transporting a variety of disease agents.

The maximum values for faecal coliforms at impact stations in the tropics are much higher than those in temperate regions. The reasons for this difference are in a large part a reflection of limited sewage treatment operations in many developing nations. Two factors that exacerbate this problem are receiving waters with ambient temperatures above 20°C for most of the year, and a steady release of decaying humic materials, both of which contribute to microbial regrowth. This situation not only causes some distortions in faecal surrogate organism densities, but also provides longer persistence of some pathogens (*Salmonella*, etc.) in these nutrient-rich waters.

Mandatory requirements for processing municipal sewage to a secondary treatment quality have produced significant improvements in surface waters. Unfortunately, wastewater treatment is not always of uniform quality due to processing difficulties and the occasional by-pass of raw waste inputs after major storm events. These interruptions cause sudden elevations in faecal pollution downstream and result in widely fluctuating bacterial densities that challenge water supply treatment. Some of these trends can be seen in a study of the Ohio River water quality data (Table IX) gathered in the continuous monitoring programme conducted by the Ohio River Valley Sanitation Commission (ORSANCO). Most of the monitoring sites selected are at locations that reflect the impact of high density human populations and industrial development. The individual high density maximum values, contrasted with geometric mean values for each period, illustrate the difficulty in achieving complete control of water quality without better management of non-point source pollution (rural and urban stormwater runoff) through protected wetlands and stormwater storage.

Table IX. Seasonal water quality data from the Ohio River mainstem*

Station name	Mile post	Faecal coliform densities per 100 ml					
		Spring		Summer		Autumn	
		Max.	Geometric mean	Max.	Geometric mean	Max.	Geometric mean
Wilkinsburg	−8·5	1180	235	2280	199	890	169
Pittsburg	4·3	1800	55	16 000	248	1200	602
Wheeling	86·8	120	25	150	56	488	361
Huntington	314·8	4700	373	2700	222	1800	134
Cincinnati	462·8	370	253	5800	177	400	253
Anderson Ferry	477·5	17 000	228	5800	269	400	90
Louisville	600·6	6200	756	37 000	1900	2300	243
West Point	625·9	4600	360	37 000	7800	7800	516
Evansville	797·3	1300	204	580	16	2000	24
Paducah	937·9	18 000	3800	11 000	1500	40 000	931

* 1993 data from ORSANCO monitoring programme

Although monitoring water quality is an important aspect, data gathering alone cannot protect public health. It can provide an alert level that should trigger the appropriate action response to protect the public from the potential presence of water-borne pathogens. The other alert mechanisms to anticipating pathogenic exposure is through expanded national surveillance for water-borne outbreaks. In each situation, it is important to characterize the aetiological agent, circumscribe occurrence, establish geographical distribution and identify control measures. In the USA, reporting of outbreaks is carried out on a voluntary basis at both the local and state level. Improvements in this reporting network have done much to bring about public health awareness. Yet the true status of the problem is biased towards a focus on large outbreaks and those occurring in community water systems. Less information is available on non-community supplies serving transient populations. The most underreported segment is private groundwater systems because they involve small clusters of cases that attract little attention (Herwaldt, et al., 1992).

Finding the aetiological agent in a water-borne outbreak is not always successful for a variety of reasons. Often the field investigation begins after an action response has inactivated the pathogen or the organism has declined in numbers below detection levels. The methodology for a few water-borne agents is poor, lacking sensitivity to cell cultivation from the water environment. As a consequence, the aetiological agent has not been found in nearly 50% of all outbreaks investigated, although epidemiological evidence and water quality data implicated water as the vehicle. These handicaps are being overcome as commitment to environmental science progresses, yielding the information voids so necessary to cope with pathogenic agents in freshwater resources.

CONCLUSIONS

Pathogenic agents (bacterial, viral and protozoan) are frequently found in the aquatic environment. Pathogens are constantly being released at variable concentrations from infected humans, animal pets, farm animals and wildlife (the warm-blooded animal population). Municipal sewage and stormwater runoff become the conduits for the passage of pathogens into surface waters and unprotected groundwater aquifers.

When water-borne outbreaks occur, illness rates increase above the norm and some people may die of exposure. Perhaps the most important lesson to be learned from outbreaks is that water-borne pathogens are a worldwide problem that needs urgent control through environmental protection to avoid further escalation of their occurrence. Since the release of pathogens is a constant threat multiple barriers are essential through management of watershed activities, treatment of sewage, control of runoff and adequate treatment of water supply. Furthermore, treatment processes must be properly operated and infrastructure deterioration of plant facilities, sewage collection networks and water distribution systems avoided to provide maximum public health protection.

REFERENCES

Akin, E. W., and Hoff, J. C. 1978. 'Human viruses in the aquatic environment: a status report with emphasis on the EPA Research Program', *Report to Congress. EPA-570/9078–006*, US Environmental Protection Agency, Cincinnati.

Alter, A. J. 1954. 'Appearance of intestinal wastes in surface water supplies at Ketchikan, Alaska' in *Proc. Fifth Alaska Sci. Conf.*, AAAS, Anchorage, 81–84.

Andersson, Y. and Stenstrom, T. A. 1986. Waterborne outbreaks in Sweden – Causes and Etiology. *Water Sci. Technol.* **18**, 185–190.

Andre, D. A., Weiser, H. H., and Maloney, G. W. 1967. 'Survival of bacterial enteric pathogens in farm pond water', *J. Am. Water Works Assoc.*, **59**, 503–508.

Anon. 1954. 'Ketchikan laboratory studies disclose gulls are implicated in disease spread', *Alaska's Health*, **11**, 1–2.

Borczyk, A. A., Karmali, M. A., Lior, H., and Duncan, L. M. C. 1987. 'Bovine reservoir for verotoxin producing *Escherichia coli* 0157:H7', *Lancet*, **i**, 98.

Bitton, G., Farrah, S. R., Montague, C. L., and Akin, E. W. 1986. 'Viruses in drinking water', *Environ. Sci. Technol.*, **20**, 216–222.

Brezenski, F. T. 1971. 'Estuary water quality and *Salmonella*' in *Proc. Nat. Spec. Conf. Disinfect.* American Society of Civil Engineers, New York. pp. 481–493, L. N. Kuzminski.

Burm, R. J., and Vaughan, R. E. 1966. 'Bacteriological comparison between combined and separate sewer discharges in southeastern Michigan', *J. Wat. Pollut. Controls Fed.*, **38**, 400–409.

Callaghan, P., and Brodie, J. 1969. 'Laboratory investigation of sewer swabs following the Aberdeen typhoid outbreak of 1964', *J. Hyg.*, **66**, 489–497.

Center for Disease Control 1973. 'Typhoid fever — Florida,' *Morbid. Mortal. Weekly Rep.*, **22**, 77–78; 85.

Center for Disease Control 1974. 'Acute gastrointestinal illness — Florida', *Morbid. Mortal. Weekly Rep.*, **23**, 134.

Center for Disease Control 1980. 'Waterborne disease outbreaks in the United States — 1978', *Morbid. Mortal. Weekly Rep.*, **29**, 46–48.

Chang, S. L. 1970. Unpublished paper presented at the *10th International Congress of Microbiologists, Mexico City.*

Clancy, J. L., Gollnitz, W. D., and Tabib, Z. 1993. 'Laboratory experience: monitoring for *Giardia* and *Cryptosporidium*' in *Proc. Water Qual. Technol. Conf.* American Water Works Association, Miami.

Claudon, D. G., Thompson, D. I., Christensen, E. H., Lawton, G. W., and Dick, E. C. 1971. 'Prolonged *Salmonella* contamination of a recreational lake by runoff waters', *Appl. Microbiol.*, **21**, 875–877.

Clemente, J., and Christensen, R. G. 1967. 'Results of a recent *Salmonella* survey of some Michigan waters flowing into Lake Huron and Lake Erie', *Proc. 10th Conf. Great Lakes Research.* Internal. Association for Great Lakes Research. pp. 1–11.

Cordano, A. M., and Virgilio, R. 1990. '*Salmonella* contamination of surface waters' in Castillo, G., Campos, V., and Herrera, L. (Eds), *Proc. Second Biennial Water Qual. Symp. Microbiological Aspects.* Editorial Universitaria, Santiago.

Craun, G. F. 1985. 'An overview of statistics on acute and chronic water contamination problems' in *Fourth Domestic Water Quality Symposium: Point-of-Use Treatment and Its Implications.* Water Quality Association, Lisle. pp. 5–15.

Craun, G. F. (Ed.) 1986. *Waterborne Diseases in the United States.* CRC Press, Boca Raton.

Craun, G. F. 1988. 'Surface water supplies and health', *J. Am. Wat. Works Assoc.*, **Feb**, 40–52.

D'Antonio, R. G., Winn, R. E., Taylor, J. P., Justafson, T. L., Current, W. L. Rhodes, M. M. Gray, G. W. and Zajac, R. A. 1985. A waterborne outbreak of crytosporidiosis in normal hosts. *Ann. Inter. Med.* **103**, 886–888.

Dolivo-Dobrovolskiy, L. B., and Rossovkaya, V. S. 1956. 'Survival of dysentery bacteria in reservoir water', *Gig. Sanit.*, **21**, 52–55.

Dragas, A.-Z., and Tradnik, M. 1975. 'Is the examination of drinkable water and swimming pools on presence of entero-pathogenic *E. coli* necessary?' *Z. Bakt. Hyg. I. Abt. Orig.*, **B160**, 60–64.

Evison, L. M., and James, A. 1973. 'A comparison of the distribution of intestinal bacteria in British and East African water sources,' *J. Appl. Bacteriol.*, **36**, 109–118.

Fennel, H., James, D. B., and Morris, J. 1974. 'Pollution of a storage reservoir by roosting gulls', *Wat. Treat. Exam.*, **23**, 5–24.

Galbraith, N. S., Barrett, N. J., and Stanwell-Smith, R. 1987. Water and disease after Croydon: A review of water-borne and water-associated disease in the U.K. 1937–86. *Jour. Inst. Water Envir. Manag.* **1**, 7–21.

Geldreich, E. E. 1966. *Sanitary Significance of Fecal Coliforms in the Environment.* US Department of the Interior WP-20–3, Federal Water Pollution Control Administration, Cincinnati.

Geldreich, E. E. 1970. 'Applying bacteriological parameters to recreational water quality', *J. Am. Wat. Works Assoc.*, **63**, 113–120.

Geldreich, E. E. 1972a. 'Waterborne pathogens' in Mitchell, R. (Ed.), *Water Pollution Microbiology.* Wiley, New York. pp. 207–241.

Geldreich, E. E. 1972b. 'Buffalo Lake recreational water quality: a study of bacteriological data interpretation', *Water Res.*, **6**, 913–924.

Geldreich, E. E. 1978. 'Bacterial populations and indicator concepts in faeces, sewage, stormwater and solid wastes' in Berg, G. (Ed.), *Indicators of Viruses in Water and Food.* Ann Arbor Science, Ann Arbor. pp. 51–97.

Geldreich, E. E. 1989. 'Pathogens in fresh water' in Meybeck, M., Chapman, D. V., and Helmer, R. (Eds), *Global Freshwater Quality. a First Assessment.* World Health Organization and United Nations Environment Programme, Geneva.

Geldreich, E. E. 1990. 'Microbiological quality of source waters for water supply' in McFeters, G. A. (Ed.), *Drinking Water Microbiology.* Springer-Verlag, New York.

Geldreich, E. E., and Bordner, R. H. 1971. 'Fecal contamination of fruits and vegetables during cultivation and processing for market, a review', *J. Mills Food Technol.*, **34**, 184–195.

Geldreich, E. E., Fox, K.R., Goodrich, J.A. Rice, E.W., Clark, R.M., and Swerdlow, D. L. (1992) 'Searching for a water supply connection in the Cabool, Missouri disease outbreak of *Escherichia coli* 0157:H7', *Wat. Res.*, **26**; 1127–1137.

Grabow, W. O. K., and Nupen, E.M. 1972. 'The load of infectious micro-organisms in the waste water of two South African hospitals', *Wat. Res.*, **6**, 1557–1563.

Greenberg, A. E., and Jongerth, J. 1966. 'Salmonellosis in Riverside, California', *J. Am. Water Works Assoc.*, **58**, 1145–1150.

Grimes, J. 1980. 'Bacteriological water quality effects of hydraulically dredging contaminated Upper Mississippi River bottom sediment,' *Appl. Environ. Microbiol.*, **39**, 782–789.

Harvey, R. W. S., Price, T. H., Foster, D. W., and Griffith, W.C. 1969. '*Salmonellas* in sewage. A study in latent human infections', *J. Hyg. (Br.)*, **67**, 517–523.

Health and Welfare, Canada 1981. 'Possible water-borne *Campylobacter* outbreak — British Columbia', *Can. Dis. Weekly Rep.*, 7, 223; 226–227.

Hendricks, C. W. 1971. 'Increased recovery rate of salmonellae from stream bottom sediments versus surface waters', *Appl. Microbiol.*, 21, 379–380.

Herwaldt, B. L., Craun, G.F. Stokes, S.L., and Juranek, D.D. 1992. 'Outbreaks of waterborne disease in the United States: 1989–90', *J. Am. Wat. Works Assoc.*, 84, 129–135.

Highsmith, A. K., Feeley, J. D., Shaliy, P., Wells, J. G., and Wood, B. T. 1977. Isolation of *Yersinia enterocolitica* from well water and growth in distilled water. *Appl. Envir. Microbio.* 34, 745–750.

Kampelmacher, E. H., and Van Noorle Jansen, L.M. 1973. '*Legionella* and thermotolerant *E. coli* in the Rhine and Meuse at their point of entry into the Netherlands', *H₂O*, 6, 199–200.

Kampelmacher, E. H., and Van Noorle Jansen, L.M. 1976. '*Salmonella* effluent from sewage treatment plants, wastepipes of butchers' shops and surface water in Walcheren', *Z. Bakteriol. Hyg. Abt. I Orig. B*, 162, 307–319.

Klein, P. D., Graham, D.Y. Gaillour, A., Opekun, A.R., and O'Brian Smith, E. 1991. 'Water source as risk factor for *Helicobacter pylori* infection in Peruvian children', *Lancet*, 337, 1503–1506.

Kowal, N. E. 1982. 'Health effects of land treatment: microbiological', *US Environmental Protection Agency, EPA-600/1–82–007*, Health Effects Research Laboratory, Cincinnati.

Lassen, J. 1972. '*Yersinia enterocolitica* in drinking water', *Scand. J. Infect. Dis.*, 4, 125–127.

Lindel, S. S., and Quinn, P. 1973. '*Shigella sonnei* isolated from well water', *Appl. Microbiol.*, 26, 424–425.

Lior, H., and Borezyk, A. A. 1987. 'False-positive identification of *Escherichia coli* 0157', *Lancet* 1 (No. 8528), 333.

Lobos, H., Garcia, J. Agiular, C., Greve, E., Olivares, A. M., Bustos, R., Valenzuela, M. E., Zapata, L., and Romero, H. 1976. 'Estudio bacteriologico comparativo de leckugas (Lactuea Sativa) provenientes de los alredeolores de Santiago y region costera', *Bol. Inst. Bacteriol. Chile.* XVIII, 33–37.

Martin, M. L., Shipman, L. D., and Wells, J. G. 1986. 'Isolation of *Escherichia coli* 0157:H7 from dairy cattle associated with two cases of haemolytic uraemic syndrome', *Lancet* ii, 276.

McMullen, L. D. 1994. 'Surviving the flood: teamwork pays off in Des Moines', *J. Am. Water Works Assoc.*, 86, 68–72.

Mentzing, L. O. 1981. 'Waterborne outbreaks of *Campylobacter enteritis* in central Sweden', *Lancet*, 8, 352–354.

Ohio River Valley Water Sanitation Commission (ORSANCO) 1993. *Quality monitor: An Appraisal of Conditions in the Ohio River and some of its Tributaries*. ORSANCO, Cincinnati.

Oliveri, V. P., Kruse, C. W., and Kawata, K. 1977. 'Microorganisms in urban stormwater', Environ Protect. Technol. Series, EPA-600/2–77–087, MERL, US Environmental Protection Agency, Cincinnati.

Popp, L. 1974. '*Salmonella* and natural purification of polluted waters', *Z. Bakteriol. Ab. I*, 158,432–445.

Prost, A. 1987. Health risks stemming from wastewater reutilization. *Water Qual. Bull.* 12, (Part 11) 73–78.

Reid, J. 1994. 'Overcoming the flood: how midwestern utilities managed disaster', *J. Am. Water Works Assoc.*, 86, 58–67.

Rose, J. B. 1990. 'Emerging issues for the microbiology of drinking water', *Wat./Engr. Manage.*, 23–29.

Rosenberg, M. L., Hazlet, K. K., Schaefer, J., Wells, J. G., and Pruneda, R. C. 1976. 'Shigellosis from swimming', *J. Am.. Med. Assoc.*, 236, 1849–1852.

Schieman, D. A. 1978. 'Isolation of *Yersinia enterocolitica* from surface and well waters in Ontario', *Can. J. Microbiol.*, 24, 1048–1052.

Schroeder, S. A., Caldwell, J. R., Vernon, T.M., White, P.C., Granger, S.I., and Bennett, J.V. 1968. 'A water-borne outbreak of gastroenteritis in adults associated with *Escherichia coli*', *Lancet*, 6, 737–740.

Seyfried, P. L., Tobin, R.S., Brown, N.E., and Ness, P.F. 1985. 'A prospective study of swimming-related illness. II. Morbidity and the microbiological quality of water', *Am.. J. Public Health*, 75, 1071–1075.

Slanetz, L. W., Bartley, C.H., and Metcalf, T.G. 1965. 'Correlation of coliform and faecal streptococcal indices with the presence of *Salmonella* and enteric viruses in seawater and shellfish', *Adv. Water Poll. Res., 2nd Int. Conf., Tokyo, Japan*, 3, 27–35.

Slanetz, L. W., Bartley, C.H., and Stanley, K.W. 1968. 'Coliform, faecal streptococci and *Salmonella* in seawater and shellfish', *Health Lab. Sci.*, 5, 66–78.

Smith, H. V. 1992. *Cryptosporidium* and Water: A review. *Jour. Ins. Water Envir. Manag.* 6, 443–451.

Taylor, D. N., McDermott, K. T., and Little, J. R. 1983. '*Campylobacter enteritis* associated with drinking water in back country areas of the Rocky Mountains', *Ann. Intern. Med.*, 99, 38–40.

Thornsberry, C., Balows, A., Feeley, J. C., and Jakubowsky, W. 1984. *Legionella: Proceedings of the 2nd International Symposium*. American Society for Microbiology, Washington. 369 pp.

US Department of Health, Education and Welfare 1965. *Report on Pollution of Interstate Waters of the Red River of the North (Minnesota, North Dakota)*. Public Health Service, Field Investigation Branch, Cincinnati.

US Environmental Protection Agency 1971. *Report on Missouri River Water Quality Studies*. Regional Office, Kansas City.

Vander Velde, T. L., and Mack, W.M. 1973. 'Poliovirus in water supply', *J. Am. Water Works Assoc.*, 65, 345–348.

Van Donsel, D. J., and Geldreich, E. E. 1971. 'Relationships of *Salmonella* to faecal coliforms in bottom sediments', *Water Res.*, 5, 1079–1087.

Vogt, R. L., Sours, H.E., Barrett, T., Feldman, R.A., Dickinson, R.J., and Witherall, L. 1982. '*Campylobacter enteritis* associated with contaminated water,' *Ann. Intern. Med.*, 96, 292–296.

Walter, W. G., and Boatman, R.P. 1967. 'Microbiological and chemical studies of an open and closed watershed,' *J. Environ. Health*, 30, 157–163.

West, A. P., Millar, M. R., and Tompkins, D. S. 1992. 'Effect of physical environment on survival of *Helicobacter pylori*', *J. Clin. Pathol.*, 45, 228–231.

Woodward, W. E., Hirschhorn, N., Sack, R. B., Cash, R. A., Brownlee, I., Chickadonz, G. H., Evans, L.K., Shephard, R. N., and Woodward, R. C. 1974. 'Acute diarrhoea in an Apache Indian Reservation', *Am. J. Epidemiol.*, 99, 281–290.

WRI (World Resources Institute) 1986. *World Resources 1986: An assessment of the Resource Base that Supports the Global Economy*. Basic Books, New York.

9

TRENDS IN GROUNDWATER QUALITY

JIM C. LOFTIS

Department of Chemical and Bioresource Engineering, Colorado State University, Fort Collins, CO 80523, USA

ABSTRACT

The term 'trend' takes on a variety of meanings for groundwater quality in both a temporal and spatial context. Most commonly, trends are thought of as changes over time at either a regional or localized spatial scale. Generally water quality managers are most interested in changes associated with some form of human activity. Carefully defining what is meant by 'trend' is a critical step in trend analysis and may be accomplished by formulating a statistical model which includes a trend component. Although there are a great many regional groundwater studies which provide a 'snapshot' description of water quality conditions over an area at one point in time, there are relatively few which consider changes over time and fewer still which include a statistical analysis of long-term trend. This review covers both regional and localized studies of groundwater quality around the world, including a few snapshots, but focusing primarily on those studies which include an evaluation of temporal changes in groundwater quality. The studies include national assessments, agricultural case studies (the largest group, mostly regional in scope), urban case studies, and point source and hazardous waste case studies.

INTRODUCTION: WHY ARE GROUNDWATER TRENDS DIFFERENT?

Trend detection is often stated as an objective of groundwater quality monitoring programmes. It is much less common, however, to carefully define what is meant by the term 'trend'. The meaning of trends in groundwater may be considerably different from the meaning of trends in surface water and may vary considerably from situation to situation.

For example, let us assume that Figure 1 represents a time series of available water quality observations at a given location. If we were asked whether there was a trend in the data, most people would look at the entire period of record and say 'yes, there is an increasing trend'. However, the period of record displayed in the figure might or might not correspond to the period of interest to water quality managers. If the period of record was 20 days, then it is probably too short to represent a change of interest in groundwater quality, but could be interesting in surface water quality or treatment plant effluent quality. If the period of record was 20 years, then we might actually be interested in a subset of the data corresponding to the implementation of some management practice. One such subset, the time period from year 13 to year 20 in Figure 1, appears to have a decreasing trend, even though the overall trend is upward.

It makes sense to define trends as 'changes in water quality over some period of time (and/or some region of space) which is of interest from a management (or scientific) standpoint'. The period of interest could be longer or shorter than the period of record. In groundwater studies, the period of interest is most likely to be longer than the period of record, thus the difficulty in performing useful trend analyses and the shortage of thorough studies performed to date.

There are at least three major differences between groundwater quality and surface water quality monitoring for trend detection and analysis: (1) the speed of development of trends; (2) the difficulty and cost of ascertaining spatial patterns; and (3) difficulty in determining cause and effect relationships.

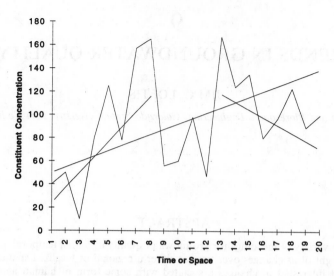

Figure 1. Hypothetical groundwater quality time series. Several different meanings may be associated with the term 'trend'

Obviously, the biggest difference between trends in groundwater quality and surface water quality is the speed with which trends typically develop and evolve. Surface water can travel hundreds of miles in a matter of days, whereas groundwater may travel only a few feet per year. Thus groundwater changes will occur more slowly and persist for a much greater period of time. There are exceptions, of course, in both cases. Some groundwater systems respond to surface inputs (such as agricultural chemicals applied in irrigation water) in a matter of days or even hours, whereas some surface water systems may be affected for many decades by the contamination of sediments.

Another major difference between groundwater and surface water trends is that, although spatial patterns are important in understanding both systems, they are much harder to characterize in groundwater owing to the high cost of drilling wells. This is true of spatial patterns in both quality and flow. In the case of surface water, we can see where the water comes from and where it is going. In the case of groundwater and vadose zone interactions with groundwater, we must rely on estimates of flow patterns which typically have a good deal of uncertainty associated with them. Thus interpreting the physical significance of apparent trends by identifying cause and effect relationships is particularly difficult. Without such an interpretation, the identification of trends is not of much value.

OBJECTIVES OF TREND STUDIES

Why do managers want to identify groundwater quality trends and how are their objectives for trend analysis changing as water quality management evolves? Water quality managers are most interested in water quality changes which are related to human activities and seek to identify trends which would at some point result in reduced utility of the water supply for its intended uses. The identification of trends, along with the identification of their causal mechanisms, can lead to appropriate water quality management decisions. For example, if a trend of increasing nitrate concentrations in the groundwater of a given region coincides with increased fertilizer use, then an education programme or even a regulatory programme for implementing best management practices (BMPs) for nutrient applications in the region might be warranted.

This paper concentrates on changes in groundwater quality over time, at a given location or over a given region, which are related to water quality or land-use management. To limit the scope of this discussion, less time is spent on characterizing spatial patterns, even though both temporal and spatial patterns are important in characterizing, understanding, and managing groundwater quality processes. Much work has been published in the field of characterizing spatial patterns in groundwater and other environmental measurements. Recent works often deal with applications of geostatistics (for example, Isotok *et al.*, 1993) and geographical information systems for characterizing spatial distributions and managing spatial information.

Several regional studies are reviewed in this paper and the reader should keep in mind a problem which arises in characterizing regional trends. In regional groundwater studies, a 'trend' is determined as a change in average groundwater quality at a defined set of points (i.e. wells) at which water quality is observed. However, for management purposes, this description of trend is adequate only if the set of points at which water quality is actually measured adequately characterizes the spatial distribution of water quality over the entire region of interest. If so, then kriging or other techniques can be used to develop adequate maps. However, in many instances, the spatial variability of groundwater quality and aquifer characteristics is large (Hoyle, 1989; Rogers, 1989; Pucci *et al.*, 1992; Smith and Ritzi, 1992) and the number of points required to map groundwater quality in a useful way would also be large, perhaps much larger than the available number of monitoring wells. In any case, it is best to be cautious in making statements about regional groundwater quality conditions and in interpreting the results of regional studies.

The spatial patterns or trends of most interest are those in which a change in groundwater quality over space is obviously related to a particular human activity. For example, groundwater quality is often measured both up-gradient and down-gradient of a waste disposal site. Any deterioration of water quality from up-gradient to down-gradient might be attributed to leakage from the site. Natural variations of groundwater quality in space complicates the problem. For naturally occurring constituents, more definitive evidence of contamination might be provided by the long-term observation of down-gradient wells, starting in the pre-operation period. However, for facilities already in operation, pre- operation data are usually not available. Thus a change in space must be taken as evidence of a change in time as well (Loftis *et al.*, 1987).

With an objective of protecting drinking water supplies, one of the most common types of groundwater quality monitoring programmes has been that associated with routine sampling of drinking water from municipal well fields. In the USA, this type of monitoring is becoming more intensive and more costly under the wellhead protection provision of the Safe Drinking Water Act Amendments of 1987. These data have not typically been analysed for trend, only for compliance with drinking water standards. Several intensive studies of individual basins have been conducted to assess the effects of human activities, such as mining, urbanization, agriculture or silviculture. Very few of these have been long-term studies with a trend analysis component, however. Spatial patterns have generally received more attention than temporal patterns. Thus relatively few data sets exist which are suitable for time trend analysis.

Globally, the picture is much the same as in the USA. Only a few studies have examined long-term data records for time trends; and, unfortunately, in most of those studies a database was compiled from existing data collected for other purposes. Thus the data are not well suited for the objective of trend analysis and the results of trend analysis must be viewed sceptically. It is therefore not yet possible to put together a good picture of groundwater quality trends around the globe.

Nevertheless, interesting work is being conducted and, just as water quality itself changes, the state of the art is changing in the field of water quality monitoring. This paper aims to provide an adequate overview of existing work and a point of reference from which to evaluate the (expected to be dramatic) progress of the next 10 years.

DEFINING TRENDS WITH A STATISTICAL MODEL

As mentioned earlier, the concept of trend in water quality means different things to different people. Therefore, to allow objective and quantitative analysis of trends, a statistical definition of trend is needed. This suggests a statistical model which includes a trend component and which describes the behaviour of observed values over the time period and spatial period of interest. For water quality, including groundwater applications, the following form is often useful:

$$\bar{y}(t) = \bar{\mu}(t) + \bar{\varepsilon}(t) + \bar{b}(t) \tag{1}$$

In the above, $\bar{y}(t) = [y_1(t), y_2(t), ..., y_n(t)]$, a vector of n groundwater quality observations at time t, each element of which corresponds to a particular sampling location, such as a particular depth within a particular well. $[\bar{\mu}(t) = \mu1(t), \mu2(t), ... , \mu n(t)]$, a set of n stationary mean functions, specifying the mean value of the water quality observations, in the absence of trend, at each sampling location at time t. Each mean function describes the predictable seasonal variation at a given point (well location and depth). Each point has a separate mean function, though all could be similar or even identical. For shallow alluvial aquifers, the seasonal variation may be pronounced. In other cases it may be zero. $\bar{\varepsilon}(t) = [\varepsilon_1(t), \varepsilon_2(t), ..., \varepsilon_n(t)]$, a vector of n noise terms with specified variance-covariance structure which describes correlation between observations. In the most general form of the model, correlation in both time and space is present, meaning that for closely spaced observations, high values tend to follow high values and low values tend to follow low values. The covariance or correlation structure may change seasonally within the year, but the seasonal pattern remains the same from year to year. As the model describes water quality observations rather than true values of water quality, the sampling error variance is included in the covariance structure of the noise terms. $\bar{b}(t) = [b_1(t), b_2(t), ..., b_n(t)]$, a set of n deterministic temporal trend functions specifying the change in the mean value of water quality observations (relative to the seasonal mean function) at each location as a function of time.

The trend functions, $\bar{b}(t)$, may be of arbitrary shape, but they are most commonly modelled as simple linear functions or as step changes occurring at particular points in time. The trend functions consider changes only in the mean. In a more general model, changes in the variance or covariance structure over time are also possible. Any set of observed values can be viewed as a realization of the stochastic process represented by the above model. Without the trend component, $\bar{b}(t)$, the model is stationary or unchanging over time. In reality, however, all processes are changing to some degree over time. Thus the trend component is never precisely zero for a real process.

For a given functional form of $\bar{b}(t)$, such as a linear or step trend, statistical trend analysis can proceed along two paths (Figure 2): (1) performing a statistical hypothesis test regarding a parameter of $\bar{b}(t)$ (for example, testing whether the slope of a trend line is significantly different from zero); or (2) estimating one or more parameters of $\bar{b}(t)$ (for example, obtaining a point estimate or confidence interval for the trend slope). Hypothesis testing is probably the more common approach as water quality managers often want a simple answer of 'yes, there is a significant trend' or 'no, there is not a significant trend.' Unfortunately, the question being answered is probably not the right one. A test for *statistical* significance of trend does not consider the *practical* significance of a trend, and it is practical significance that managers really care about. Therefore, the alternative path, estimation of trend magnitude, will probably yield information of real value to managers more often than will tests of statistical hypotheses. McBride *et al.* (1993) examine in some depth the question of 'What do significance tests really tell us about the environment?'

As illustrated in Figure 1, the trend component can be different over different portions of the time domain. It could be increasing for a portion, zero for a portion and decreasing for another portion, and so on. The way in which a period of record is divided for a study of trend should match our understanding of the physical behaviour of the system and of the way in which the system responds to external inputs. It is

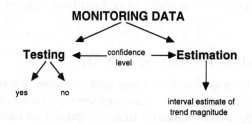

Figure 2. Estimation and testing are statistical alternatives for trend analysis

not logical to analyse groundwater records for monotonic trend over the entire period of record if the period of interest or the period over which a trend is expected is a subset of the period of record.

As mentioned earlier, in groundwater, the time segments during which a trend would be expected to develop and persist in response to an input would be longer than for surface water, probably of the order of years as opposed to weeks or months. Therefore, a common problem is that the period of interest is much longer than the period of record. However, groundwater quality changes in response to surface inputs can sometimes be rapid, and changes of interest can be detected with a short period of record or with a small segment of a long record.

TEMPORAL AND SPATIAL SCALE EFFECTS

Serial correlation versus trend

In Equation (1), the trend functions, $\bar{b}(t)$, are assumed to apply over the entire time domain, representing deterministic or fully predictable changes in the mean. However, changes in average conditions are rarely fully predictable and might sometimes be better described as stochastic persistence or serial correlation of the noise terms $\bar{\varepsilon}(t)$ (Harris *et al.*, 1987; Montgomery *et al.*, 1987; Close, 1989). Most real situations probably include changes of both types. The degree to which changes as trend versus correlation are modelled will depend on our understanding of the physical system—physically explainable changes related to human activities might be better modelled as trend—and the time-scale of interest (Loftis *et al.*, 1991). From a purely statistical viewpoint, the persistence of duration much shorter than the time-scale of interest might be better modelled as correlation than as trend.

For an example, refer back to Figure 1 and assume yearly time steps. The general increase over the period of record might be modelled as trend, whereas the short-term changes from years 1–8, upward, and from years 13–20, downward, could be considered as serial correlation. However, if we imagine a larger time-scale of, say, 100 years, the 20 year upward trend shown in the graph might be only one of several upward and downward shifts, with little change in the long-term average over the 100 year period. Thus at the larger scale, the 20 year trend might be better modelled as serial correlation. Serial correlation may be thought of as short-term trend and the distinction between the two depends on the time-scale of interest and the time-scale for which data are available, as well as the degree to which the physical system is understood.

Sample size considerations

When sufficient data are available, the statistical techniques described in the quoted references can provide some objective help in identifying an appropriate model. However, because of the limitations imposed by a finite number of measurements over a finite time period for representing the real world situation, the concept of a single correct model is probably fallacious.

As the cited references point out, persistence in a groundwater system will generally be greater than for a surface water system, as the water moves and changes more slowly. Repeated samples a few days apart might sample the same few cubic metres of water. This persistence poses some interesting problems for statistical trend analysis. If we view the persistence as stochastic, i.e. as correlation and therefore temporary, then it will take longer to statistically detect 'real' deterministic—and therefore permanent or at least predictable—trends than it would if there were no serial correlation in the series. Close (1989) presents the calculations for sampling frequency requirements for detecting permanent changes in a groundwater system with correlated observations.

However, if we are really interested in the changes over a time period shorter than that for which data are available (and do not care whether they are permanent or not), then those changes could be modelled as trend. In that instance, fewer observations will be required to characterize the changes in a correlated series than would be required in an independent series; see Loftis *et al.* (1991) for a more complete discussion.

Design considerations for regional studies

Alley (1993) discusses issues of scale, both temporal and spatial, in the context of general design considerations for regional groundwater studies. The author contrasts low density survey sampling over a broad area against targeted sampling at a finer range of spatial scales, often including nested monitoring wells screened at different depths. Targeted sampling is much more useful for characterizing the chemical, physical and biological processes which determine water quality and for developing cause and effect relationships between management or land-use alternatives and water quality.

Alley (1993) also discusses the importance of distinguishing between short-term effects—such as seasonality, recharge and drawdown due to pumping—from long-term trends. 'Any evaluation of trends in groundwater quality should be done to the extent possible in the context of the groundwater flow system.'

Hydrological and land-use factors must also be considered. An example is included from the former Czechoslovakia in which data for precipitation, groundwater nitrate concentration, wheat yield and inorganic fertilizer application are compared over a 20 year period from 1960 to 1981. Over the entire period, both fertilizer applications and groundwater nitrate concentrations generally increase. However, during a period of low rainfall and high wheat yields for 1980–1985, nitrate concentrations in groundwater declined, probably due to the increased crop uptake of nitrogen.

Another important factor discussed by Alley (1993) is the use of existing wells compared with specially constructed monitoring wells. A related paper by Hamilton et al. (1993) discusses the uses and limitations of existing monitoring data. There is a place in regional groundwater analyses for both existing wells and existing data. However, we must always keep in mind the fact that the comparison of data over time or space for describing either processes or trends requires a knowledge of how the samples were collected and analysed and what part of the aquifer, vertically and horizontally, is represented by each sample. A simple knowledge of the latitude and longitude and the visual appearance of a well will not suffice when comparisons between wells are made. We must also know the type of well and construction methods, the pumping rate and the depth and length of the screened interval. Otherwise we are comparing 'apples and oranges' in analysing groundwater trends.

Alley (1993) identifies a related confounding factor in determining and interpreting regional trends as the analysis of samples from wells producing water of significantly different ages. Age-dating of groundwater (Plummer et al., 1993) can be an asset in interpreting such data.

CATEGORIZING MONITORING AND TREND STUDIES

A large number of studies have examined the impact of agriculture, mining, urbanization, waste disposal and other human activities on regional groundwater quality. As mentioned earlier, these include relatively few studies of long records, suitable for time series analysis or detailed time trend analysis. Most are snapshots of groundwater quality at a particular time which may use data collected in several different studies for a variety of purposes. Things are changing, however, with the implementation of more well designed, comprehensive monitoring programmes which should produce high quality, consistent data over a long time period. Such data can then be used to reliably assess changes in water quality and to study the causes of those changes.There is, of course, no sharp dividing line between the types of groundwater pollution. Still, some generalization might help to break up our case studies into logical groups: national water quality studies; agricultural pollution studies; urban pollution studies; and studies of solid and hazardous waste disposal sites.

National studies

There are few groundwater monitoring programmes which cover entire countries. There is an increasing recognition, however, that the long-term management of groundwater at the national level is best supported by comprehensive monitoring programmes which are designed to provide a national perspective.

Agricultural studies

There is increasing concern around the globe about the impacts of agriculture on ground and surface water quality. For most practitioners water quality trends due to agriculture bring to mind fairly gradual changes occurring over a large area, especially with regard to nitrates and pesticides. Exceedences of drinking water limits for nitrates are common; exceedences for pesticides are much less common and are more site-specific.

Not all changes in groundwater quality related to agriculture are for the worse. In many parts of the world, farmers, with technical agency assistance, are attempting to modify their chemical application practices to both protect and improve groundwater quality. In the USA several programmes have arisen from the Presidential Water Quality Initiative, which started in 1987 under the Bush Administration and continues under President Clinton. Unlike waste disposal, there are no legal requirements for monitoring surface water and groundwater in agricultural areas. As the operating budgets of farms are a tiny fraction of those of major landfills or hazardous waste facilities, regulatory monitoring of agriculture is fiscally unrealistic. Only a few special studies related to the Rural Clean Water Program (US Environmental Protection Agency, 1992a) and other targeted federal or state programmes have as yet resulted in data sets adequate for the analysis of temporal trends. Most of those studies deal primarily with surface water.

Regional projects which are receiving special federal funds for water quality improvements related to agriculture are generally required to perform monitoring to document their success. However, this has not been easy for either surface water or groundwater because changes must be rather large and/or persist over a long time period to be observable in the presence of natural variability in water quality.

The National Water Quality Evaluation Project (NWQEP Notes, 1994) is a good source of current information on US Department of Agriculture water quality projects and the monitoring activities associated with them. Most monitoring efforts reported thus far have concentrated on surface water, but groundwater projects are also considered. The NWQEP is also a good source of information on the statistical methods for analysis of non-point source monitoring data; but, here again, the primary focus has been on surface water.

Urban studies

In the case of urban contamination of groundwater, we generally think of impacts which occur more rapidly and over a smaller area than agricultural impacts. Changes may occur from improper waste disposal practices, a lack of sewers within urbanizing areas, fuel and chemical spills and many other causes. Urban contamination could be viewed as non-point sources or collections of many point sources.

Hazardous and solid waste studies

The final category to be discussed is the detection of changes in groundwater quality at hazardous waste sites, regulated facilities for solid waste disposal, and other 'point' sources. These changes occur over a still smaller area than the agricultural and urban cases, although large or multiple facilities or sources can cause a regional effect.

The objectives of monitoring regulated facilities are the protection of aquifers from the impacts of solid waste disposal facilities and documentation of the effectiveness of remediation efforts. Changes in water quality occur from leaks or poor disposal practices as opposed to normal agricultural activities or urban development. Therefore we often encounter more extreme pollution—greater concentrations of toxic substances—and we find more immediate threats to public health. The degree of regulation of such sources is usually much greater and the amount of money spent in management and clean-up is, thus far, many times greater than for urban and agricultural sources.

As, in the case of waste disposal, the system is highly regulated, a regulatory definition of trend is needed. An appropriate definition might be 'a change in groundwater quality caused by the disposal of waste.' We think of such trends occurring in the form of contaminant plumes moving down-gradient of one or more point sources. Thus the contamination would occur abruptly as a leak or spill, but it might take many years to reach any down-gradient water supply wells. If the contaminant is not naturally occurring, then trend detection might simply amount to finding the contaminant in a well down-gradient from the suspected source. However, it is generally not that simple. Infrequent detection of trace-level contaminants can occur even for uncontaminated wells due to sampling and analysis errors, accidental sample contamination, etc. Therefore a statistical approach to trend analysis is warranted for both naturally occurring constituents and those which should be present only under contaminated conditions.

The objectives of monitoring for waste disposal sites will require long-term monitoring and trend analysis of the resulting data. The US Resource Conservation and Recovery Act (RCRA) regulations have fairly rigid monitoring requirements which will ensure that monitoring programmes are designed and operated consistently with those objectives.

Statistical problems in monitoring regulated facilities

As mentioned earlier, the trend analysis problem often amounts to the estimation of parameters of the trend function $\bar{b}(t)$ or testing hypotheses about those parameters. In studying cause and effect relationships, the estimation of model parameters might be more useful, whereas in a regulatory setting, hypothesis testing is more commonly used. As mentioned earlier, hypothesis testing presents some problems for sorting out important water quality responses from unimportant ones: 'Under the Resource Conservation and Recovery Act, the U.S. Environmental Protection Agency has established guidelines both for monitoring of solid and hazardous waste facilities and for statistical analysis of monitoring data (US Environmental Protection Agency, 1989; 1992b)'.

For those facilities at which contamination has not occurred at present, the goal of monitoring and data analysis is to make future determinations of whether or not the contamination of groundwater has occurred as a result of the facility, i.e. to detect a release. If contamination has occurred, the goal of monitoring will change to assessing the nature and extent of the contamination and determining whether the contamination is getting better or worse over time as a result of management or remediation efforts.

In the detection phase of monitoring, the overall strategy is to obtain a background data set representing uncontaminated conditions and to statistically compare future observations with those background data. Background data may be obtained from up-gradient wells which presumably cannot be affected by the facility or from down-gradient wells which have not yet been affected. If the mean or median of a group of future observations is significantly (in a statistical sense) greater than the mean or median of the background samples, this is taken as evidence that a release has occurred.

A statistical hypothesis test is used to provide a 'yes' or 'no' answer to the question of whether the means or medians are different. The 'yes' or 'no' answer is desirable from a regulatory point of view as it leaves no grey area requiring a subjective decision. However, there will always be *some* difference in water quality from up-gradient to down-gradient wells and from one time period to another. Thus the statistical tests really answer not the question of whether there is a difference, but rather the question of whether the difference is large enough to cause a rejection of the null (no trend) hypothesis given the available sample size. If the sample size is large enough, we always reject a null hypothesis of no difference between groups.

If the sample size is too large, the hypothesis test will be overly sensitive to changes that are too small to be of concern. If the sample size is too small, the test will be insensitive to changes that are of concern. Rarely in regulatory monitoring is there sufficient statistical analysis to make sure that the sample size is 'just right' to detect changes of real concern and ignore those of no consequence. One possible solution to this problem is to place a greater emphasis on the estimation, particularly the interval estimation, of parameters rather than on the testing of hypotheses. Thereby, we avoid the question of whether or not there has been a change in water quality over time or space and try to answer the questions of how large are those changes, and how precise are the estimates of change.

NATIONAL WATER QUALITY ASSESSMENTS

US Geological Survey National Water Quality Assessment Program

As stated earlier, there seems to be an increased emphasis on long-term comprehensive studies which are properly designed for identifying changes in water quality and for evaluating cause and effect relationships behind those changes. One of the larger studies of this type which has a major emphasis on ground water is the US Geological Survey's National Water Quality Assessment Program (NAWQA), which began in 1991. The purpose of NAWQA is to describe the status and trends of a large, representative part of the nation's surface water and groundwater resources and to provide an understanding of both the natural and human factors affecting the quality of those resources. The study area consists of 60 large hydrological units (thousands of square kilometres each) covering about 60–70% of the USA's water use and population served by a public water supply. Forty of the regional studies have been started thus far, 20 in fiscal year 1991 and 20 in 1994, with the remainder to start in 1997 (Figure 3). Preliminary results of the studies begun in 1991 are now becoming available.

The studies are designed to be long term, with three years of intensive data collection followed by six years of low intensity sampling, with the cycle repeated into the future at each regional study. Statistical comparison of data across the intensive study periods provides the basis for the identification of temporal trends. As the studies are interdisciplinary and comprehensive, including chemical, biological, hydrological, and land-use factors, trend analysis will include an assessment of the causes and practical significance of changes in water quality. Pesticides and nutrients are an initial focus of the study (Leahy et al., 1993).

In 1993 an entire issue of *Water Resources Bulletin* was devoted to the NAWQA programme. The papers therein provide an excellent overview of the study (Leahy et al., 1993) and a more detailed discussion of five of the individual study areas: the Central Nebraska River Basins (Huntzinger and Ellis, 1993), the Red River of the North Basin (Stoner et al., 1993), the Rio Grande Valley (Ellis et al., 1993), the South Platte River Basin, (Dennehy et al., 1993) and the Trinity River Basin (Ulery et al., 1993).

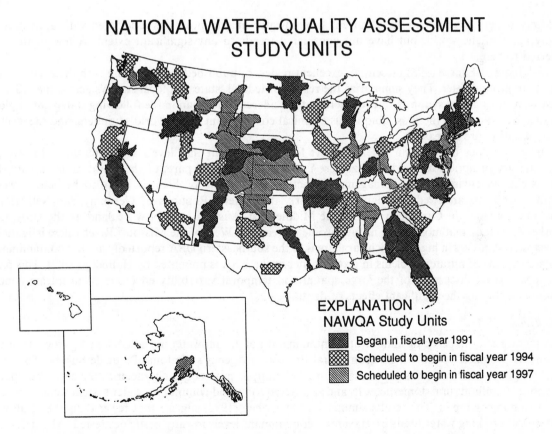

Figure 3. NAWQA study units (Leahy et al., 1993, with permission from the American Water Resources Association)

Australian Ground-Water Quality Assessment Program

According to Evans and Bauld (1993), Australia, like many countries, is becoming increasingly concerned about the sustainability of its water resources management. Natural resources management agencies at both the state and national levels are concerned that management decisions are based on adequate scientific information.

To address the need for scientifically based and consistent water quality management, the Australian Water Resources Council and the Australia and New Zealand Environment and Conservation Council have jointly developed a National Water Quality Management Strategy. As part of the strategy, the Australian Geological Survey Organization is developing an Australian Ground-Water Quality Assessment Program.

The objectives of the programme are to: support groundwater protection guidelines; provide information on the current state and trends in groundwater systems as a foundation for sustainable resource management; identify existing point and non-point sources of pollution; and provide information on spatial and temporal patterns in groundwater quality as a result of natural processes.

The main tasks are to: establish protocols for data collection and storage; analyse existing data to establish suitability for the above objectives; perform detailed catchment studies to assess management performance; and perform detailed research on groundwater quality processes.

The high priority areas identified for detailed studies are the Murray–Darling Basin, the populated region of south-eastern Australia, the populated regions of south-western Australia and northern Australia.

AGRICULTURAL CASE STUDIES

Overview

Many papers have documented current groundwater quality problems associated with agriculture in many parts of the world, but have not considered trends to any significant extent. A few of these are referred to here.

Spalding and Exner (1993) present an excellent review of published work dealing with the occurrence of nitrate in groundwater. They summarize the results of federal, state and local surveys across the USA and provide a short global synopsis. Trends were not considered other than in the following statement of global perspective: 'Nitrate is the most ubiquitous chemical contaminant in the world's aquifers, and the levels of contamination are increasing.'

The paper does not contain much support for the statement that levels are increasing. However, as the intensity of agriculture and fertilizer use is increasing in most parts of the world more rapidly than the associated environmental management practices, the nitrate problem is likely to be getting worse. Nitrate problems are cited in Belgium, Denmark, the Federal Republic of Germany, the Netherlands, England, Canada, the Caribbean, Africa, the Middle East, Australia and New Zealand. In the USA, these workers conclude that most of the nitrate problem areas 'are West of the Missouri River where irrigation is a necessity. Aquifers in highly agricultural areas in the southeastern USA reportedly are not contaminated'.

An overview of nitrate problems in groundwater in the USA is presented by Hallberg (1989). This paper includes a good discussion of the large spatial and temporal variability encountered in nitrate concentrations, including the significant effects of depth.

Snapshots, USA

A picture of both pesticide and nitrate contamination of groundwater in the USA at a particular instant in time is provided by the US Environmental Protection Agency's National Pesticide Survey. This study used a rigorous statistical design and sampling and analysis protocol to characterize the entire population of about 10 million rural domestic wells and a hundred thousand community water system wells using only a few hundred samples. The results summarized by Cohen (1992) suggest that fewer than 1% of all wells exceed the drinking water limits or maximum contaminant levels for any pesticides tested. About 2.4% of the rural wells and 1.2% of the community wells exceed the drinking water limits for nitrate.

Snapshots of particular regions of the USA were given by Burkhart and Kolpin (1993) and Koterba *et al.* (1993), to mention only two. Burkhart, and Kolpin (1993) describe the 'hydrogeologic spatial, and seasonal distribution' of selected herbicides, especially atrazine and its residues, and nitrate based on samples from 303 wells in 1991. Nitrate exceeded the drinking water limit of 10 mg/l as nitrogen (N) in 6% of the samples. At least one herbicide or atrazine metabolite was detected (with a detection limit of 0.05 μg/l) in 24% of the samples. Koterba *et al.* (1993) provides a similar snapshot of pesticides in shallow groundwater in the Delmarva Peninsula, in the north-eastern USA between Chesapeake Bay and the Atlantic, based on 100 wells sampled by the US Geological Survey in 1988–1990. As in most studies of this type, herbicide residues were detected at low levels, mostly in the shallowest aquifers (less than 10 m to the water-table). Neither of these studies considered trends.

Snapshot, Australia

Bauld *et al.* (1992) describe a snapshot monitoring programme involving 14 observation wells in the Shepparton Region Irrigation Area of the Murray Basin in south-eastern Australia, north of Melbourne. The water-table in the area is shallow, with a median depth to water of about 1 m. The wells were sampled once and samples were analysed for nutrients, major ions, metals, microbial contamination and over 30 pesticides. Half of the wells showed contamination by triazine herbicides, none above health advisory levels, and half exhibited contamination by fecal indicator bacteria, probably from livestock, above World Health Organization drinking water standards. Interestingly, nitrate concentrations were generally below drinking water standards. Only one sample exceeded 10 mg/l nitrate-N concentration. There was no observed correlation between nitrate concentrations and pesticide contamination.

Snapshot, Egypt

El Ghandour *et al.* (1985) discussed the spatial distribution of carbonate, hydrogencarbonate and pH in groundwater for the Nile Delta region in Egypt. They discussed the implications regarding the suitability of the water for irrigation but did not discuss trends, temporal patterns or land-use impacts on water quality.

Trends, New Zealand

There have been few attempts at formal trend analysis of groundwater quality records in an agricultural setting. An excellent attempt, however, was reported by Close (1987). In this study of irrigation impacts on groundwater quality in New Zealand, monthly data from 15 wells in a shallow (1–20 m depth to water-table) alluvial aquifer were analysed for the period 1977 to 1984. The study area was in the north-eastern part of South Island, one of the drier parts of the country. Surface irrigation was begun on about 30% of the study area midway through the study period. This highly desirable situation of several years of record before such a significant change in agricultural practices is also highly unusual, making this study unique.

Several water quality variables were studied, the most important being nitrate concentration. Formal statistical tests for trend, including the non-parametric Mann–Whitney test and a test based on Spearman's rho, were applied to examine the effects of irrigation. Corrections for serial correlation were used. Surprisingly, nitrate concentrations tended to increase over the before-irrigation period and decreased during the period after irrigation was started. Physical explanations were postulated, such as the dilution effect of excess irrigation water percolating below the root zone. Leaching a greater mass of nitrate to the groundwater would not increase the concentration of nitrate in groundwater if the mass of percolating water were sufficiently large.

Physical insight into the important causal mechanisms of water quality changes was enhanced by sampling two wells at discrete vertical intervals during part of the study period. As the vertical distribution of nitrates in the aquifer is strongly governed by the leaching process, this is an important, but sometimes overlooked, component of groundwater monitoring programmes to assess agricultural impacts on water quality. In this study, as expected, the depth of 'drainage affected groundwater' increased during irrigation seasons and after winter recharge by precipitation.

Management-induced change, USA

A few other studies have attempted some trend analysis, though generally not as detailed as that by Close (1987). One monitoring study related improvements in groundwater quality to improved nutrient management through tightly controlled manure applications (Hall, 1992). Two years of pre-practice data were compared with three years of post-practice data at five wells. Statistically significant differences in concentrations were observed using the Mann–Whitney test. This represents one of few studies in which a 'statistically significant' improvement in groundwater quality was measured in response to an improvement in agricultural practices such as nutrient management.

Wall *et al.* (1992) report on groundwater monitoring activities in the Garvin Brook watershed and recharge area in Minnesota, north-central USA. A 15-well network was established in 1981 to create baseline data and to evaluate changes in groundwater quality from implementation of the Rural Clean Water Program BMPs, including pesticide and nutrient management. Beginning in 1983, 80 domestic wells were sampled to raise public awareness of water quality concerns as well as to track changes in water quality. The study included statistical tests (Kendall's tau) for trend in nitrate concentrations using series of five to eight annual observations per well. Decreasing concentrations were found more often than increasing concentrations over the study period. However, it was not possible to relate changes in water quality to management practices. For the original 15-well network, age dating of groundwater revealed that most of the wells yielded water that entered the ground over 30 years earlier and did not reflect recent land-use practices. This illustrates the importance of characterizing the hydrogeological setting when designing monitoring programmes and interpreting data therefrom. As aquifer response times to management inputs vary greatly from site to site, it is impossible to generalize their magnitudes.

Owens and Edwards (1992) discuss long-term changes in groundwater quality in response to a one-time surface application of bromide. They reported the highest concentrations of bromide in groundwater three years after application, with increased levels persisting 10 years after application.

Modelling management-induced change, USA

Several studies have used modelling as opposed to monitoring approaches to predict the impact of improved agricultural management on groundwater quality. Shirmohammadi *et al.* (1991), for example, describe the use of CREAMS to simulate the long-term effects of alternative BMPs on nitrate loadings to a shallow groundwater system, using two representative watersheds in the Coastal Plain region of Maryland. Studies of this type are, no doubt, a useful part of planning water quality management programmes for a given region. However, there does not appear to be a great deal of experimental evidence, as yet, that the actual or even the relative magnitudes of improvement in groundwater quality predicted by a model for alternative management practices will be realized.

Regional trends, Midwestern USA

One of the regions which has received the most attention regarding agricultural chemical impacts on water quality is the 'Corn Belt'. Oddly referred to as the midwest, this region is actually located in the northern half of the central part of the USA. In most of this region, corn and other crops are grown under rain-fed conditions, without irrigation. Most of the studies of the region have looked at fairly short data records.

Hallberg and Keeney (1993) provide an overview of the issue of nitrate contamination of groundwater, discussing nitrogen cycling, sources of nitrate to groundwater and nitrate distribution and variability. Using the Big Spring Basin in Iowa (heart of the USA's Corn Belt) as an example, they note significant increases in nitrate concentrations in groundwater from less than 1 mg/l nitrate-N in the 1930s to annual averages of about 10 mg/l in the 1980s. This increase in groundwater nitrate corresponded to an increase of 30% in manure nitrogen application and almost 300% in inorganic fertilizer nitrogen application between the late 1950s and 1980s (Figure 4). Similar or more dramatic increases were noted in Nebraska (another large corn-producing state in the USA), England and the former USSR.

McDonald and Splinter's (1982) study of nitrate concentrations in Iowa water supplies is one of the few which have analysed long-term trends. Not surprisingly, the study found increasing concentrations of

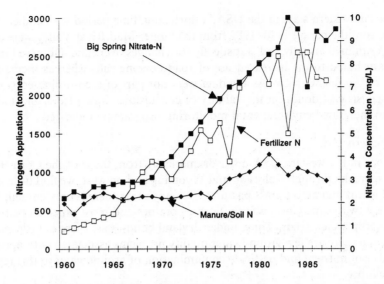

Figure 4. Nitrogen applications in the Big Spring Basin and groundwater nitrate-N concentrations at Big Spring Iowa, USA (modified from Hallberg and Keeney, 1993)

nitrate in shallow wells, probably due to the increasing use of anhydrous ammonia fertilizer. Deeper wells had much lower nitrate concentrations.

Rajogopal (1987) described the integration of two large groundwater quality databases for Iowa from the US Geological Survey WATSTORE (Water Data Storage and Retrieval) and the US Environmental Protection Agency MSIS (Model State Information System), and discussed the application of the data for water quality management decisions and for improving future monitoring activities.

Using the Iowa data with 52 years of record, Rajogopal (1986) investigated the estimation of water quality population parameters for nitrate and fluoride concentrations using samples of various sizes from 2000 down to 50. The parameters studied were the minimum, maximum and the following quantiles: 0.01, 0.05, 0.10, 0.25, 0.50, 0.75, 0.90, 0.95 and 0.99. As the distributions were positively skewed, lower quantiles were estimated more closely than upper quantiles, but good estimates were obtained with sample sizes down to 100 for all quantiles. Not surprisingly, the smallest sample size did poorly in estimating the 0.99 quantile. Still, a sample size of 50 would be considered large if we were interested, not in regional conditions, but in the distribution at a single well. Unfortunately, trend analysis was not discussed.

Fedkiw (1991) summarizes a few localized studies of long-term trends in groundwater nitrate concentrations. The largest of these was a US Geological Survey study (Chen and Druliner, 1988) of six hydrologically distinct areas of the High Plains Aquifer in Nebraska. A total of 2308 groundwater samples was available for the period 1936 to 1983. In only one of the six areas did an upward trend in nitrate concentration appear to be statistically significant at the 95% level. In the other areas, concentrations appeared to be increasing, but were not statistically significant.

The most commonly reported problem associated with pesticides in the Corn Belt is the presence of atrazine—a widely used herbicide on corn—in both groundwater and surface water supplies. As atrazine is moderately mobile in soils, it can move from alluvial groundwater into surface water supplies. Blum et al. (1993) document atrazine transport in the Platte River near Ashland, Nebraska and the movement of atrazine from the river into the alluvial aquifer. They estimated that atrazine transported down the Platte River in 1989, 1403 kg, increased to 10 157 kg in 1990 and 15 057 kg in 1991, a large upward trend (we might question whether it is meaningful to estimate these annual mass loadings of atrazine to the nearest kilogram). The upward trend in atrazine transported in the river suggests an upward trend in atrazine present in the aquifer. These workers attribute the increase to changes in the amount and timing of rainfall over the period as agricultural practices were relatively stable during those years. Atrazine moves from the river into the aquifer with an observed delay of 21 days from the river to wells 3 m from the bank.

In Ohio, another midwestern state in the USA, a fairly long time period was studied by Pennino (1984). Groundwater data from the period 1940–1983 from the Upper Mad River Valley were collected by several Ohio agencies and examined visually and statistically for changes over time. Observed increases in chloride and sodium were attributed to the increased use of road de-icing salt, whereas increases in sulphate and potassium were linked to fertilizer use. As the data were not part of a co-ordinated study, inconsistencies in sampling and analysis cast doubt on the validity of conclusions drawn from the study. This was noted and the need for an integrated environmental monitoring programme indicated.

Regional trends, Western U.S.

Climatic conditions in the western USA differ dramatically from those of the Corn Belt. From western-most Kansas and Nebraska through Colorado and Wyoming and on west, we find little rain-fed agriculture other than dryland wheat and native grass pastures. In many areas there is only enough rainfall to produce one wheat crop every two years. Corn, alfalfa, beans, potatoes and most other crops require irrigation to supply most of the crop water needs. Thus, under dryland conditions, there is little leaching of moisture past the root zone, and there are few groundwater quality problems associated with agricultural chemicals. Essentially, all studies of nutrient and pesticide contamination of groundwater in this region are associated with irrigated agriculture.

The most intensive and extensive irrigated agriculture in the USA is found in California. Letey and Pratt (1984) used monitoring data for agricultural tile drainage waters from the San Joaquin Valley (in the northern half of the state, centrally located from east to west) as a case study of agricultural pollutants and groundwater quality. The concentrations and mass loads of constituents in drainage waters probably represent a worst-case scenario for what could reach groundwater if no drains were present, as some reduction in concentrations and loads can occur below the root zone by denitrification. Over an 11 year period from 1971 to 1981, both the concentrations and loads of nitrate were stable in the southern part of the valley. In the central region of the valley, both concentrations and loads were higher than in the south. In this region, the concentrations tended to decrease over time, whereas the loads remained at about the same level, suggesting that the drainage volume increased over the period. No statistical analyses for the significance of trend were performed.

Thompson (1993) conducted a study of nitrate contamination of the shallow aquifer of the San Luis Valley of south-central Colorado (in the west-central USA.) This intensively farmed region is a high (2300 m elevation) inter-mountain valley characterized by coarse soils, shallow water-tables (1–3 m below the ground surface in many areas), cold winters and little rainfall.

Groundwater quality data were available from three sources: Agro Engineering, a local crop management and irrigation consulting firm for 1984–1992, the Colorado Department of Health for 1990 and the US Geological Survey for 1973 and 1981. The data quality is variable among the various data sources, and no wells had nice time series of data, though a few had multiple observations over time. As mentioned earlier, any results from an analysis such as this, using existing wells and existing data, must be viewed sceptically. That does not mean, however, that the data are useless for any kind of regional analysis.

Although all of the data were used to map nitrate contamination, only the Agro Engineering data (consisting of 645 data points from 339 wells in a study area of 1455 km^2)were used in trend analysis. These data were from irrigation wells which are screened over a large fraction of the saturated zone and have pumping rates of the order of 60–70 l/s. This type of well should provide a good indication of average water quality conditions over the saturated depth, but would not be the optimum for evaluating leaching from the surface.

Trends were examined using three simple procedures. Firstly, the area which exceeded the drinking water standard for nitrate was estimated for each year using a geographical information system (GIS) and graphics software package. This fraction was plotted versus year. Next, trend slopes were calculated for each well that was sampled more than once over the study period and the number of positive slopes was statistically compared with the number of negative slopes using the binomial distribution. Finally, a similar approach was taken using estimated areas of increasing trend and decreasing trend, the areas again delineated using the GIS and graphics software. All three methods used by Thompson (1993) suggested an increasing trend in nitrate concentrations over the study area during the study period. Although the approach is not statistically rigorous, agreement among the three methods of data analysis suggests that the overall conclusions of a general increase in nitrate concentrations over the region during the study period are reasonable.

Regional trends—France, Denmark, Germany

Meybeck et al. (1989) describe trends in groundwater nitrate concentrations in France and Denmark. The distributions of nitrate concentrations in roughly 300 samples taken in France were compared for the years 1965 (318 samples), 1971 (292 samples) and 1977 (266 samples). In 1965, 100% of the samples were below 20 mg/l nitrate as NO_3^-. The fraction dropped to 60% in 1971 and to 26.5% in 1977. The fraction exceeding 40 mg/l as NO_3^- was 3.4% in 1971 and 14.7% in 1977. Nitrate levels in the Petite Traconne Spring in the Brie region of France rose from 6 mg/l during the 1930s and 1940s to about 16 mg/l in 1980. In Denmark, five year mean nitrate concentrations for 11 000 groundwater samples from wells greater than 10 m depth more than doubled over the period 1940–1982. The increase, from less than 1 mg/l to more than 2 mg/l, corresponds to increased agricultural use of nitrogen compounds.

Meybeck et al. (1989) also present an interesting case study of the Upper Rhine Aquifer, which is situated between the Vosges of France and the Black Forest of Germany. This aquifer, the most important water resource in the region, is overlain with shallow, permeable soils and is vulnerable to pollution from both agricultural and industrial development. Water quality problems include nitrate contamination from agriculture, chloride contamination from potassium and salt mining, and chlorinated hydrocarbon (solvent) contamination of the aquifer in industrial areas. 'On average, nitrate levels have doubled during the last ten years'. A continuation of this trend would result in half of the water resources exceeding drinking water standards for nitrate by the year 2000.

Regional trends, Israel

Ronen et al. (1984) discussed 'the importance of the unsaturated zone as a buffering link between the top soil and the water table' using a nitrogen balance for the Coastal Plain aquifer of Israel. They suggest that

1. The average nitrate concentration in the aquifer increased from less than 5 mg/l (as NO_3^-), assumed background concentration, in 1930 to over 50 mg/l in 1980.
2. The nitrate flux to groundwater across the unsaturated–saturated interface decreased from about 3300 tons N per year during the 1930s to about 300 tons per year in the 1970s.
3. The rate of increase in average nitrate concentration in groundwater decreased from 1.45 mg/l yr during the 1940s to 0.13 mg/l yr in the 1970s. Thus the nitrogen balance in the aquifer appears to be approaching steady state.

URBAN CASE STUDIES

The literature is rich with studies of urban groundwater pollution, but, here again, few longer term data records are available, and few trend analyses have been performed.

Urban growth and development

A list of 36 references through 1990 covering 15 countries is presented by Somasundaram et al. (1993). These workers provide a detailed description of groundwater pollution of the Madras urban aquifer in India, reporting 'gross pollution of ground and surface waters by a range of species including nitrate, heavy metals, and microorganisms.' The study did not consider changes in groundwater quality over time. In fact, not many studies have related urbanization to changes in groundwater quality over time. However, Sulam and Ku (1977) related changes in groundwater quality in south-east Nassau County, New York from 1910 to 1975 to population growth accompanied by the installation of cesspools and septic tanks and increased fertilizer use.

Very few studies examine trends in urban groundwater quality over both time and space. In a relatively early work, however, Kaufman (1977) did just that for groundwater quality in the Las Vegas Valley in Nevada, USA, a region heavily affected by development. Kaufman found that groundwater quality at depths less than 50 ft was highly variably in space, probably reflecting spatially variable land-use and water-use practices. At depths greater than 50 ft, changes were more gradual and explainable by geological conditions. Time trends over the period 1912–1968 were not statistically significant, but the number of available data was small.

De-icing salt

Urban groundwater problems can result from the use of de-icing salts on roads. The study by Pennino (1984), mentioned earlier under agricultural impacts, suggested that trends in chloride and sodium in Ohio, USA were related to the use of road de-icing salt. Meybeck et al. (1989) report a dramatic example from Burlington, Massachusetts, USA (after Todd, 1980) in which groundwater chloride concentrations increased from around 10 mg/l in 1956 to over 150 mg/l in 1970 due to roadway de-icing. Concentrations started to decrease after the use of de-icing salts was discontinued.

Acid precipitation

Another environmental impact of urbanization and industrial growth is acid precipitation and the associated acidification of soils and water resources, including eventually groundwater. Jacks et al. (1984) document the 'Effect of acid rain on soil and groundwater in Sweden'. These workers studied hydrochemical data from 'about 400 municipal wells in six counties in different parts of Sweden . . . as well as water analysis from about 550 private wells and springs in six test areas.' The study concentrated in the southern half of Sweden and included resampling of about 250 wells and springs for which earlier data were available to provide a comparison over time. The available period of record ranged from 10 to 25 years. The data analysis included time series plots and statistics, but the statistical methods were not described. Drawing conclusions from the analysis was difficult because of the complexity and variability of the physical and chemical system, including the effect of wet and dry years and of pumping. However, some general observations were made regarding mainly municipal wells in sand and gravel aquifers.

1. 'Real' acidification or 'significantly' lowered pH cannot be proved, but shallow wells and springs in south-eastern and western Sweden show apparent downward trends in pH.

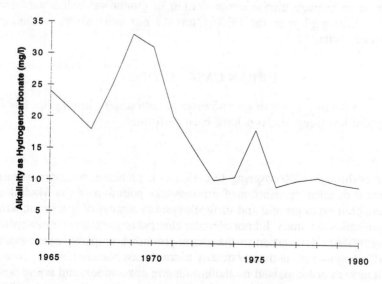

Figure 5. Groundwater alkalinity time series for Kronoberg county, Sweden: an example of decreases in response to acid precipitation where bedrock consists of weathering-resistant materials (modified from Jacks et al., 1984)

2. In the same regions alkalinity tended to decrease in areas where the bedrock consisted of weathering-resistant minerals (Figure 5), but increased where the bedrock was easily weathered.
3. Total hardness, the hardness to alkalinity ratio and sulphate tended to increase over time.

Jacks (1993) states that pH is usually not a good indicator variable for changes in groundwater caused by environmental acidification. The pH is difficult to measure in poorly buffered waters, which are most likely to be affected, and carbon dioxide may be lost before and during measurement, leading to questionable measurements.

SOLID AND HAZARDOUS WASTE CASE STUDIES

In the introductory issue of *Ground Water Monitoring Review*, Sgambat and Stedinger (1981) provided an overview of the difficulties in obtaining useful regulatory information from groundwater monitoring of waste disposal sites. They discuss problems associated with sampling procedures, natural variability in data and statistical analysis. They mention trend analysis as a monitoring objective, but discuss statistical methods only for a comparison of means across time or space.

Although there are hundreds of groundwater monitoring programmes in place for waste disposal sites, not too many have been documented. One obvious explanation is that many examples are politically sensitive or tied up in court. Also the primary objective of these programmes is meeting regulatory criteria, which might or might not include tracking long-term trends.

In the past, regulatory monitoring programmes were not often rigorously designed from an overall systems perspective and have often not provided the information needed to support effective management decisions. An example programme in which problems were corrected is the IBM Ground-water Quality Monitoring Program at East Fishkill, New York, USA described by Bell (1991).

Before the recent downsizing of IBM Corporation, the East Fishkill Facility was one of the world's largest semiconductor manufacturing plants, with a 300 ha site, roughly 400 000 m^2 of building floor space and over 10 000 employees at its peak. A groundwater quality monitoring programme was begun in 1979 to aid in the characterization and remediation of solvents in groundwater below the site. Over 470 different wells (some off-site) have been sampled as part of this programme. In the first few years of operation the data produced were highly variable and did not provide the information needed for effective management of existing groundwater programmes.

Over the course of several years, beginning in 1982, a complete redesign of the monitoring programme was effected, in which information goals were clearly defined and a data quality improvement programme was instituted, which included: resampling when outliers occur; in-house laboratory analysis of water quality samples; elimination of censoring of data by the laboratory, i.e. recording of non-detects (Porter, 1986; Porter *et al.*, 1988); and institution of a data analysis protocol which specifies statistical methods to be used in given situations (Ward *et al.*, 1988).

The data analysis protocol addresses three statistical information objectives: (1) determination of average conditions; (2) determination of trends; and (3) assessment of compliance with regulations.

The protocol emphasizes non-parametric and graphical approaches to maximize its applicability for data from a variety of statistical distributions. The complete redesign and implementation of the revised monitoring programme has resulted in improved data quality, reduced variability (Figure 6) and an improved information product provided cost-effectively.

In addition to the data analysis protocol developed for the IBM East Fishkill facility, similar protocols which supplement the RCRA guidance are being developed for other sites. For example, Chou and Jackson (1992) discuss data handling methods, graphical evaluation techniques, statistical tests and reporting procedures for the Hanford Site, Washington, a US Department of Energy facility. A general framework for the development of such protocols is presented by Adkins (1993).

Long-term trend analyses of groundwater quality at waste disposal sites are scarce to date. Shorter term trends were qualitatively examined by Rivett *et al.* (1990) for chlorinated solvents in the sandstone aquifer underlying the city of Birmingham, UK. Solvent concentrations were observed at 59 sampling points

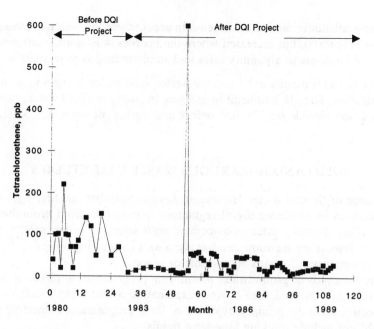

Figure 6. Time series of tetrachloroethene concentrations from monitoring well at IBM East Fishkill, New York, USA. A data quality improvement (DQI) project in 1982 resulted in greatly reduced variability in water quality observations beginning in 1983 (adapted from Bell, 1990)

(abstraction wells) within an area of roughly 100 km^2 over a 20 month period during 1987 and 1988. Variations in water quality over the period were related to pumping patterns and contaminant movements. According to these workers: 'Data obtained during an organic water quality survey of the Birmingham Aquifer have indicated that abstraction wells may be used to provide valuable data on the degree and trend of ground water contamination by chlorinated solvents'.

There are no statistical analyses in the paper. It is implied that the term 'trend' here refers to spatial and temporal patterns resulting from pumping and plume movement, with temporal scales smaller than the 20 month period of record.

Plumb (1987) provides a comparison of groundwater quality monitoring data generated under two different US federal programmes for waste disposal regulation and hazardous waste clean-up, the RCRA and the Comprehensive Environmental Response Compensation and Liability Act (CERCLA), commonly known as the Superfund. In comparing data from 334 sites, Plumb found that RCRA sites generally had more wells sampled more frequently, whereas samples from CERCLA sites were analysed for a larger number of constituents. There was no discussion of length of record available or of any trend analyses.

SUMMARY AND FUTURE DIRECTIONS

In summary, the term 'trend' takes on several different meanings in both spatial and temporal contexts in relation to groundwater quality. We perhaps think first of groundwater quality trends as gradual changes over large areas as a result of changing land-use practices. As temporal and spatial scale effects dominate the interpretation of trends, this definition should be amended, emphasizing changes that occur over a defined region of interest in time and space. This review also considered a purely statistical definition of trend, based on a statistical model, and a regulatory definition of trend arising from the US RCRA dealing with solid and hazardous wastes.

Several groundwater quality studies were reviewed under the headings of national assessments, agricultural impacts, urban impacts and impacts of waste disposal. These studies illustrated the importance of defining the meaning of 'trend' in terms of monitoring and management objectives of each specific study and interpreting groundwater quality observations in terms of the specific hydrogeological setting, including the age, depth and flow path of groundwater being sampled. These criteria are particularly difficult to meet when using existing data and/or existing wells for trend studies.

Unfortunately, there are relatively few data records available for adequately characterizing long-term trends in groundwater quality. Instead we find many snapshots of groundwater quality and studies of short-term changes related to land use and management. The greatest number of such studies deal with agricultural impacts, especially from nitrate fertilizers. There are also several studies characterizing urban and waste disposal impacts.

There is a perception that groundwater quality is getting worse in many areas of the world. Although this is logical, direct evidence of time trends towards poorer groundwater quality is limited and localized. In the few instances where longer term records are examined, data are often obtained from a variety of sources and may not be comparable. Statistical methods of trend analysis have been sparingly applied.

For the future, however, we can expect more resources to be devoted to comprehensive studies of groundwater quality over longer study periods. Even though the pace of implementing such studies may be slowed by economic conditions in both the USA and Europe, the need for usable information on the status and trends of water quality is becoming ever more apparent. The NAWQA programme is an excellent beginning for a 'trend' towards more carefully designed monitoring programmes of both groundwater and surface water quality. Programmes such as the Australian Ground-Water Quality Assessment Program demonstrate that commendable efforts for linking groundwater monitoring and management are not limited to the USA.

ACKNOWLEDGEMENTS

Funding support for this review was provided by Colorado State University and by the Colorado Agricultural Experiment Station, Project #723, 'Modeling and Monitoring Agricultural Chemical Transport in the Subsurface Environment.'

REFERENCES

Adkins, N.C. 1993. 'A framework for development of data analysis protocols for ground water quality monitoring', *Tech. Rep. No. 60*, Colorado Water Resources Research Institute, Colorado State University, Fort Collins.

Alley, W. M. 1993. 'General design considerations' in Alley, W. M. (Eds), *Regional Ground-Water Quality*. Van Nostrand Reinhold, New York. pp. 3–22.

Bauld, J., Evans, W. R., and Sandstrom, M. W. 1992. 'Groundwater quality under irrigated agriculture: Murray Basin, southeastern Australia' in Fei Jin (Ed.), *Groundwater and Environment. Proceedings of International Workshop on Groundwater and Environment, Beijing, China*. Seismological Press, Beijing. pp. 447–457.

Bell, H. F. 1991. 'IBM ground water quality monitoring program at East Fishkill, New York' in Ward, R. C., Loftis, J. C., and McBride, G. B. (Eds), *Design of Water Quality Monitoring Systems*. Van Nostrand Reinhold, New York. pp. 130–147.

Blum, D. A., Carr, J. D., Davis, R.K., and Pederson, D.T., 1993. 'Atrazine in a stream–aquifer system: transport of atrazine and its environmental impact near Ashland, Nebraska', *Ground Water Monitoring Rev.*, **13**(2), 125–133.

Burkhart, M. R. and Kolpin, D. W. 1993. 'Hydrologic and land-use factors associated with herbicides and nitrate in near-surface aquifers', *J. Environ. Qual.*, **22**, 646–656.

Close, M. E. 1987. 'Effects of irrigation on water quality of a shallow unconfined aquifer', *Wat. Resour. Bull.*, **23**, 793–802.

Close, M. E., 1989. 'Effect of serial correction on ground water quality sampling frequency', *Wat. Resour. Bull.*, **25**, 507–515.

Chen, H. and Druliner, A. D. 1988. 'Agricultural chemical contamination of ground water in six areas of the High Plains Aquifer, Nebraska, National Water Summary 1986—hydrologic events and ground water quality', *US Geol. Surv., Wat.-Supply Pap.*, **2325**, 103–106.

Chou, C. J. and Jackson, R. L. 1992. 'Analysis protocol for the Hanford Site, Washington', presented at the *American Nuclear Society Meeting RCRA Groundwater Data Spectrum '92, Boise, Idaho, August 23–27*. Westinghouse Hanford, Richland.

Cohen, S. 1992. 'Agricultural chemical news', *Ground Water Monitoring Rev.*, **12**(2), 90–91.

Dennehy, K. F., Litke, D. W., Tate, C. M., and Heiny, J. S. 1993. 'South Platte River Basin—Colorado, Nebraska, and Wyoming,' *Wat. Resour. Bull.*, **29**, 647–684.

El Ghandour, M. F. M., Khalil, J. B., and Atta, S. A. 1985. 'Distribution of carbonate, bicarbonate, and pH values in ground water of the Nile Delta region, Egypt', *Ground Water*, **23**, 35–41.

Ellis, S. R., Levings, G. W., Carter, L. F., Richey, S. F., and Radell, M. J. 1993. 'Rio Grande Valley, Colorado, New Mexico and Texas', *Wat. Resour. Bull.*, **29**, 617–646.

Evans, W. R. and Bauld, J. 1993. 'Towards an Australian groundwater quality assessment program', *AGSO J. Aust. Geol. Geophys.*, **14**, 307–311.

Fedkiw, J. 1991. *Nitrate Occurrence in U.S. Waters (and Related Questions): a Reference Summary of Published Sources from an Agricultural Perspective*. United States Department of Agriculture, Washington.

Hall, D. W. 1992. 'Effects of nutrient management on nitrate levels in ground water near Ephrata, Pennsylvania', *Ground Water*, **30**, 720–730.

Hallberg, G. R. 1989. 'Nitrate in ground water in the United States' in Follett, R. F. (Ed.), *Nitrogen Management and Ground Water Protection*. Elsevier Science, New York. 395 pp. pp 35–69.

Hallberg, G. R. and Keeney, D. R. 1993. 'Nitrate' in Alley, W.M. (Ed.), *Regional Ground-Water Quality*. Van Nostrand Reinhold, New York. pp. 297–322.

Hamilton, P. A., Welch, A. H., Christenson, S. C., and Alley, W. H. 1993. 'Uses and limitations of existing groundwater-quality data' in Alley, W. M. (Ed.), *Regional Ground-Water Quality*. Van Nostrand Reinhold, New York. pp. 613–622.

Harris, J., Loftis, J. C., and Montgomery, R. H. 1987. 'Statistical methods for characterizing ground water quality', *Ground Water*, **25**, 176–184.

Hoyle, B. L. 1989. 'Ground-water quality variations in a silty alluvial soil aquifer, Oklahoma', *Ground Water*, **27**, 540–549.

Huntzinger, T. L. and Ellis, M. J. 1993. 'Central Nebraska River Basins, Nebraska', *Wat. Resour. Bull.*, **29**, 533–574.

Isotok, J. D., Smyth, J. D., and Flint, A. L. 1993. 'Multivariate geostatistical analysis of groundwater contamination: a case history', *Ground Water*, **31**, 63–73.

Jacks, G. 1993. 'Acid precipitation' in Alley, W. M. (Ed.), *Regional Ground-Water Quality*. Van Nostrand Reinhold, New York. pp. 405–421.

Jacks, G., Knutsson, G., Maxe, L., and Flykner, A. 1984. 'Effect of acid rain on soil and groundwater in Sweden' in *Pollutants in Porous Media, Ecological Studies*. Vol. 47. Springer-Verlag, New York. pp. 94–114. B. Yaron, G. Dagan and J. Goldshmid (Eds).

Kaufmann, R. F. 1977. 'Land and water use impacts on groundwater quality in Las Vegas Valley' *Ground Water*, **15**, 81–89.

Koterba, M. T., Banks, W. S. and Shedlock, R. J. 1993. 'Pesticides in shallow groundwater in the Delmarva Peninsula', *J. Environ. Qual.*, **22**, 500–518.

Leahy, P.P., Ryan, B. J., and Johnson, A. I. 1993. 'An introduction to the U.S. Geological Survey's National Water-Quality Assessment Program', *Wat. Resour. Bull.*, **29**, 529–532.

Letey, J. and Pratt, P. F. 1984. 'Agricultural pollutants and groundwater quality' in *Pollutants in Porous Media, Ecological Studies*. Vol. 47. Springer-Verlag, New York. pp. 211–222. B. Yaron, G. Dagan and J. Goldshmid (Eds).

Loftis, J. C., Harris, J., and Montgomery, R. H. 1987. 'Detecting changes in ground water quality at regulated facilities', *Ground Water Monitoring Rev.*, **7**(1), 72–76.

Loftis, J. C., McBride, G. B., and Ellis, J. C. 1991. 'Considerations of scale in water quality monitoring and data analysis', *Wat. Resour. Bull.*, **27**, 255–264.

Meybeck, M., Chapman, D. V., and Helmer, R. (Eds) 1989. *Global Freshwater Quality, A First Assessment*. World Health Organization and the United Nations Environment Programme, Basil Blackwell, Oxford. 306 pp.

McBride, G. B., Loftis, J. C., and Adkins, N. C. 1993. 'What do significance tests really tell us about the environment?' *Environ. Manage.*, **17**, 423–432.

McDonald, D. B. and Splinter, R. C. 1982. 'Long-term trends in nitrate concentrations in Iowa's water supplies', *J. Am. Wat. Works Assoc.*, **74**, 437–440.

Montgomery, R. H., Loftis, J. C., and Harris, J. 1987. 'Statistical characteristics of ground water quality variables', *Ground Water*, **25**, 185–193.

NWQEP Notes 1994. *Newsletter of the National Water Quality Evaluation Project*. NCSU Water Quality Group, North Carolina State University, 615 Oberlin Rd., Suite 100, Raleigh, NC 27605–1126, USA.

Owens, L. B. and Edwards, W. M. 1992. 'Long-term groundwater quality changes from a one-time surface bromide application', *J. Environ. Qual.*, **21**, 406–410.

Pennino, J. D. 1984. 'Ground water quality in the aquifer of the Upper Mad River Valley 1940–1983', *Ground Water Monitoring Rev.*, **4**(3), 27–38.

Plumb, R. H. 1987. 'A comparison of ground water monitoring data from CERCLA and RCRA sites', *Ground Water Monitoring Rev.*, **7**(4), 94–100.

Plummer, L. N., Michel, R. L., Thurman, E. M., and Glynn, P. D. 1993. 'Environmental tracers for age dating young ground water' in Alley, W. M. (Ed.), *Regional Ground-Water Quality*. Van Nostrand Reinhold, New York. pp. 255–294.

Porter, P. S. 1986. 'Statistical analysis of water quality data affected by limits of detection', *PhD Dissertation*, Department of Agricultural and Chemical Engineering, Colorado State University, Fort Collins.

Porter, P. S., Ward, R. C., and Bell, H. F. 1988. 'The detection limit', *Environ. Sci. Technol.*, **22**, 856–861.

Pucci, A. A., Ehlke, T. A., and Owens, J. P. 1992. 'Confining unit effects on water quality in the New Jersey Coastal Plain', *Ground Water*, **30**, 415–427.

Rajagopal, R. 1986. 'The effect of sampling frequency on groundwater quality characterization', *Ground Water Monitoring Rev.*, **6**(4), 65–73.

Rajagopal, R. 1987. 'Large data bases and regional ground-water quality assessments—an Iowa case study', *Ground Water*, **25**, 415–426.

Rivett, M. O., Lerner, D. N., and Lloyd, J. W. 1990. 'Temporal variations of chlorinated solvents in abstraction wells', *Ground Water Monitoring Rev.*, **10**(4),127–133.

Rogers, R. J. 1989. 'Geochemical comparison of ground water in areas of New England, New York, and Pennsylvania', *Ground Water*, **27**, 690–711.

Ronen, D., Kanfi, Y., and Magaritz, M. 1984. 'Nitrogen presence in groundwater as affected by the unsaturated zone' in *Pollutants in Porous Media, Ecological Studies*. Vol. 47. Springer-Verlag, New York. pp. 223–236. B. Yaron, G. Dagan and J. Goldshmid (Eds).

Sgambat, J. P. and Stedinger, J. R. 1981. 'Confidence in groundwater monitoring', *Ground Water Monitoring Rev.*, **1**(1), 67–69.

Shirmohammadi, A., Magette, W. L. and Shoemaker, N. L. 1991. 'Reduction of nitrate loadings to ground water', *Ground Water Monitoring Rev.*, **11**(1), 12–118.

Smith, R. T. and Ritzi, R. W. 1992. 'Designing a nitrate monitoring program in a heterogeneous, carbonate aquifer', *Ground Water*, **31**, 576–582.

Somasundaram, M. V., Ravindram, G. and Tellam, J. H. 1993. 'Ground-water pollution of the Madras urban aquifer, India', *Ground Water* **31**, 4–11.

Spalding, R. F. and Exner, M. E. 1993. 'Occurrence of nitrate in groundwater—a review', *J. Environ. Qual.*, **22**, 392–402.

Stoner, J. D., Lorenza, D. L., Wiche, G. J., and Goldstein, R. M. 1993. 'Red River of the North Basin, Minnesota, North Dakota, and South Dakota', *Wat. Resour. Bull.*, **29**, 575–616.

Sulam, D. J. and Ku, H. F. H. 1977. 'Trends of selected groundwater constituents from infiltration galleries, southeast Nassau County, New York', *Ground Water*, **15**, 439–445.

Thompson, K. L. 1993. 'Nitrate contamination in the unconfined aquifer of the San Luis Valley', *Senior Honors Thesis*, Colorado State University, Fort Collins.

Todd, D. K. 1980. *Groundwater Hydrology*. 2nd Edn. Wiley, New York.

US Environmental Protection Agency 1989. *Statistical Analysis of Ground-Water Monitoring Data at RCRA Facilities—Interim Final Guidance*. Office of Solid Waste, Waste Management Division, USEPA, Washington.

US Environmental Protection Agency. 1992a. *Proceedings of the National RCWP Symposium*. US Environmental Protection Agency, Cincinatti.

US Environmental Protection Agency. 1992b. *Statistical Analysis of Ground- Water Monitoring Data at RCRA Facilities—Addendum to Interim Final Guidance*. Office of Solid Waste, Permits and State Programs Division, US EPA, Washington.

Ulery, R. L., Van Metre, P. C. and Crossfield, A. S. 1993. 'Trinity River Basin, Texas', *Wat. Resour. Bull.*, **29**, 685–711.

Wall, D. B., Evenson, M. G., Regan, C. P., Magner, J. A., and Anderson, W. P. 1992. 'Understanding the groundwater system: the Garvin Brook experience' in *Proceedings of the National RCWP Symposium*. US Environmental Protection Agency, Cincinatti. pp. 59–70.

Ward, R. C., Loftis, J. C., DeLong, H. P., and Bell, H. F. 1988. 'Ground water quality: data analysis protocol', *J. Wat. Pollut. Control Fed.*, **60**, 1938–1945.

SECTION II

GEOCHEMICAL MASS BALANCE

10

ATMOSPHERIC DEPOSITION TO WATERSHEDS IN COMPLEX TERRAIN

GARY M. LOVETT[1], JONATHAN J. BOWSER[2] AND ERIC S. EDGERTON[2]

[1]*Institute of Ecosystem Studies, Box AB, Millbrook, NY 12545, USA*
[2]*Environmental Science and Engineering, Inc., 4915 Prospectus Drive, Suite J, Durham, NC 27713, USA*

ABSTRACT

Single collection stations for wet or bulk deposition are generally inadequate to describe atmospheric inputs to watersheds in complex terrain. Atmospheric deposition is delivered by wet, dry and cloud deposition processes, and these processes are controlled by a wide range of landscape features, including canopy type and structure, topographic exposure, elevation and slope orientation. As a result, there can be a very high degree of spatial variability within a watershed, and a single sampling point, especially at low elevation, is unlikely to be representative. Atmospheric inputs at the watershed scale can be calculated from the whole watershed mass balance if the outputs and within-watershed sources and sinks are known with sufficient accuracy. Alternatively, indices of atmospheric deposition such as Pb accumulation in the forest floor and SO_4^{2-} flux in throughfall can be used to characterize patterns of total deposition, and these indices can be used to model deposition to the entire watershed based on known landscape features such as elevation and canopy type. © 1997 by John Wiley & Sons, Ltd.

INTRODUCTION

Studies of chemical mass balance of watersheds require accurate information on atmospheric inputs. Too often in watershed studies the input information comes from a single collector of precipitation or bulk deposition (the material collected by a continuously exposed collector such as a funnel or a bucket). Recent research has shown that this approach can produce serious inaccuracies in deposition measurements because: (1) atmospheric deposition is delivered by three mechanisms — wet, dry and cloud water deposition, and the precipitation or bulk deposition collector does not adequately sample the latter two; and (2) rates of atmospheric deposition by wet, dry and cloud processes can vary tremendously within a watershed, especially if the terrain is mountainous, and the location of the single collector may not be representative of the watershed as a whole. Because of these problems, the accurate estimation of atmospheric inputs to watersheds requires a more comprehensive measurement effort.

There have been several recent reviews of deposition mechanisms and measurement methods (Fowler, 1980; Hicks *et al.*, 1990; Lovett, 1994), and the reader is referred to these papers as background. This paper will summarize briefly the deposition processes, emphasizing the problems that are encountered in trying to estimate atmospheric input to a watershed in mountainous terrain.

DEPOSITION MECHANISMS

Wet Deposition

The delivery of atmospheric substances to the earth dissolved in falling precipitation, either liquid or frozen, is called wet deposition. Wet deposition is usually collected with a sampling device that exposes a collection vessel (usually a bucket or a funnel and bottle) automatically when the precipitation begins, and seals the vessel closed to prevent contamination during non-precipitation periods. These 'wet-only' collectors are used by the major precipitation monitoring networks in the USA, Canada and Europe. Generally these collectors are used to sample precipitation chemistry only, while precipitation amount is measured in a standard rain gauge nearby.

The amount of wet deposition depends on the amount of precipitation and the chemical composition of that precipitation. These two factors are related, in that concentration of many substances shows a hyperbolic relationship to precipitation amount (e.g. Hicks and Shannon, 1979). Within the context of a watershed study, both precipitation amount and chemistry are likely to show significant variation within a watershed if there is substantial topographic relief. Precipitation generally increases with elevation because of orographic effects, and rain shadowing may cause differences in precipitation between slopes with different orientations (Dingman, 1981; Barros and Lettenmaier, 1994). The amount of increase of precipitation with elevation averages about 75 cm per km of elevation in the mountains of the eastern US (Lovett and Kinsman, 1990), although the rates of increase appear to be less than that mean at elevations below 400 m and greater at elevations above 400 m (Ollinger *et al.*, 1993).

Precipitation concentrations change with elevation in some areas and not in others. In the eastern US, elevational gradients in precipitation concentrations are generally weak or non-existent (Lovett and Kinsman, 1990, Ollinger *et al.*, 1993). However, several conditions can result in significant changes in precipitation concentrations with elevation. For example: (1) a temperature inversion can lock ground level emissions into a lower layer of the atmosphere, resulting in less concentrated precipitation at higher elevations. A classic example of this situation is in the Los Angeles basin, in which precipitation concentrations of H^+, SO_4^{2-} and NO_3^- are lower at high elevation sites (Munger *et al.*, 1983). (2) Material emitted to the atmosphere at low elevation may not be carried long distances in the atmosphere, and thus would be more dilute in precipitation at high elevation sites. For example, the concentrations of base cations associated with dust decrease with elevation on the east slope of the Rocky Mountains (Lewis *et al.*, 1984). (3) Raindrops falling at high elevation may pass through a mountain cap cloud and accrete cloud droplets of higher chemical concentration. This is part of the 'seeder–feeder' phenomenon, which has been reported for rain in the hills of northern England (Fowler *et al.*, 1988) as well as for fine aerosols in the eastern US (Graustein and Turekian, 1989).

Variation in wet deposition in mountainous terrain can also be caused by the shifting proportions of snow and rain along an elevational gradient. Snow often has a different chemical composition than rain, even if both emanate from the same clouds, because snowflakes and raindrops have different scavenging efficiencies for atmospheric particles and gases (Raynor and Hayes, 1983; Topol, 1986; Dasch, 1987). Snow is problematic for two other reasons. First, snow is more difficult than rain to sample accurately, because it is more subject to collection errors caused by wind over the gauge. Secondly, because snow is redistributed by wind after deposition, the amount ultimately deposited may be highly variable, even on the scale of metres, making the choice of sampling locations especially important.

The result of these various processes can be much variation in wet deposition within a watershed if that watershed contains appreciable topographic relief. This implies that a single wet deposition monitoring site is insufficient in mountainous watersheds, and an accurate measurement requires a network of sites.

Dry Deposition

Processes. The direct deposition of gases and particles to canopy, ground or water surfaces is called dry deposition. The processes involved in dry deposition are physically more complicated than the gravitational delivery of raindrops or snowflakes. While gravitational sedimentation can be important for larger particles (>2 μm diameter), most gases and smaller particles are deposited by turbulent transport into the canopy and then diffusion or inertial impaction through the leaf boundary layers (Figure 1) (Lovett, 1994). The turbulence is generated by the drag imposed by canopy surfaces on the air passing over them, and is dependent on the wind speed and the structure of the canopy. At the surface of the leaf or branch, the site

DRY DEPOSITION PROCESSES

H.120 B.125

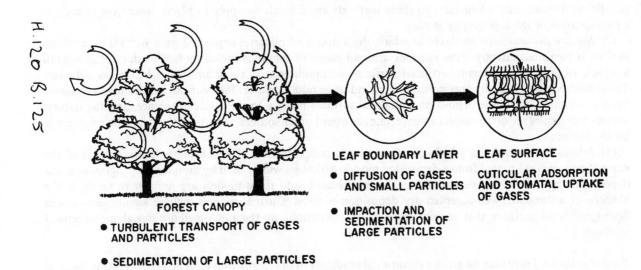

FOREST CANOPY

- **TURBULENT TRANSPORT OF GASES AND PARTICLES**

- **SEDIMENTATION OF LARGE PARTICLES**

ADAPTED FROM FOWLER (1980)

LEAF BOUNDARY LAYER

- **DIFFUSION OF GASES AND SMALL PARTICLES**

- **IMPACTION AND SEDIMENTATION OF LARGE PARTICLES**

LEAF SURFACE

CUTICULAR ADSORPTION AND STOMATAL UPTAKE OF GASES

Figure 1. Schematic depiction of processes of dry deposition to foliage. Deposition involves transport through three regimes: (1) the air within the canopy; (2) the boundary layer around the leaf; and (3) external and internal surfaces of the leaf. Transport through each regime relies on specific physical processes that depend on variables such as the structure of the canopy, wind speed, leaf shape, stomatal aperture, and the size of the depositing particle. From Lovett (1994)

and mode of deposition depends on the size of the particle, the chemistry of the receptor surface and the physiological state of the plant, among other factors. Small particles are deposited directly to all sides of canopy surfaces. Some reactive gas species, for instance HNO_3 vapour, also deposit efficiently to external plant surfaces. Less reactive species, like NO and NO_2, have their primary site of deposition within the substomatal cavity, thus their deposition rate is partially dependent on the opening of the stomata. Sulfur dioxide deposition is dependent on stomatal opening if the canopy surfaces are dry, but will deposit efficiently to external surfaces if those surfaces are wet with a film of water, such as rain or dew (Fowler and Unsworth, 1979).

As a result of these various depositional processes, many factors interact to control the rate of dry deposition. These factors range from meteorology and air chemistry, to canopy structure and leaf physiology. These multiple controls make dry deposition both difficult to measure and extremely variable in space and time. For instance, significant differences in dry deposition rates can be expected between coniferous and deciduous canopies (Weathers, 1993), between topographically exposed and sheltered locations (Lindberg and Owens, 1993) and between edges and interiors of forests (Hasselrot and Grennfelt, 1987; Draaijers et al., 1988; Beier and Gundersen, 1989; Weathers et al., 1995).

Methods. Measurement methodology for dry deposition is an active area of research (Hicks et al., 1990). Methods generally fall into three categories, each of which has strengths and weaknesses (Hicks et al., 1987; Lovett, 1994).

(1) *Micrometeorological methods,* such as eddy correlation and flux gradient measurements, in which the deposition to a surface is calculated from the vertical gradient or temporal variation in concentration of the depositing substance above the surface. These methods are well developed and quite accurate for gases such as SO_2 and O_3, and because of the short measurement time, are useful for evaluation of meteorological and plant physiological controls on deposition rates. However, the methods assume a constant flux zone above

the canopy, in which the vertical flux of the depositing substance is equal to the depositional flux to the canopy. Development of this constant flux layer requires adequate upwind fetch in which the wind moves parallel to a homogeneous surface, so these methods are difficult to apply in places where the canopy is heterogeneous or the topography is steep.

(2) *Surface accumulation methods*, in which the amount of material deposited on a natural or artificial surface is measured directly. This includes a broad range of methods including leaf washing, throughfall analysis, isotope measurements, artificial surfaces, accumulation on snow and watershed mass balances. These methods are generally applicable in any kind of terrain, however, the artificial surface methods suffer from uncertainty about the deposition to natural surfaces relative to the artificial surface, and the natural surface techniques are often subject to uncertainties about other sources or sinks of the depositing substance on the surface.

(3) *Inferential methods*, in which the atmospheric concentration of a substance is measured and the deposition rate is calculated from a deposition model. Although widely used by monitoring programmes, the deposition rates are highly dependent on the model used, and these models are difficult to verify in the absence of a standard, well-accepted dry deposition method. Current models include assumptions about homogeneity of surfaces that require similar site restrictions as the micrometeorological measurement methods.

Spatial patterns. There may be greater elevational gradients in dry deposition than in wet deposition because of elevational gradients in wind speed, canopy type and other controlling factors (Lovett and Kinsman, 1990). Air concentrations of common pollutant gases may also change with elevation. Because deposition to ground and canopy surfaces is a major sink for many atmospheric substances, air at lower elevations, which has greater contact with the ground than air at high elevations, can be more depleted in the concentrations of reactive substances. Preliminary results from the US Environmental Protection Agency's Clean Air Status and Trends Network sites at the Hubbard Brook Experimental Forest in central New Hampshire show that a lower elevation site (250 m elevation) has similar concentrations of fine SO_4^{2-} particles, but lower concentrations of HNO_3 vapour and SO_2, compared with a higher elevation site (650 m) (Figure 2). The high elevation site is on a south-facing slope about 150 m elevation below the ridge top, while the low elevation site is 3·5 km away in a broad valley. Model estimates (see Clark and Edgerton, 1992) of the rate of transfer of these substances to the forest canopy at the two sites indicate that the dry deposition flux is greater at the high elevation site than at the low elevation site by factors of 1·9, 3·6 and 2·1 for SO_4^{2-}, HNO_3 and SO_2, respectively (Table I).

Cloud Water Deposition

Deposition of chemicals dissolved in cloud droplets can be substantial in areas that are frequently immersed in clouds, such as mountain tops and sea coasts. Cloud droplets are generally in the diameter range 5–100 μm, and as such are deposited by gravitational sedimentation and inertial impaction (Lovett, 1984). Because cloud droplets tend to be more highly concentrated than raindrops (Weathers *et al.*, 1988), chemical deposition via cloud water can be the dominant form of atmospheric input at high elevation sites (Lovett and Kinsman, 1990; Mohnen *et al.*, 1990; Vong *et al.*, 1991; Lovett *et al.*, 1982). The deposition mechanisms for cloud water are similar to those for dry atmospheric particles, and, as a result, the deposition rates can vary as much as or more than dry deposition rates. Cloud water deposition is particularly sensitive to the chemical concentration of the droplets, the liquid water content of the cloud, the amount of time the vegetation is immersed in cloud and the wind speed (Lovett, 1984). Secondary controlling factors include the droplet size distribution and the structure of the forest canopy. As a result of these controlling factors, the rate of cloud deposition varies greatly as a function of elevation, slope orientation, topographic exposure, canopy type and the presence of canopy edges (Weathers, 1993, Weathers *et al.*, 1995).

Measurement methods for cloud water deposition are essentially similar to those for dry deposition. Collection of throughfall is the most commonly used method, because cloud water deposition (minus

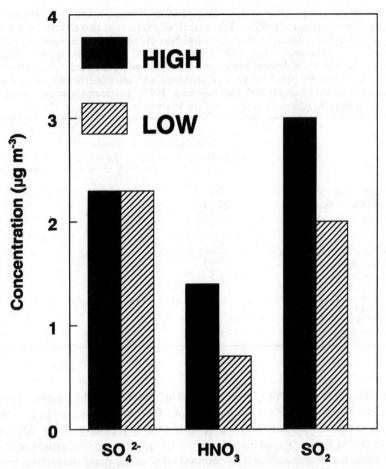

Figure 2. Mean annual concentrations of fine SO_4^{2-} particles, HNO_3 vapour, and SO_2 at two sites in the Hubbard Brook Experimental Forest in central New Hampshire. Measurements made during calendar year 1993 using the filter pack method described by Edgerton and Lavery (1991). The high site is on a south-facing slope at 650 m, and the low site is in a valley at 250 m. Filter packs were run continuously and analysed weekly, using identical methods at both sites

evaporative loss) can be measured directly as drip from the canopy during rain-free periods of cloud immersion. The cloud water deposition rate is then multiplied by a cloud chemical concentration measured above the canopy or in the open.

ATMOSPHERIC DEPOSITION AT THE WATERSHED SCALE

Given the inherent variability in the processes depositing atmospheric substances, how does one evaluate deposition at the scale of an entire watershed? While networks of precipitation collectors are feasible, networks of stations to measure dry and cloud deposition would require more expense and logistical effort than the typical watershed study could bear. There are two approaches that can be used to develop watershed-scale input estimates. The first is to use the watershed mass balance itself to estimate total atmospheric input. The best example of this is the sulfur deposition estimates made at the Hubbard Brook Experimental Forest, where stream water outputs and sources and sinks of sulfur in the watershed were used to estimate total S deposition, and measurements of S in precipitation were used to partition this input into

Table I. Concentration (μg m^{-3}), deposition velocities (cm s^{-1}), and fluxes (kg S or N ha^{-1} yr^{-1}) of fine particle SO$_4^{2-}$, SO$_2$, and HNO$_3$ vapour at two elevations in the Hubbard Brook Experimental Forest in central New Hampshire. Concentration data are means of continuously run filter packs, analysed weekly throughout 1993. Deposition velocities are derived from a modified version of the Hicks *et al.* (1987) dry deposition model, based on canopy structure and micrometeorological parameters measured at the sites (Clark and Edgerton, 1992). The high site is on a south-facing slope at 650 m elevation, and the low site is in a valley at 250 m

	Site	
	High	Low
SO$_4^{2-}$		
Concentration	2·3	2·3
Deposition velocity	0·13	0·05
Flux (kg S ha^{-1} yr^{-1})	0·33	0·17
SO$_2$		
Concentration	3·0	2·0
Deposition velocity	0·33	0·24
Flux (kg S ha^{-1} yr^{-1})	1·5	0·7
HNO$_3$		
Concentration	1·4	0·7
Deposition velocity	1·5	0·9
Flux (kg N ha^{-1} yr^{-1})	1·4	0·4

wet and dry deposition (Eaton *et al.*, 1978; Likens *et al.*, 1990). This method is on the appropriate scale for watershed studies, but it requires the accurate knowledge of other sources and sinks for the element under study. For elements that have a large weathering source within the watershed (e.g. Ca, Mg, Na, K) or a large sink in the vegetation or soil (e.g. N or P) the inaccuracies in the source and sink estimations may overwhelm the calculation of the atmospheric deposition as a residual term in the mass balance equation. Even in the case where the inputs and outputs of the element are much larger than the known sources and sinks within the watershed, such as for S in the north-eastern US, the watershed method may disagree with other input estimates. Lovett *et al.* (1992) reported a fourfold difference in S dry deposition estimates made using the watershed method, the throughfall method (Lindberg and Lovett, 1992) and the inferential method (Hicks *et al.*, 1991). Of these three estimates, the watershed method gave the highest estimates, but the reason for the discrepancy is unclear. Rustad *et al.* (1994) also found the watershed mass balance to give a higher estimate than the throughfall method for a watershed in Maine, but Hultberg and Grennfelt (1992) found the two methods to compare quite closely for watersheds in Sweden.

Another way to estimate deposition on the watershed scale is to measure the deposition at a single point and model relative deposition rates throughout the watershed as a function of landscape features. Unfortunately, models based on first principles are not yet sufficiently well developed to be applicable to complex terrain and heterogeneous canopies, so it is necessary to develop empirical models of variation in deposition rate for each site under study. There are several indices of total deposition that can be used to develop models of spatial patterns. The most frequently used index is the accumulation of Pb in the forest floor. Lead is deposited as wet, dry and cloud water deposition, and, once deposited, tends to bind to organic matter and remain relatively immobile in the organic layers of the soil (Reiners *et al.*, 1975). Thus, the Pb content in the forest floor is an index of the accumulated amount of Pb deposition at that site, and the pattern of Pb content can be used as an index of the pattern of atmospheric Pb deposition. Because Pb is deposited by wet, dry and cloud water processes, it may have a similar deposition pattern to S, N and other substances deposited by similar means. Recent evidence suggests that Pb in the forest floor can be transferred to lower soil horizons, probably in association with the leaching of dissolved organic carbon compounds (Friedland *et al.*, 1992; Johnson *et al.*, 1995). This may, over time, obscure the patterns of Pb accumulation, because Pb inputs have declined to very low levels since the advent of lead-free gasoline (Johnson *et al.*, 1995). None the less, soil Pb content has been used to evaluate patterns of atmospheric deposition along an elevational gradient (Reiners *et al.*, 1975), and across the north-eastern region (Johnson *et al.*, 1982). Weathers (1993) used this method to evaluate patterns of deposition on Hunter Mountain in the Catskill Mountains of New York State. She measured patterns of Pb accumulation associated with changes in elevation, slope orientation, canopy type and presence of forest edges, and used these data to develop a landscape model of patterns of deposition. The model was implemented in the framework of a geographic information system (GIS) that included data on topography and canopy type for the mountain landscape.

Another technique for estimating patterns of total deposition in complex terrain, especially for S, is the use of throughfall. Throughfall integrates wet deposition, cloud deposition and dry deposition, to the extent that the latter can be washed from the canopy (Lindberg and Garten, 1988). Use of throughfall to estimate total deposition will be inaccurate if the canopy retains some of the deposited chemicals or releases chemicals from the plant into throughfall solution. However, if the amount of deposition of an element is large compared with the storage within the plant tissues of the canopy (which is frequently the case for S, Na and Cl), throughfall (including stemflow) probably gives a fairly accurate estimate of total deposition. Throughfall has been used to estimate patterns of deposition at forest edges (Potts, 1978; Draaijers *et al.*, 1988; Beier and Gundersen, 1989; Weathers *et al.*, 1995), in different canopy types (Hultberg and Grennfelt, 1992) and in forests of different age and size (Draaijers and Erisman, 1993; Lindberg and Owens, 1993). Our recent data on an elevational gradient of S in throughfall from Slide Mt. in the Catskill Mountain range of south-eastern New York State shows a doubling of throughfall S deposition between 750 and 1380 m during the growing season of 1993 (Figure 3). Bulk deposition collected in the open did not increase over this elevational gradient, so the increase in total deposition as measured in throughfall is a result of increased dry and/or cloud deposition at higher elevations. The ability to measure throughfall at a large number of points allows the generation of mathematical functions that can be used to describe patterns and model deposition as a function of landscape characteristics.

Whatever method is used to describe deposition patterns within the watershed, scaling the input estimate up to the entire watershed requires knowledge of the distribution of controlling landscape features such as slope orientation, elevation and canopy type. Many of the most important data are available as publicly accessible databases and can be integrated and displayed within a GIS. Application of the equations that describe the deposition patterns and summation of inputs over the entire watershed can also be done within a GIS.

CONCLUSIONS

Atmospheric deposition can vary considerably within a watershed because of variation in meteorology and air chemistry, topography, exposure, canopy type and canopy structure. As a result, single monitoring sites can give only a very crude estimate of watershed inputs. If the measurement site is in the lower (generally the most accessible) part of the watershed, it is likely to yield a significant underestimate of deposition to the watershed as a whole, because wet, dry and cloud deposition often increase with elevation. Networks of monitoring sites for wet deposition are possible, but networks of monitoring sites for dry and cloud deposition are prohibitively expensive for most watershed mass balance studies. In addition, meteorological methods and models that are based on the establishment of a constant flux layer over the vegetation are not, in principle, applicable to the complex terrain and heterogeneous canopies that typify most mountain watersheds. One approach to this problem is use of a watershed mass balance to estimate the atmospheric inputs, but this requires accurate information on the outputs and all the sources and sinks of the element within the watershed. Another approach is the use of an empirical index of total deposition, such as Pb in the forest floor or SO_4^{2-} in throughfall, to describe patterns of deposition. The patterns can be described

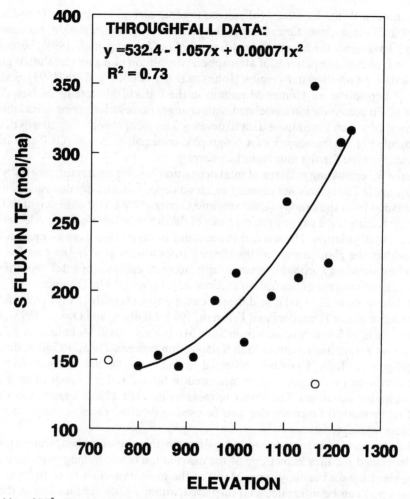

Figure 3. Bulk deposition of SO_4^{2-}-S in throughfall (filled circles) and in the open (open circles) along an elevational gradient on a north west-facing slope on Slide Mountain in the Catskills. Data are summed deposition for the period 1 June to 30 September 1993. Each point represents the mean of four throughfall collectors or two open collectors at that elevation

mathematically as a function of controlling landscape features such as elevation and canopy type, and then used to extrapolate point measurements to an entire watershed. This requires a substantial initial effort to obtain the pattern information, but the result is both an improved whole watershed deposition estimate, and knowledge of the key sources of variation of deposition within the watershed.

ACKNOWLEDGEMENTS

We thank Andrew Thompson and Scott Nolan for help with the field work in the Catskills and at Hubbard Brook, respectively. This project was supported by the US Geological Survey (Grant # 1434-92-G2247), the National Science Foundation (through the project 'Long-Term Ecological Research at the Hubbard Brook Experimental Forest', BSR-87002331 to Cornell University), and by the Environmental Protection Agency's Clean Air Status and Trends Network. (EPA contract 68-02-D134 to Environmental Science and Engineering, Inc.). This is a contribution to the Hubbard Brook Ecosystem Study and to the programme of the Institute of Ecosystem Studies.

REFERENCES

Barros, A. P. and Lettenmaier, D. P. 1994. 'Dynamic modeling of orographically induced precipitation', *Rev. Geophys.*, **32**, 265–284.

Beier, C. and. Gunderson, P. 1989. 'Atmospheric deposition to the edge of a spruce forest in Denmark', *Environ. Pollut.*, **60**, 257–272.

Clark, J. F. and Edgerton, E. S. 1992. 'Dry deposition flux calculations for the National Dry Deposition Monitoring Network', *Report of contract 68-02-D134*. US Environmental Protection Agency, Atmospheric Research and Exposure Assessment Laboratory, Research Triangle Park, NC.

Dasch, J. M. 1987. 'On the difference between SO_4^{2-} and NO_3^- in wintertime precipitation', *Atm. Environ.*, **21**, 137–141.

Dingman, S. L. 1981. 'Elevation: a major influence on the hydrology of New Hampshire and Vermont', *Hydrol. Sci. Bull.*, **26**, 399–413.

Draaijers, G. P. J. and Erisman, J. W. 1993. 'Atmospheric sulphur deposition onto forest stands: throughfall estimates compared to estimates from inference', *Atm. Environ.*, **27A**, 43–55.

Draaijers, G. P. J., Ivens, W. P. M. F., and Bleuten, W. 1988. 'Atmospheric deposition in forest edges measured by monitoring canopy throughfall', *Water Air Soil Pollut.*, **42**, 129–136.

Eaton, J. S., Likens, G. E. and Bormann, F. H. 1978. 'The input of gaseous and particulate S to a forest ecosystem', *Tellus B*, **30**, 546–551.

Edgerton, E. S. and Lavery, T. F. 1991. 'National Dry Deposition Network', *Fourth Annual Progress Report (1990)*, Contract # 68-02-4451. US Environmental Protection Agency, Atmospheric Research and Exposure Assessment Laboratory, Research Triangle Park, NC. 102 pp.

Fowler, D. 1980. 'Removal of sulphur and nitrogen compounds from the atmosphere in rain and by dry deposition, in Drablos, D. and Tolan, A. (Eds), *Ecological Impact of Acidic Precipitation*. SNSF Project, Oslo. pp. 22–32.

Fowler, D. and Unsworth, M. H. 1979. 'Turbulent transfer of sulphur dioxide to a wheat crop', *Q. J. R. Meteorol. Soc.*, **105**, 767–783.

Fowler, D., Cape, J. N., Leith, I. D., Choularton, T. W., Gay, M. J., and Jones, A. 1988. 'The influence of altitude on rainfall composition at Great Dun Fell', *Atm. Environ.*, **22**, 1355–1362.

Friedland, A. J., Craig, B. W., Miller, E. K., Herrick, G. T., Siccama, T. G., and Johnson, A. H. 1992. 'Decreasing lead levels in the forest floor of the northeastern USA', *Ambio*, **21**, 400–403.

Graustein, W. C. and Turekian, K. K. 1989. 'The effects of forests and topography on the deposition of sub-micrometer aerosols measured by lead-210 and cesium-137 in soils', *Agric. Forest Meteorol.*, **47**, 199–220.

Hasselrot, B. and Grennfelt, P. 1987. 'Deposition of air pollutants in a wind-exposed forest edge', *Water Air Soil Pollut.*, **34**, 135–143.

Hicks, B. B. and Shannon, J. D. (1979). 'A method for modeling the deposition of sulfur by precipitation over regional scales', *J. Appl. Meteorol.*, **18**, 1415–1420.

Hicks, B. B., Baldocchi, D. D., Meyers, T. P., Hosker, R. P. Jr., and Matt, D. R. 1987. 'A preliminary multiple resistance routine for deriving deposition velocities from measured quantities', *Water Air Soil Pollut.*, **36**, 311–330.

Hicks, B. B., Draxler, R. R., Albritton, D. L., Fehsenfeld, F. C., Dodge, M., Schwartz, S. E., Tanner, R. L., Hales, J. M., Meyers, T. P., and Vong, R. J. 1990. 'Atmospheric processes research and process model development', *State of Science/Technology Report 2*. National Acid Precipitation Assessment Program, Washington, D.C.

Hicks, B. B., Hosker, R. P. Jr., Meyers, T. P., and Womack, J. D. 1991. 'Dry deposition inferential measurement techniques. I. Design and tests of a prototype meteorological and chemical system for determining dry deposition', *Atm. Environ.*, **25A**, 2345–2359.

Hultberg, H. and Grennfelt, P. 1992. 'Sulphur and seasalt deposition as reflected by throughfall and runoff chemistry in forested catchments', *Environ. Pollut.*, **75**, 215–222.

Johnson, A. H., Siccama, T. G., and Friedland, A. J. 1982. 'Spatial and temporal patterns of lead accumulation in the forest floor in the northeastern United States', *J. Environ. Qual.*, **11**, 577–580.

Johnson C. E., Siccama, T. G., Driscoll, C. T., Likens, G. E., and Moeller, R. E. 1995. 'Changes in lead biogeochemistry in response to decreasing atmospheric inputs', *Ecol. Applic.*, **5**, 813–822.

Lewis, W. M. J., Grant, M. C., and Saunders, I. F., III 1984. 'Chemical patterns of bulk atmospheric deposition in the state of Colorado', *Wat. Resour. Res.*, **20**, 1691–1704.

Likens, G. E., Bormann, F. H., Hedin, L. O., Driscoll, C. T., and Eaton, J. S. 1990. 'Dry deposition of sulfur: a 23-year record for the Hubbard Brook Forest ecosystem', *Tellus B*, **42**, 319–329.

Lindberg, S. E. and Lovett, G. M. 1992. 'Deposition and canopy interactions of airborne sulfur: results from the Integrated Forest Study', *Atm. Environ.*, **26A**, 1477–1492.

Lindberg, S. E., and Owens, J. G. 1993. 'Throughfall studies of deposition to forest edges and gaps in montane ecosystems', *Biogeochemistry*, **19**, 173–194.

Lovett, G. M. 1984. 'Rates and mechanisms of cloud water deposition to a subalpine balsam fir forest', *Atm. Environ.*, **18**, 361–371.

Lovett, G. M. 1994. 'Atmospheric deposition of nutrients and pollutants to North America: an ecological perspective', *Ecol. Applic.*, **4**, 629–650.

Lovett, G. M. and Kinsman, J. D. 1990. 'Atmospheric pollutant deposition to high-elevation ecosystems', *Atm. Environ.*, **24A**, 2767–2786.

Lovett, G. M., Reiners, W. A., and Olson, R. K. 1982. 'Cloud droplet deposition in subalpine balsam fir forests: hydrological and chemical inputs', *Science*, **218**, 1303–1304.

Lovett, G. M., Likens, G. E., and Nolan, S. S. 1992. 'Dry deposition of sulfur to the Hubbard Brook Experimental Forest: a preliminary comparison of methods', in Schwarz, S. E. and Slinn, W. G. N. (Eds), *Precipitation Scavenging and Atmosphere-Surface Exchange*, Vol. 3. Hemisphere Publ. Co., Washington, D.C. pp. 1391–1402.

Mohnen, V. A., Aneja, V., Bailey, B., Cowling, E., Goltz, S. M., Healey, J., Kadlecek, J. A., Meagher, J., Mueller, S. M., and Sigmon, J. T. 1990. 'An assessment of atmospheric exposure and deposition to high-elevation forests in the eastern United States', *Report EPA/600/3-90/058 Edition*. Environmental Protection Agency, Office of Research and Development, Washington, D.C.

Munger, J. W., Waldman, J. M., Jacob, D. J., and Hoffman, M. R. 1983. 'Vertical variability and short term temporal trends in precipitation chemistry', in Pruppacher, H., Semonin, R. G., and Slinn, W. G. N. (Eds), *Precipitation Scavenging, Dry Deposition, and Resuspension*, Vol. 1. Elsevier, New York. pp. 275–281.

Ollinger, S. V., Aber, J. D., Lovett, G. M., Millham, S. E., Lathrop, R. G., and Ellis, J. M. 1993. 'A spatial model of atmospheric deposition in the Northeastern U.S.', *Ecol. Applic.*, 3, 459–472.

Potts, M. J. 1978. 'The pattern of deposition of airborne salt of marine origin under a forest canopy', *Plant Soil*, 50, 233–236.

Raynor, G. S. and Hayes, J. V. 1983. 'Differential rain and snow scavenging efficiency implied by ionic concentration differences in winter precipitation', in Pruppacher, H., Semonin, R. G., and Slinn, W. G. N. (Eds), *Precipitation Scavenging, Dry Deposition and Resuspension*, Vol. 1. Elsevier, New York. pp. 249–264.

Reiners, W. A., Marks, R. H., and Vitousek, P. M. 1975. 'Heavy metals in subalpine and alpine soils of New Hampshire', *Oikos*, 26, 264–275.

Rustad, L. E., Kahl, J. S., Norton, S. A., and Fernandez, I. J. 1994. 'Underestimation of dry deposition by throughfall in mixed northern hardwood forests', *J. Hydrol.*, 162, 319–336.

Topol, L. E. 1986. 'Differences in ionic compositions and behavior in winter rain and snow. *Atm. Environ.*, 20, 347–355.

Vong, R. J., Sigmon, J. T., and Mueller, S. F. 1991. 'Cloud water deposition to Appalachian forests', *Environ. Sci. Technol.*, 25, 1014–1021.

Weathers, K. C. 1993. 'The effect of four landscape features on atmospheric deposition to Hunter Mountain, New York', *Ph.D. Dissertation*, Rutgers University.

Weathers, K. C., Likens, G. E., Bormann, F. H., Bicknell, S. H., Bormann, B. T., Daube, B. C., Eaton, J. S., Galloway, J. N., Keene, W. C., Kimball, K. D., McDowell, W. H., Siccama, T. G., Smiley, D., and Tarrant, R. A. 1988. 'Cloud water chemistry from ten sites in North America', *Environ. Sci. Technol.*, 22, 1018–1026.

Weathers, K. C., Lovett, G. M., and Likens, G. E. 1995. 'Cloud water deposition to a spruce forest edge', *Atmos. Environ.*, 29, 665–672.

11

COMPARISON OF METHODS FOR CALCULATING ANNUAL SOLUTE EXPORTS FROM SIX FORESTED APPALACHIAN WATERSHEDS

BRYAN R. SWISTOCK[1], PAMELA J. EDWARDS[2], FREDERICA WOOD[2]
AND DAVID R. DEWALLE[3]

[1] School of Forest Resources and Environmental Resources Research Institute, The Pennsylvania State University, 132 Land and Water Research Building, University Park, PA 16802, USA
[2] USDA, Forest Service, Northeastern Forest Experiment Station, Timber and Watershed Laboratory, PO Box 404, Parsons, WV 26287, USA
[3] School of Forest Resources and Environmental Resources Research Institute, The Pennsylvania State University, 107 Land and Water Research Building, University Park, PA 16802, USA

ABSTRACT

Six methods were compared for calculating annual stream exports of sulfate, nitrate, calcium, magnesium and aluminum from six small Appalachian watersheds. Approximately 250–400 stream samples and concurrent stream flow measurements were collected during baseflows and storm flows for the 1989 water year at five Pennsylvania watersheds and during the 1989–1992 water years at a West Virginia watershed. Continuous stream flow records were also collected at each watershed. Solute exports were calculated from the complete data set using six different scenarios ranging from instantaneous monthly measurements of stream chemistry and stream flow, to intensive monitoring of storm flow events and multiple regression equations. The results for five of the methods were compared with the regression method because statistically significant models were developed and the regression equations allowed for prediction of solute concentrations during unsampled storm flows. Results indicated that continuous stream flow measurement was critical to producing exports within 10% of regression estimates. For solutes whose concentrations were not correlated strongly with stream flow, weekly grab samples combined with continuous records of stream flow were sufficient to produce export estimates within 10% of the regression method. For solutes whose concentrations were correlated strongly with stream flow, more intensive sampling during storm flows or the use of multiple regression equations were the most appropriate methods, especially for watersheds where stream flows changed most quickly. Concentration–stream flow relationships, stream hydrological response, available resources and required level of accuracy of chemical budgets should be considered when choosing a method for calculating solute exports. © 1997 by John Wiley & Sons, Ltd.

INTRODUCTION

There has been considerable interest in calculating import–export budgets for specific chemical solutes from watersheds. Early budget studies were concerned primarily with nutrient cycling (Likens *et al.*, 1967; Johnson *et al.*, 1976) and land use mangement effects on nutrient cycling (Likens *et al.*, 1970). More recently, budget calculations have been used to evaluate the effects of deposition of atmospheric sulfur (Lynch and Corbett, 1989; Rochelle and Church, 1987) and nitrogen (Johnson, 1992) on stream chemistry.

The inherent difficulty in estimating chemical budgets lies in the need to determine annual integrated import and export loads based upon data collected at discrete time intervals. Over the past decade, input estimates have become more accurate as methods of measuring dry deposition have improved (Shepard *et al.*,

1989; Hicks *et al.*, 1991; Lindberg and Lovett, 1992). Numerous methods for estimating export budgets also have been developed, ranging from using intermittent grab sample chemistries and associated stream flows, to complex linear regression models. Theoretically, grab samples collected on an evenly spaced time interval will include samples from a range of stream flow conditions. In turn, the chemistry of these samples will be representative of that range of stream flows for the time period of sampling. However, if storms and subsequent high stream flows tend to occur in predictable patterns (Patric and Studenmund, 1975), or if the sampling interval is large in comparison to the frequency and duration of storm flows, routinely timed grab samples may not be sufficient to estimate exports accurately. Additionally, if the concentration of a solute undergoes large changes during storms, intensive storm flow monitoring may be necessary to quantify the chemical changes adequately so annual exports are not poorly estimated. The reliability of chemical budgets will increase with increased frequency of stream flow and chemistry measurement. However, the increased costs of more frequent sampling may not result in significantly more accurate estimates of export; thus, the benefits may not justify the costs.

In this paper, we examine the importance of stream sampling and stream flow measurement intervals in determining annual exports of several solutes. Continuous stream flow measurements and frequent stream samples were collected for six forested, headwater Appalachian watersheds. Solute exports were calculated from this extensive field data set using six scenarios ranging from simple monthly measurements (12 samples per year) to intensive storm sampling and regression relationships (300–400 samples per year). Export budgets were calculated for five solutes that are commonly of interest in budget studies and that showed a range of sensitivity to changes in stream flow. We hypothesized that solutes whose concentrations were correlated well with stream flow would show a wider range of calculated exports, depending upon the method used, than those that were correlated poorly with stream flow.

METHODS

Stream flow and stream chemistry data from six first- and second-order Appalachian watersheds were used for analysis. The physical characteristics of each of the study watersheds are given in Table I. One small watershed (WS4) was located on the Fernow Experimental Forest in north-central West Virginia, and the other five larger basins were on the Appalachian Plateau in Pennsylvania (Figure 1). Data for WS4 in West Virginia were collected as part of an EPA-funded watershed acidification study initiated in 1987 in which WS4 was used as a control watershed. The five Pennsylvania basins (BNR, RBS, STN, BWN and LNN) were part of the EPA-sponsored Episodic Response Project in which effects of episodic acidification in several regions of the north-eastern US were studied (Wigington *et al.*, 1993).

The five Pennsylvania watersheds are representative of previously unglaciated forested basins in this region. Approximately 100 to 135 cm yr^{-1} of precipitation falls on average, which is distributed uniformly on a monthly basis, and the mean annual air temperature is approximately 7°C. Snowpack accumulation has

Table I. Drainage basin locations and physical characteristics of each study watershed

Stream	Abbreviation	Latitude	Longitude	Basin area (ha)	Stream order*	Max. elevation (m)
Benner Run	BNR	40°56′04″N	78°01′22″W	1134	2	749
Roberts Run	RBS	41°10′12″N	78°24′22″W	1070	2	733
Stone Run	STN	41°05′52″N	78°26′48″W	1156	2	701
Baldwin Creek	BWN	40°21′05″N	79°03′04″W	535	2	832
Linn Run	LNN	40°08′40″N	79°12′37″W	1000	2	893
Watershed 4	WS4	39°03′00″N	79°41′00″W	39	1	854

* Stream orders defined based upon US Geological Survey 7·5′ quadrangles.

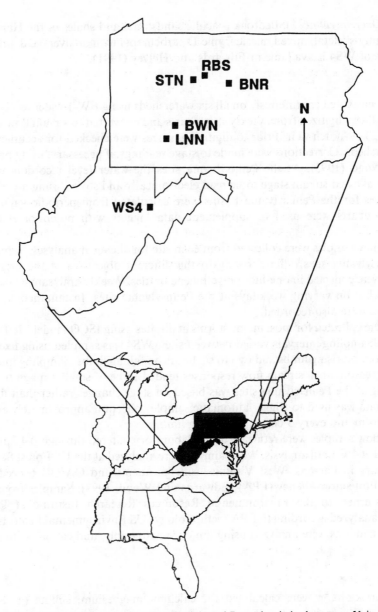

Figure 1. Study watershed locations in West Virginia and Pennsylvania in the eastern United States

occurred only intermittently in the last five years. Overstory vegetation is dominated by oak species (*Quercus* spp.), red maple (*Acer rubrum* L.) and black cherry (*Prunus serotina* Ehrh.). These basins are underlain primarily by sandstone and shale bedrock from the Pocono Group (BWN and BNR) and the Pottsville Group (STN, RBS and LNN). Soils are shallow (0·5–2·0 m), extremely rocky, sandy loam, Typic Dystrochrepts from the Hazelton and Dekalb series. More detailed descriptions of the Pennsylvania sites are available in DeWalle *et al.* (1993) and Wigington *et al.* (1993).

The West Virginia watershed (WS4) receives 150 cm yr^{-1} of precipitation on average, which is evenly distributed throughout the year, and has a mean annual air temperature of 9°C. Overstory vegetation is dominated by sugar maple (*Acer saccharum* Marsh.), American beech (*Fagus grandifolia* Ehrh.) and northern red oak (*Quercus rubra* L.). Bedrock is acidic sandstones and shales of the Hampshire Formation. Soils are Calvin loamy-skeletal, mixed, mesic Typic Dystrochrepts with an average depth of 1·5 m. A more detailed description of WS4 is available in Edwards and Helvey (1991).

Data collection

Stream stage was measured continuously on all six watersheds using FW-1 water level recorders with strip charts. Stage at WS4 was digitized from weekly strip charts and converted to stream flow using a rating curve developed from a 120° V-notch weir. Prior to digitizing, stages were checked for accuracy using strip charts from adjacent watersheds. Corrections were made to stage readings, if necessary, using procedures described by Edwards and Wood (1993). At the Pennsylvania sites the water level recorders were outfitted with potentiometers that allowed stream stage to be read electronically and stored using a Campbell CR-10 data logger. Rating curves for the Pennsylvania basins were calculated from periodic gaging over a range of stream flows. Strip charts were used to supplement data loggers with potentiometers when the latter malfunctioned.

Routine stream grab samples were collected from each site for chemical analyses. Sampling schedules for WS4 and the Pennsylvania sites differed owing to the different objectives of the original studies. Grab samples at all sites were collected in one-liter polyethylene bottles. Weekly grab samples were collected every Tuesday from WS4 but on varying weekdays at the Pennsylvania sites. Instantaneous stream flows at the time of grab samples were also recorded.

Storm samples were collected for most major storms at all sites using ISCO model 2700 samplers housed in instrument shelters. Sampling intervals varied between sites. WS4 was sampled using fixed time intervals of 15 minutes to one hour on rising limbs and two to six hours on falling limbs. Sampling intervals were shorter during the growing season when stream flow responses tended to be of smaller magnitudes and durations. Storm flow sampling at the Pennsylvania sites was based on stage changes rather than fixed time intervals. A data logger program was used to trigger automatic sample collection approximately every 1·5 cm on the rising limb of the storm and every 3 cm on the falling limb.

Grab and storm flow samples were returned to the laboratory, filtered through 0·45 μm pore filters, acid preserved and stored at 4°C until analysis. WS4 samples were analyzed at the US Forest Service's Timber and Watershed Laboratory in Parsons, West Virginia using protocols and QA/QC procedures approved by US Environmental Protection Agency (EPA) (Edwards and Wood, 1993). Samples from the Pennsylvania watersheds were returned to the Environmental Resources Research Institute at Pennsylvania State University and also analyzed according to EPA methodology (US Environmental Protection Agency, 1983). At both laboratories, anions were analyzed using ion chromatography and cations with atomic absorption spectrophotometry.

Export calculations

Annual exports in kg ha^{-1} were calculated for calcium, magnesium, sulfate (as SO_4^{2-}) and nitrate (as NO_3^-) at each site. These solutes were chosen because they are frequently the primary nutrients of interest on forested watersheds and/or their cycling is affected by atmospheric deposition and land use changes. Aluminum export was also calculated for the Pennsylvania sites because it has been shown to be correlated well with stream flow (Sharpe *et al.*, 1984; Swistock *et al.*, 1989) and, therefore, was most likely to be affected by variations in sampling interval. Aluminum data were not available for the WS4 site during the study period.

Export budgets were calculated based on a 1 October–30 September water year. Data for water years 1989 to 1992 were available for the WS4 site, but only water year 1989 was available for the five Pennsylvania sites. Four water years of data for WS4 permitted comparison of methods across a range of meteorological conditions, including two dry years (1991 and 1992), a near-normal year (1990) and a wet year (1989).

Six of the most commonly used export computation methods were compared in this study. For each method, solute export (kg ha^{-1}) for a given time period was calculated as the product of stream flow

(m^3 s^{-1}), solute concentration (mg l^{-1}) and a conversion factor. The conversion factor used depended upon the time interval between samples and the watershed area. Export of each solute on an annual basis was calculated by summing export quantities estimated for the successive time increments. The primary difference between the six methods was the time interval between stream samples (i.e. solute concentrations) and stream flow measurements. A detailed description of each of the six export calculation methods follows and is summarized in Table II.

Method 1: monthly grab samples/instantaneous stream flow.   Exports were determined by applying the mean solute concentration of grab samples collected on two consecutive sampling periods to the instantaneous stream flow at the time the first sample of the pair was taken. In this case, the sampling interval was once per month. Since actual sampling in our data sets was more frequent than this, the first grab sample collected during the month was used as the monthly grab sample. For example, the calcium export for January was determined by calculating the mean calcium concentration from the January and February grab samples and applying this concentration to the instantaneous stream flow measured during collection of the January grab sample. The twelve monthly exports were then summed for each water year. This method is attractive because it only requires twelve stream visits per year and no continuous stream flow recording equipment. Because of the limited sampling frequency and the probability of missing storm flows, however, this method is likely to be the least accurate of the six compared here.

Method 2: monthly grab samples/continuous stream flow.   This method is identical to method 1 except that mean monthly solute concentrations were applied to total stream flow for the month calculated from continuously collected stream flow measurements. Although only monthly grab samples are needed for this method, data collection is more intensive because continuous stream flow records are necessary.

Method 3: weekly grab samples/instantaneous stream flow.   This method is identical to method 1 except that the sampling for solute concentrations was increased to a weekly interval. Instantaneous stream flow measurements at the time of the weekly grab samples were used in conjunction with the mean weekly solute concentration to calculate solute exports.

Method 4: weekly grab samples/continuous stream flow.   This method is the same as method 2 except mean weekly solute concentrations were applied to total stream flow for the week calculated from continuous stream flow measurements. This method is perhaps the most frequently used for estimating nutrient exports.

Method 5: grab samples and storm flow samples.   In addition to routine grab samples, all six of the study watersheds were sampled intensively during storm flow events. A total of 250–400 samples were collected at each site during each water year. Approximately 60–70% of the 1989 water year storms that occurred at the Pennsylvania sites were intensively sampled. At the WS4 site, 42% of 1989 water year storms were sampled while 62–92% of storms were monitored during the 1990–1992 water years. Solute export was determined by first calculating the mean solute concentrations from two consecutive stream samples (now including both grab and storm flow samples). The time period between consecutive samples varied from as long as a week during non-storm periods, to as short as 15 minutes during storm flows. The mean solute concentration for a given time interval was applied to the total stream flow that occurred during that interval, as determined from continuous stream flow records, to determine the solute export that occurred during each time interval. The solute export from each time interval was summed to produce the annual solute export in kg ha^{-1}. This calculation method parallels method 4, with the addition of samples collected during storm flow periods.

Method 6: multiple regression equations.   Numerous studies have used linear regression models to predict solute concentrations from stream flow measurements (Likens *et al.*, 1967; Johnson, 1979; Dann *et al.*, 1986; Dow, 1992). The regression method is advantageous because it allows for complete coverage of the annual

Table II. Summary of sampling schemes and calculation techniques for each of the six solute export methods

Method	Stream sampling interval	Number of samples per year	Stream flow measurement interval	Numer of stream flow measurements per year	Calculation method for annual solute export
1	Monthly	12	Monthly	12	Σ (monthly concentration $\times$ stream flow at time of sampling)
2	Monthly	12	Continuous	Variable	Σ (monthly concentration $\times$ continuous stream flow values)
3	Weekly	52	Weekly	52	Σ (weekly concentration $\times$ stream flow at time of sampling)
4	Weekly	52	Continuous	Variable	Σ (weekly concentration $\times$ continuous stream flow values)
5	Baseflow and storm flows	250–400	Continuous	Variable	Σ (baseflow and storm flow concentrations $\times$ continuous stream flow values)
6	Baseflow and storm flows used for regression equations	Hourly values predicted using regression equations	Hourly	8760	Σ (predicted hourly concentration from regression equations $\times$ hourly stream flow value)

stream hydrograph by transforming continuous stream flow records into continuous solute concentrations and exports. The disadvantage of this method is that reasonably predictive regression equations often require frequent sampling across all stream flow regimes and times of year. The development of statistically valid regression equations seemed reasonable for this data set because 250–1820 stream samples had been collected over 99·8–100% of the observed stream flows in each stream. Using the data from the five Pennsylvania watersheds, numerous regression models were investigated to determine the best predictive models for solute concentrations. The model developed from the Pennsylvania data was also used to predict solute concentrations at the WS4 site in West Virginia. The best predictive model (highest R^2) included stream flow, time of year harmonic terms and interaction terms (Bliss, 1970) as shown in Equation (1):

$$\begin{aligned}
\text{concentration (mg l}^{-1}) = a + \log Q + \cos(ct) + \sin(ct) + \cos(2ct) + \sin(2ct) + \cos(3ct) + \sin(3ct) \\
+ \log Q[\cos(ct)] + \log Q[\sin(ct)] + \log Q[\cos(2ct)] + \log Q[\sin(2ct)] \\
+ \log Q[\cos(3ct)] + \log Q[\sin(3ct)]
\end{aligned} \quad (1)$$

where a is the intercept, $\log Q$ is the $\log_{10}$ of stream flow in $m^3\ s^{-1}\ km^{-2}$, $c = 2\pi/365$ and t is the Julian day. A stepwise regression procedure was used for each watershed and solute to determine which terms in Equation (1) were statistically significant ($\alpha = 0·05$) and should be included in the model (Neter et al., 1985). The resulting regression equations for each solute and watershed were used to predict hourly solute concentrations from hourly stream flow data. The predicted hourly solute concentrations were applied to hourly stream flows and the resulting loads were summed over the water year.

The residuals from each regression model were plotted against predicted solute concentrations to determine if errors (residuals) associated with each regression model had a constant variance (Neter et al., 1985). A non-constant variance is indicative of a model that is biased towards specific stream flow ranges and might consistently underestimate or overestimate solute exports.

RESULTS AND DISCUSSION

Strong correlations between solute concentrations and stream flow have previously been shown to be an important cause of variations between exports calculated by different methods (Johnson, 1979). A wide range of solute–stream flow correlations were found among the solutes chosen for this study (Table III), and nearly all of the correlations were statistically significant ($p \leqslant 0·05$). As expected, aluminum was correlated most strongly with stream flow, with positive Pearson correlation coefficients (r) ranging from $+0·68$ to $+0·91$. Other strong correlations included calcium and magnesium on LNN, sulfate on BNR and STN and nitrate on WS4. Each of the solutes, except aluminum, was also weakly correlated with stream flow on at least one of the study streams. These solute–streamflow correlations were important in explaining observed differences between exports calculated from the six methods.

Table III. Pearson correlation coefficients between $\log_{10}$ of solute concentration and $\log_{10}$ of stream flow for each study watershed. Values with a 'ns' indicate correlations that were not statistically significant ($p \leqslant 0·05$)

Watershed	N*	Calcium	Magnesium	Sulfate	Nitrate	Aluminum
BNR	277	−0·30	0·03 ns	0·56	−0·17	0·68
RBS	319	−0·62	−0·56	0·01 ns	0·02 ns	0·85
STN	362	−0·27	0·15	0·65	−0·18	0·91
BWN	300	−0·32	0·08 ns	0·37	0·29	0·81
LNN	256	−0·81	−0·64	−0·16	0·22	0·89
WS4	1828	0·10	0·16	0·29	0·39	†

* Number of stream samples collected.
† Aluminum was not measured on WS4.

Table IV. R^2 values from multiple regression models used to predict hourly solute concentrations from measured hourly stream flow, harmonic time variables and interaction terms [Equation (1)]. All models are statistically significant ($p \leqslant 0.05$)

Watershed	N*	Calcium	Magnesium	Sulfate	Nitrate	Aluminum
BNR	277	0·40	0·13	0·51	0·31	0·42
RBS	319	0·69	0·63	0·53	0·40	0·52
STN	362	0·56	0·18	0·61	0·25	0·72
BWN	300	0·38	0·26	0·28	0·62	0·48
LNN	256	0·74	0·54	0·27	0·47	0·71
WS4	1828	0·28	0·31	0·23	0·34	†

* Number of stream samples collected and used for regression analysis.
† Aluminum was not measured on WS4.

Addition of six harmonic time variables and six interaction terms to the stream flow correlations [Equation (1)] provided significant improvement in describing the behavior of each solute. The stepwise regression procedure generally found between six and nine of the 13 independent variables to be statistically significant ($\alpha = 0.05$). The resulting multiple regression equations, with R^2 values ranging from about 0·2 to 0·7 (Table IV), were used to calculate exports from the regression method (method 6). All of the regression models were statistically significant ($p \leqslant 0.05$) but the least predictive models (lowest R^2) were generally found at the WS4 watershed in West Virginia. This may be because the regression model [Equation (1)] was developed from the five Pennsylvania watersheds and the same independent variables were applied to the WS4 site in West Virginia with no attempt to explore other equation forms.

Residual analysis indicated that, with two exceptions, the regression models were not biased towards high or low stream flows, and, therefore, should not consistently underestimate or overestimate solute exports. Regression residuals for nitrate at WS4 and aluminum at BNR were correlated negatively with predicted concentrations revealing a consistent overestimation of nitrate export from WS4 and aluminum export from BNR.

In order to compare the annual solute export results from the six calculation methods, the percentage difference between results from each of the grab sampling methods (methods 1–5) and results from the regression technique (method 6) were computed (Table V). The regression method was assumed to be the most accurate of the six methods because it allowed for characterization of all storm flows (not just those that were sampled), and because all of the models were statistically significant ($p \leqslant 0.05$). Previous work by Johnson (1979) also reported that regression methods developed from sampling over the observed stream flow regime are the most satisfactory for budget studies where solute concentrations are correlated with stream flow. Because the WS4 nitrate and BNR aluminum regression equations were biased, export results for these cases were compared to the storm flow sampling method (method 5) rather than the regression method (method 6).

Comparison of methods

Calculated annual exports for each of the solutes are shown in Figures 2–6. The six bars (from left to right) for each stream and water year correspond to the six export methods (1–6, respectively). Methods 1 and 3, which require only monthly or weekly site visits and instantaneous stream flow observations, represent the least complicated techniques for measuring solute exports. Not surprisingly, both methods consistently produced the most variable export results, often 20–50% different from those from the other four methods (Figures 2–6). Increasing the sampling frequency from monthly (method 1) to weekly (method 3) only slightly improved the export estimates relative to the regression results in most cases (Table V).

Since grab sampling generally characterizes low stream flows better than high stream flows (Johnson, 1979; Dann *et al.*, 1986), methods 1 and 3 would be expected to overestimate exports for solutes negatively correlated to stream flow and underestimate exports for solutes positively correlated to stream flow.

Table V. Percentage differences between solute exports calculated from each of the grab sampling methods (methods 1–5) and outputs from the regression method (method 6). Negative values indicate export estimates were lower than those from regression

Watershed	Water year	Method				
		1	2	3	4	5
Calcium						
BNR	1989	−18·7	0·5	−12·5	−1·3	−1·9
STN	1989	−5·8	0·5	−13·7	2·2	−0·3
RBS	1989	−1·4	3·1	−2·2	2·6	2·0
BWN	1989	−2·2	−4·6	3·2	−6·7	−6·0
LNN	1989	21·8	7·7	35·0	6·6	0·8
WS4	1989	35·2	−9·2	−0·1	−6·2	−6·7
WS4	1990	−21·6	−0·8	−16·4	−0·3	4·2
WS4	1991	−21·4	4·1	−9·7	4·2	5·1
WS4	1992	−25·2	5·2	−15·5	4·7	6·0
Magnesium						
BNR	1989	−20·6	−1·9	−13·2	−1·6	−2·6
STN	1989	−7·2	−3·0	−18·9	−3·6	−4·7
RBS	1989	−4·0	0·0	−3·1	1·0	0·7
BWN	1989	0·8	−4·9	6·3	−5·5	−4·9
LNN	1989	18·0	7·5	30·7	4·7	1·7
WS4	1989	30·5	−11·7	−2·5	−7·7	−6·9
WS4	1990	−21·6	−0·9	−17·3	−0·9	2·9
WS4	1991	−22·1	3·7	−9·7	5·8	7·0
WS4	1992	−25·7	5·0	−16·0	4·2	4·7
Sulfate						
BNR	1989	−25·5	−8·8	−21·6	−8·8	−7·9
STN	1989	−5·6	−0·3	−18·3	−2·3	−2·8
RBS	1989	−2·3	0·7	−2·9	0·5	−0·0
BWN	1989	2·4	0·9	8·5	−3·4	−3·5
LNN	1989	8·9	−4·8	21·4	−5·8	−8·5
WS4	1989	49·0	0·8	9·4	1·4	3·6
WS4	1990	−27·0	−6·2	−22·4	−6·6	−4·1
WS4	1991	−32·4	−9·8	−22·4	−10·3	−9·1
WS4	1992	−27·9	0·8	−17·5	1·5	2·6
Nitrate						
BNR	1989	−4·0	17·6	−2·6	8·6	2·7
STN	1989	18·3	40·9	4·3	10·8	6·4
RBS	1989	−18·7	−3·1	−0·4	1·9	2·7
BWN	1989	−9·8	−5·0	4·5	−7·2	−5·5
LNN	1989	7·4	14·6	33·3	10·6	14·4
WS4	1989	42·0	−2·7	4·3	−1·5	*
WS4	1990	−24·7	−9·5	−22·5	−7·4	*
WS4	1991	−22·3	1·4	−17·2	−1·1	*
WS4	1992	−31·0	−0·7	−20·5	−0·6	*
Aluminum						
BNR	1989	−38·8	−31·4	−32·3	−16·1	*
STN	1989	−14·6	−17·1	−22·0	−4·7	−0·4
RBS	1989	−8·4	−11·5	−14·2	−11·9	−1·6
BWN	1989	−2·4	−11·4	−6·2	−14·1	0·3
LNN	1989	−44·7	−56·3	−26·7	−47·4	−34·6

* Percentage differences for nitrate at WS4 and aluminum at BNR were calculated compared to method 5 because the regression equations were found to be biased.

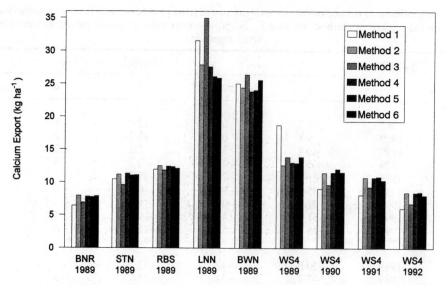

Figure 2. Calcium export estimates (kg ha^{-1}) for each of the study sites and water years. The six bars from left to right represent export estimates using methods 1–6, respectively. Refer to Table II for a summary of the six methods

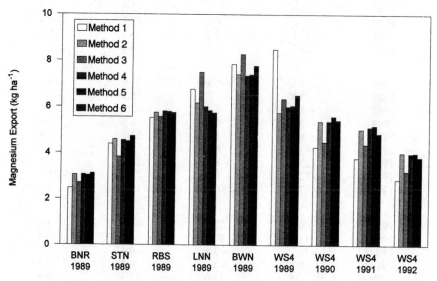

Figure 3. Magnesium export estimates (kg ha^{-1}) for each of the study sites and water years. The six bars from left to right represent export estimates using methods 1–6, respectively. Refer to Table II for a summary of the six methods

Comparison of export estimates (Figures 2–6) with solute–stream flow correlations (Table III) indicated that this pattern was generally true. For example, on Linn Run where calcium was negatively correlated with stream flow ($r = -0.81$), methods 1 and 3 overestimated calcium exports by 22–35% compared with exports using the regression method (Figure 2, Table V). Conversely, methods 1 and 3 frequently underestimated aluminum export (Figure 6, Table V) which was positively correlated with stream flow at the Pennsylvania sites ($r = +0.68$ to $+0.91$).

The magnitude of errors associated with grab sampling methods also appeared to be influenced by the hydrological response of the streams to rainfall. Of the six watersheds, LNN and WS4 were the most 'flashy'

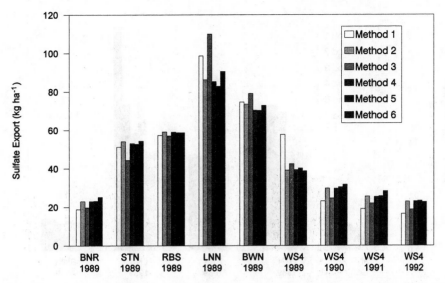

Figure 4. Sulfate (as SO_4^{2-}) export estimates (kg ha^{-1}) for each of the study sites and water years. The six bars from left to right represent export estimates using methods 1–6, respectively. Refer to Table II for a summary of the six methods

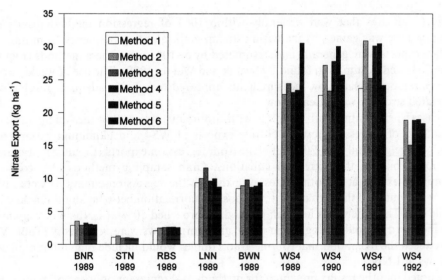

Figure 5. Nitrate (as NO_3^-) export estimates (kg ha^{-1}) for each of the study sites and water years. The six bars from left to right represent export estimates using methods 1–6, respectively. Refer to Table II for a summary of the six methods

with more rapid responses to precipitation, shorter storm flow durations and larger flows per unit area. During the 1989 water year, for example, peak stream flows at LNN and WS4 exceeded 0·8 and 0·4 m^3 s^{-1} km^{-2}, respectively, while peak flows at the other four sites were less than 0·3 m^3 s^{-1} km^{-2}. Shorter storm flow durations on LNN and WS4 make it less likely that routine grab samples would be taken during moderate or high flows. As a result, errors associated with methods 1 and 3 are enhanced at LNN and WS4 (Figures 2–6).

Increasing the frequency of stream flow measurement was critical to improving export estimates. Hourly stream flow values in conjunction with monthly (method 2) or weekly (method 4) solute concentrations

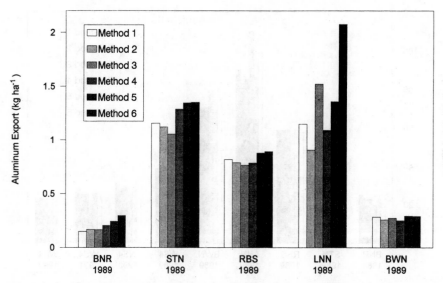

Figure 6. Aluminum export estimates (kg ha^{-1}) for each of the study sites and water years. The six bars from left to right represent export estimates using methods 1–6, respectively. Refer to Table II for a summary of the six methods

produced export estimates that were generally within 10% of regression method results (Figures 2–6, Table V). Overall, little was gained by increasing stream grab sampling frequency from monthly to weekly intervals. Sulfate exports were generally underestimated by 5–10% using these methods (Figure 4), which is better than the 20% underestimation found by Swank and Waide (1988). The use of weekly or monthly grab samples with continuous stream flow measurements appeared to provide adequate results for solutes that were not correlated strongly with stream flow.

There were cases where routine grab sampling with continuous stream flow measurements did not produce exports within 10% of regression results. Nitrate exports at WS4 and aluminum exports at BNR using methods 2 and 4 were generally 20–30% different from regression exports (Figure 5), primarily because of the aforementioned bias of the regression equations. Grab sampling methods also consistently underestimated aluminum export at the other sites even though the regression equations were unbiased. These differences were attributed to the extremely strong positive correlation between aluminum and stream flow at each site (Table III). The differences between methods approached 50% at LNN, where aluminum concentrations often increased 10-fold during storm flows, and storm flows were very flashy (Table V). Combined, these factors reduced the likelihood that routine grab samples would include high aluminum concentrations during high flows at LNN.

It should be noted that the results presented for methods 1–4 represent only one of many routine monthly and weekly sampling scenarios that could have been selected from the complete data sets for each stream. The timing of sampling in comparison to the occurrence of storm flows and their durations could have an effect on how these methods compare with more precise procedures. Comparison of results between the four water years at WS4 (Figures 2–6), however, indicated that the methods produced consistent results relative to one another. Similarity of results between methods across the four water years also suggests that differences between methods were not influenced strongly by wet vs. dry water years.

The storm flow sampling and regression methods (methods 5 and 6) represent more complex calculation methods. For both, intensive storm flow sampling was coupled with routine grab sampling and continuous stream flow data. These methods included a significantly larger database since 60–80% of all samples collected at the study sites were storm flow samples. The results from the storm flow sampling method (method 5) were generally within 7% of regression results (Table V). Exports calculated from the regression method tended to be slightly higher than from method 5 for solutes that showed a strong positive correlation with stream flow, and slightly lower than method 5 exports for solutes with a strong negative stream flow correlation. Dann *et al.* (1986) also found that the regression method produced the highest export estimate for sulfate, which was correlated positively with stream flow. The slightly lower or higher exports from storm flow sampling (method 5) compared with the regression technique (method 6) probably occurred because only 60–70% of the storm flow events were sampled during most water years.

The addition of storm flow samples greatly improved aluminum export estimates compared with regression results at most sites (Figure 6, Table V). Thus, for solutes that are strongly correlated with stream flow, the storm flow sampling or regression approaches are clearly preferable to the less intensive grab sampling methods. At LNN, however, storm flow sampling still underestimated aluminum export by 35% compared with regression results. On hydrologically flashy streams, such as LNN, the regression method is probably the only useful method for solutes that vary greatly with stream flow because obtaining samples that are representative of all stream flow conditions is difficult.

Although the regression equations were all statistically significant, 17 of the 29 equations had R^2 values less than 0.5 (Table IV). The predominance of equations with low R^2 values suggests that the storm flow sampling method (method 5) may be more accurate and more appropriate as a standard for comparison of the other methods rather than the regression method, especially since it uses measured rather than predicted solute concentrations. However, this approach depends even more on intensive sampling than the regression method which is probably not cost effective. The fact that calculated solute exports using methods 5 and 6 nearly always agreed to within 10%, and usually agreed to within 5% (Table 5), suggests that either of these methods could have been used for comparisons. Furthermore, differences between exports calculated from methods 5 and 6 were no larger for solutes with low R^2 regression equations than for those with high R^2 regression equations (Table IV).

An additional concern regarding the regression method involves the number of stream samples that must be collected to produce reasonable regression equations. The regression equations developed in this study were based upon more than 250 samples per stream which may be too costly or impractical in many cases. Johnson (1979) compared phosphorous export calculated using the regression method with 281 stream samples and with subsets of 16–136 stream samples. Regressions from the subsets generally underestimated P export by 10–50% compared with the 'best estimate' (regression with all 281 samples). This consistent underestimation was attributed to a bias towards lower stream flows when using subsets of data. These results suggest that regression equations from small data sets (< 50 samples) may only be useful if they are collected over the range of observed stream flows, and even then may only be as accurate as methods coupling routine grab samples with continuous stream flow measurements (methods 2 and 4).

The similarity between grab sampling methods (methods 2 and 4) and methods including storm flow samples (methods 5 and 6) implies that storms are generally not substantial contributors to solute export from these watersheds. Wilson *et al.* (1991) reported that shallow subsurface discharge, generally associated with storm flows (Christopherson *et al.*, 1982; Shanley, 1992), had little effect on annual nutrient exports on Walker Branch watershed. Instead, solute export by deeper originating baseflow was predominant. Apparently, groundwater discharge accounts for much of the chemical export from the catchments used in this analysis. Aluminum was the only solute studied that illustrated a strong case for storm-controlled export. Previous studies have suggested that aluminum increases during storm flows are caused by displacement of stored soil water from the upper organic and soil layers in response to rapid hydrological changes (Swistock *et al.*, 1989).

CONCLUSIONS

Results from this study suggest that measurements of solute export can be influenced strongly by the frequency of data collection. The common practice of using fixed interval grab samples with instantaneous stream flows at the time of sampling generally produced export estimates that were 10–30% different from methods using continuous stream flow records and more frequent stream samples. Fixed interval grab sampling was adequate for solutes that were not correlated strongly with stream flow provided that stream flow was measured continuously. Using continuous stream flow data was more helpful to producing improved export estimates than increasing sampling frequency. Exports calculated using routine grab samples and continuous stream flow measurements generally could be estimated to within 10% of the more precise methods, especially for solutes that were not particularly responsive to stream flow changes. The relatively low cost associated with continuous stream flow recorders coupled with periodic grab sampling makes these methods especially attractive when accuracy to within $\pm 10\%$ is adequate.

Intensive storm flow monitoring produced results similar to complex multiple regression equations. Intensive monitoring and regression approaches were clearly the best methods for solutes that were strongly correlated with stream flow, such as aluminum. Intensive storm flow sampling was unnecessary for solutes that were weakly correlated with stream flow. Regression was the preferable method at sites where hydrological responses were flashy and difficult to document without very intensive sampling. However, care must be taken when using regression equations to ensure that they are unbiased and that they are predictive over the range of observed flows. Basin hydrology, solute–stream flow correlations, available resources and required level of accuracy of solute budgets should all be considered when choosing a method for estimating solute exports.

ACKNOWLEDGEMENTS AND DISCLAIMER

Although partial funding for data collection for this study was provided by the US Environmental Protection Agency under agreements DW12932656-01-1 and CR814566-02, this manuscript has not been subjected to review by the EPA and, therefore, does not necessarily reflect the views of that agency and no official endorsement should be inferred. In addition, the use of trade, firm or corporation names in this paper is for the information and convenience of the reader. Such use does not constitute an official endorsement or approval by the US Department of Agriculture or the Forest Service of any product or service to the exclusion of others that may be suitable.

REFERENCES

Bliss, C. I. 1970. *Statistics in Biology*, Vol. 2. McGraw-Hill Book Co, New York. pp. 219–287.

Christopherson, N., Seip, H. M., and Wright, R. F. 1982. 'A model for streamwater chemistry at Birkenes, Norway', *Wat. Resour. Res.*, **18**, 977–996.

Dann, M. S., Lynch, J. A., and Corbett, E. S. 1986. 'Comparison of methods for estimating sulfate export from a forested watershed', *J. Environ. Qual.*, **15**, 140–145.

DeWalle, D. R., Swistock, B. R., Dow, C. L., Sharpe, W. E., and Carline, R. F. 1993. 'Episodic Response Project-Northern Appalachian Plateau: site description and methodology', *EPA 600/R-93/023*. US Environmental Protection Agency, Washington, D.C. 55 pp.

Dow, C. L. 1992. 'Sulfur and nitrogen budgets on five forested Appalachian plateau basins', *M.S. Thesis*, The Pennsylvania State University, University Park, Pennsylvania. 133 pp.

Edwards, P. J. and Helvey, J. D. 1991. 'Long-term ionic increases from a central Appalachian forested watershed', *J. Environ. Qual.*, **20**, 250–255.

Edwards, P. J. and Wood, F. 1993. 'Field and laboratory quality assurance/quality control protocols and accomplishments for the Fernow Experimental Forest watershed acidification study', *GTR-NE-177*, Northeastern Forest Experiment Station. USDA Forest Service, Radnor, PA.

Hicks, B. B., Hosker, R. P. Jr., Meyers, T. P., and Womack, J. D. 1991. 'Dry deposition inferential measurement techniques — I. Design and tests of a prototype meteorological and chemical system for determining dry deposition', *Atm. Environ.*, **25A**, 2345–2359.

Johnson, A. H. 1979. 'Estimating solute transport in streams from grab samples', *Wat. Resour. Res.*, **15**, 1224–1228.

Johnson, D. W. 1992. 'Nitrogen retention in forest soils', *J. Environ. Qual.*, **21**, 1–12.

Johnson, A. H., Bouldin, D. R., Goyette, E. A., and Hedges, A. M. 1976. 'Phosphorous loss by stream transport from a rural watershed: quantities, processes and sources', *J. Environ. Qual.*, **5**, 148–157.

Likens, G. E., Bormann, F. H., Johnson, N. M., and Pierce, R. S. 1967. 'The calcium, magnesium, potassium, and sodium budgets for a small forested ecosystem', *Ecology*, **48**, 772–785.

Likens, G. E., Bormann, F. H., Johnson, N. M., Fisher, D. W., and Pierce, R. S. 1970. 'Effects of forest cutting and herbicide treatment on nutrient budgets in the Hubbard Brook Watershed ecosystem', *Ecol. Monogr.*, **40**, 23–47.

Lindberg, S. E., and Lovett, G. M. 1992. 'Deposition and forest canopy interactions of airborne sulfur: Results from the Integrated Forest Study', *Atm. Environ.*, **26A**, 1477–1492.

Lynch, J. A. and Corbett, E. S. 1989. 'Hydrologic control of sulfate mobility in a forested watershed', *Wat. Resour. Res.*, **25**, 1695–1703.

Neter, J., Wasserman, W., and Kutner, M. H. 1985. *Applied Linear Statistical Models*, 2nd edn. Richard D. Irwin Inc., Homewood, Illinois. pp. 111–121.

Patric, J. H. and Studenmund, W. R. 1975. 'Some seldom-reported statistics on precipitation at Elkins, West Virginia', *West Virginia Agric. Forest.*, **6**, 14–16.

Rochelle, B. P. and Church, M. R. 1987. 'Regional patterns of sulfur retention in watersheds of the eastern U.S.', *Water Air Soil Pollut.*, **26**, 61–73.

Shanley, J. B. 1992. 'Sulfate retention and release in soils at Panola Mountain, Georgia', *Soil Sci.*, **153**, 499–508.

Sharpe, W. E., DeWalle, D. R., Leibfried, R. T., Dinicola, R. S., Kimmel, W. G., and Sherwin, L. S. 1984. 'Causes of acidification of four streams on the Laurel Hill in southwestern Pennsylvania', *J. Environ. Qual.* **13**, 619–631.

Shepard, J. P., Mitchell, M. J., Scott, T. J., Zhang, Y. M., and Raynal, D. J. 1989. 'Measurements of wet and dry deposition in a northern hardwood forest', *Water Air Soil Pollut.*, **48**, 225–238.

Swank, W. T. and Waide, J. B. 1988. 'Characterization of baseline precipitation and stream chemistry and nutrient budgets for control watersheds', in Swank, W. T. and Crossley, D. A., Jr. (Eds), *Forest Hydrology and Ecology at Coweeta*. pp. 57–79.

Swistock, B. R., DeWalle, D. R., and Sharpe, W. E. 1989. 'Sources of acidic storm flow in an Appalachian headwater stream', *Wat. Resour. Res.*, **25**, 2139–2147.

US Environmental Protection Agency, 1983. 'Methods for chemical analysis of water and wastes', *EPA-600/4-79-020*, Environ. Monitoring and Support Lab., Office of Res. and Develop. USEPA, Cincinnati, OH.

Wigington, P. J., Baker, J. P., DeWalle, D. R., Kretser, W. A., Murdoch, P. S., Simonin, H. A., Van Sickle, J., McDowell, M. K., Peck, D. V., and Barchet, W. R. 1993. 'Episodic acidification of streams in the northeastern United States: chemical and biological results of the episodic response project', *EPA/600/R-93/190*. US Environmental Protection Agency, Environmental Research Laboratory, Corvallis, OR. 337 pp.

Wilson, G. V., Jardine, P. M., Luxmoore, R. J., Zelazny, L. W., Todd, D. E., and Lietzke, D. A. 1991. 'Hydrogeochemical processes controlling subsurface transport from an upper subcatchment of Walker Branch watershed during storm events. 2. Solute transport processes', *J. Hydrol.*, **128**, 317–336.

12

APPLICATION OF THE GEOGRAPHICAL INFORMATION SYSTEMS APPROACH TO WATERSHED MASS BALANCE STUDIES

PAUL L. RICHARDS AND LEE R. KUMP

Department of Geosciences and Earth System Science Center, The Pennsylvania State University, University Park, PA 16802, USA

ABSTRACT

This study was undertaken to test the utility of a geographical information systems (GIS) approach to problems of watershed mass balance. This approach proved most useful in exploring the effects that watershed scale, lithology and land use have on chemical weathering rates, and in assessing whether mass balance calculations could be applied to large multilithological watersheds. Water quality data from 52 stations were retrieved from STORET and a complete GIS database consisting of the watershed divide, lithology and land use was compiled for each station. Water quality data were also obtained from 7 experimental watersheds to develop a methodology to estimate annual fluxes from incomplete data sets. The methodology consists of preparing a composite of daily flux data, calculating a best fit sinusoid and integrating the equation to obtain an annual flux. Comparison with annual fluxes calculated from high resolution data sets suggests that this method predicts fluxes within about 10% of the true annual flux.

Annual magnesium fluxes (moles km^{-2} yr^{-1}) were calculated for all stations and adjusted for fluxes from atmospheric deposition. Magnesium flux was found to be a strong function of the amount of carbonate in the watershed, and silica fluxes were found to increase with the fraction of sandstone present in the watershed. All fluxes were strongly influenced by mining practices, with magnesium fluxes from affected watersheds being 6–10 times higher than fluxes from comparable pristine watersheds. Mining practices enhance chemical weathering by increasing the surface area of unweathered rock to which water has access and by increasing acidity and rate of mineral weathering. Fluxes were also found to increase with watershed size. This scale dependence is most likely caused by the sensitivity of weathering fluxes to even minor quantities of carbonates, which are likely to be found in all lithologies at larger scales.

Mass balances were carried out in watersheds where gauged sub-watersheds made up more than 95% of the area. The calculations show large magnesium flux and water balance discrepancies. These errors may be a result of significant groundwater inputs to streams between gauges. The results suggest that improvements in how we measure discharge and estimate fluxes may be required before we can apply mass balance techniques to larger scales. © 1997 by John Wiley & Sons, Ltd.

INTRODUCTION

Mass balance models are commonly used to estimate rates of weathering in watersheds (Cleaves *et al.*, 1970; Paces, 1983; Katz *et al.*, 1985; Velbel, 1985). Despite their common use, several questions remain to be explored. The most important of these is how to apply the results at increasingly larger scales. If we carried out a mass balance calculation on two watersheds upstream of a stream junction, and then carried out a mass balance calculation downstream of the junction, could we account for all the mass flux coming from the overall watershed by adding up the fluxes from the contributing watersheds? Another aspect of mass balance calculations that is not typically addressed is the manner in which annual fluxes are estimated from point measurements of concentration. Discharge is commonly measured with automatic stage recorders, where data are collected frequently enough to define accurately the history of runoff. Chemical sampling is not usually obtained as frequently, however. Because flux is the product of concentration and discharge, one must devise a way to match these time-scales of observation through averaging.

The influence of lithology in controlling rates of chemical weathering is also poorly understood because it is difficult to characterize quantitatively the lithology of a watershed. Although strong advances have been made in watersheds composed of specific rock types such as basalt (Peters, 1984; Meybeck, 1987; Bluth and Kump, 1994), watersheds that have heterogeneous lithologies have not been studied as much (e.g. Stallard and Edmond, 1983; Probst et al., 1992). Almost all of the studies of lithological control of weathering fluxes have been made in very small catchments where the lithology is uniform. Since the rivers that contribute the highest fluxes to the oceans are large multilithological watersheds (Meybeck, 1976), we need to determine what lithologies exert the strongest control in weathering within them. In this study we use geographical information systems (GIS) technology to calculate accurately the fractional area of each component lithology. Decisions on how to classify a particular region then fall to the scale at which geological formations are mapped, where lithology is better defined.

Land use is probably the least understood factor in chemical weathering. Although many studies have suggested that land use practices can significantly increase total dissolved load in streams (Meade, 1969; Lystrom et al., 1978), relatively little work has been done to quantify the effect. April et al. (1986) computed major cation weathering rates for a particular watershed in the Adirondacks using water chemistry data. They discovered that recent rates are three times the long-term rates of weathering estimated from soil mineralogy. They interpret this recent increase in weathering to be a result of acid rain deposition. Dissolved loads for strip mining regions have been calculated to be three times 'normal' concentrations (Feltz and Wark, 1962). Paces (1983, 1986) suggests that agricultural practices can accelerate weathering rates by as much as five times the normal rate. In addition to land use effects on chemical weathering, land use practices can also introduce solutes that may hinder the accurate estimation of weathering fluxes (Janda, 1971). The role of land use in chemical weathering will need to be addressed if we are to predict accurately how fluxes might change in the future.

This study examines the relationship of lithology and land use to weathering from multilithological watersheds in the Susquehanna River Basin. We also explore the relationship of these weathering fluxes to watershed scale to see how larger scale watersheds integrate the fluxes of smaller sub-watersheds.

METHODOLOGY

Our technique consists of retrieving historical chemical data from 52 gauge stations in the Susquehanna River Basin in Pennsylvania and Maryland (Table I, Figure 1), defining the watershed divide associated with each and examining the relationship between flux, lithology and land use for each station. Data were also obtained for four watersheds from the Episodic Response Project (DeWalle et al., 1993), and for three watersheds from the Susquehanna River Basin Experiment (P. L. Richards, unpubl.), to examine these relationships in watersheds at smaller scales (Table II). Magnesium was chosen as a proxy for weathering over calcium and alkalinity because its source is primarily from silicate minerals, at least in carbonate-poor units (e.g. Holland, 1978; Berner and Berner, 1987). Agricultural practices commonly introduce large quantities of low magnesium lime to watersheds, which can greatly increase the alkalinity and calcium in runoff. The general lack of silica data makes interpretations from silica flux estimates uncertain.

Study area

The study area (Figure 1) is the Susquehanna River Basin in Pennsylvania and Maryland. Historic data associated with 52 stations in the study area were obtained electronically from the STORET database. The STORET database is a repository of chemical data collected by private and government agencies in the United States. The database is managed by the Environmental Protection Agency and is accessible to the public. Table I lists the stations retrieved in this study. The data come from the United States Geological

Table I. Location and STORET search parameters of water quality sampling stations considered in this study

Station name	ID	Latitude (N)	Longitude (W)	Agency	Code	Station ID
Baldeagle Creek @ Blanchard	1	41 03 06	77 36 17	USGS	112wrd	01547500
Baldeagle Creek @ Curtin	2	40 58 31	77 44 35	DER	21pa	wqn0413
Beech Creek @ Monument	3	41 06 42	77 42 09	USGS	112wrd	01547950
Beech Creek @ US 220	4	41 04 29	77 35 32	DER	21pa	wqn0423
Clearfield Creek	5	40 59 09	78 24 22	DER	21pa	wqn0422
Conodoguinet Creek	6	40 15 31	77 04 39	DER	21pa	wqn0213
Conodoguinet Creek @ Enola	7	40 16 38	76 57 00	DER	21pa	wqn0240
Conestoga Creek @ Lancaster	8	40 03 00	76 16 39	DER	21pa	wqn0205
				USGS	112wrd	01576500
Driftwood Branch Sinnemahoning Ck.	9	41 24 48	78 04 50	DER	21pa	wqn0420
First Fork Sinnemahoning Creek	10	41 19 12	78 04 51	DER	21pa	wqn0419
Fishing Creek	11	40 59 42	76 28 25	DER	21pa	wqn0308
Frankstown Branch Juniata River	12	40 28 34	78 10 39	DER	21pa	wqn0224
Juniata River @ Huntingdon	13	40 29 05	78 01 09	DER	21pa	wqn0216
				USGS	112wrd	01559000
Juniata River @ Newport	14	40 28 42	77 07 46	DER	21pa	wqn0214
				USGS	112wrd	01567000
Kettle Creek	15	41 19 10	77 52 25	DER	21pa	wqn0434
Lackawana River @ Archbald	16	41 30 16	75 32 33	USGS	112wrd	01534500
Lackawana River @ Old Forge	17	41 21 33	75 44 41	USGS	112wrd	01536000
Little Juniata River @ Spruce Creek	18	40 36 33	78 08 01	DER	21pa	wqn0217
Loyalsock Creek	19	41 19 31	76 54 43	DER	21pa	wqn0408
Mahantango Creek @ Dalmatia	20	40 36 40	76 54 44	DER	112wrd	01555500
Moshannon Creek	21	40 50 58	78 16 05	DER	21pa	wqn0421
Penns Creek	22	40 52 00	77 02 55	USGS	112wrd	01555000
Pine Creek @ Ramsey	23	41 16 59	77 19 21	DER	21pa	wqn0410
Raystown Branch Juniata River	24	40 25 35	78 01 47	USGS	112wrd	01562000
Raystown Branch Juniata @ Saxton	25	40 12 57	78 15 56	DER	21pa	wqn0223
Sherman Creek @ Duncannon	26	40 22 49	77 04 56	DER	21pa	wqn0243
				USGS	112wrd	01568200
Sinnemahoning Creek	27	41 19 02	78 06 12	USGS	112wrd	01543500
Sinnemahoning Creek @ wqn0418	28	41 15 41	77 54 16	DER	21pa	wqn0418
South Branch Codorus Creek	29	39 55 14	76 44 57	DER	21pa	01575000
Spring Creek @ Bellfont	30	40 53 23	77 47 40	DER	21pa	wqn0415
Susquehanna River @ Conowingo, MD	31	39 39 26	76 10 31	USGS	112wrd	01578310
Susquehanna River @ Dansville	32	40 57 30	76 37 10	USGS	112wrd	01540500
Susquehanna River @ Harrisburg	33	40 15 27	76 53 12	DER	21pa	wqn0202
		40 15 10	76 52 27	USGS	112wrd	01570500
Susquehanna River @ Marietta	34	40 03 16	76 31 52	DER	21pa	wqn0201
Susquehanna River @ Sunbury	35	40 51 15	76 51 20	DER	21pa	wqn0203
Swatara Creek @ Inwood	36	40 28 38	76 31 26	USGS	112wrd	01572200
Swatara Creek @ Middleton	37	40 11 28	76 43 52	DER	21pa	wqn0211
Tioga River	38	41 54 30	77 07 47	DER	21pa	wqn0319
Tioga River @ Tioga	39	41 57 27	77 06 58	DER	21pa	wqn0324
				USGS	112wrd	01518000
Toby Creek	40	41 16 51	75 53 46	DER	21pa	wqn0312
Wapwollen Creek	41	41 04 17	76 08 02	DER	21pa	wqn0310
West Conewago Creek @ Rte 181	42	40 04 52	76 43 07	DER	21pa	wqn0210
West Branch Codorus	43	39 53 14	76 50 09	DER	21pa	wqn0208
West Branch Susq. @ Karthus	44	41 07 03	78 06 33	DER	21pa	wqn0404
West Branch Susq. @ Lewisburg	45	40 58 05	76 52 25	USGS	112wrd	01553500
West Branch Susq. @ Lockhaven	46	41 08 20	77 26 30	USGS	112wrd	01545800
West Branch Susq. @ PR453	47	40 57 41	78 31 10	DER	21pa	wqn0405
West Branch Susq. @ Renova	48	41 19 26	77 45 02	DER	21pa	wqn0403
				USGS	112wrd	01545500
West Branch Susq. @ T41B	49	40 53 49	78 40 38	DER	21pa	wqn0406
West Branch Susq. @ Williamsport	50	41 13 14	77 01 09	DER	21pa	wqn0402
Yellow Breeches Creek	51	40 13 27	76 51 38	DER	21pa	wqn0212
Youngwomans Creek	52	41 19 28	77 41 28	USGS	112wrd	01545600

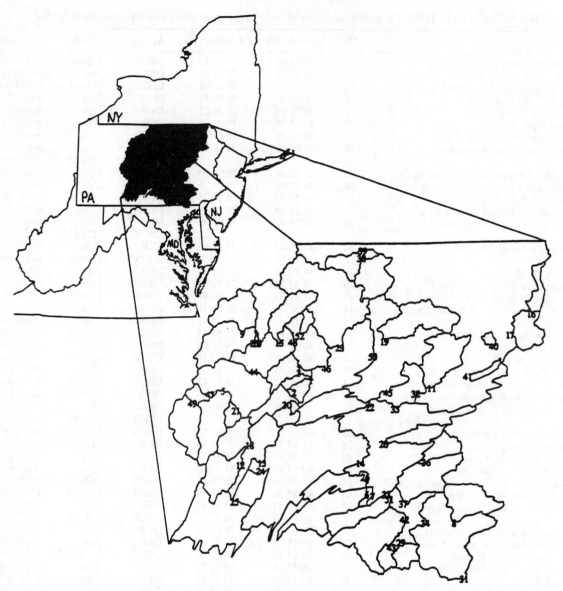

Figure 1. Watershed divides and locations of STORET stations. The numbers refer to the ID# in Table I.

Survey (USGS) and the Pennsylvania Department of Environmental Resources (DER). The agency codes and station numbers listed in Table I are the search parameters that were used to perform the retrieval. The data were collected by these agencies on a relatively infrequent basis depending on local stream use and land use. Short periods of more frequent sampling were carried out for some of these watersheds for specific projects. Detailed summaries of the chemical sampling programmes for the DER and the USGS can be found in Shertzer and Schreffler (1993) and the USGS Water Supply Papers, respectively. Data for most of the stations are collected quarterly, but stations associated with urban areas are collected monthly. Large gaps in the data exist in the records of almost all stations. The water quality parameters assessed in this study are dissolved Mg and total Mg.

Table II. Locations and references for small experimental watersheds considered in this study

Watershed name	Latitude	Longitude	Agency/project	Reference
Baldwin Creek	40 21 05	79 03 04	EPA	De Walle *et al.* (1993)
Benner Run	40 56 04	78 01 22	EPA	De Walle *et al.* (1993)
Roberts Run	41 10 12	78 24 22	EPA	De Walle *et al.* (1993)
Stone Run	41 05 52	78 26 48	EPA	De Walle *et al.* (1993)
W.2	40 43 48	76 36 02	SRBX	Richards (unpublished data)
WD-38	40 42 44	76 35 08	SRBX	Richards (unpublished data)
WE-38	40 42 15	76 35 17	SRBX	Richards (unpublished data)

Watershed hierarchy

The 52 gauge stations evaluated in this study form a series of nested watersheds with a total of 29 *primary* watersheds and 23 *composite* watersheds, composed of one or more *primary* watersheds (Figure 2). A *primary* watershed is defined as the smallest scale of watershed in the nested watershed hierarchy. Thus, only those watersheds with no sampling stations upstream are primary watersheds. Evaluation of how fluxes from contributing watersheds add up and compare with the larger composite watershed encompassing them enables us to test the accuracy of the mass balance approach at larger scales.

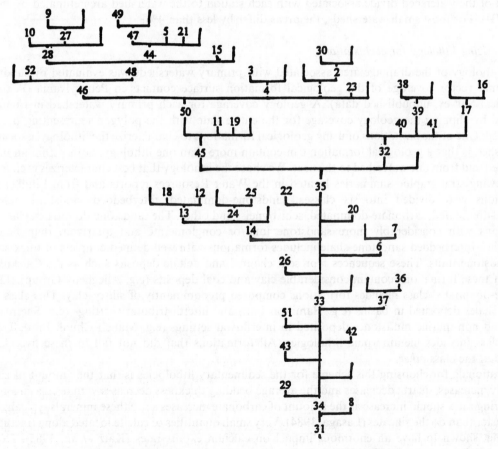

Figure 2. Stream hierarchy. The numbers refer to the ID# in Table I

Climate — geography

The study area encompasses portions of the Plateau and Valley and Ridge provinces of the Appalachian Mountains. Average annual temperatures vary from 7·8°C in the northern part of the study area to 11·1°C in the southern tip of the study area (Cuff, 1989). Average rainfall varies little, from 96·5 cm in the Valley and Ridge portion of the study area to 106·7 cm along the Appalachian Plateau, located in the northern and western part of the study area. Soils are typically thin and residual. Thicker oils can be found in valleys containing carbonate deposits. Unconsolidated sediments can also be found in the study area as colluvium that mantles the sides of ridges, glacial sediments that occur in the northern part of the study area and floodplain deposits in some valleys.

Estimating watershed divides for the sampling stations

Watershed divides for all of the stations in Table I were delineated by plotting the locations of the stations on 1 : 250 000 scale topographic maps and judging the position of topographic highs from contours. A total of nine such maps were needed to cover the study area. The divide maps were then scanned with a 600 dpi scanner to create raster images. The raster images were then converted to vector coverages and projected to 'real world' coordinates using the GIS software package ARC/INFO. These vector coverages were then joined together to create one vector coverage encompassing the entire study area. The coverage was checked by overlaying the watershed divides on a 1 : 100 000 scale stream coverage compiled for Pennsylvania. Watershed boundaries were edited so that they did not intersect streams. Errors were assessed by comparing the area of the watershed divide associated with each station to the watershed area etimated by the USGS (Table III). For most small watersheds, the areas differ by less than 4%.

Characterizing lithology for each station

The lithology of the drainage area associated with primary watersheds was evaluated by subdividing a provisional vector coverage of the geological formation surface contacts in Pennsylvania (Pennsylvania Geological Survey, unpublished data). A geology coverage for each primary watershed in Figure 1 was obtained by clipping the geology coverage for the entire state with the polygon representing the primary watershed. One of the limitations of using geological formations to characterize the lithology associated with a watershed is that a geological formation can contain more than one lithology, both within an individual watershed and from one watershed to the next. We chose the lithology that best characterizes each formation by reviewing stratigraphic summaries located in the Water Resources reports and from Lindberg (1984). Formations were divided into six classes: sandstone-dominated, interbedded sandstone-shale, shale/siltstone-dominated, carbonate-dominated, schist/gneiss and other. The sandstone-dominated class includes formations with considerably more sandstone (and/or conglomerate and quartzite) than fine-grained lithologies. Interbedded sandstone-shale includes formations with well-defined couplets of thick sandstone and mudstone units. These sequences represent channel and deltaic deposits such as the Catskill Group. Some of these formations contain considerable clay and coal deposits (e.g. Allegheny Group). The shale/siltstone-dominated class includes formations composed predominantly of silt or clay. This class includes marine shales deposited in basins (e.g. Hamilton Fm.) and interdistributary settings (e.g. Sherman Creek Fm.), and non-marine mudstones deposited in interfluvial settings (e.g. Mauch Chunk Fm.). The schist/gneiss class includes metamorphic lithologies. All formations that did not fall in these five classes are grouped in the class *other*.

The rationale for choosing this scheme for the sedimentary lithologies is that the amount of carbonate generally increases, quartz decreases and the average bedding thickness decreases with decreasing grain size. Weathering rates should increase as the amount of carbonate increases since these minerals typically weather much faster than do the silicates (Lasaga, 1984). Very small quantities of calcite located along fracture planes have been shown to have an enormous impact on calcium export rates (Katz et al., 1985). Decreasing bedding thickness will also favour weathering since it has been shown that fracture spacing generally

Table III. Watershed areas calculated for the stations considered in this study

Station name	Area USGS (km^2)	Area ARC/INFO (km^2)	Relative error (%)
Baldeagle Creek @ Blanchard	879·1	864·4	−1·67
Baldeagle Creek @ Curtin	687·2	703·5	+2·38
Beech Creek @ Monument	394·2	391·6	−0·65
Beech Creek @ US 220		438·7	
Clearfield Creek	962·0	941·3	−2·15
Conodoguinet Creek	1218·8	1109·1	−8·90
Conodoguinet Creek @ Enola	1299·1	1183·3	−8·90
Conestoga Creek @ Lancaster	840·2	827·5	−1·51
Driftwood Branch Sinnemahoning Ck.	705·3	767·4	+8·80
First Fork Sinnemahoning Creek		696·8	
Fishing Creek		980·5	
Frankstown Branch Juniata River		796·2	
Juniata River @ Huntingdon	2116·0	2169·6	+2·54
Juniata River @ Newport	8697·3	8661·8	−0·41
Kettle Creek	604·2	611·3	+1·18
Lackawana River @ Archbald	280·1	287·6	+2·69
Lackawana River @ Old Forge	860·9	865·2	+0·50
Little Juniata River @ Spruce Creek	518·6	562·7	+8·40
Loyalsock Creek	1148·7	1111·9	−3·21
Mahantango Creek @ Dalmatia	420·1	422·6	+0·60
Moshannon Creek		174·7	
Penns Creek	780·5	810·0	+3·78
Pine Creek @ Ramsey	2447·9	2470·5	+0·92
Raystown Branch Juniata River	2481·6	2396·0	−3·45
Raystown Branch Juniata @ Saxton		1917·8	
Sherman Creek @ Duncannon	518·6	620·6	+19·66
Sinnemahoning Creek	1776·3	1792·4	+0·91
Sinnemahoning Creek @ wqn0418	2411·6	2509·0	+0·91
South Branch Codorus Creek	303·4	301·0	−0·79
Spring Creek @ Bellfont	226·1	222·1	−1·78
Susquehanna River @ Conowingo, MD		70 350·8	
Susquehanna River @ Dansville	29 094·6		
Susquehanna River @ Harrisburg	62 493·7		
Susquehanna River @ Marietta	63 394·7		
Susquehanna River @ Sunbury	47 453·7		
Swatara Creek @ Inwood	433·0	416·1	−3·91
Swatara Creek @ Middleton		1463·2	
Tioga River		718·4	
Tioga River @ Tioga	1156·5	1153·1	−0·30
Toby Creek		74·8	
Wapwollen Creek	113·6	122·1	+0·75
West Conewago Creek @ Rte 181	1322·5	1337·5	+1·14
West Branch Codorus		212·4	
West Branch Susq. @ Karthus		3614·1	
West Branch Susq. @ Lewisburg	17 754·0	17 364·7	−2·20
West Branch Susq. @ Lockhaven	8653·2	8660·8	+0·08
West Branch Susq. @ PR453	951·7	978·6	+2·83
West Branch Susq. @ Renova	7714·5	7708·4	+0·08
West Branch Susq. @ T41B	816·8	806·3	−1·29
West Branch Susq. @ Williamsport	14 734·0	14 675·7	−0·40
Yellow Breeches Creek	567·9	591·7	+4·19
Youngwomans Creek	119·8	123·0	+2·67

Small scale watersheds	Area (km^2)
Baldwin Creek	5·35
Benner Run	11·34
Roberts Run	10·70
Stone Run	11·56
W.2	0·15
WD-38	0·63
WE-38	7·41

Table IV. Lithological classification for the geological formations present in the study area

Sandstone/quartzite-dominated		Carbonate dominated	
Antietam Fm.	Ca	Annville Fm.	Oan
Baldeagle Fm.	Obe	Axeman Fm.	Oa
Burgoon Sandstone	Mb	Beekmantown Group	Ob
Chickies Fm.	Cch	Bellfont Fm.	Obf
Gettysburg Conglomerate	Trgc	Benner-Loysburg Fm. undivided	Obl
Hammer Creek Fm.	Trh	Buffalo Springs Fm.	Cbs
Hardystown Fm.	Cha	Coburn-Nealmont Fm. undivided	Ocn
Harpers Fm.	Ch	Conestoga Fm.	OCc
Huntley Mountain Fm.	MDhm	Epler Fm.	Oe
Juniata Fm.	Oj	Hershey-Myerstown Fm. undivided	Ohm
Montalto Member of the Harpers Fm.	Chm	Kinzers Fm.	Ck
New Oxford conglomerate	Trnc	Ledger Fm.	Cl
Pocono Fm.	Mp	Leithsville Fm.	Clv
Pottsville Group	Pp	Limestone fanoglomerate	Trfl
Quartz Fanoglomerate	Trfq	Millbach Fm.	Cm
Specht Kopf Fm.	MDsk	Nittany Fm.	On
Stockton Fm.	Trs	Onondaga-Old Port Fm	Doo
Tuscorora Fm.	St	Otelaunee Fm.	Oo
Weaverton and Loudoun Fm. undivided	Cwl	Richland Fm.	Cr
		Snitz Creek Fm.	Csc
Interbedded sandstone-shale		Stonehenge Fm.	Os
Alleghany Group	Pa	Vintage Fm.	Cv
Catskill Fm. undivided	Dck	Wakefield Marble of Wissahickon Fm.	Xww
Duncannon member of the Catskill Group	Dcd	Zooks Corner Fm.	Czc
Glenshaw Fm.	Pcg		
Irish Valley member of the Catskill Group	Dciv	Schist/Gneiss	
Llewellyn Fm.	Pl	Granitic Gneiss	gn
New Oxford Fm.	Trn	Granodiorite and Granodiorite Gneiss	ggd
Rockwell Fm.	Mdr	Greenstone Schist	vs
Shenango-Oswayo Fm. undivided	MDso	Hornblende Gneiss	hg
		Marsburg Schist	Xwm
Shale/siltstone-dominated		Wissahickon Fm.	Xwc
Casselman Fm.	Pcc		
Clinton Group	Sc	Other	
Cocallico Fm.	Oco	Diabase	Trd
Gettysburg Fm.	Trg	Metabasalt	mb
Hamilton Group	Dh	Metarhyolite	mr
Heidlersburg Fm.	Trgh	Metavolcanics	Xwv
Lockhaven Fm.	Dlh		
Martinburg Fm.	Om		
Mauch Chunk Fm.	Mmc		
Reedsville Fm.	Or		
Sherman Creek member of the Catskill Group	Dcsc		
Trimmers Rock Fm.	Dtr		
Wills Creek Fm.	Swc		

increases with bedding thickness (Narr and Suppe, 1991) and that most of the groundwater in the north-east travels through fractures (Heath, 1984). Decreased fracture spacing leads to more fractures per unit area, and more potential conduits where water can react with mineral surfaces. As a consequence we might expect that weathering fluxes will increase as: sandstone-dominated < interbedded sandstone/shale < shale/ siltstone-dominated < carbonates. Table IV summarizes the formations associated with each class. The fraction of watershed area encompassed by each lithological class was calculated for all of the stations. The results are presented in Table V and Figure 3.

Table V. Fraction of watershed area composed of a particular lithology

Station name	Sandstone-dominated	Interbedded sandstone-shale	Shale-dominated	Carbonate-dominated	Schist-gneiss	Other
Baldeagle Creek @ Blanchard	0·16	0·10	0·28	0·46	0·00	0·00
Baldeagle Creek @ Curtin	0·17	0·12	0·25	0·45	0·00	0·00
Beech Creek @ Monument	0·60	0·17	0·23	0·00	0·00	0·00
Beech Creek @ Rte 220	0·64	0·12	0·24	0·00	0·00	0·00
Clearfield Creek	0·12	0·80	0·08	0·00	0·00	0·00
Conodoguinet Creek	0·10	0·01	0·56	0·33	0·00	0·00
Conodoguinet Creek @ Enola	0·12	0·00	0·52	0·36	0·00	0·00
Conestoga Creek @ Lancaster	0·39	0·04	0·08	0·43	0·02	0·04
Driftwood Branch Sinnemahoning Ck.	0·51	0·45	0·04	0·00	0·00	0·00
First Fork Sinnemahoning Creek	0·53	0·46	0·01	0·00	0·00	0·00
Fishing Creek	0·16	0·30	0·50	0·01	0·00	0·00
Frankstown Branch Juniata River	0·19	0·23	0·33	0·25	0·00	0·00
Juniata River @ Huntingdon	0·18	0·10	0·35	0·37	0·00	0·00
Juniata River @ Newport	0·18	0·11	0·49	0·22	0·00	0·00
Kettle Creek	0·54	0·44	0·02	0·00	0·00	0·00
Lackawana River @ Archbald	0·18	0·82	0·00	0·00	0·00	0·00
Lackawana River @ Old Forge	0·20	0·79	0·01	0·00	0·00	0·00
Little Juniata River @ Spruce Creek	0·25	0·16	0·36	0·23	0·00	0·00
Loyalsock Creek	0·57	0·30	0·13	0·00	0·00	0·00
Mahantango Creek @ Dalmatia	0·17	0·28	0·55	0·00	0·00	0·00
Moshannon Creek	0·24	0·56	0·20	0·00	0·00	0·00
Penns Creek	0·38	0·00	0·37	0·25	0·00	0·00
Pine Creek @ Ramsey	0·54	0·42	0·05	0·00	0·00	0·00
Raystown Branch Juniata River	0·19	0·18	0·46	0·16	0·00	0·00
Raystown Branch Juniata @ Saxton	0·22	0·30	0·29	0·19	0·00	0·00
Sherman Creek @ Duncannon	0·15	0·06	0·65	0·15	0·00	0·00
Sinnemahoning Creek	0·54	0·44	0·02	0·00	0·00	0·00
Sinnemahoning Creek @ wqn0418	0·54	0·44	0·02	0·00	0·00	0·00
South Branch Codorus Creek	0·13	0·00	0·00	0·02	0·77	0·00
Spring Creek @ Bellfont	0·11	0·00	0·05	0·84	0·00	0·00
Susquehanna River @ Dansville*	0·13	0·52	0·34	0·00	0·00	0·00
Susquehanna River @ Harrisburg*	0·27	0·32	0·33	0·08	0·00	0·00
Susquehanna River @ Marietta*	0·25	0·30	0·34	0·09	0·01	0·01
Susquehanna River @ Sunbury*	0·31	0·42	0·24	0·03	0·00	0·00
Swatara Creek @ Inwood	0·18	0·28	0·52	0·02	0·00	0·00
Swatara Creek @ Middleton	0·08	0·09	0·65	0·16	0·00	0·02
Tioga River	0·39	0·35	0·27	0·00	0·00	0·00
Tioga River @ Tioga	0·34	0·31	0·35	0·00	0·00	0·00
Toby Creek	0·05	0·93	0·02	0·00	0·00	0·00
Wapwollen Creek	0·04	0·53	0·43	0·00	0·00	0·00
West Conewago Creek @ Rte 181	0·13	0·23	0·34	0·06	0·03	0·21
West Branch Codorus Creek	0·41	0·00	0·00	0·19	0·40	0·00
West Branch Susq. @ Karthus	0·29	0·62	0·09	0·00	0·00	0·00
West Branch Susq. @ Lewisburg	0·45	0·35	0·15	0·05	0·00	0·00
West Branch Susq. @ Lockhaven	0·49	0·45	0·06	0·00	0·00	0·00
West Branch Susq. @ PR453	0·04	0·88	0·08	0·00	0·00	0·00
West Branch Susq. @ Renova	0·46	0·49	0·05	0·00	0·00	0·00
West Branch Susq. @ T41B	0·02	0·89	0·09	0·00	0·00	0·00
West Branch Susq. @ Williamsport	0·47	0·37	0·11	0·05	0·00	0·00
Yellow Breeches Creek	0·31	0·00	0·18	0·37	0·00	0·14
Youngwomans Creek	0·83	0·16	0·01	0·00	0·00	0·00
Baldwin Creek	0·70	0·20	0·10	0·00	0·00	0·00
Benner Run	0·75	0·00	0·25	0·00	0·00	0·00
Roberts Run	0·95	0·05	0·00	0·00	0·00	0·00
Stone Run	0·95	0·05	0·00	0·00	0·00	0·00
W.2	1·00	0·00	0·00	0·00	0·00	0·00
WD-38	0·00	0·20	0·80	0·00	0·00	0·00
WE-38	0·20	0·50	0·30	0·00	0·00	0·00

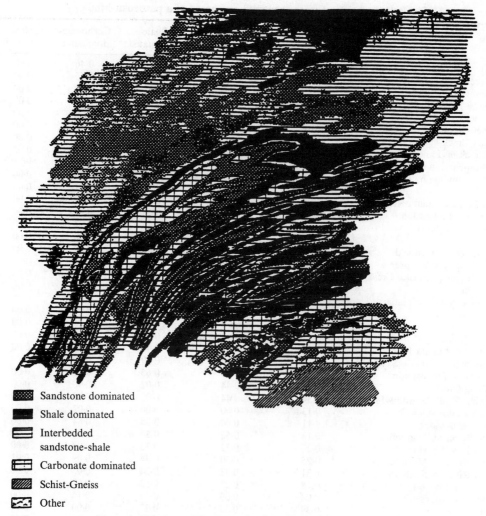

Sandstone dominated

Shale dominated

Interbedded
sandstone-shale

Carbonate dominated

Schist-Gneiss

Other

Figure 3. Lithology of the Pennsylvania portion of the study area

Characterizing land use for each station

The land use associated with each station was evaluated by subdividing a land use database (LUDA) obtained from the USGS. The LUDA database is an Anderson level II land use classification of Pennsylvania, which consists of 26 categories that were mapped from aerial photographs (Anderson *et al.*, 1976). Categories were grouped into six classes: urban, agriculture, forest, wetland, mining and other (Table VI). The areal percentage of each watershed composed of each land use class is presented in Table VII.

COMPUTING ANNUAL FLUXES

Flux is calculated as the product of discharge and concentration. However, stream samples for water chemistry are not typically collected as frequently as discharge data. Several approaches have been used to rectify this problem. They differ in how discharge and concentration data are mathematically reduced to obtain a simple product between discharge and concentration. Dann *et al.* (1986) review some of the advantages and disadvantages between the various techniques, including the period-weighted method

Table VI. Land use classification scheme used in this study

Land use classification	USGS LUDA VAT CODE	Description
Urban	11	Residential
	12	Commercial and services
	13	Industrial
	15	Transportation, communication and utilities
	16	Mixed urban or built-up land
	17	Other urban or built-up land
Agriculture	21	Cropland and pasture
	22	Orchards, groves, vineyards, nurseries and ornamental horticultural areas
	23	Confined feeding operations
	24	Other agricultural land
	31	Herbaceous range land
	32	Shrub and brush range land
	33	Mixed range land
Forest	41	Deciduous forest land
	42	Evergreen forest land
	43	Mixed forest land
Wetland	51	Streams and canals
	52	Lakes
	53	Reservoirs
	54	Bays and estuaries
	61	Forested wetland
	62	Non-forested wetland
Mining	75	Strip mines, quarries and gravel pits
	76	Transitional areas
Other	0	Unclassified
	74	Bare exposed rock

(Likens *et al.*, 1977; Probst *et al.*, 1992), discharge-weighted method (Paces, 1983; Probst *et al.*, 1992), estimating concentration directly from discharge (Johnson, 1979; Probst *et al.*, 1992) and simple averaging (Langbein and Dawdy, 1964; Peters, 1984).

The sampling interval (monthly at best) of the STORET data set precludes the period- and discharge-weighted methods. The period of record is long for many of the stations so there is an abundance of data. We decided to convert the data to daily solute fluxes and to overlay all of it to gain an impression of the temporal variability of solute fluxes.

Variation of daily fluxes

Much of the previous work (e.g. Langbein and Dawdy, 1964; Holland, 1978; Peters, 1984; Bluth, 1990; Bluth and Kump, 1994) demonstrates that fluxes for most water quality parameters are highly correlated with runoff. These observations suggest that the range of variation of concentration is not as great as the range of variation of runoff. One possible exception may be sulfate (Dann *et al.*, 1986) which has a large contribution from precipitation. Since flux is autocorrelated with runoff, we might expect a time series of fluxes to resemble a hydrograph, with the peaks attenuated. The attenuated peaks represent high discharge periods when overland and shallow through-flow paths are contributing dilute water to stream flow. These relationships are well demonstrated by a daily magnesium flux time series for the Benner Run (Episodic Response) watershed (Dewalle *et al.*, 1993).

Figure 4b shows all the instantaneous flux data for the Susquehanna River at Harrisburg gauge station, with a time series of flux data for 1980 superimposed (solid line). The time series plots within the envelope of historical data. The total yearly flux for the Susquehanna River at Harrisburg 1980 can be estimated by

Table VII. Percentage of watershed area composed of a particular land use

Station name	Urban	Agriculture	Forest	Wetland	Mining	Other
Baldeagle Creek @ Blanchard	7·36	32·62	58·46	1·04	0·53	0·00
Baldeagle Creek @ Curtin	8·12	32·72	58·64	0·03	0·50	0·00
Beech Creek @ Monument	0·91	0·69	93·87	0·02	4·51	0·00
Beech Creek @ US 220	0·85	2·21	92·83	0·02	4·09	0·00
Clearfield Creek	1·82	17·21	78·62	0·99	1·36	0·00
Conodoguinet Creek	6·84	62·22	30·61	0·17	0·16	0·00
Conodoguinet Creek @ Enola	6·88	63·40	29·42	0·16	0·15	0·00
Conestoga Creek @ Lancaster	11·01	59·30	28·98	0·34	0·37	0·00
Driftwood Branch Sinnemahoning Ck.	1·08	3·01	94·81	0·00	1·10	0·00
First Fork Sinnemahoning Creek	0·67	4·53	94·67	0·13	0·00	0·00
Fishing Creek	0·98	37·57	61·01	0·41	0·03	0·00
Frankstown Branch Juniata River	8·78	26·22	63·43	0·29	1·28	0·00
Juniata River @ Huntingdon	5·22	28·92	64·94	0·24	0·69	0·00
Juniata River @ Newport	2·91	28·51	67·33	0·73	0·53	0·00
Kettle Creek	0·17	5·76	93·65	0·16	0·25	0·00
Lackawana River @ Archbald	8·15	14·77	68·42	2·07	6·58	0·00
Lackawana River @ Old Forge	18·36	9·59	63·29	1·82	6·94	0·00
Little Juniata River @ Spruce Creek	5·46	18·83	74·73	0·19	0·79	0·00
Loyalsock Creek	0·35	12·60	85·93	0·93	0·19	0·00
Mahantango Creek @ Dalmatia	1·85	51·04	43·64	0·04	3·42	0·00
Moshannon Creek	2·73	2·60	75·76	0·00	18·91	0·00
Penns Creek	1·24	31·08	67·59	0·09	0·00	0·00
Pine Creek @ Ramsey	0·51	14·28	84·66	0·17	0·38	0·00
Raystown Branch Juniata River	2·40	31·24	64·61	1·02	0·73	0·00
Raystown Branch Juniata @ Saxton	2·75	34·61	61·89	0·16	0·59	0·00
Sherman Creek @ Duncannon	0·64	33·01	66·29	0·04	0·02	0·00
Sinnemahoning Creek	0·85	2·94	94·4	0·01	1·77	0·00
Sinnemahoning Creek @ wqn0418	0·82	3·37	94·51	0·04	1·26	0·00
South Branch Codorus Creek	3·88	77·73	17·34	0·77	0·25	0·02
Spring Creek @ Bellfont	16·72	50·28	32·44	0·00	0·56	0·00
Susquehanna River @ Conowingo, MD*	4·20	32·36	60·00	1·27	1·84	0·31
Susquehanna River @ Dansville*	4·69	35·49	55·80	1·74	1·67	0·61
Susquehanna River @ Harrisburg*	3·34	27·53	65·79	1·16	1·99	0·18
Susquehanna River @ Marietta*	3·94	30·34	62·40	1·19	1·93	0·20
Susquehanna River @ Sunbury*	3·29	24·99	69·98	1·19	0·25	0·28
Swatara Creek @ Inwood	4·77	26·14	60·08	0·36	8·65	0·00
Swatara Creek @ Middleton	10·59	52·28	30·49	0·32	6·33	0·00
Tioga River	1·67	42·73	54·17	0·22	1·22	0·00
Tioga River @ Tioga	1·29	45·66	51·93	0·33	0·80	0·00
Toby Creek	26·57	17·30	53·64	2·28	0·20	0·00
Wapwollen Creek	13·45	24·83	60·75	0·08	0·88	0·00
West Conewago Creek @ Rte 181	0·56	96·06	3·13	0·05	0·07	0·13
West Branch Codorus	6·60	68·06	21·34	3·30	0·00	0·70
West Branch Susq. @ Karthus	2·53	13·06	72·09	0·48	11·84	0·00
West Branch Susq. @ Lewisburg	2·04	14·61	79·45	0·68	3·21	0·00
West Branch Susq. @ Lockhaven	1·41	6·98	85·13	0·58	5·89	0·00
West Branch Susq. @ PR453	1·92	20·20	68·24	0·46	9·19	0·00
West Branch Susq. @ Renova	1·50	7·74	83·72	0·47	6·57	0·00
West Branch Susq. @ T41B	2·20	22·23	67·92	0·02	7·63	0·00
West Branch Susq. @ Williamsport	1·85	12·19	81·65	0·57	3·74	0·00
Yellow Breeches Creek	8·03	41·60	49·81	0·20	0·36	0·00
Youngwomans Creek	0·00	0·00	100·00	0·00	0·00	0·00

* Estimate; does not include regions outside Pennsylvania.

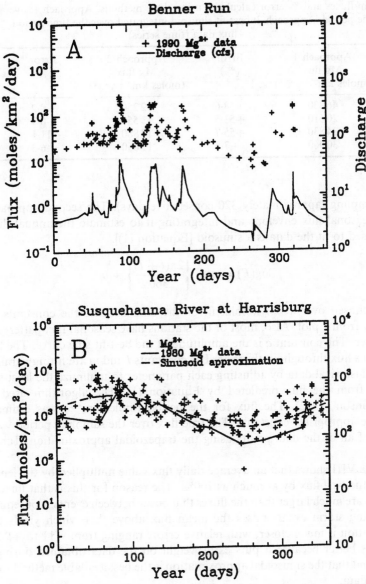

Figure 4. Daily flux time series plotted against Julian day. (a) Benner Run (1980). (b) Susquehanna River @ Harrisburg (1980) and the best-fit sinusoid overlaid on all historical data

integrating a time series of daily fluxes over the year. This is essentially the same as obtaining the area under the 1980 time series curve. If we did not have a complete set of time series data, we could fit a sinusoid through a composite of historic flux data and calculate the area underneath it. The integral of a best fit sinusoid is a reasonable approximation of the annual flux.

Testing the sinusoidal approximation method

Two empirical approaches were tested with data from the Episodic Response Project (Dewalle *et al.*, 1993). The first approach involves calculating an average daily flux from the data and multiplying by the number of

Table VIII. Magnesium fluxes and % error calculated using three methods. Approach 1: average daily concentration integrated over the time period. Approach 2: best-fit sinusoid integrated over the time period. 'True': Integral of daily flux data (time series)

Watershed name	Approach 1 Mg flux (moles km^{-2} yr^{-1})	Error (%)	Approach 2 Mg flux (moles km^{-2} yr^{-1})	Error (%)	'True' Mg flux (moles km^{-2} yr^{-1})
Baldwin Creek	46 830	+58·6	32 250	+9·2	29 530
Benner Run	20 610	+53·5	13 550	+0·9	13 430
Roberts Run	34 830	+55·7	19 880	−11·1	22 370
Stone Creek	30 960	+7·8	15 850	+6·3	14 900

days of intensive sampling (approximately 320 consecutive days). The second approach involves fitting a curve through the seasonal flux envelope and integrating it to estimate the annual flux. The equation of the curve that was used to fit the data is a sinusoid [Equation (1)].

$$\log(F) = a \sin\left\{\frac{2\pi(t + l)}{360}\right\} + m \tag{1}$$

This equation was chosen over a polynomial fit of the data because the constants a, l and m can be estimated graphically from a plot. Also, most of the stations have seasonal flux patterns that are essentially sinusoidal in character. The constant a is the amplitude of the best-fit sine curve. The larger a is, the more variable is the stream's flux throughout the year. The constants l and m are fitting parameters. Equation (1) was fitted to station historical data by adjusting each parameter to minimize the least-squares deviation of the daily flux data from the flux predicted by Equation (1). The algorithm used was developed by Bevington and Robinson (1992). The flux for the period (moles km^{-2}) was estimated by integrating Equation (1) with the appropriate constants. The true flux over the sampling period was estimated to be a straight integration of all of the data points using the trapezoidal approximation. Table VIII summarizes the results.

Inspection of Table VIII shows that an average daily flux value multiplied by the length of the sampling period over predicts the true flux by as much as 100%. The reason for this is that fluxes during infrequent high discharge events are much larger than the fluxes that occur between events. Consequently, inclusion of a few data points during storm events raises the mean flux above that which yields the 'true' flux. The sinusoidal approximation is much closer, with relative errors ranging from −11 to +9·2%. The sinusoidal approximation works better because it puts extra weight on the data obtained during normal discharge conditions. We contend that the sinusoidal approximation is the best available method for estimating annual fluxes with sporadic data.

Estimating atmospheric deposition

Previous authors (Gibbs, 1970; Goudie, 1970; Janda, 1971; Stallard and Edmond, 1983; Drever, 1988) have observed that the atmospheric flux of some solutes is large and that these fluxes must be accounted for to obtain realistic chemical erosion rates from stream data. Atmospheric deposition comes in two forms, wet and dry deposition. Wet deposition is the process whereby dissolved solutes and insoluble particles are carried to the ground by rainfall. Dry deposition is the settling of particulates and aerosols from the atmosphere without rain as a transport medium. Some of this material accumulates on the forest canopy where it is washed out during storm events.

To quantify deposition rates in this study, we obtained annual atmospheric deposition rate data for magnesium from Lynch *et al.* (1993). They measured the rates for 15 monitoring stations in Pennsylvania from 1982 to 1992 (shorter periods for certain stations). An average deposition rate was calculated for each

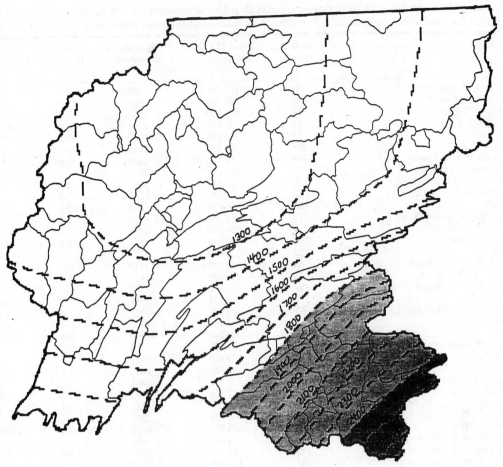

Figure 5. Atmospheric deposition map of magnesium for the field area. Contours are in moles km^{-2} yr^{-1}

station and a surface was fit through the data to interpolate atmospheric fluxes between stations. The calculations suggest that magnesium fluxes from the atmosphere range from 1200 to 2700 moles km^{-2} yr^{-1} (Figure 5). The highest fluxes were associated with the south-east portion of the field area. The total atmospheric flux for each watershed was estimated by gridding this surface at 1 km resolution and calculating the mean for all cells falling within each watershed. It is important to note that the manner with which the original deposition rate data were collected is biased against dry deposition. Consequently, the calculated fluxes underestimate the true atmospheric deposition rate.

Estimating the weathering rate

The parameters of Equation (1) were evaluated for all stations with at least 30 or more data points covering the entire year to obtain an average yearly flux. Flux data were evaluated from both dissolved magnesium and total magnesium. No discernible difference was found between them. The magnesium flux into the watershed from atmospheric deposition was subtracted from the total annual flux to obtain the weathering flux (Table IX). Table IX shows that the atmospheric flux of magnesium only contributes a minor portion (0·5–6%) of the total annual flux.

Table IX. Atmospheric deposition fluxes of magnesium calculated for watersheds, % of total watershed flux contributed from the atmosphere and effective denudation flux. Obs. is the total number of data points used to calculate the total flux

Station name	Obs. (#)	Total Mg flux (moles km^{-2} yr^{-1})	Mg atmos. deposition (moles km^{-2} yr^{-1})	Percent of total flux	Net Mg flux (moles km^{-2} yr^{-1})
Baldeagle Creek @ Blanchard	66	141 100	1257	0·9	139 843
Baldeagle Creek @ Curtin	140	308 760	1264	0·4	307 496
Beech Creek @ Monument	39	120 700	1232	1·02	119 468
Beech Creek @ US 220	55	150 610	1231	0·8	149 379
Clearfield Creek	59	235 050	1370	0·6	233 680
Conodoguinet Creek	209	84 300	1680	2·0	82 620
Conodoguinet Creek @ Enola	169	84 270	1686	2·0	82 584
Conestoga Creek @ Lancaster	187	169 110	2292	1·4	166 818
Driftwood Branch Sinnemahoning Ck.	128	27 410	1285	4·7	26 125
First Fork Sinnemahoning Creek	142	23 810	1247	5·2	22 563
Fishing Creek	105	26 130	1368	5·2	24 762
Frankstown Branch Juniata River	158	126 300	1514	1·2	124 786
Juniata River @ Huntingdon	134	101 860	1433	1·4	100 427
Juniata River @ Newport	290	92 840	1511	1·6	91 329
Kettle Creek	126	28 070	1223	4·4	26 847
Lackawana River @ Archbald	66	260 370	1470	0·6	258 900
Lackawana River @ Old Forge	90	117 570	1471	1·3	116 099
Little Juniata River @ Spruce Creek	159	138 100	1384	1·0	136 716
Loyalsock Creek	211	21 440	1311	6·1	20 129
Mahantango Creek @ Dalmatia	63	69 500	1652	2·4	67 848
Moshannon Creek	48	318 490	1327	0·4	317 163
Penns Creek	38	53 260	1262	2·4	51 998
Pine Creek @ Ramsey	228	24 400	1214	5·0	23 186
Raystown Branch Juniata River	48	73 100	1657	2·3	71 443
Raystown Branch Juniata @ Saxton	151	87 140	1687	1·9	85 453
Sherman Creek @ Duncannon	72	38 470	1601	4·2	36 869
Sinnemahoning Creek	85	42 280	1295	3·1	40 985
Sinnemahoning Creek @ wqn0418	72	55 320	1281	2·3	54 039
South Branch Codorus Creek	136	38 620	2261	5·9	36 359
Spring Creek @ Bellfont	158	249 860	1272	0·5	248 588
Susquehanna River @ Conowingo, MD	174	174 620?	1494	—	—
Susquehanna River @ Dansville	165	99 100	1373	1·4	97 727
Susquehanna River @ Harrisburg	201	118 530	1385	1·2	117 145
Susquehanna River @ Marietta	166	98 450	1440	1·5	97 010
Susquehanna River @ Sunbury	80	99 470	1314	1·3	98 156
Swatara Creek @ Inwood	71	105 700	1858	1·8	103 842
Swatara Creek @ Middleton	143	96 250	1945	2·0	94 305
Tioga River	77	69 200	1237	1·8	67 963
Tioga River @ Tioga	139	56 150	1230	2·2	54 920
Toby Creek	97	49 090	1421	2·9	47 669
Wapwollen Creek	205	39 760	1446	3·6	38 314
West Conewago Creek @ Rte 181	156	52 880	1981	3·7	50 899
West Branch Codorus	146	50 810	2160	4·3	48 650
West Branch Susq. @ Karthus	50	288 720	1339	0·5	287 381
West Branch Susq. @ Lewisburg	167	106 360	1270	1·2	105 090
West Branch Susq. @ Lockhaven	95	119 810	1290	1·1	118 520
West Branch Susq. @ PR453	134	182 400	1383	0·8	181 017
West Branch Susq. @ Renova	137	145 860	1300	0·9	144 560
West Branch Susq. @ T41B	151	187 450	1393	0·7	186 057
West Branch Susq. @ Williamsport	161	125 140	1265	1·0	123 875
Yellow Breeches Creek	205	124 290	1814	1·5	122 476
Youngwomans Creek	189	14 740	1210	8·2	13 530
Baldwin Creek	164	32 250	1590	4·9	30 660
Benner Run	188	13 550	1260	9·3	12 290
Roberts Run	229	23 680	1306	5·5	22 374
Stone Run	188	14 890	1308	8·8	13 582
W.2	55	29 730	1576	5·3	28 154
WD-38	55	59 420	1605	2·7	57 815
WE-38	42	62 720	1605	2·6	61 115

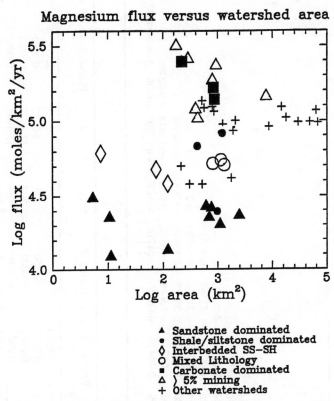

Figure 6. Log Mg flux versus log watershed area for a variety of lithologies. Sandstone watersheds have >0.5 sandstone-dominated formations and no carbonate-dominated formations. Shale watersheds have >0.5 shale-dominated formation. Interbedded SS-SH have >0.5 interbedded sandstone-shale formations. Carbonate watersheds have >0.4 carbonate-dominated formations. Mixed watersheds have approximately 0·2 to 0·3 of all of the sedimentary lithology classes.+ symbols are all other watersheds

EFFECTS OF LITHOLOGY, LAND USE AND WATERSHED SIZE

Pristine watersheds

Magnesium fluxes from watersheds not heavily influenced by mining ($<1\%$ of fractional area as mining) varied from 13 000 to 320 000 moles km^{-2} yr^{-1}; Table IX. Magnesium fluxes generally increase with the sequence of lithological classes hypothesized earlier: sandstone-dominated $<$ interbedded sandstone-shale $<$ siltstone $<$ carbonate-dominated. Low fluxes (13 000 to 35 000 moles km^{-2} yr^{-1}) were associated with sandstone-dominated watersheds defined as watersheds containing greater than 50% sandstone (Figure 6). Relatively pristine watersheds composed predominantly ($>50\%$) of interbedded sandstone-shale yield annual magnesium fluxes from 35 000 to 50 000 moles km^{-2} yr^{-1}. The two mixed lithology watersheds had magnesium fluxes around 55 000 moles km^{-2} yr^{-1}. Shale-dominated watersheds composed of greater than 50% shale had fluxes between 60 000 and 105 000 moles km^{-2} yr^{-1}. The one notable exception to this was Fishing Creek, with a flux of 24 760 moles km^{-2} yr^{-1}. Fishing Creek differs from the other shale watersheds in that all of its shale is comprised of the Mauch Chunk Formation. Since the Mauch Chunk Formation is a non-marine mudstone deposited in an interfluvial setting, it may not contain as much carbonate as other marine shale units.

The highest magnesium fluxes were associated with the predominantly ($>50\%$) carbonate watersheds. Conestoga Creek, which contains only 43% carbonate-dominated formations has a flux of 167 000 moles km^{-2} yr^{-1}. Spring Creek, with 84% carbonate-dominated functions, has a flux 249 000 moles km^{-2} yr^{-1}.

Table X. Statistical results

Area	Mg flux	Urban	Agric.	Forest	Wetland	Mining	Other	SS	SS-SH	SH	Carb.	Schist	Other
						Pearson correlation coefficients (R) 58 total observations							
Mg flux	1												
Urban	0.24770	1											
Agric.	−0.03171	0.25135	1										
Forest	−0.11327	−0.45436	−0.96549	1									
Wetland	−0.08973	−0.42168	0.10148	−0.21617	1								
Mining	0.63263	−0.00666	−0.26379	0.10494	0.02699	1							
Other	−0.06961	−0.02877	0.25181	−0.25116	0.64021	−0.11054	1						
SS	−0.38175	−0.50681	−0.60897	0.71178	−0.23390	−0.17070	−0.07157	1					
SS-SH	0.23864	0.15402	−0.22892	0.11915	0.23555	0.42047	−0.05583	−0.28709	1				
SH	−0.06729	0.01664	−0.40885	−0.37828	−0.16262	−0.05775	−0.00871	−0.47564	−0.37045	1			
Carb.	0.33265	0.42111	0.38216	−0.40993	−0.05673	−0.25354	−0.02466	−0.31066	−0.54720	0.23891	1		
Schist	0.01070	0.02325	0.38145	−0.36062	0.31167	−0.11020	−0.30776	−0.08287	−0.23018	−0.20543	−0.01902	1	
Other	−0.02669	0.01844	0.40995	−0.37477	−0.10315	−0.10601	0.05704	−0.10689	−0.16508	0.07087	0.13618	−0.0798	1
Area	0.04064	−0.06153	−0.01671	0.01984	0.31458	−0.01599	0.44420	−0.06240	0.04716	0.12206	−0.07303	−0.06112	−0.04277

Stepwise regression

Variable	Parameter estimate	Standard error	T for H0 parameter = 0	Probability > T	Partial R^2	C(p)	F	Probability > F
Intercept	937·86	11 803·73	0·079	0·9370				
Urban	−2350·59	1335·37	−1·760	0·0841	0·0150	1·571	3·0985	0·0001
Mining	13 963·00	1631·90	8·556	0·0001	0·4002	61·702	37·3672	0·0001
Interbedded sandstone-shale	121 880·00	29 548·48	4·125	0·0001	0·0680	2·469	13·4998	0·0005
Carbonate	357 869·00	47 231·71	7·577	0·0001	0·2598	13·586	42·0266	0·0001

Adjusted R^2 for model: 0·72.

Overall, yields from carbonate-dominated watersheds are 7–18 times those from sandstone/quartzite-dominated watersheds. This is considerably greater than the 4–8-fold increase calculated by Peters (1984) for carbonate-dominated watersheds.

Statistical analysis

To isolate which lithologies and land uses are the most important factors controlling weathering fluxes, a stepwise regression was performed on the data. A correlation matrix was computed for all variables (Table X). The matrix suggests that a few of the dependent variables are highly correlated with each other. For example, forest and agriculture are strongly negatively correlated and forest and sandstone are weakly correlated. The strong negative relationship between the fractional area of agriculture and forest is arte-factual: the two together represent the majority of most watersheds, so when one is high the other is low. The weak correlation between forest and sandstone is probably because forested areas tend to be associated with ridge tops, which in the Valley and Ridge are commonly composed of sandstone-dominated formations.

Stepwise regressions are sensitive to colinear data sets, so we removed forest as a dependent variable before performing the regression. Table X presents the results. The analysis suggests that the fraction of area composed of mining (m), urban (u), carbonate (c) and interbedded sandstone-shale (s) are the factors that are most significantly correlated with magnesium flux. Equation (2) presents the model that best explained the magnesium flux:

$$\text{Flux}_{Mg} = 14\,000m + 358\,000c + 122\,000s - 2350u \tag{2}$$

The model explained 72% of the total variance in the data, with mining (m) and carbonate (c) being the most important variables. A plot of magnesium flux, fraction of watershed area composed of carbonate-dominated lithologies and percentage of area composed of mining is presented in Figure 7, together with a

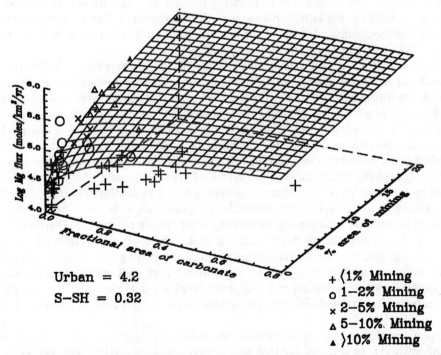

Figure 7. Log Mg flux plotted against fraction of carbonate-dominated formations and % area of mining. The surface is a representation of the model [Equation (2)]. This surface was calculated under the assumption that the % area of urban and fraction of interbedded sandstone-shale is 4·2 and 0·32 (average for the entire study area), respectively

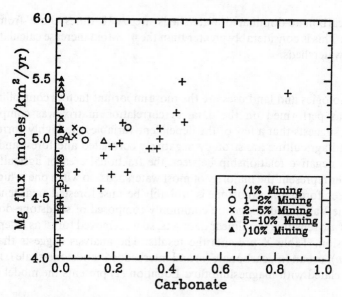

Figure 8. Log Mg flux versus fraction of (sedimentary) lithology as a function of % mining. (a) Sandstone-dominated. (b) Interbedded sandstone-shale. (c) Shale-dominated. (d) Carbonate-dominated

surface representing the model. The surface in Figure 7 was calculated by fixing the fractional area of interbedded sandstone-shale and percentage area of urban to be equal to the average for all watersheds.

Figures 7 and 8 show a strong exponential increase in magnesium flux as carbonate increases for watersheds with less than 1% mining. The relationship between the presence of carbonate and the potential for magnesium weathering partially explains why the magnesium fluxes increase with interbedded sandstone-shale < shale-dominated < carbonate-dominated. The amount of carbonate-bearing minerals generally increases in the same sequence (Holland, 1978).

Magnesium fluxes associated with watersheds having >1% mining were much higher than the fluxes from watersheds with <1% mining having similar lithologies (Figure 8). The calculations illustrate the significance of mining on watersheds. Interbedded sandstone-shale watersheds with acid mine drainage yielded magnesium fluxes ranging from 160 000 to 317 000 moles km^{-2} yr^{-1}. This represents a 6–10-fold increase in the flux of magnesium over interbedded sandstone-shale watersheds with <1% mining. Beech Creek, a predominantly sandstone watershed with 4·5% mining, has a magnesium flux of 119 500 moles km^{-2} yr^{-1}. If this watershed had a pre-mining magnesium yield equivalent to the other sandstone watersheds, mining has increased magnesium fluxes almost six times in this watershed. This is similar to the sevenfold increase in flux calculated by Feltz and Wark (1962) for a strip mining region in the North Branch of the Potomac River.

We interpret the high magnesium flux associated with watersheds with a high % (>1%) of mining to be caused by enhanced physical and chemical weathering from mining practices. Stallard and Edmond (1983) suggest that the highest denudation fluxes occur in areas where physical weathering can keep up with chemical weathering. Mining practices accelerate physical weathering by stripping soil and breaking up bedrock. This will increase the surface area of unweathered rock with which the soil water is in contact. Exposure of sulfide minerals by mining can also promote acidity in the water. The common application of limestone to remediate acid mine drainage will also increase fluxes by providing a source of easily weathered material.

The regression revealed no significant relationship between magnesium fluxes and the other land use categories. Other workers have postulated that agricultural practices could increase cation fluxes (e.g. Paces, 1983). Although acid rain has been demonstrated by April et al. (1986) to increase significantly the cation denudation rates from watersheds, no attempt was made to estimate the amount of acid rain deposition occurring for the watersheds in this study. Further work should consider the effect that acid rain deposition might have on magnesium fluxes.

Effects of watershed scale

The data suggest that the effect of increasing scale is a blending of high flux (carbonate-dominated, high mining) and low flux (sandstone, interbedded sandstone-shale) sub-watersheds. The overall Susquehanna River has a magnesium flux of 100 000 moles km^{-2} yr^{-1} (Figure 6). The fraction of carbonate-dominated and interbedded sandstone-shale formations in the Susquehanna River Basin (Pennsylvania) is 0·09 and 0·30, respectively. If there were no mining and urban land use classes present in the watershed (i.e. if it were pristine), Equation (2) predicts that the flux from the Susquehanna River Basin would be 69 700 moles km^{-2} yr^{-1}. The observed flux is much greater than one would expect for the fraction of carbonate-dominated and interbedded sandstone-shale formations present in the basin. It seems likely that mining is responsible for this difference.

Mass balances

A fundamental test of the feasibility of the mass balance method at larger scales is whether we can account for all the magnesium flux leaving a large watershed by adding up the fluxes of sub-watersheds nested within it. The nested hierarchy (Figure 2) allows us to perform this test on fourteen watersheds. Of these, six have sub-watersheds that make up better than 95% of the larger watershed. Relative errors (%) in them ranged from 45% above to 16% below the magnesium flux for the larger watershed (Table XI).

The question, then, is whether or not the observed difference is an artefact of the error inherent in this analysis, a water balance problem or an unknown source or sink of magnesium. Errors (%) in water balance range from −42% to 56%. Errors were generally in the same direction as the magnesium flux errors except for the Susquehanna River at Marietta. This suggests that the source of the error in the magnesium flux calculation is caused in part by errors in estimating runoff. The percentage water budget error is much higher, however, than the error in measuring discharge suggested by the USGS for its stream gauges (5–15%). Part of the error may be owing to the manner in which annual runoff was calculated (the sinusoidal approximation). Other factors that may contribute to the errors calculated here are spatial variations in dry deposition, analytical error and errors in estimating watershed area. It is also possible that stream water may be bypassing the stream gauge or that significant groundwater inputs exist. The results seem to indicate that problems in measuring discharge and estimating elemental fluxes in larger rivers may make mass balance calculations inaccurate at larger scales.

CONCLUSIONS

The manner in which fluxes are extrapolated from concentration and discharge data plays an important role in determining the accuracy of flux estimates. Where data are scant and obtained over short periods, mass balance models should perhaps be viewed as qualitative measures of weathering fluxes. Integration of a best fit sinusoid through sparse runoff and concentration data provides a reasonable way of estimating flux from data sets that are abundant but sporadically sampled.

The flux of magnesium is a strong function of the amount of carbonate present in the watershed. This is apparent in the strong correlation between the fractional area of carbonate coverage and magnesium flux, as well as in the increase in flux in the sequence; sandstone < interbedded sandstone-shale < shale < carbonate. The observation that magnesium flux increases with watershed size is also probably related to the increasing likelihood of encountering carbonate lithologies at larger scales. Estimates of the fraction of carbonate in a watershed made by computing the fraction of carbonate-dominated formations may not be accurate enough to capture the effects of small amounts of carbonates present in other lithologies.

Land use has had a strong impact on weathering fluxes in the Susquehanna River Basin. This is thought to be primarily a result of mining practices. Mining has probably increased the flux of magnesium from the Susquehanna River Basin at least 50% above its 'pristine' level. The effects of mining and carbonate-rock

Table XI. Percentage area composed of sub-watersheds and % relative error of magnesium and water flux between sub-watersheds nested within a watershed. A positive relative error indicates that the main watershed has higher flux than the sum of all the component sub-watersheds. A negative relative error indicates that the sum of the fluxes from component sub-watersheds is greater than the main watershed. ID numbers refer to the main watershed (Figure 2)

Stream junction station name	ID #	Area Sub-watersheds (% difference)	Water flux Main watershed (km³ yr⁻¹)	Sub-watersheds (km³ yr⁻¹)	Error (%)	Mg flux Main watershed (millions moles yr⁻¹)	Sub-watersheds (millions moles yr⁻¹)	Error (%)
Sinnemahoning Creek @ wqn0418	28	102	0·65	0·69	−5·51	133·41	91·69	+31·27
Susquehanna River @ Conowingo, MD	31	97	54·98	24·27	+55·86	12 284·66	6777·09	+44·83
Susquehanna River @ Harrisburg	33	95	33·13	21·58	+34·88	7407·38	5727·87	+22·67
Susquehanna River @ Marietta	34	99	23·95	34·17	−42·65	6635·01	7711·24	−16·22
Susquehanna River @ Sunbury	35	99	16·95	24·10	−42·23	4720·22	4771·59	−1·09
West Branch Susq. @ Lewisburg	45	95	9·26	6·77	+26·83	1888·32	1894·06	−0·30

proportion on weathering were so strong that the effects of other land use practices could not be discerned in this study.

Mass balance calculations of magnesium flux and discharge carried out in large watersheds showed large inconsistencies. These inconsistencies are thought to be owing to errors in measuring discharge and estimating watershed fluxes. Improvements in estimating these variables are necessary before we can accurately apply mass balance techniques to larger scale watersheds.

REFERENCES

Anderson, J. R., Hardy, E. E., Roach, J. T., and Witmer, R. E. 1976. 'Land use and land cover data', *US Geol. Surv. Prof. Paper*, **964**, 28.

April, R. R., Newton, R., and Coles, L. T. 1986. 'Chemical weathering in two Adirondack watersheds; past and present-day rates', *Geol. Soc. Am. Bull.*, **97**, 1322–1328.

Berner, K. B. and Berner, R. A. 1987. *The Global Water Cycle*. Prentice-Hall, Englewood Cliffs. p. 397.

Bevington, P. R. and Robinson, D. K. 1992. *Data Reduction and Error Analysis for the Physical Sciences*. McGraw-Hill, New York.

Bluth, G. J. S. 1990. 'Effects of paleogeography, chemical weathering and climate on the global geochemical cycle of carbon dioxide', *Ph.D. Thesis*, The Pennsylvania State University, University Park.

Bluth, G. J. S. and Kump, L. 1994. 'Lithologic and climatologic controls of river chemistry', *Geochim. Cosmochim. Acta*, **48**, 2341–2359.

Cleaves, E. T., Godfrey, A. E., and Bricker, O. P. 1970. 'Geochemical balance of a small watershed and its geomorphic implications', *Geol. Soc. Am. Bull.*, **81**, 3015–3032.

Cuff, D. J. 1989. *The Atlas of Pennsylvania*. Temple University Press, Philadelphia.

Dann, M. S., Lynch, J. A., and Corbett, E. S. 1986. 'Comparison of methods for estimating sulfate export from a forested watershed', *J. Environ. Water Qual.*, **15**, 140–145.

Dewalle, D. R., Swistock, B. R., Dow, C. L., Sharpe, W. E., and Carline, R. F. 1993. 'Episodic response project — northern Appalachian Plateau site description and methodology', *EPA/600/R-93/190*. Environmental Resources Research Institute, Penn State University.

Drever, J. I. 1988. *The Geochemistry of Natural Waters*. Prentice-Hall, Englewood Cliffs. 437 pp.

Feltz, H. R. and Wark, J. W. 1962. 'Solute degradation in the Potomac River basin', *US Geol. Surv. Prof. Paper*, **450-D**, D186–D187.

Gibbs, R. J. 1970. 'Mechanisms controlling world water chemistry', *Science*, **170**, 1088–1090.

Goudie, A. 1970. 'Input and output considerations in estimating rates of chemical denudation', *Earth Sci. J.*, **4**, 59–65.

Heath, R. C. 1984. 'Ground-water regions of the United States', *US Geol. Surv. Water-Supply Paper*, **2242**, 78.

Holland, H. D. 1978. *The Chemistry of the Atmosphere and Oceans*. Princeton University Press, Princeton, New Jersey.

Janda, J. J. 1971. 'An evaluation of procedures used in computing chemical denudation rates', *Geol. Soc. Am. Bull.*, **82**, 67–80.

Johnson, A. H. 1979. 'Estimating solute transport in streams from grab samples', *Wat. Resour. Res.*, **15**, 1224–1228.

Katz, B. G., Bricker, O. P., and Kennedy, M. M. 1985. 'Geochemical mass-balance relationships for selected ions in precipitation and stream water, Catoctin mountains, Maryland', *Am. J. Sci.*, **285**, 931–962.

Langbein, W. B. and Dawdy, D. R. 1964. 'Occurrence of dissolved solids in surface waters in the United States', *US Geol. Surv. Prof. Paper*, **501-D**, D115–D117.

Lasaga, A. C. 1984. 'Chemical kinetics of water-rock interactions', *J. Geophys. Res.*, **89**, 4009–4025.

Likens, G. E., Bormann, F. H., Pierce, R. S., Eaton, J. S., and Johnson, N. M. 1977. *Biogeochemistry of a Forested Ecosystem*. Springer-Verlag, New York. 146 pp.

Lindberg, F. A. 1984. 'Northern appalachian region correlation of stratigraphic units of North America (COSUNA) project', *Correlation Chart Series*. Publ: AAPG Book Store, Tulsu, Oklahoma.

Lynch, J. A., Horner, K. S., Grimm, J. W., and Corbett, E. S. 1993. 'Atmospheric deposition: spatial and temporal variations in Pennsylvania — 1992', *ER9304A*. Environmental Resources Research Institute, Penn State University. 95 pp.

Lystrom, D. J., Rinella, F. A., Rickett, D. A., and Zimmermann, L. 1978. 'Regional analysis of the effects of land use on stream water quality, methodology, and application in the Susquehanna River Basin, Pennsylvania and New York', *US Geol. Surv. Wat. Resour. Invest.*, **78-12**, 1–60.

Meade, R. H. 1969. 'Errors in using modern stream-load data to estimate natural rates of denudation', *Geol. Soc. Am. Bull.*, **80**, 1265–1274.

Meybeck, M. 1976. 'Total mineral dissolved transport by world major rivers', *Hydrol. Sci. Bull.*, **2**, 265–284.

Meybeck, M. 1987. 'Global chemical weathering of surficial rocks estimated from river dissolved loads', *Am. J. Sci.*, **287**, 401–428.

Narr, W. and Suppe, J. 1991. 'Joint spacing in sedimentary rocks', *J. Struct. Geol.*, **13**, 1037–1048.

Paces, T. 1983. 'Rate constants of dissolution derived from the measurements of mass balance in hydrological catchments', *Geochim. Cosmochim. Acta*, **47**, 1855–1863.

Paces, T. 1986. 'Rates of weathering and erosion derived from mass balance in small drainage basins', in Colman, S. M. and Dethier, D. P. (Eds), *Rates of Chemical Weathering of Rocks and Minerals*. pp. 531–550. Academic Press, Orlando, FL.

Peters, N. E. 1984. 'Evaluation of environmental factors affecting yields of major dissolved ions in streams in the United States', *US Geol. Surv. Water-Supply Paper*, **2228**.

Probst, J. L., Nkounkou, R. R., Krempp, G., Bricquet, J. P., Thiebaux, J. P., and Olivry, J. C. 1992. 'Dissolved major elements exported by the Congo and the Ubangi rivers during the period 1987–1989', *J. Hydrol.*, **135**, 237–257.

Shertzer, R. H. and Schreffler, T. L. 1993. 'Pennsylvania's surface water quality monitoring network (WQN)', *Commonwealth of Pennsylvania Department of Environmental Resources Publication*, Vol. 33, DER #636-9/91. Published by the Pennsylvania Dept. of Environmental Protection, Harrisburg, PA.

Stallard, R. F. and Edmond, J. M. 1983. 'Geochemistry of the Amazon 1. Precipitation chemistry and the marine contribution to the dissolved load at the time of peak discharge', *J. Geophys. Res.*, **86**, 9844–9858.

Velbel, M. A. 1985. 'Geochemical mass balances and weathering rates in forested watersheds of the Southern Blue Ridge', *Am. J. Sci.*, **285**, 904–930.

13

CHEMICAL MODELLING ON THE BARE ROCK OR FORESTED WATERSHED SCALE

R. L. BASSETT

Department of Hydrology and Water Resources, The University of Arizona, Tucson, AZ 85721, USA

ABSTRACT

The simulation of weathering, solute distribution or acidification at the catchment scale is predominantly done with either a mass balance or process level model. The former redistributes total elemental concentrations between known points with measured total concentration, but does not explicitly include catchment hydrology. The latter includes compartmental hydrological models and detailed descriptions of spatially averaged chemical reactions. Interestingly, the model applications tend towards hydrologically different watershed structures: mass balance modelling favours bare rock watersheds similar to the Apache Leap Research Site, whereas process level models are applied most often to forested watersheds, among which the Hubbard Brook Experimental Forest is an example. Although constrained either by mineral or water compositions on the one hand, or calibrated against stream or lake water chemistry on the other, both approaches basically fit parameters to the geochemical circumstances of the specific watershed of interest. Limited success is attained if the hydrological conditions remain within the circumstances of the parameter fitting. The principal differences in model formulation and approach to mass balance modelling are discussed. Without advancements in model calibration and rigorous model testing, and the development of methods for optimizing the important reactions and pathways, the transportability of models between watersheds or the simulation of extreme events will continue to be inadequate.

INTRODUCTION

Chemical modelling on the watershed scale in recent years has been principally driven by two foci of scientific inquiry: (1) catchment response to acidification from atmospheric sources; and (2) mass balance studies of chemical weathering. Certainly, modelling of chemical processes in watersheds extends beyond these two principal applications into many other subdisciplines, for example, elemental redistribution in watershed soils, ecosystem models, infiltration, pedogenesis and stream–subflow quantitation; however, acidification and weathering are clearly the dominant issues at present.

The term 'chemical modelling' is frequently used in the literature, and has become an ambiguous expression because of the large number of chemical modelling approaches, the range of complexity in mathematical formulations and the variety of assumptions inherent within any given model (Bassett and Melchior, 1990; Mangold and Tsang, 1991). Chemical modelling approaches include: static material balance, mass balance with mass transfer, mass balance with mass transport and statistical analysis. The application of a chemical model may be constrained to the single solute case, or may be robust enough to include multicomponent and multicompartment systems. Finally, the more complex chemical models may consider more than the basic inorganic solute compositions and may also include atmospheric gases and particulates, organic compounds and both radioactive and stable isotopes. Distinctions between these principal mass balance model types, as well as the defining equations are described in subsequent sections.

Rigorous calibration and verification studies of watershed models are rarely done. Consequently, the utility of a watershed model for a potential user depends entirely on his ability to evaluate the model based on existing field studies. This obviously restricts evaluations to either a comparison of a few models for the same watershed, or the comparison of a single model for several watersheds. Even though such model comparisons and evaluations can be made, and generally it is agreed that this is a necessary and useful exercise, the evaluation will unfortunately still be superficial because of unevenness of model assumptions and the wide variety of field circumstances. Variability at the field scale is unavoidable; calibration must be done at a smaller scale with repeatable circumstances in order to gain confidence in the predictability of chemical throughput and the transportability of chemical models between different watersheds.

The literature related to chemical modelling is large and the applications numerous. For example, the mobility of elements in a watershed has been simulated on a multitude of *length scales*, ranging from ionic flux through centimetres of snowpack to chemical changes over kilometres of stream channel. Similarly, *time frames* for periods of simulations vary from static *snapshot* characterizations, using a mass balance–mass transfer approach to multiyear *time series* analysis using process level models calibrated with chemical hydrographs.

Regarding chemical and mathematical details, there are both implicitly and explicitly stated assumptions that a user must examine for any chemical modelling application. For example, one should confront the implicit assumption that reactions are rapid and that local equilibrium circumstances prevail. Unfortunately, equilibrium is generally assumed and thermodynamic equilibrium constants are used for many of the computations in watershed chemical models, rather than employing rate-limited reactions, or approximating non-thermodynamic processes with purely empirical constants. Similarly, explicitly stated assumptions require scrutiny. As an example, many models come with a predetermined input structure. How does one justify the number of components explicitly chosen for a specific model? The specific number and type of components directly affects the complexity of the model, owing to the fact that few components are truly conservative, and therefore, essentially all components require a definition of the chemical processes responsible for the variation in that component, and distribution of each within the watershed.

The complexity of many watershed chemical models fosters an 'off the shelf approach' by users to avoid writing new models. However, the criteria employed in selecting a model for use in a given watershed, and how well the chosen model fits with the hydrological setting, are generally not clearly articulated. The two extreme 'consequent evils' to be avoided when using complex watershed models are: (1) avoid excessive expenditures for predetermined chemical components required as input for a given model, but which may have little significance in the watershed being investigated (data-driven model); and (2) avoid a complex model if too little data are available and numerous approximations and estimates will be required to run the code (model-driven project).

The simplest case of mass balance modelling is the distribution or cycling in a watershed of a single element or component, e.g. sulfate, aluminium, calcium or various nutrients. This has been simulated by ignoring the effect of other elements or approximating all other reactions as lumped parameters (Likens *et al.*, 1977; Bischoff *et al.*, 1984). At the other extreme, the most complex case is the multicompartment and multicomponent process model which includes, essentially: (1) all major ions, nutrients and gases; (2) capacity terms such as alkalinity, acid-neutralizing capacity and ion exchange capacity; (3) a mathematical designation explicitly defining the chemical processes such as aqueous speciation, ion exchange, adsorption, solubility, decomposition, nitrification and rate-limited reactions; and (4) coupling of the chemical process to a separately calibrated hydrological model.

This value of the increased complexity in modelling is possibly the most contentious issue in the catchment modelling literature, with proponents for multiparameter and multicomponent approaches set apart from those for lumped parameter and simplified mass balance models. It is the opinion here that one should be as objective as possible regarding this issue. In the most rigorous sense, neither approach at present provides true, predictive capability. Rather, both approaches essentially fit the equations in the model to each catchment being investigated with a non-unique set of parameters and consequently provide historical fits to monitored events.

The objective of this summary paper is to discuss the principal approaches, review the general equations and assumptions, and identify areas of promising advancement. No attempt is made to evaluate critically the validity of any approach; however, an effort is made to clarify the modelling terminology and to explain the modelling strategies used by research groups in their efforts to simulate the cycling of chemical components in watersheds.

On a broader environmental scale beyond the watershed, environmental regulators and scientists are discussing similar issues related to the use of mass balance models for evaluating the fate of pollutants and the consequent human exposure (Renner, 1995). Chemical mass balance models are being increasingly considered because of their simplicity of operation and the ability to distribute chemicals between different 'media', e.g. air, water and soil phases. These multimedia models are increasingly favoured by regulators because they lend themselves to rapid intercomparison studies that can identify inconsistencies (Renner, 1995). Even though the pressure is on for simpler mass balance models for cost-effective regulations and efficient risk analysis, the counter arguments of oversimplification and lack of validation for the models are strong and are remarkably similar to the issues regarding watershed scale research.

One additional observation is made here as a result of the analysis of model applications, and that is the now obvious circumstance that the choice of modelling methodology is strongly influenced by the hydrological characteristics of the watershed. In other words, what most influences, if not determines, the modelling approach is the type and quantity of data one can most readily obtain from a specific watershed. The net result is that almost all major watershed modelling activities are directed towards only two distinctively different watershed hydrogeological types, here referred to as the bare rock watershed type (BRWT) and the forested watershed type (FWT). In general, chemical redistribution within these two principal watershed types has been simulated with significantly different mathematical formulations. These two cases are best described as either: (1) 'static' material balance calculations (BRWT); or (2) more integrated 'process driven' mass balance approaches (FWT). This classification of two watershed types will be used here in discussing and characterizing the modelling assumptions.

WATERSHED HYDROGEOLOGICAL TYPES

One can take a broad operational view of the watershed classification scheme, which, in general, is difficult to describe anyway. For example, a watershed is generally classified by its climatic, geological, hydrological, vegetative, land use and topographic characteristics (Woolridge, 1990; Gordon et al., 1992). The broad view, however, would note that the majority of the watershed modelling studies described in the literature are, surprisingly, limited to the bare rock catchments frequently studied in the western United States and the heavily vegetated or forested experimental watersheds that have emerged in the eastern United States (Figure 1). Further, many of the investigations of acid sulfate precipitation have been done in regions that contain significant woody or agricultural vegetation. These catchments often have the following similar characteristics that can affect chemical cycling: significant soil cover, trees, organic decomposition, throughfall, seasonal snow cover, perennial streams, lakes and wetlands, advanced soil development and shallow water table. Many of the research groups performing these investigations have large institutional commitments to continuous monitoring and data collection such as the Hubbard Brook Experimental Forest (Likens et al., 1977), or the ILWAS Project (Gherini et al., 1985; Goldstein et al., 1985) with participants from several institutions. Certainly, a watershed like Hubbard Brook would serve only as one example of the many catchment sites in the broad category of forested watersheds, but will at least be useful in demonstrating the extreme contrast between forested and bare rock circumstances. For illustration, a photograph of a typical stream landscape at Hubbard Brook is shown in Figure 2a.

Complex chemical models designed for the forested watershed sites generally combine elemental mass balance constraints with time series hydrological and chemical data, principally because of high precipitation, permanent lakes and perennial stream flow. This use of integrated data is referred to as 'process level mass balance modelling' and has become the dominant modelling approach in acidification studies for at least four main reasons. First, the regions most affected by acidification are in the eastern United States and are *dominated by forested watersheds* receiving high precipitation, are drained by gaining streams and have

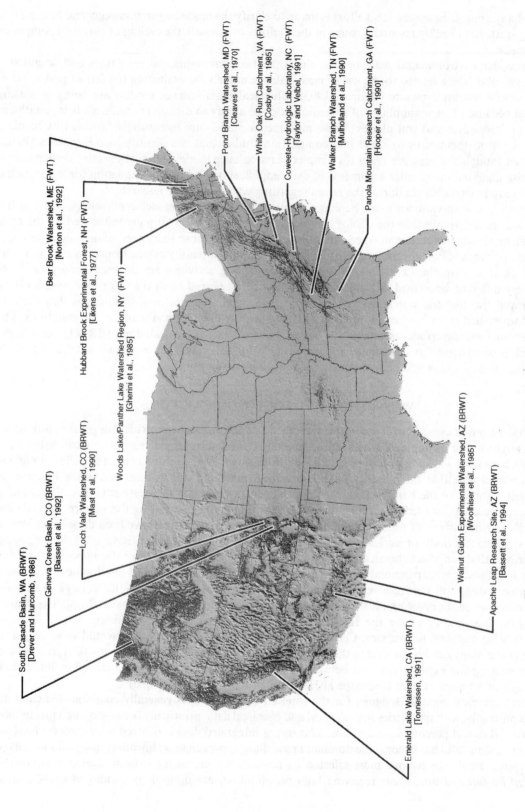

Figure 1. Locations of the major watershed modelling studies described in the literature

Bear Brook Watershed, ME (FWT)
[Norton et al., 1992]

Hubbard Brook Experimental Forest, NH (FWT)
[Likens et al., 1977]

Woods Lake/Panther Lake Watershed Region, NY (FWT)
[Gherini et al., 1985]

Pond Branch Watershed, MD (FWT)
[Cleaves et al., 1970]

White Oak Run Catchment, VA (FWT)
[Cosby et al., 1985]

Coweeta-Hydrologic Laboratory, NC (FWT)
[Taylor and Velbel, 1991]

Walker Branch Watershed, TN (FWT)
[Mulholland et al., 1990]

Panola Mountain Research Catchment, GA (FWT)
[Hooper et al., 1990]

South Casade Basin, WA (BRWT)
[Drever and Hurcomb, 1986]

Geneva Creek Basin, CO (BRWT)
[Bassett et al., 1992]

Loch Vale Watershed, CO (BRWT)
[Mast et al., 1990]

Walnut Gulch Experimental Watershed, AZ (BRWT)
[Woolhiser et al., 1985]

Apache Leap Research Site, AZ (BRWT)
[Bassett et al., 1994]

Emerald Lake Watershed, CA (BRWT)
[Tonnessen, 1991]

Figure 2. (a) Hubbard Brook, a typical stream landscape. (b) The Apache Leap Reasearch Site, a bare rock watershed

shallow water tables. Consequently, the hydrological component of the system can be monitored and is easily integrated into the analysis. Secondly, changes in elemental concentration are readily noticeable in surface water, and specific *solute flow paths become important* because of different reaction sequences in soil, ground water or biological compartments encountered. Thus, there is at least the perceived requirement for increased chemical complexity. Thirdly, in such regions the *soil, vegetation and even canopy cover are quite important* components in the chemical cycling, and the processes inherent in elemental attenuation and release in these subsystems are often available and are being investigated collaboratively by other research groups, and can be integrated into the elemental cycling studies. Finally, a single metric is available for gauging the success for many modelling efforts and that is a *direct comparison of simulated chemical hydrographs* with the continuous record of stream water chemical composition. As a result, many research groups in the United States, Canada and elsewhere, that are dependent on the data from forested watersheds (FWT) have relied on the more complex process level models used to describe the multiplicity of reaction pathways. The watershed modelling literature is quite large and only a few of the US research watershed sites are identified here (Figure 1). Similarly, the modelling of forested watersheds outside the US is equally active, with a well-established history; a few of the more commonly referenced field sites are Afon Hafron Catchment, Wales (Christopherson *et al.*, 1990); Allt a' Mharcaidh Catchment, Scotland (Stone *et al.*, 1990); Birkenes, Norway (Wright, 1982); Duddon Catchment, England (Tipping, 1989); Elbe and Trnavka River Basin, Czech Republic (Paces, 1983); Funäsdalen, Nolsjön, and Assman Basins, Sweden (Bergström *et al.*, 1985); Harp Lake, Ontario, Canada (Rustad *et al.*, 1986); Turkey Lakes Watershed, Ontario, Canada (Bobba and Lam, 1988).

The second archetypal watershed is the bare rock watershed type (BRWT). This is a common catchment circumstance, especially in the high alpine regions of the western United States, and as a group, has antithetical characteristics to the forested watershed. Bare rock watersheds often contain significant areas of exposed and fractured bedrock, sparse soil, minimal vegetation, seasonal snow cover, ephemeral or intermittent streams, deeper permanent water tables, significant stretches of losing streams and minimal long-term monitoring is done because many locations are remote. As one would expect, much less time series data are available because of the intermittent stream flow conditions. Continuous monitoring of the flashy hydrological events has not been common in these regions, but has been done, for example, at Emerald Lake (Kattelman and Elder, 1991; Tonnessen, 1991) and certainly will be pursued more frequently with advanced technology for remote sites.

Detailed rock weathering studies are more commonly carried out in the bare rock watersheds, in which complications from vegetation are minimal, the geological terrain is visible and weathering products are accessible. Statistical analysis of the data from the National Surface Water Survey (Stauffer, 1990; Stauffer and Wittchen, 1991) and the mass balance studies using field data from specific watersheds (Roth *et al.*, 1985; Drever and Hurcomb, 1986; Barron and Bricker, 1987; Mast *et al.*, 1990; Williams and Melack, 1991; Bassett *et al.*, 1992) indicated that alpine lake water and stream water chemistry is dominated by classical weathering reactions such as silicate hydrolysis and carbonate mineral dissolution. The weathering process is not mitigated significantly by the factors often seen in forested watersheds, such as soil ion exchange processes and buffering effects from bioaccumulation (Velbel, 1985).

Simulations of the FWT setting favour level modes, with defined individual reaction pathways, batch-type reactor reservoirs and predictions of the stream chemical hydrographs. In contrast, the BRWT environments are more compatible with a static mass balance approach. Surface water flow events are episodic and the static mass balance concept relies on a 'snapshot' description of the elemental mass distribution at one point in time, or an average value over a defined increment of time. Total mass involved in the reactions depends on average precipitation and average stream or ground water compositions as end-members. It is noteworthy that many potential sites for hazardous waste storage facilities, low level radioactive waste facilities and even potentially the high level waste repository at Yucca Mountain, will be constructed amongst these bare rock and arid bare soil alluvial watersheds. The appeal of these western geographical settings, is the paucity of water, minimal vegetation, little anthropogenic land use and low infiltration rates (Conrad, 1993).

The Apache Leap Research Site (Figures 1 and 2b) is an example of a bare rock watershed, located on Miocene-age tuft, east of Superior, Arizona, and is the research site cosponsored by the US Nuclear Regulatory Commission and the University of Arizona (Bassett *et al.*, 1994; 1996). This instrumented watershed is analogous to conditions at the proposed nuclear repository site at Yucca Mountain, Nevada, and serves as an example of the 'typical' bare rock watershed model for this discussion.

Most acidification studies have been done in environments similar to either FWT or BRWT, or fit hydrologically in the spectrum between these two generalized models. Figure 3 is a representation of these

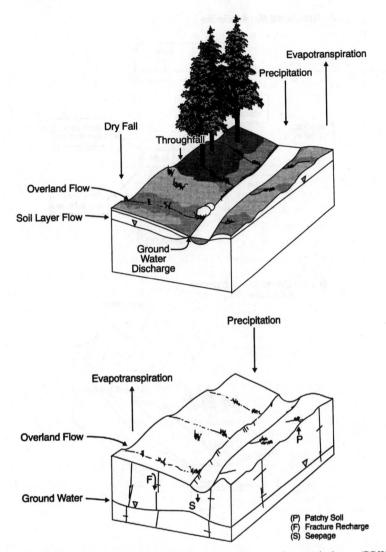

Figure 3. (Top) The forested type (FWT). (Bottom) The bare rock watershed type (BRWT)

two broad watershed types and this depiction, however oversimplified, will be used here in subsequent discussions about chemical modelling and applications. Note in particular, the hydrological distinctions. In the FWT, catchments are drained predominantly by gaining streams and stream flow is generally perennial and fed by surface runoff, drainage from soil horizons or groundwater discharge. Consequently, the specific solute flow path is important because of different reaction scenarios, and numerous schemes to separate the hydrograph into compartments have emerged using either chemical or isotopic tracers (commonly δD and $\delta^{13}C$).

In the BRWT the hydrological characteristics are generally quite different. Rather than the stream flow increasing from groundwater discharge, local groundwater may be below the stream bed, the unsaturated zone is often quite extensive and the streams are intermittent or ephemeral, and lose water through seepage. Rain storms and snowmelt result in extreme runoff events in bare rock watersheds, often producing flashy discharge that scours dry alluvial stream beds and rapidly recharges fracture networks

A. Distributed Mass Balance

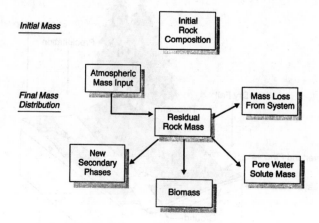

B. Flux Driven Mass Balance
(After Paces, 1983)

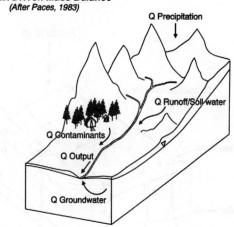

C. Element Cycle Mass Balance with Flux Between Compartments

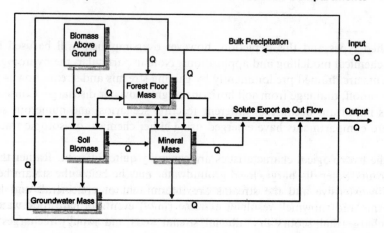

D. Reaction Pathway Mass Balance
(After Bassett et al., 1992)

Optional Path for the Formation of Clay Minerals

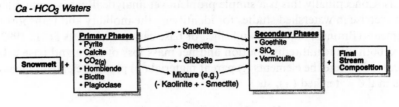

Optional Path for the Dissolution of Secondary Silicates

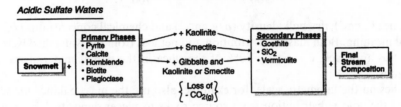

E. End-member Mass Balance
(After Hooper et al., 1990)

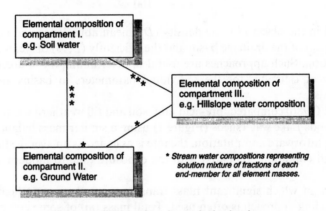

Figure 4. Models of catchment hydrochemistry

(Bassett *et al.*, 1994; Woodhouse *et al.*, 1995a,b). Process level models are difficult to apply to the BRWT without having the perennial flow for comparison. Consequently, static mass balance approaches dominate the literature.

GOVERNING EQUATIONS

Mass Balance

Models of catchment hydrochemistry typically include some form of elemental m terial balance. The simplest accounting is a distribution of some original or historical elemental mass balance over all 'chemical compartments' that represent the current system of interest, e.g. soil, minerals, surface or groundwater, biomass, etc. (Figure 4a). Even though the process is simple, important assumptions are inherent: (1) homogeneous concentrations within a given compartment; (2) steady-state system without material production or loss during the time frame of the assessment; and (3) the total number of significant chemical compartments has been identified.

An example of this material balance accounting is the redistribution of elements from parent rock to secondary minerals. Conceptually this is a simple problem yet analytically it is quite difficult; nevertheless, it is an important exercise in watershed studies for identifying the mobility and transport of specific species, e.g. rare earth elements (Braun et al., 1993) and silica or base cations (Merritts et al., 1992). Chadwick et al. (1990) investigated both elemental redistribution and net mass loss of silica and base cations in a carefully done mass balance study. Here the elements (i) parent material (p) plus the elemental mass flux (m_i) were set equal to the mass in the watershed (w), such that:

$$\frac{V_p \rho_p C_{i,p}}{100} + m_{i,\text{flux}} = \frac{V_w \rho_w C_{i,w}}{100}$$

In this case, volume (V, cm^3), dry bulk density (ρ, g cm^{-3}) and chemical concentration (C, wt%) of element i, have their usual meaning. Both mass redistribution and net loss from the watershed were determined to be significant using this approach, e.g. nearly 50% of the silica in the parent material was leached from the surface soil horizons.

Watershed studies on the kilometre scale, for example, also use the mass balance model approach. Loen (1992) determined the gold redistribution from source rocks to placer deposits. The material balance was computed as follows:

$$P = \frac{(D_x A_d \cdot C \cdot T \cdot R) \cdot E}{100}$$

Here, the mass of gold in the placer (P), the density (D), mean abundance and denudation rate (R) of the source rocks, the area (A_d) of the drainage basin and the efficiently (E,%) of weathering are used to describe the material redistribution. Such approaches are useful in determining the sources of the elements and for evaluating the validity of estimates for the individual parameters in basins and watersheds for which significant field data exist.

Similarly, April et al. (1986) compared 'pre-acid rain' soil and till weathering rates with current rates in the Panther Lakes and Woods Lake watersheds (Figure 1) using a similar mass balance approach. The element redistribution is a straightforward computation, the rate is assigned as an average time period for till and soil existence in the pre-acid condition compared with the exposure time to modern atmospheric acidification (Figure 4a.)

For the circumstance in which significant mass transfer occurs between 'chemical compartments' in a watershed, the mass cycling approach is often used. Total mass (m) of some component (i) in the system is changing with time (t) and can be accounted for by including the production (P) or consumption C terms as well as input and output flux terms (F_i and F_0):

$$\frac{dm_i}{dt} = F_i - F_0 + P - C$$

This allows for mass transfer between compartments and is the approach used in representations of elemental cycling on the global scale (Garrels et al., 1975), and is extended to elemental cycling within watersheds. This method is generally restricted to cycling of only a few elements at a time in order to focus on elemental source and mechanisms of redistribution especially, for acid precipitation-related investigations. As examples, in an early study, Likens et al. (1977) defined nutrient cycles for the Hubbard Brook Experimental Watershed (Figure 4b). Bischoff et al. (1984) examined sulfur and nitrogen cycles and were able to identify sources and establish fluxes for these elements for the entire eastern United States. Paces (1983) used the flux of elements involved in the weathering process to establish mineral dissolution rates assuming the mass balance for the elements in the watershed was defined (Figure 4c). Lynch and Corbett (1989) investigated the export rate of sulfate from the Leading Ridge Experimental Watershed, Central Pennsylvania, using the best method selected from the seven variations of the mass balance approaches presented by Dann *et al.* (1986) for the same watershed. The results of this analysis revealed that discrepancies between input and export rates of sulfate could be explained by the dryfall contribution. In a related study, the mass balance calculations of Fu and Winchester (1994a,b) compared the import/export approach with a multivariate analysis to identify the atmospheric versus terrestrial sources of nitrogen in rivers draining three watersheds. Similarly, the mass balance approach was used to advance the understanding of aluminium cycling (Cronan *et al.*, 1990). The interaction of aluminium with biomass involves a multiplicity of reactions and their identity provided a better definition of the flow system. As these authors point out, as the complexity of the chemical interactions and flow system increases, the inadequacy of this simple mass balance approach becomes evident. There is the tendency to shift from the single element cycling mass accounting approach to a process level model with more components, hydrological process descriptions and specifically defined chemical reactions if the data are available.

Mass balance — inverse modelling

A variation of the 'global cycling' material balance model was described initially by Garrels and McKenzie (1967) in their pioneering study of Sierra Nevada springs. In this work specific weathering reactions were defined that described the spring compositions. This material balance approach, often referred to as the 'balance sheet method' has been used repeatedly and successfully in groundwater and watershed studies (see Taylor and Velbel, 1991, for an exhaustive tabulation of related studies).

Changes in water composition between two points, e.g. wells in a groundwater study, or atmospheric input and stream discharge in catchment studies, can be described by identifying all the phases that could explain increase and loss of each element, yet still honour the stoichiometry of a selected combination of the plausible minerals. This mass balance process is referred to as 'inverse modelling' by Plummer *et al.* (1983). Using the following formulation:

$$\sum_{\rho=1}^{P} \alpha_\rho b_{P,k} = \Delta m_{T,k} \quad k = 1, j$$

Here, the change in total elemental concentrations ($\Delta m_{T,k}$) between two points can be accounted for by mass transfer between minerals, solid phases, surfaces or gasses by identifying the phases (ρ) with stoichiometry (b) and net mass transferred (α) for each element (k) of interest.

Mass balance on hydrogen and oxygen are not considered, but the process is constrained by using a chemical equilibrium program to compute aqueous speciation, saturation states of minerals and partial pressures of gases. Similarly, electron transfer and isotopic exchange can be included in the mass transfer calculations, to identify further the dominant reactions.

$$\sum_{\rho=1}^{\rho} \mu_\rho \alpha_P = \Delta RS$$

$$\sum_{\rho=1}^{P} \alpha_\rho b_{P,I} SI_\rho = \Delta m_T I_T$$

where ΔRS is the mass balance accounting for the difference in the redox state using the operational valence (μ), which is known for all components entering into redox reactions. Similarly, the stable isotopic concentration (δI) in per mille notation, must be known for the phases and solute of interest.

Although mass balance/equilibrium model approaches may identify elemental sources and sinks, the final result does not provide a prediction of the chemical hydrograph. For example, the term 'inverse mass transfer

Table I. Comparison of the principal features of the major process level models

	[MAGIC] Model of acidification in catchments	[BIM] Birkenes model	[ILWAS] Integrated lake–watershed acidification study	[ETD] Enhanced trickle down
Reference	Cosby et al. (1985a–c)	Christopher and Wright (1981); Stone et al. (1990)	Gherini et al. (1985)	Nikolaidus et al. (1988)
Catchment	White Oak Run, VA, USA	Birkenes, Norway	Adirondack Mtns, New York	Adirondal Punte, New York
Principal objective	Catchment response to acid precipitation	Catchment response to acid precipitation	Catchment response to acid precipitation	Catchment response to acid precipitation
Hydrological submodel	Surface, soil, ground water and stream flow [often determined externally to the model]	Yes: 2 or 3 reservoir Piston flow [Precip, ET]	Darcy's Law and stage flow [precipitation, canopy interception, ET, snowpack melting, percolation, lake volume balance]	Darcy flow between compartments, [ET, ppt, and precipitation snowmelt, lake seepage, drainage]
a. Chemical model				
Gases	Carbon dioxide	Carbon dioxide	CO_2, NO, SO_x	No
Speciation	Aluminium only	No	Yes	No
Solubility	Cations and Al	Yes	Al	No
Ion exchange	Sulphate	Yes	Yes	No, kinetic expression
Adsorption	Anion retention	No	Sulfate, phoshate	Sulfate
Organic acids	No	No	Yes	No
b. Mass balance		—		
Single element	—	—	Yes	Sulfate and alkalinity
Base cations	Yes	Yes	Yes	Balance
Strong acids	Yes	Yes	Yes	Yes
Aluminium	Yes	Yes	Yes	Yes
Mineral sources	Al	–	Yes	Yes
Amos. input	Yes	Yes	Yes	
c. Kinetic expression				
Mineral wealth	No	No	Yes	Yes
Litter decay	No	No	Yes	No
Nutrient uptake	No	No	Yes	No
Respiration	No	No	Yes	No
Nitrification	No	No	Yes	No
Mineralization	No	Sulfate	Yes	No
d. Output prediction	ANC, sulfate, base cations, acids anions, stream/lake chemistry	ANC, stream/lake chemistry	ANC, sulfate, base cations, acid anions, stream/lake chemistry	ANC, sufate, stream/ lake chemistry
e. Data requirements				
Equil. consts	Yes	No	Yes	No
CEC, select.	Yes	Yes	Yes	No
Rate constants	No	Yes	Yes	Yes
Solute in/out soil	Yes	Yes	Yes	Yes

modelling' is used to describe the process because a stream or groundwater composition is required for the mathematical solution. Essentially all plausible reaction pathways are examined by *repetitively* solving for *all possible combinations* of the specified minerals, surface reactions and gases, which could, stoichiometrically, contribute to or remove mass from the solution. The identified reactions must not violate thermodynamic, redox or isotopic constraints, and should have some physical evidence that they are present in the system.

The resulting calculation is a 'snapshot' of the mass transfer pathway that would result in the observed mass distribution in the system. This approach is essentially a computationally enhanced version of the method introduced by Garrels and McKenzie (1967) for either groundwater or surface water compositions. This approach is described in detail by Plummer *et al.* (1983) and Plummer *et al.* (1990) and can be easily implemented using the computer program NETPATH (Plummer *et al.*, 1991).

Recent studies have used this or similar mass balance approaches to determine the solute source in streams draining 'bare rock watersheds' (Drever and Hurcomb, 1986; Mast *et al.*, 1990; Williams and Melack, 1991). These alpine watersheds are dominated by rock weathering processes, contain little soil or vegetation and the solute source can be attributed to atmospheric input and mineral degradation. The suspected chemical processes, mineral source and atmospheric contribution have been identified and 'ground truth' is provided by field inspection of weathered phases and positive identification of suspected minerals. Plausible reaction pathways are presented.

This approach was extended by Bassett *et al.* (1992) to a bare rock watershed in which acid sulfate water was naturally generated from weathering of a sulfide intrusive in Geneva Creek, Colorado (Figure 4d). This watershed is potentially an analogue of a bare rock watershed that might be affected by severe acid sulfate precipitation, which created a strongly acidic surface water and accompanying rapid degradation of exposed minerals.

Even in early work in forested watersheds underlain by igneous rocks, chemical weathering was determined to be a significant source of solute, and can be modelled with the mass balance approach (Cleaves *et al.*, 1970; Wood and Low, 1986). Similarly, watersheds with high solute loads from mining activities or other similar anthropogenic activities, can be analysed with this approach, which assists in identifying the dominant processes (Filipek *et al.*, 1987).

Mass balance and rate-determined reaction models

An entire subdiscipline in chemical mass balance modelling, largely for bare rock watersheds, has emerged, which focuses on the rates of mineral weathering. These studies are essential to the understanding of solute flux and this process is often integrated into more complex catchment acidification modelling (Table I).

Early laboratory studies of the kinetics of silicate dissolution created significant controversy regarding the mechanism of mass transfer, pH dependence, mode of formation of secondary products and transportability of rate constants (for detailed reviews see Berner, 1981; Lasaga, 1981; Velbel, 1985). Comparisons between laboratory- and field-derived constants have not been conclusive and the reasons for the large differences are not clear; however, constant ratios between the two indicate that field-related factors may yet be identified (Velbel, 1993).

Recent simulations of the field weathering rates with models such as PROFILE (Sverdrup and Warfvinge, 1992; 1993) indicate advancement can be made by focusing on key minerals and adjusting the dissolution rate to the amount of exposed mineral surface. Mass balance models average solute flux over time, and for bare rock watersheds the mineral weathering rates will affect the mass accounting when longer time frames, e.g. seasonal and annual, are considered.

Mass balance (end-member mixing)

A simplification of the above-mentioned mass balance approaches is presented by Woolhiser *et al.* (1982; 1985) for Walnut Gulch, Arizona, and Hooper *et al.* (1990) and Christopherson *et al.* (1990) for Panola Mountain Research Catchment (Figure 1). This end-member mixing analysis procedure (EMMA; Hooper *et al.*, 1990) assumes that for a perennial catchment stream, the measured stream chemical concentration (S) of any species (i) can be described as a linear combination of j number of water types having fixed compositions (e.g. groundwater, soil solutions from separate horizons, etc.). The concentration C of these same species (i) in j number of end-member waters can be summed as contributing fractions (x_j), represented as:

$$\sum_j C_{i,j} = S_i \quad i = 1, m$$

Furthermore, it is a requirement that all end-members be identified such that

$$\sum_j x_j = 1$$

Rather than include specific mineral or gas phases as solute sources, only water sources are used for the end-members. The end-member solutions are measurements of water compositions from physical locations within the watershed, e.g. soil water from different horizons, groundwater, etc. If the end-member solutions have been correctly identified, then the plot of variation in stream water composition for any element will be contained within a geometric shape generated by the plot of elemental composite of the n number of end-members. In some circumstances, stream water analysis will align primarily along a boundary, indicating binary mixing (Figure 4e).

If all the points representing stream water compositions plot within the diagram, it is assumed that conservative mixing dominates, that the correct end-members are identified, that other necessary water types have not been omitted and that the hydrological regime is in steady, state. The authors indicate that this approach, although useful in limited circumstances, assists in identifying the principal contributors to stream water composition and in testing the assumption about conservative mixing, before proceeding to more complex hydrochemical modelling.

Mass balance — process models

Catchment response to atmospheric acid sulfate chemical stress has been monitored in many watersheds; consequently, data for comparing the atmospheric chemical load with the observed chemical changes in the catchment stream waters are readily available. A simple comparison between atmospheric input and stream water solute load is uninformative because sulfur mobility through the watershed is a complex hydrological and chemical pathway. More than any other factor, the acidification process in forested watersheds has been responsible for the development of complex process level models that generally couple the hydrological simulations with some detailed level of chemical reactions and chemical transformations. Four of the more frequently referenced models, MAGIC (Cosby *et al.*, 1985a–c), BIM (Christopherson and Wright, 1981; Wright, 1982; Stone *et al.*, 1990), ILWAS(Gherini *et al.*, 1985), and ETD (Nikolaidus *et al.*, 1988), are described in more detail in Table I, and a generic flow chart is given in Figure 5. For reference, one may also benefit from examining variations of these models, among which are the following: BICK1 (Eshleman, 1985), EEM (Nikolaidis *et al.*, 1988), PULSE (Bergstrom *et al.*, 1985), RAINS (Kamari, 1985), TMWAM (Bobba and Lam, 1988), and WAM (Booty and Kramer, 1984).

One fundamental assumption behind the development of a process level model is that once it is properly 'calibrated', output from this modelling process can be used predictively, e.g. to assess the consequences of unregulated sulfur deposition or evaluate the improvement of stream quality through management of the sulfur deposition. Although sulfur is the environmental stressing agent receiving the most attention, there are offending contaminants as a result of derivative effects like pH change. For example, the sulfate load overwhelms the natural acid-neutralizing capacity and acidity increases. Simultaneously, aluminium concentrations are elevated in soil water, lakes and streams, and weathering rates increase, mobilizing nutrients and base cations.

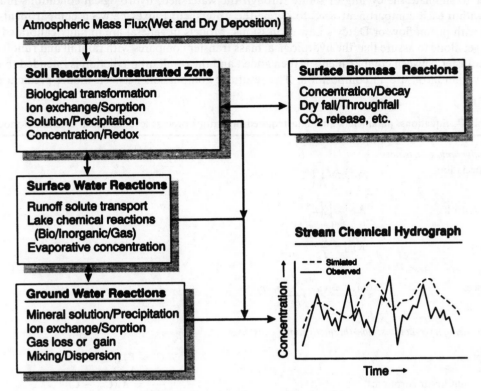

Process Level Model with Output Simulating Chemical Hydrograph with Chemical Mass Balance Coupled to Water Balance (not shown)

Figure 5. A flow chart for process level models

Simple mass balance models address the sources and sinks of elements but do not directly include the actual speciation of aqueous components and their redistribution in response to changes in pH. To address aqueous solution speciation, mass action expressions with the appropriate thermodynamic data are required. The aqueous speciation will allow for a more precise depiction of gas and solid phase solubility effects, but will not describe rigorously the non-thermodynamic ion exchange reactions, nor will it much improve the kinetically constrained reactions. Each chemical process also requires the definition of an 'aqueous compartment' with defined aqueous compositions. This approach, if applied rigorously to an entire watershed, results in models with multicompartment and multispecies representations (Figure 5). The number of compartments will never be sufficient, nor are the components within a compartment homogeneously distributed; however, the assumption is that they are 'representative enough' of the watershed to predict the composition of the stream water.

Complex process models require substantial data from numerous sources, among which are requirements for hydrological information on rain and snowfall quantity, distribution and intensities, as well as records from stream flow gauging. Similarly, required chemical data include elemental compositions from atmospheric precipitation and particulates; solute concentration in soil, surface and groundwater contributions, and their spatial and temporal distribution; and, lastly, the solid phase contribution may be constrained by mineral, soil and biomass compositions.

The interaction between these water and solute sources is defined by mass action equations requiring thermodynamic equilibrium constants, non-thermodynamic exchange constants and rate constants for dissolution, decomposition, etc. A summary of the general equations and attendant constants is given in Table II.

In order to simulate the cycling of solute through the watershed, hydrological conditions must first be specified within each compartment used to define the watershed. Fluxes between compartments are then simulated with piston flow or Darcy's Law expressions. Models in general are initially calibrated using the water budget alone to assure that the hydrological mass transfer compares with rainfall and runoff data. The chemical transfer between compartments is then added and the resultant prediction is intended to match the temporal variation of stream composition. The resultant comparison between model prediction and the

Table II. Equations, parameters and data required in various process level chemical catchmant models

A. *Thermodynamic constraints*

Aqueous speciation:
$$a_j = K_j \prod_{i=1}^{I} a_i^{n_{i,j}}$$

Solubility control:
$$K_P = \prod_{i=1}^{I} a_j^{n_{i,p}}$$

Henry's Law:
$$K_{H,i} = \frac{a_i}{f_i}$$

Mass balance:
$$m_{T,i} = m_i + \sum_{j=1}^{J} n_{i,j} m_j$$

B. *Surface chemistry reactions: non-thermodynamic constants*

Linear sorption:
$$C_i = K_i S_i$$

Non-equilibrium linear sorption:
$$\frac{dC_i}{dt} = K_i(C_{eq} - C_i)$$

Ion exchange between divalent cations and H^+, Gapon formulation:
$$K_G = \frac{[H^+]}{[m^{+2}]^{1/2}}$$

Ion exchange selectivity coefficients:
$$S_{12} = \frac{a_1^{n_1} x_{z_1}}{a_2^{n_2} x_{z_2}}$$

C. *Rate expressions*

$$\frac{-dM}{dt} = kM[H^+] \quad M = \text{minerals mass}; \quad k = \text{specific rate constant}$$

$$\frac{dm}{dt} = k([Al_{T_{sat}}] - [Al_T]) \quad Al_{T_{sat}} = \text{Total conc. Al (at equilibrium)}$$

$$R_w = \sum_{j=1}^{\text{minerals}} r_j A_w x_j \theta z \quad R_w = \text{weathering rate}; \quad A_w = \text{surface area}; \quad x_j = \text{surface area fraction};$$

$\theta = $ moisture saturation; $z = $ soil layer thickness

D. *Commonly required parameters*

CEC, cation exchange capacity; Ads_{SO}, adsorbed sulfate; w, uptake and release functions (undefined rate constants); k, specific rate constants (mineral or organic deposition); M, mineral concentration or surface area

E. *Data requirements*

Chemical analysis of water sources, atmospheric input
Partial pressure of CO_2
Chemical hydrographs
Mineral compositions
Exchangeable ions on clays

measured data used to construct the chemical hydrograph is generally similar in historical trend without significant match to the detail of the chemical changes.

The difference between the four process models shown in Table I is both in hydrological and chemical complexity. Among the differences are the number of compartments, the inclusion and omission of specific reactions, the number of components and the explicit representation of aqueous speciation using thermodynamic constraints or lumped parameter representations controlled by rate equations. A systematic comparison was made of the ILWAS, MAGIC and ETD models by Rose et al. (1991a,b) and Cook et al. (1992) by mapping the same equations and assumptions through each model as well as possible. Similarly, Booty et al. (1992) compared TMWAM, ETD, ILWAS and RAINS in the Turkey Lakes Watershed, Canada. For the watersheds investigated, the models were similar in performance when used to predict acid-neutralizing capacity, and the sums of acid anions as well as base cations in lakes in two watersheds in the north-east with 15 years of data.

PREDICTIVE QUALITY

It is interesting to note that the approach used to model a given catchment is directly related to the kind of data available and the questions posed. For example, questions of, 'How much mass is derived from weathering?' or 'How is a specific element compartmentalized within the watershed?' or 'What is the magnitude of atmospheric contribution versus other sources?', are generally answered with mass balance models. Conversely, questions such as, 'How does the stream composition change if more sulfate is added, or more carbonate, or if hydrological conditions change?' are process level questions. These questions focus on the temporal variations of stream water chemical composition: because it is measurable, monitorable, known to vary and represents the net effect of all reactions in the watershed system. It is precisely this focus that has driven the development of models that incorporate all 'significant chemical reactions' and 'dominant hydrological pathways' required to simulate the changes in this stream water chemistry.

There is a need to be clear about the predictions these models are making. The output from these models is useful only after significant manipulation of parameters in a 'calibration' step. In actual fact, the model is neither calibrated nor validated. Validation would imply that true circumstances were known and the model was independently making predictions of chemical change that reproduced this truth. This, of course, is not the case, neither are these models actually calibrated, that is, tested against standard circumstances that are completely known and controlled. The most appropriate definition of this process is probably 'parameterization'. These process models are 'fitted' against the chemical hydrographs under several steady state, low flow or better understood circumstances, such that as the conditions become more complex the model can still make predictions that match the hydrograph. As pointed out by Eshleman et al. (1992), most parameters that govern the mixing or interaction are not measured but, rather, are obtained in the calibration of the predicted chemical hydrograph against the measured stream composition.

Setting aside the uncertainty related to flow quantity and pathway, there are important points related to the chemical circumstances that must be considered. First, while there are fundamental thermodynamic constants in many models which should be valid for all conditions of equilibrium, in many circumstances equilibrium may not be attained. Secondly, most of the parameters and constants are not thermodynamic and are therefore determined experimentally, estimated, or fit; e.g. selectivity constants, exchange capacity and decay or reaction rate constants. Finally, even if the non-thermodynamic parameters used in the models were experimentally derived or measured in the field, they represent a measurement for a specific location and time within a large watershed, and homogeneity most certainly is a poor assumption over the expanse of any watershed. Certainly, the clay mineralogy, clay mineral and organic matter surface area and soil thickness vary widely.

The net effect of all this uncertainty is that these empirical parameters, however detailed in their chemical application, are treated as lumped parameters describing processes extending over large areas, perhaps even over an entire hypothetical watershed. The more reactions one describes for the watershed and the more individual compartments that are used to account for the mass being transferred, then the more parameters one has to fit against the chemical hydrograph. This is especially apparent when the models are used for circumstances beyond the bounds of their parameterization and they fail to predict extreme events. Testing small components of a watershed independently, and experimentally stressing the chemical conditions of the watershed, adds a level of detail, but does not make the process any less of a lumped parameter approach. The most obvious evidence that process models are basically parameterizations is the fact that they do not transport between watersheds without new parameterization.

The distribution of species in the aqueous phase and the calculation of solubility are most likely the only chemical portions of the process model that are rigorously defensible, because they rely on thermodynamic constants. However, even the application of these equations depends on the concentrations of the species at a given location, which unfortunately is known only for the monitored points. This information is extrapolated over the entire compartment, e.g. soil or groundwater, and therefore the uncertainty remains.

Details often considered in groundwater studies such as activity coefficients, ionic strength, the effects of omitted elements, ignored redox conditions, amorphous oxyhydroxide compounds controlling solubility, etc., are representative of a level of complexity that is essentially lumped into the uncertainty of the fitted parameters in watershed studies.

Nevertheless, it is remarkable that these models are well-enough defined that they can indeed be used in limited predictive capacities, within a given watershed, over the range of conditions against which they were fit.

CONFIDENCE BUILDING

In both watershed types, existing procedures of field sampling, monitoring and analysis will undoubtedly be transferable to other watersheds, and many new case studies will emerge, but for any given watershed, there will be the continual dilemma of an infinite number of hydrological pathways and chemical reaction scenarios. The number of mathematical solutions depends on the minerals selected, and how one chooses to distribute the elemental mass between compartments. If investigators hope to build credibility for this approach, more evidence is required to demonstrate two levels of confidence. First, it must be shown that complex process level models *can be calibrated* under numerous controlled conditions on the laboratory or well-controlled small field scale. This may require that the relationship between hydrological calibration and the attendant chemical reactions be well defined. It may be true that the reason extreme rainfall events are not predicted well in these models is because the predominant flow paths and residence times are so different from the ambient circumstances.

Secondly, it must be shown that the *minimum number of significant reactions* are considered. Implausible reactions must be identified and eliminated; whether implausible because of physical absence of material on the site, kinetic constraint or other restriction, it should be clear if and when these reactions will not be influential to the simulation process. The reverse situation must also be clarified, namely the discovery of which reactions do proceed, and do so in an understandable sense, supportable with sufficient field data, evidence for the presence of all reactants, and experimental evidence to support the probability that the reaction will proceed in the time-scale of the study. It is unfortunate that so many reactions are hypothesized but not confirmed.

Models of the BRWT are dominated by weathering reactions and atmospheric import of solute; increased credibility of mass balance models for the BRWT environment will depend on answers to questions such as the following. Are the selected minerals, gases or other solid phases that are assumed to be responsible for the observed chemical changes actually present in the system? Does direct physical evidence exist for the decomposition of these primary phases and the formation of suspected secondary phases. This 'ground truth' in actual fact is seldom done. How important are the rates of individual reactions? If the rates are important, how can the rate constants be verified and what are the variables that attract them, e.g. surface area, moisture content, pH, etc., and how do field-derived rates correlate to laboratory-derived rates of weathering?

For both watershed types, are the hydrological assumptions reasonable for the watershed, e.g. steady state, no seepage, evapotranspiration rates, etc? Similarly, can the predominant flow path be identified, especially as it depends on rainfall frequency, duration and intensity?

What are the variances for the above-mentioned components in the system? What are the ranges in mineral compositions, reaction rates, spatial distribution, hydrological conditions, etc., and can the modelled process be represented as a bounded prediction? Mass balance computations are inherently dependent on knowing a total mass or end-point concentration, but the variance in the quantities must be considered.

There is of course a significant scepticism that either the various mass balance model forms or the process level models represent anything more than multiparameter fitting. They obtain unique mathematical solutions, can be made to describe historical events, but have little utility in prediction. How can the model development proceed for these watersheds to provide confidence in prediction? It is suggested that procedures such as those discussed in the next section could advance the credibility in this environment.

ADVANCES AND FUTURE APPROACHES

Regarding advancing the state of the science, one can see that the watershed type will most likely still determine the modelling approach. Especially in terms of the BRWT, advancements in technology providing real-time monitoring of chemical and hydrological conditions will undoubtedly improve predictability in the immediate future.

Beyond this, perhaps a forensic approach will supplement the understanding of the system. For example, can key solute components with indicator capabilities offer additional constraint on the hydrological component of the system to define better the flow path? Stable solute isotopes such as the common $\delta^{34}S$, $\delta^{13}C$, δD, $\delta^{18}O$, have been useful in hydrograph separation but are either affected by redox reactions and antecedent conditions or have too little variation to be definitive? Newly emerging isotopic tracers such as $\delta^{11}B$, δ^7Li or radiogenic isotopes of noble gases like ^{85}Kr and ^{39}Ar are less affected by chemical reactions like redox, and have large source-related signatures (Bassett, 1990; Bassett et al., 1995). In particular, extreme events may carry enhanced cosmogenic isotopes that have diminished importance in ambient circumstances. Forensic isotope hydrology may offer signatures distinctive enough to dissect the system behaviour.

A probabilistic approach to the spatial distribution of catchment characteristics may be useful. As discussed by Mulholland et al. (1990), chemical compositions were different in type and magnitude depending on the subcatchment, flow path and runoff volume, etc. In general, this variability is ignored and is possibly a primary contributor to the inability of simulating extreme conditions, or even the details of more common conditions.

Chemical indicators are needed that can be continuously monitored remotely with sensors or probes placed in numerous locations throughout the watershed. Advancements in chemical microsensors, satellite monitoring and data transmission will most likely revolutionize watershed modelling. Current methods of sample collection and analysis are much too slow and spatially restricted.

What is the separate error associated with the hydrological model? In some cases the chemical error is independent of flow, e.g. thermodynamic constants, rate constants, variance in composition or capacity over time even at a given location, whereas other errors are inseparable, such as specific reactions that occur only in certain parts of the watershed, and consequently the flow path will significantly affect the stream water composition. In some cases the error in the hydrological prediction is additive to the error of the chemical components and can be minimized separately. Studies to determine the dominant chemical reactions must be correlated with hydrological studies. Preferential pathways control the chemical composition of catchment output. This is seen in a study of a grass-covered watershed (Muscutt et al., 1990) in which soil macropores controlled the flow and consequently the water composition, or, similarly, in bare rock watersheds, with fracture-dominated recharge, similar effects on subsurface water composition are observed (Bassett et al., 1995). Additionally, infrequently accessed but ill-defined solute reservoirs containing compounds such as sulfate, are apparently not detectable until hydrological conditions are right; such as flooding. Long-term monitoring will undoubtedly be needed to account for this kind of variability which depends both on antecedent conditions and extreme events (Lynch and Corbett, 1989).

A true calibration should be attempted with process models using small scale, highly instrumented watersheds and laboratory studies using tracers. This would be invaluable in both model intercomparison and in identifying the sensitivity of individual components of each model. If sufficient variability in the conditions of the tests and levels of solute can be created on this small scale, then probabilistic approaches can be used to evaluate the accuracy as well as precision of the models, and the importance of each chemical predictor.

True predictive modelling is not achieved in any environmental field project. The goal, however, is to minimize the arbitrary fitting and maximize the ability to predict independently the processes that have the greatest probability of effecting the important chemical changes in the system. The appropriate watershed model and modelling approach must be used.

CONCLUSIONS

Mass balance models decoupled from a hydrological model have been used widely in all fields of geoscience and several variants of the approach are used to simulate chemical processes in catchments. They are used almost exclusively in bare rock watershed circumstances and, in fact, this field environment is common enough to serve as an archetype end-member watershed (BRWT).

One could think of this mass balance process as a closed form solution in which either the total concentration of some element is known, or the concentration at specific end-points has been measured, such that the total mass can be distributed between selected compartments and the mass gained or lost can be quantified. The error of this approach is difficult to quantify even though the solutions are constrained by defined mineral compositions, water compositions and input or output concentrations. Sources of error include the assumptions of average compositions for given compartments, or that the amount and spatial distribution of minerals and soils water types is known, or that the temporal changes at any given location are not important. Solutions to the mass balance models may be conceptually correct but they are difficult to verify. Ground truth can be done to check randomly for mineral presence or intermediate water compositions, but because the elemental distribution is an average for a given compartment, the random checks will require large sample sizes to offer a meaningful comparison. This is seldom, if ever, done and therefore simulations are estimates with little verification. Because of the structure of the equations, there is almost always a solution for the distribution of mass, but not an evaluation of the error.

Similarly, *process level* models are used almost exclusively in the forested watershed type (FWT) environment, which also serves as the second archetype environment. These complex process models incorporate the additional complication of the details of flow; in fact, the mathematical description of the flow system in the model is first fit against the stream hydrograph. In reality there are an infinite number of hydrological models for any given watershed, and, most importantly, the flow path changes with storm intensity and distribution. Chemical compartments are defined with average compositions that are reservoirs of solute moved about by the constraints of the defined flow system. Predictions of stream water chemistry are made and the appropriate fitting parameters are adjusted until the prediction matches the stream water chemistry as well as possible. Obviously the fit over wide ranges of conditions will require sufficient parameterization over those circumstances.

The users should not be overly confident regarding the application of these models to new catchments. Until sufficient research is done to identify process-related constants, including the probability of occurrence over certain ranges, the model parameterization will be useful only for the catchment of interest.

Similarly, it is now recognized that although numerous process level models exist, most can be similarly configured such that they all fit the chemical hydrograph to the same general degree of accuracy. Technological advancements in rapid chemical analysis or field monitoring with inexpensive datalogging devices and transmitting equipment will undoubtedly greatly assist the model calibration process within a given watershed.

There is actually no strong evidence at present that increasingly complex models are required to answer the acidification or weathering questions being posed. Optimization of the major hydrological pathways and verification of the principal chemical reactions may be a better approach, requiring less intensive field sampling, and simpler models. Forensic hydrological studies and real-time monitoring of processes on a smaller scale will most likely take these models to a higher level of confidence and improve the transferability between watersheds.

ACKNOWLEDGEMENTS

The research at the Apache Leap Research Site is funded by the United States Nuclear Regulatory Commission. Assistance in creating the graphics was provided by Dan K. Braithwaite, Department of Hydrology and Water Resources and Dave Cantrell, Biomedical Communications, University of Arizona.

REFERENCES

April. R., Newton, R., and Coles, L. T. 1986. 'Chemical weathering in two Adirondack watersheds: past and present day roles', *Geol. Soc. Am. Bull.*, **97**, 1232–1238.

Baron, J. and Bricker, O. P. 1987. 'Hydrologic and chemical flux in Loch Vale Watershed, Rocky Mountain National Park', in McKnight, D. and Averett, R. C. (Eds), *Chemical Quality of Water and the Hydrologic Cycle*. Lewis Publishers, Chelsea, MI. pp. 141–156.

Bassett, R. L. 1990. 'A critical evaluation of the available measurements for the stable isotopes of boron', *Applied Geochemistry*, **5**, 541–554.

Bassett, R. L. and Melchior, D. C. 1990. 'Chemical modeling in aqueous systems: an overview', in Melchior, D. C. and Bassett, R. L. (eds), *Chemical Modeling in Aqueous Systems II*, ACS Symposium Series 416. American Chemical Society, Washington, D.C. pp. 1–14.

Bassett, R. L., Miller, W. R., McHugh, J., and Catts, J. G. 1992. 'Simulation of natural acid sulfate weathering in an alpine watershed', *Wat. Resour. Res.*, **28**, 2197–2209.

Bassett R. L., Neuman S. P., Rasmussen, T. C., Guzman, A., Davidson, G. R., and Lohrstorfer, C. F. 1994. 'Validation studies for assessing unsaturated flow and transport through fractured rock', *Rep. NUREG/CR-620. US Nuclear Regulatory Commission*, Washington, D.C.

Bassett, R. L., Buszka, P., Davidson, G. R., and Diaz-Chong, D. 1995. 'The identification of groundwater sources using boron isotopic composition', *Environ. Sci. Technol.*, **29**, 2915–2922.

Bergström, S., Carlsson, B., Sandberg, G., and Maxe, L. 1985. 'Integrated modelling of runoff, alkalinity, and pH on a daily basis', *Nord. Hydrol.*, **16**, 89–104.

Berner, R. A. 1981. 'Kinetics of weathering and diagenesis', in Lasaga, A. C. and Kirkpatrick, R. J. (Eds), *Kinetics of Geochemical Processes*, Min. Soc. Am. Reviews in Mineralogy, 8. Mineralogical Society of America. pp. 111–135.

Bischoff, W. D., Peterson, V. L., and MacKenzie, F. T. 1984. 'Geochemical mass balance for sulfur- and nitrogen-bearing acid components: eastern United States', in Bricker, O. P. (Ed.), *Geological Aspects of Acid Deposition*, Acid Precipitation Series, 7. Butterworth. pp. 1–22.

Bobba, A. G. and Lam, D. C. L. 1988. 'Application of a hydrological model to the acidified Turkey Lakes Watershed', *Can. J. Fish. Aquat. Sci.*, **45**, 81–87.

Booty, W. G. and Kramer, J. R. 1984. 'Sensitivity analysis of a watershed acidification model', *Phil. Trans. R. Soc. Lond.*, **B305**, 441–449.

Booty, W. G., Bobba, A. G., Lam, D. C. L., and Jeffries, D. S. 1992. 'Application of four watershed acidification models to Batchawana Watershed, Canada', *Environ. Pollut.*, **77**, 243–252.

Braun, J. J., Pagel, M., Herbillon, A., and Rosin, C. 1993. 'Mobilization and redistribution of REES and thorium in a syenitic lateritic profile; a mass balance study', *Geochim. Cosmochim. Acta*, **57**, 4419–4434.

Chadwick, O. A., Brimhall, G. H., and Hendricks, D. M. 1990. 'From a black to a gray box — a mass balance interpretation of pedogenesis', *Geomorphology*, **3**, 369–390.

Christophersen, N. and Wright, R. F. 1981. 'Sulfate budget and a model for sulfate concentrations in stream water at Birkenes, a small forested catchment in Southernmost Norway', *Wat. Resour. Res.*, **17**, 377–389.

Christophersen, N., Neal, C., Hooper, R. P., Vogt, R. D., and Andersen, S. 1990. 'Modelling streamwater chemistry as a mixture of soilwater end-members — a step toward second generation acidification models', *J. Hydrol.*, **116**, 307–320.

Cleaves, E. T., Godfrey, A. E., and Bricker, O. P. 1970. 'Geochemical balance of a small watershed and its geomorphic implications', *Geol. Soc. Am. Bull.*, **81**, 3015–3032.

Conrad, S. H. 1993. 'Using environmental tracers to estimate recharge through an arid basin', *Proceedings, Fourth International Conference on High-Level Radioactive Waste Management, Las Vegas, NV*. American Nuclear Society and American Society of Civil Engineers. 132–137.

Cook, R. B., Rose, K. A., Brenkert, A. L., and Ryan, P. F. 1992. 'Systematic comparison of ILWAS, MAGIC, and ETD watershed acidification models: 3. Mass balance budgets for acid neutralizing capacity', *Environ. Pollut.*, **77**, 235–242.

Cosby, B. J., Hornberger, G. M., Galloway, J. N., and Wright, R. F. 1985a. 'Time scales of catchment acidification', *Environ. Sci. Technol.*, **19**, 1144–1149.

Cosby, B. J., Hornberger G. M., Galloway, J. N., and Wright, R. F. 1985b. 'Modeling the effects of acid deposition: assessment of a lumped parameter model of soil water and streamwater chemistry', *Wat. Resour. Res.*, **21**, 51–63.

Cosby, B. J., Wright, R. F., Hornberger, G. M., and Galloway, J. N. 1985c. 'Modeling the effects of acid deposition: estimation of long-term water quality responses in a small forested catchment', *Wat. Resour. Res.*, **21**, 1591–1601.

Cronan, C. S., Driscoll, C. T., Newton, R. M., Kelly, J. M., Schofield, C. L., Bartlett, R. J., and April, R. 1990. 'A comparative analysis of aluminum biogeochemistry in a Northeastern and a Southeastern forested watershed', *Wat. Resour. Res.*, **26**, 1413–1430.

Dann, M. S., Lynch, J. A., and Corbett, E. S. 1986. 'Comparison of methods for estimating sulfate export from a forested watershed', *J. Environ. Qual.*, **15**, 140–145.

Drever, J. I. and Hurcomb, D. R. 1986. 'Neutralization of atmospheric acidity by chemical weathering in an alpine drainage basin in the North Cascade Mountains', *Geology*, **14**, 221–224.

Eshleman, K. N. 1985. 'Hydrochemical response of a New England watershed to acid deposition,' *Ph.D. Thesis*, Massachusetts Institute of Technology, Cambridge, 180 pp.

Eshleman, K. N., Wigington, P. J., Jr., Davies, T. D., and Tranter, M. 1992. 'Modelling episodic acidification of surface waters: the state of the science', *Environ. Pollut.*, **77**, 287–295.

Filipek, L. H., Nordstrom, D. K. and Ficklin, W. H. 1987. 'Interaction of acid mine drainage with waters and sediments of West Squaw Creek in the West Shasta Mining District, California', *Environ. Sci. Technol.*, **21**, 388–396.

Fu, J. and Winchester, J. W. 1994a. 'Sources of nitrogen in three watersheds of northern Florida, USA: mainly atmospheric deposition', *Geochim. Cosmochim. Acta*, **59**, 1581–1590.

Fu, J. and Winchester, J. W. 1994b. 'Inference of nitrogen cycling in three watersheds of northern Florida, USA, by multivariate statistical analysis', *Geochim. Cosmochim. Acta*, **58**, 1591–1600.

Garrels, R. M. and MacKenzie, F. T. 1967. 'Origin of the chemical composition of some springs and lakes', in Gould, R. F. (Ed.), *Equilibrium Concepts in Natural Water Systems*. Advances in Chemistry Series, 67. American Chemical Society, Washington, D.C. pp. 222–242.

Garrels, R. M., MacKenzie, F. T., and Hunt, C. 1975. *Chemical Cycles and the Global Environment*. William Kaufman, Inc., Los Altos, California.

Gherini, A. S., Mok, L., Hudson, R. J. M., Davis, G. F., Chen, C. W., and Goldstein, R. A. 1985. 'The ILWAS model: formulation and application', *Water Air Soil Pollut.*, **26**, 425–459.

Goldstein, R. A., Chen, C. W., and Gherini, S. A. 1985. 'Integrated lake-watershed acidification study: summary', *Water Air Soil Pollut.*, **26**, 327–337.

Gordon, N. D., McMahon, T. A., and Finlayson, B. L. 1992. *Stream Hydrology: an Introduction for Ecologists*. John Wiley & Sons Ltd., Chichester.

Kamari, J. 1985. 'A model for analyzing lake water acidification on a regional scale', *Report No. CP-85-48*. International Institute for Applied Systems Analysis, 2361, Laxenberg, Austria.

Kattelmann, R. and Elder, K. 1991. 'Hydrologic characteristics and water balance of an alpine basin in the Sierra Nevada', *Wat. Resour. Res.*, **27**, 1553–1562.

Hooper, R. P., Christophersen, N., and Peters, N. 1990. 'Modeling streamwater chemistry as a mixture of soilwater end-members — an application to the Panola Mountain Catchment, Georgia, U.S.A., *J. Hydrol.*, **116**, 321–343.

Lasaga, A. C. 1981. 'Dynamic treatment of geochemical cycles: global kinetics', in Lasaga, A. C. and Kirkpatrick, R. J. (Eds), *Reviews of Mineralogy: Kinetics of Geochemical Processes*, Vol. 8. pp. 69–110.

Likens, G. E., Bormann, F. H., Pierce, R. S., Eaton, J. S., and Johnson, N. M. 1977. *Biogeochemistry of a Forested Ecosystem*. Springer-Verlag, New York.

Loen, J. 1992. 'Mass balance constraints on gold placers: possible solutions to source area problems', *Economic Geol.*, **87**, 1624–1634.

Lynch, J. A. and Corbett, E. S. 1989. 'Hydrologic control of sulfate mobility in a forested watershed', *Wat. Resour. Res.*, **25**, 1695–1703.

Mangold, D. C. and Tsang, C. 1991. 'A summary of subsurface hydrological and hydrochemical models', *Rev. Geophys.*, **29**, 51–79.

Mast, M. P., Drever, J. I., and Baron, J. 1990. 'Chemical weathering in the Loch Vale Watershed, Rocky Mountain National Park, Colorado', *Wat. Resour. Res.*, **26**, 2971–2978.

Merritts, D. J., Chadwick, O. A., Hendricks, D. M., Brimhall, G. H., and Lewis, C. J. 1992. 'The mass balance of soil evolution on late Quaternary marine terraces, Northern California', *Geol. Soc. Am. Bull.*, **104**, 1456–1470.

Mulholland, P. J., Wilson, G. V., and Jardine, P. M. 1990. 'Hydrochemical response of a forested watershed to storms: effects of preferential flow along shallow and deep pathways', *Wat. Resour. Res.*, **26**, 3021–3036.

Mucutt, A. D., Wheater, H. S., and Reynolds, B. 1990. 'Stormflow hydrochemistry of a small Welsh upland catchment', *J. Hydrol.*, **116**, 239–249.

Nikolaidis, N. P., Rajaram, H., Schnoor, J. L., and Georgakakos, K. P. 1988. 'A generalized soft water acidification model', *Wat. Resour. Res.*, **24**, 1983–1996.

Norton, S. A., Wright, R. F., Kahl, J. S., and Scofield, J. P. 1992. 'The MAGIC simulation of surface water acidification at, and first year results from, the Bear Brook Watershed manipulation, Maine, USA', *Environ. Pollut.*, **77**, 279–286.

Paces, T. 1983. 'Rate constants of dissolution derived from the measurements of mass balance in hydrological catchments', *Geochim. Cosmochim. Acta*, **47**, 1855–1863.

Plummer, L. N., Parkhurst, D. L., and Thorstenson, D. C. 1983. 'Development of reaction models for groundwater systems', *Geochim. Cosmochim. Acta*, **47**, 665–685.

Plummer, L. N., Busby, J. F., Lee, R. W., and Hanshaw, B. B. 1990. 'Geochemical modeling of the Madison Aquifer in parts of Montana, Wyoming, and South Dakota', *Wat. Resour. Res.*, **26**, 1981–2014.

Plummer, L. N., Prestemon, E. C., and Parkhurst, D. L. 1991. 'An interactive code (NETPATH) for modeling net geochemical reactions along a flow path', *US Geol. Surv. Water Resour. Invest. Rep., 91-4078*.

Renner, R. 1995. 'Predicting chemical risks with multimedia fate models', *Environ. Sci. Technol.*, **29**, 556A–559A.

Rose, K. A., Cook, R. B., Brenkert, A. L., Gardner, R. H., and Hettelingh, J. P. 1991a. 'Systematic comparison of ILWAS, MAGIC, and ETD watershed acidification models 1. Mapping among model inputs and deterministic results', *Wat. Resour. Res.*, **27**, 2577–2589.

Rose, K. A., Brenkert, A. L., Cook, R. B., Gardner, R. H., and Hettelingh, J. P. 1991b. 'Systematic comparison of ILWAS, MAGIC, and ETD watershed acidification models 2. Monte Carlo analysis under regional variability', *Wat. Resour. Res.*, **27**, 2591–2603.

Roth, P., Blanchard, C., Harte, J., Michaels, H., and El-Ashry, M. 1985. 'The American west's acid rain test', *Rep. I*. World Resour. Inst.-Res., Washington, D.C.

Rustad, S., Christophersen, N., Seip, H. M., and Dillon, P. J. 1986. 'Model for streamwater chemistry of a tributary to Harp Lake, Ontario', *Can. J. Fish. Aquat. Sci.*, **43**, 625–633.

Stauffer, R. E. 1990. 'Granite weathering and the sensitivity of alpine lakes to acid deposition', *Limnol. Oceanogr.*, **35**(5), 1112–1134.

Stauffer, R. E. and Wittchen, B. D. 1991. 'Effects of silicate weathering on water chemistry in forested, upland, felsic terrane of the USA', *Geochim. Cosmochim. Acta*, **55**, 3253–3271.

Stone, A., Seip, N. M., Tuck, S., Jenkins, A., Ferrier, R. C., and Harriman, R. 1990. 'Simulation of hydrochemistry in a Highland Scottish Catchment using the Birkenes Model', *Water Air Soil Pollut.*, **51**, 239–259.

Sverdrup, H. and Warfvinge, P. 1992. 'PROFILE: A mechanistic geochemical model for calculation of field weathering rates', in Kharaka, Y. K. and Maest, A. S. (Eds), *Water–Rock Interaction*, Vol. 1. pp. 585–590.

Sverdrup, H. and Warfvinge, P. 1993. 'Calculating field weathering rates using a mechanistic geochemical model PROFILE', *Appl. Geochem.*, **8**, 273–283.

Taylor, A. B. and Velbel, M. A. 1991. 'Geochemical mass balances and weathering rates in forested watersheds of the Southern Blue Ridge: II. Effects of botanical uptake terms', *Geoderma*, **51**, 29–50.

Tipping, E. 1989. 'Acid-sensitive waters of the English Lake District: a steady-state model of streamwater chemistry in the upper Duddon catchment,' *Environ. Pollut.*, **60**, 181–208.

Tonnessen, K. A. 1991. 'The Emerald Lake Watershed study: introduction and site description', *Wat. Resour. Res.*, **27**, 1537–1539.

Velbel, M. A. 1985. 'Geochemical mass balances and weathering rates in forested watersheds of the Southern Blue Ridge', *Am. J. Sci.*, **285**, 904–930.

Velbel, M. A. 1993. 'Constancy of silicate-mineral weathering-rate ratios between natural and experimental weathering: implications for hydrologic control of differences in absolute rates', *Chem. Geol.*, **105**, 89–99.

Williams, S. and Melack, J. M. 1991. 'Solute chemistry of snowmelt and runoff in an alpine basin, Sierra Nevada', *Wat. Resour. Res.*, **27**, 1575–1588.

Wood, W. W. and Low, W. H., 1986. 'Aqueous geochemistry and diagenesis in the eastern Snake River Plain aquifer system, Idaho', *Geol. Soc. Am. Bull.*, **97**, 1456–1466.

Woodhouse, E. G., Bassett, R. L., Orr, T. W., and Thompson, D. L. 1995a. 'Apache Leap Research Site weather data Feb. 1, 1994 through July 31, 1994', *UA DATA REP. 95-001*. University of Arizona, Tucson.

Woodhouse, E. G., Bassett, R. L., Shuttleworth, W. J., and Thompson, D. L. 1995b. 'Field measurement of evapotranspiration and infiltration at the Apache Leap Research Site, Superior, AZ', *EOS, 76, no. 46, F251*.

Woolhiser, D. A., Gardner, H. R., and Olsen, S. R. 1982. 'Estimation of multiple inflows to a stream reach using water chemistry data'. *Trans. A.S.A.E.*, **25**, 616–622.

Woolhiser, D. A., Emmerich, W. E., and Shirley, E. D. 1985. 'Identification of water sources using normalized chemical ion balances: a laboratory test', *J. Hydrol.*, **76**, 205–231.

Woolridge, D. D. 1990. *Manual on Land Use Survey and Capability Classification for Upland, Watersheds* ASEAN-US Watershed Project, College, Laguna, Phillippines.

Wright, R. F. 1982. 'A model for streamwater chemistry at Birkenes, Norway', *Wat. Resour. Res.*, **18**, 977–996.

14

USE OF GEOCHEMICAL MASS BALANCE MODELLING TO EVALUATE THE ROLE OF WEATHERING IN DETERMINING STREAM CHEMISTRY IN FIVE MID-ATLANTIC WATERSHEDS ON DIFFERENT LITHOLOGIES

ANNE K. O'BRIEN, KAREN C. RICE, OWEN P. BRICKER, MARGARET M. KENNEDY AND R. TODD ANDERSON

US Geological Survey, Trenton, NJ, USA

ABSTRACT

The importance of mineral weathering was assessed and compared for five mid-Atlantic watersheds receiving similar atmospheric inputs but underlain by differing bedrock. Annual solute mass balances and volume-weighted mean solute concentrations were calculated for each watershed for each year of record. In addition, primary and secondary mineralogy were determined for each of the watersheds through analysis of soil samples and thin sections using petrographic, scanning electron microscope, electron microprobe and X-ray diffraction techniques. Mineralogical data were also compiled from the literature. These data were input to NETPATH, a geochemical program that calculates the masses of minerals that react with precipitation to produce stream water chemistry. The feasibilities of the weathering scenarios calculated by NETPATH were evaluated based on relative abundances and reactivities of minerals in the watershed. In watersheds underlain by reactive bedrocks, weathering reactions explained the stream base cation loading. In the acid-sensitive watersheds on unreactive bedrock, calculated weathering scenarios were not consistent with the abundance of reactive minerals in the underlying bedrock, and alternative sources of base cations are discussed.

INTRODUCTION

Mass balance techniques have been used widely for several decades as a means to infer watershed processes. Many studies compare solute mass balances between adjacent watersheds on similar bedrock types. Some address specifically the nutrient fluxes (Likens *et al.*, 1977; Swank and Waide, 1988), while others focus on solutes of primarily atmospheric origin (Campbell and Turk, 1988; Campbell *et al.*, 1991; Probst *et al.*, 1992). Several studies document the effects of different vegetation on watershed solute mass balances (Johnson and Swank, 1973; Durand *et al.*, 1992). Mass balance techniques have also been used to approximate geochemical transformations from precipitation to surface water. The method has proven particularly useful for the analysis of stream flow and chemical data collected over a long period of time in intensively studied watersheds throughout the US (Garrels and Mackenzie, 1967; Cleaves *et al.*, 1970; Bricker *et al.*, 1968; Shaffer and Galloway, 1982; Katz *et al.*, 1985; Velbel, 1985; Swank and Waide, 1988; Mast *et al.*, 1990; Bassett *et al.*, 1992). The primary objectives of this research were to: (1) calculate annual solute mass balances for five watersheds in the Appalachian Valley and Ridge and Blue Ridge physiographic provinces in the mid-Atlantic region of the US, and (2) examine the role of mineral weathering in producing observed stream water chemistry in each of the watersheds. This study compares long-term stream solute outputs in watersheds receiving similar atmospheric loadings but situated on different bedrock types. NETPATH

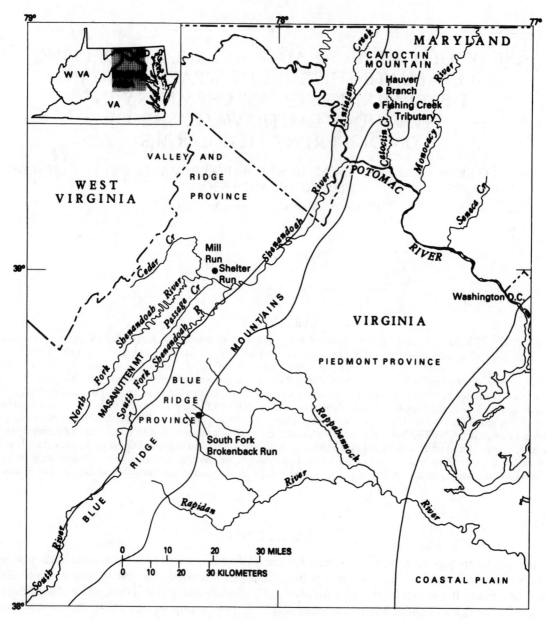

Figure 1. Site map: locations of five mid-Atlantic watersheds

(Plummer *et al.*, 1991), a code for modelling geochemical mass balance along a flow path, was used to calculate masses of individual minerals weathered in each of the watersheds to produce observed stream water chemistry.

DESCRIPTION OF STUDY AREAS

The watersheds selected for this study are in the mid-Atlantic region of the United States and contribute to headwater streams of the Chesapeake Bay (Figure 1, Table I): (1) Mill Run and (2) Shelter Run, located on Massanutten Mountain in the Appalachian Valley and Ridge Physiographic Province of Virginia; (3) South

Table I. Site characteristics

Physiographic Province	Mill Run	Shelter Run	S.F. Brokenback Run	Hauver Branch	Fishing Creek Tributary
	Valley and Ridge	Valley and Ridge	Blue Ridge	Blue Ridge	Blue Ridge
Area (ha)	303	36	279	550	104
Bedrock	Silurian sandstone/shale	Silurian–Devonian sandstone/shale	Precambrian granite	Precambrian metabasalt	Cambrian quartzite
Range in elevation (feet)	1250–2000	1250–1800	1600–3200	1030–1870	550–1283
Vegetation					
Deciduous	75	75	90	90	100
Coniferous	25	25	10	10	0
Stream order	First	First	First	Second	First
Range in mean daily discharge ($l\ s^{-1}$)	0–187	0·06–21	3–184	0·6–2860	1·4–311
Maximum instantaneous discharge ($l\ s^{-1}$)	844	48	198	23050	934
Baseflow pH range	4·1–5·9	6·5–8·2	6·7–8·0	6·5–7·5	5·5–6·5

Table II. Mean annual inputs and outputs of H^+, NO_3^- and SO_4^{2-} in moles $ha^{-1}\ yr^{-1}$ in five mid-Atlantic streams

Watershed	Period of record		(cm)	H^+	NO_3^-	SO_4^{2-}	Alkalinity
Massanutten Mt.	1983–1992	Precipitation	87·7	425	201	216	
		Mill Run	37·7	85	5	435	9
		Shelter Run	33·9	0	3	253	2065
Old Rag Mt.	1983–1992	Precipitation	114·9	595	278	282	
		S.F. Brokenback	39·5	1	9	78	263
Catoctin Mt.	1982–1992	Precipitation	113·8	534	259	315	
		Hauver Branch	52·2	1	154	478	1498
		Fishing Creek Trib.	47·1	3	44	112	193

Fork (S.F.) Brokenback Run, located on Old Rag Mountain in the Blue Ridge Physiographic Province of Virginia; (4) Hauver Branch and (5) Fishing Creek Tributary, located on Catoctin Mountain in the Blue Ridge Physiographic Province of Maryland. The bedrock types underlying these watersheds represent the range of abundant bedrock types in the physiographic provinces studied.

The mid-Atlantic region receives some of the most acidic precipitation in the US (National Atmospheric Deposition Program, 1990). Table II shows the mean annual precipitation volume and mean annual wet inputs of H^+, NO_3^- and SO_4^{2-} measured at the three US Geological Survey precipitation collection stations, Massanutten Mountain (1983–1992), Old Rag Mountain (1983–1992) and Catoctin Mountain (1982–1992). Old Rag Mountain and Catoctin Mountain receive similar amounts of precipitation annually (*ca.* 114 cm), while Massanutten Mountain, which lies in a rain shadow, receives an average of only 88 cm of rain per year. Inputs of H^+, NO_3^- and SO_4^{2-} are similar for all three sites, Massanutten Mountain receiving only slightly lower inputs. Despite similar inputs, different outputs are observed at each of the five watersheds (Table II).

Mill Run and Shelter Run, Massanutten Mountain, Virginia

Mill Run and its tributary Shelter Run are located in a structurally controlled synclinal basin on the eastern limb of the Massanutten Mountain in northern Virginia (Figure 1). These adjacent watersheds occupy 339 ha and lie within the easternmost part of the Valley and Ridge Physiographic Province of the

Appalachian Highlands. The streams are headwaters of the North Fork of the Shenandoah River, which lies in the Potomac River basin.

Mill Run drains to the south-west along the asymmetric axis of the valley floor fault; in the lower part of the basin, the stream abruptly changes course and flows towards the north-west through a narrow fault-controlled gap in the valley sidewall (Olson and Hupp, 1986). Shelter Run flows from the south-east and parallels the regional joint system. Most of the watershed is underlain by the Silurian Massanutten sandstone, consisting of more than 99% quartz sand cemented by interlocking quartz overgrowths (E. K. Rader, personal communication). A member of the Devonian Bloomsburg Formation underlies about 40% of the watershed, although only about 10% lies within the hydrological flow paths of the unsaturated zone (Olson, 1988). In the Mill Run/Shelter Run watershed, the Bloomsburg is primarily ferruginous silty mudstones cemented with haematite or locally contained iron concretions. Mineralogically, it is dominated by an illite mica (Rader and Biggs, 1976). Sandstone and shale colluvium cover most of the basin. The most abundant soils in both the basins are moderately well-drained profiles developed in sandstone colluvium; base saturation values are typically less than 20% (Olson, 1988). Major clay minerals are hydroxy–interlayer–vermiculite and kaolinite (Olson, 1988). Although chlorite was not identified by Olson (1988), recent X-ray analyses of samples collected during a drilling operation indicate the presence of a small amount of chlorite. Mill Run stream water is very dilute, reflecting the unreactive substrate upon which it is situated; the pH varies annually from 4·5 to 5·9 and the acid-neutralizing capacity (ANC) ranges from below 0 to 200 μeq l^{-1}.

The upper reach of the Shelter Run valley is underlain by the same bedrock as the Mill Run valley. However, field observations suggest that the lower reach of Shelter Run flows on a down-dropped section of the Lower Devonian/Upper Silurian rock unit. These strata are mainly calcareous sandstones, shales and limestones. Perennial flow is confined to the lower reach where groundwater discharges as springs along the 200–300 m reach of the channel just upstream from the gauge. In contrast to that of Mill Run, Shelter Run pH ranges from 6·5 to 8·2 and ANC ranges from 100 and 2000 μeq l^{-1}. The upper reach of Shelter Run flows only during storms and is similar in composition to the dilute acidic Mill Run.

South Fork Brokenback Run, Old Rag Mountain, Virginia

S.F. Brokenback Run watershed is located on the southern flank of Old Rag Mountain within Shenandoah National Park in the Blue Ridge Physiographic Province in central Virginia (Figure 1). The 279 ha watershed lies in the Rappahannock River basin. The watershed is underlain by Old Rag granite, a coarse-grained porphyritic rock composed mainly of perthitic microcline and quartz (Gathright, 1976). Saprolite, having a potentially large water and solute storage capacity, covers the bedrock. The saprolite thickness was found to be approximately 20 m on the side slopes; however, bedrock is exposed at lower elevations in the watershed, including the stream bed. This saprolite distribution is similar to that described by Pavich (1986) in a granitic basin in the Piedmont province, Virginia. The stream pH ranges annually from 6·7 to 8 and ANC from 50 to 120 μeq l^{-1}.

Hauver Branch and Fishing Creek Tributary, Catoctin Mountain, Maryland

Hauver Branch is a 550 ha watershed located on Catoctin Mountain in the Blue Ridge Physiographic Province in north-central Maryland (Figure 1). The stream lies within the Potomac River basin. The watershed is underlain by the Catoctin Formation, a metabasalt consisting of plagioclase feldspar, amphibole, chlorite, epidote, clinopyroxene, biotite, sericite-muscovite, quartz and minor amounts of orthoclase and calcite (Katz *et al.*, 1985; Badger, 1989). Soils range in depth from 0 to 50 cm, depending on location, and are thickest in the valley bottoms. Soils are underlain by saprolite derived from and grading into metabasalt (Katz *et al.*, 1985). The depth to unweathered bedrock in the watershed ranges from 0 (exposed bedrock) to 10 m (data on file at Maryland Geologic Survey). Kaolinite is the major clay mineral, although interstratified clays and vermiculite also are present (Olson, 1988). The stream flows perennially with pH ranges from 6·5 to 7·5 and ANC ranges from 150 to 650 µeq l^{-1}. Hauver Branch stream chemistry is affected by the application of deicing salt on roads in the watershed.

Fishing Creek Tributary lies in a 104 ha watershed located on Catoctin Mountain approximately 10 km south of Hauver Branch (Figure 1) within the Potomac River basin. The watershed is underlain by the sedimentary Weverton Formation, which consists primarily of quartz grains cemented by quartz overgrowths. The bedrock also contains feldspathic horizons, local concentrations of magnetite and thin interbedded phyllitic layers which contain chlorite and sericite (Stose and Stose, 1946). Major clay minerals are hydroxy–interlayer–vermiculite and kaolinite (Webster, personal communication). The stream flows perennially with pH ranges from 5·5 to 6·5 and ANC from 0 to 70 µeq l^{-1}.

METHODS

Field data

Wet-only precipitation samples were collected weekly (whenever possible) at Massanutten Mountain, Old Rag Mountain and Catoctin Mountain precipitation stations with Aerochem Metrics wet/dry collectors. Precipitation amounts were recorded with a 5–780 series Belfort weighing bucket and tipping bucket rain gauges connected to CR-21 data loggers.

Stage levels in each of the streams were measured continuously using a float and counterweight apparatus installed within a stilling well. The data were recorded by Campbell CR-10 electronic data loggers at 15-minute intervals. Parshall flumes were installed in Mill Run, Shelter Run and S.F. Brokenback Run, and rating curves were developed accordingly. Rating curves for Hauver Branch and Fishing Creek Tributary are continuously updated using monthly stage–discharge measurements. Discharge was computed from the stage data based on rating curves developed for each of the streams. Grab samples were collected periodically from each of the streams. Hauver Branch has been sampled since 1982, S.F. Brokenback Run, Mill Run and Shelter Run since 1983, and Fishing Creek Tributary since 1987.

Specific conductance and pH of precipitation and stream water samples were measured on unfiltered waters according to recommended procedures for low conductivity waters (Busenburg and Plummer, 1987). Samples were filtered through 0·1 µm (in Maryland) and 0·2 µm (in Virginia) Nuclepore polycarbonate membrane filters. One aliquot was acidified with Baker's Instra-analysed nitric acid to a pH less than 2·0 for cation and silica analysis; the other aliquot was refrigerated at 4°C for anion, ANC and pH analyses.

Laboratory analysis

Analytical methods for determining cation and anion concentrations have changed and improved during the past decade. Anion concentrations (Cl^-, NO_3^-, SO_4^{2-} and NO_2^-) were determined by ion chromatography. Cation concentrations (Na^+, K^+, Mg^{2+}, Ca^{2+}) and SiO_2 were analysed by direct current plasma atomic emission spectrometry (DCP-A). Total aluminum determinations were performed by DCP-A on samples with a pH of less than 5·4. Radiometer's low ionic strength titration system (LIST) was used to determine ANC. Bicarbonate was assumed equal to the ANC, if ANC was greater than zero; if ANC was less than zero, bicarbonate was assumed to be zero. Analytical precision for all analyses is less than 2% relative standard deviation (%RSD) except for Na^+ and K^+, which is less than 3%RSD.

Mineralogy

The mineralogical and chemical composition of the rocks and soils underlying each watershed were compiled from the literature and analysed in the laboratory. A scanning electron microscope, optical microscope and electron microprobe were used to analyse representative samples collected from outcrops in the watershed and from other localities. Table III shows a complete list of minerals, chemical composition and sources of information for formations present in each watershed. Modal abundance is reported for the Old Rag Formation and the Catoctin Formation, the two mineralogically diverse formations included in this study.

Table III. Mineralogy of the formations underlying Mill Run, Shelter Run and S.F. Brokenback, Virginia, and Fishing Creek Tributary and Hauver Branch, Maryland. Unless otherwise noted, generalized formulae are provided. Italicized entries are minerals not important in weathering reactions

Formation	Sources	Mineral	Chemical formula	Modal %
Massanutten Mountain — Mill Run and Shelter Run				
Massanutten	(Harris, 1972;	Biotite	$KMg_3AlSi_3O_{10}(OH)_2$	
Sandstone	Rader and Biggs,	Chlorite	$Mg_5Al_2Si_3O_{10}(OH)_8$	
	1976; Olson, 1988;	Epidote	$Ca_2FeAl_2Si_3O_{12}(OH)$	
Silurian	A. K. O'Brien,	Goethite	$FeO \cdot OH$	
	unpublished data)	*Haematite*	Fe_2O_3	
		Illite*	$K_{0.6}Mg_{0.25}Al_{2.3}Si_{3.5}O_{10}(OH)_2$	
		K-spar	$KAlSi_3O_8$	
		Kaolinite	$Al_2Si_2O_5(OH)_4$	
		Monazite	$(Ce, La, Y, Th)PO_4$	
		Muscovite	$KAl_3Si_3O_{10}(OH)_2$	
		Plagioclase†	$Na_{0.95}Ca_{0.05}Al_{1.05}Si_{2.95}O_8$	
		Quartz	SiO_2	
		Rutile	TiO_2	
		Vermiculite	$Mg_{4.62}Al_{4.667}Si_{5.44}O_{20}(OH)_4$	
		Zircon	$ZrSiO_4$	
Old Rag Mountain — S.F. Brokenback Run — General				
Old Rag	(Gathright, 1976;	*Apatite*	$Ca_5(PO_4)OH$	
Granite	Lukert, 1982;	Actinolite	$Ca_2Mg_5Si_8O_{22}(OH)_2$	
	Piccoli, 1987;	Biotite	$KMg_3AlSi_3O_{10}(OH)_2$	
Precambrian	A. K. O'Brien,	Chlorite	$Mg_5Al_2Si_3O_{10}(OH)_8$	
	unpublished data;	Epidote	$Ca_2FeAl_2Si_3O_{12}(OH)$	
	Jorgensen,	*Garnet*	$Fe_3Al_2(SiO_4)_3$	
	unpublished data)	Goethite	$FeO \cdot OH$	
		Ilmenite	$FeTiO_3$	
		Kaolinite	$Al_2Si_2O_5(OH)_4$	
		K-spar	$KAlSi_3O_8$	
		Muscovite	$KAl_3Si_3O_{10}(OH)_2$	
		Plagioclase	$Na_{0.74}Ca_{0.26}Al_2Si_3O_8$	
		Rutile	TiO_2	
		Sphene	$CaTiO(SiO_4)$	
		Quartz	SiO_2	
		Zircon	$ZrSiO_4$	
Catoctin Mountain — Hauver Branch — General				
Catoctin	(Katz *et al.*, 1985;	Actinolite	$Ca_2Mg_5Si_8O_{22}(OH)_2$	
Formation	Badger, 1989)	Albite	$NaAlSi_3O_8$	
		Biotite‡	$KMg_3AlSi_3O_{10}(OH)_2$	
		Calcite	$CaCO_3$	
		Chlorite	$Mg_5Al_2Si_3O_{10}(OH)_8$	
		Diopside	$CaMgSi_2O_6$	
		Epidote	$Ca_2Al_3SiO_{12}(OH)$	
		Garnet	$Fe_3Al_2(SiO_4)_3$	
		K-spar	$KAlSi_3O_8$	
		Kaolinite	$Al_2Si_2O_5(OH)_4$	
		Mag/Haem	Fe_3O_4, Fe_2O_3	
		Quartz	SiO_2	

Table III. Continued

Formation	Sources	Mineral	Chemical formula	Modal %
Old Rag Mountain — S.F. Brokenback Run				
Old Rag	(Gathright, 1976;	*Apatite*	$Ca_5(PO_4)OH$	tr
Granite	Lukert, 1982;	Actinolite‡	$(Ca_{1.59}Na_{0.01}K_{0.04})(Mg_{2.57}Mn_{0.05}Fe_{2.13}Ti_{0.01}$	tr
	Piccoli, 1987;		$Al_{0.22})Si_8O_{20}(OH)_4$	
Precambrian	A. K. O'Brien,	Biotite‡	$(K_{1.58}Na_{0.03}Ca_{0.01})(Fe_{2.71}Mg_{1.17}Ti_{0.23}Mn_{0.02}$	1·4
	unpublished data;		$Al_{0.52})(Al_{2.63}Si_{5.37})O_{20}(OH)_4$	
	Jorgensen,	Chlorite	$Mg_5Al_2Si_3O_{10}(OH)_8$	1·2
	unpublished data)	Epidote	$Ca_2FeAl_2Si_3O_{12}(OH)$	0·9
		Garnet‡	$(Fe_{2.34}Mg_{0.3}Ca_{0.25}Mn_{0.05}Na_{0.01})Al_{1.99}Si_{3.02}O_{12}$	0·5
		Goethite	$FeO \cdot OH$	
		Ilmenite	$FeTiO_3$	tr
		Kaolinite	$Al_2Si_2O_5(OH)_4$	
		K-spar‡	$(K_{0.57}Na_{0.37}Ca_{0.04}Mg_{0.01}Fe_{0.01})Al_{1.07}Si_{2.93}O_8$	43·9
		Muscovite‡	$(K_{1.39}Na_{0.05}Ca_{0.01})(Mg_{0.1}Fe_{0.16}Ti_{0.01}Al_{3.86})$	1·4
			$Al_{1.57}Si_{6.43}O_{20}(OH)_4$	
		Plagioclase‡	$(Na_{0.76}Ca_{0.18}K_{0.04}Fe_{0.02}Mg_{0.01})Al_{1.16}Si_{2.82}O_8$	17·2
		Rutile	TiO_2	tr
		Sphene	$CaTiO(SiO_4)$	
		Quartz	SiO_2	32·5
		Zircon	$ZrSiO_4$	tr
Catoctin Mountain — Hauver Branch — *Microprobe analyses of Badger (1989)				
Catoctin	(Katz *et al.*, 1985;	Amphibole‡	$(Ca_{1.701}Na_{0.552}K_{0.157})Mg_{2.202}Mn_{0.044}Fe_{2.801}$	2·4
Formation	Badger, 1989;		$Ti_{0.174}Al_{0.908}Si_{7.02}Cl_{0.06}H_{1.711}F_{0.229}O_{24}$	
	R. L. Badger, personal	Albite‡	$Na_{0.936}Ca_{0.023}K_{0.013}Mg_{0.004}Fe_{0.007}Mn_{0.002}$	36·9
	communication)		$Al_{1.053}Si_{2.955}O_8$	
		Biotite‡	$K_{1.871}(Na_{0.022}Ca_{0.014}Mg_{0.515}Fe_{4.347}Ti_{0.217}Mn_{0.048}$	tr
			$Al_{0.532})Al_{2.466}Si_{5.534}O_{20}(OH)_4$	
		Calcite	$CaCO_3$	1·6
		Chlorite‡	$Mg_{2.017}Fe_{2.225}Mn_{0.048}Al_{1.332}Na_{0.004}Ca_{0.032}$	20·6
			$K_{0.197}Al_{0.903}Si_{3.097}O_{10}(OH)_8$	
		Clinopyroxen‡	$(Ca_{0.787}Na_{0.02}K_{0.001}Mg_{0.881}Fe_{0.257}Ti_{0.022}$	1·1
			$Mn_{0.007}Al_{0.036})Al_{0.083}Si_{1.917}O_6$	
		Epidote‡	$(Ca_{1.994}K_{0.006}Ba_{0.002})Mg_{0.055}Fe_{1.089}$	23·9
			$Mn_{0.032}Ti_{0.007}Al_{2.034}Si_{3.128}O_{12}(OH)$	
		Garnet‡	$K_{0.002}Ca_{1.304}Mg_{0.048}Fe_{1.295}Mn_{0.341}$	tr
			$Fe_{0.099}Al_{1.901}Si_{3.024}O_{12}$	
		K-spar	$KAlSi_3O_8$	tr
		Kaolinite	$Al_2Si_2O_5(OH)_4$	
		Mag/Haem	Fe_3O_4, Fe_2O_3	4·6
		Plagioclase‡	$Na_{0.441}Ca_{0.525}K_{0.018}Mg_{0.01}Fe_{0.031}Ti_{0.005}$	tr
			$Al_{1.46}Si_{2.502}O_8$	
		Quartz	SiO_2	1·5
Catoctin Mountain — Fishing Creek Tributary				
Weverton	(Stose and Stose,	Albite	$NaAlSi_3O_8$	
Formation	1946; A. K. O'Brien,	*Apatite*	$Ca_5(PO_4)OH$	
	unpublished data)	Chlorite	$Mg_5Al_2Si_3O_{10}(OH)_8$	
		Epidote	$Ca_2Al_3Si_3O_{12}(OH)$	
		Illite	$K_{0.6}Mg_{0.25}Al_{2.3}Si_{3.5}O_{10}(OH)_2$	
		Muscovite	$KAl_3Si_3O_{10}(OH)_2$	
		Rutile	TiO_2	
		Quartz	SiO_2	
		Sphene	$CaTiO(SiO_4)$	
		Zircon	$ZrSiO_4$	

* General formula provided in NETPATH (Plummer *et al.*, 1991).

† Formula estimated from qualitative energy dispersive X-ray analysis (A. K. O'Brien, unpublished data).

‡ Microscope analyses of Piccoli (1987).

Mass balance calculations and geochemical modelling

Annual solute mass balances were calculated for each of the watersheds for all years of record available by the period-weighted method. Watershed inputs were calculated by multiplying the concentration of the sample representing the period by the corresponding precipitation volume. The sum of the products was divided by the total annual precipitation volume to determine the watershed inputs in mol ha^{-1} yr^{-1}. Watershed outputs were calculated by multiplying the concentration of the sample representing the period by the corresponding runoff volume. The sum of the products was divided by the total annual runoff volume to determine the watershed outputs in mol ha^{-1} yr^{-1}.

Differences between inputs and outputs of base cations are equal to the weathering contribution of solutes plus the change in composition of the exchange complex and biomass:

$$\text{Watershed inputs} - \text{Watershed outputs} = \text{Weathering inputs} + \Delta\text{Exchange pool} + \Delta\text{Biomass}$$

Because we do not have quantitative data on the elemental release and uptake of biomass and the forest floor, nor do we have data concerning dynamics of the exchange pool, we make the null hypothesis that mineral weathering can explain the entire cation loading in each of the streams. By assessing the feasibility of the weathering reactions as determined by NETPATH and evaluating independent inputs such as road salt, we indirectly explore the potential sources of each cation in each of the streams.

The mean volume-weighted annual mean concentrations of cations (Na^+, K^+, Mg^{2+}, Ca^{2+}) and SiO_2 in precipitation and seam water were input to NETPATH along with the primary and secondary minerals compiled in Table III. With these inputs NETPATH finds a plausible set of chemical reactions constrained to fit a set of analytical data; i.e. balanced reactions in the form:

$$\text{Initial water} + \textit{Reactants} = \text{Final water} + \textit{Products}$$

Mean volume-weighted precipitation concentration represents the initial water, and mean flow-weighted stream water concentration represents the final water. Since precipitation inputs are better quantified than throughfall in each watershed, we chose to use mean annual precipitation concentrations as input to the model. We explored, in a preliminary fashion, the use of volume-weighted throughfall concentrations to approximate wet and dry deposition concentrations, and found that weathering amounts determined by NETPATH changed only slightly for the minerals containing solutes enhanced in throughfall. Reactants and products are mineral phases found to dissolve or precipitate in each of the watersheds (Table III). For S.F. Brokenback and Hauver Branch, more detailed mineralogical information was available. The models generated using microprobe data are referred to as specific models, and models generated using general mineral formulae are referred to as general models. From the constraints and phases input, NETPATH calculated models, or 'masses of a set of plausible minerals and gases that must enter or leave the initial solution in order to exactly define a set of selected elemental and isotopic (not used here) constraints observed in a final (evolutionary) water' (Plummer *et al.*, 1991).

We used volume- and discharge-weighted mean concentrations averaged over the entire period of study (10 years) as input to NETPATH. An evaporation factor representing the mean rainfall–runoff ratio was incorporated into each model. In S.F. Brokenback Run, the range in rainfall–runoff ratio was so large (2·0– 6·4) that we allowed NETPATH to calculate the evaporation factor assuming the conservation of chloride. NETPATH output is masses of minerals in mol l^{-1} of input water. To obtain an estimate of relative weathering rates, we multiplied the NETPATH output by mean annual runoff in each of the watersheds.

Mean volume-weighted concentrations and fluxes of Cl^-, SO_4^{2-}, Na^+, Ca^{2+} and Mg^{2+} at Hauver Branch were adjusted to account for the ion contributions of deicing salts. Corrections were applied according to the method of Shanley (1994): the method assumes conservative transport of Cl^-, and small amounts of SO_4^{2-}, Ca^{2+} and Mg^{2+} derived from the salt. The net Cl^{-1} flux, equal to the difference between Cl^{-1} inputs in precipitation and Cl^{-1} outputs in the stream, was assumed to be entirely attributable to deicing salts. The net Cl^{-1} flux was balanced by the Na^+ contribution from salt, equal to the net Na^+ flux minus the Na^+ flux from weathering, the minor amounts of Ca^{2+} and Mg^{2+} in deicing salts and the Ca^{2+} and Mg^{2+} from ion exchange. The Na^+ contribution from weathering was calculated based on the Na^+ to SiO_2 ratio in groundwater. Ca^{2+} and Mg^{2+} from exchange were calculated based on the relative proportions of Ca^{2+} and Mg^{2+} in the stream. Calculated Ca^{2+} and Mg^{2+} contributions from deicing salt and ion exchange were subtracted from the total fluxes measured. SO_4^{2-} in deicing salts was subtracted from the total SO_4^{2-} flux measured (see Table VIII later). Adjusted fluxes of Ca^{2+} and Mg^{2+} and the calculated Na^+ flux from weathering were converted to mean annual volume-weighted concentrations in $\mu eq\ l^{-1}$ and used as input to NETPATH (see Table V later).

RESULTS

The results of this study are presented in three sections. The first section describes the volume-weighted precipitation and stream chemistry for each of the five watersheds. The second section compares solute mass balances for the five watersheds during the entire study period, and the third presents the results of the NETPATH geochemical modelling.

Precipitation and stream chemistry

Table IV compares the mean annual volume-weighted chemical composition of precipitation at the three collection stations. The composition varied little between the sites, although Na^+ and Cl^- concentrations were slightly lower at Massanutten Mountain than at the other two stations. Calcium and Mg^{2+} concentrations were slightly higher at Catoctin Mountain than the other two stations. Differences in NO_3^- and SO_4^{2-} concentrations in precipitation between the sites were less than 3% relative standard deviation.

Table V shows the volume-weighted mean chemical composition of each of the five streams studied. Sulfate was the dominant anion and Mg^{2+} the dominant cation in Mill Run, Virginia and Fishing Creek Tributary, Maryland. Concentrations of Ca^{2+}, Mg^{2+} and SO_4^{2-} were much lower at Fishing Creek Tributary than Mill Run, but NO_3^- and ANC concentrations were higher. In contrast, the dominant anion and cation at Shelter Run, Virginia and Hauver Branch, Maryland were bicarbonate and Ca^{2+}. At S.F. Brokenback Run, Virginia, bicarbonate was the dominant anion, but Na^+ was the dominant cation.

Solute mass balances

Figure 2 shows the annual precipitation and runoff volumes at each of the five mid-Atlantic watersheds. As stated previously, Old Rag and Catoctin Mountain received similar amounts of precipitation;

Table IV. Mean annual precipitation volume and volume-weighted concentrations in $\mu eq\ l^{-1}$

	Period of record	cm	H^+	Ca^{2+}	Mg^{2+}	Na^+	K^+	Cl^-	NO_3^-	SO_4^{2-}
Massanutten	1983–1992	87·7	49·5	7·0	1·6	4·1	1·7	9·5	23·2	51·3
Old Rag	1983–1992	114·9	52·1	6·7	2·0	6·4	2·1	11·6	24·4	49·8
Catoctin	1982–1992	113·8	44·5	8·9	2·7	6·4	1·8	10·8	22·9	51·6

Table V. Mean annual runoff volume and volume-weighted stream concentrations in $\mu eq\ l^{-1}$

	Period of record	cm	H^+	Ca^{2+}	Mg^{2+}	Na^+	K^+	Cl^-	NO_3^-	SO_4^{2-}	HCO_3^-	SiO_2
Mill Run	1983–1992	37·7	21·9	61·6	95·9	27·2	26·9	32·4	1·4	231·8	2·9	83·0
Shelter Run	1983–1992	33·9	0·1	553·2	230·4	38·1	31·2	29·2	0·9	147·8	691·4	102·2
S.F. Brok. R.	1983–1992	39·5	0·3	46·4	23·4	59·0	10·7	26·0	2·9	39·7	68·8	154·9
Hauver Br.	1982–1992	52·2	0·2	281·0	215·4	114·4	7·1	137·6	29·7	182·8	287·9	168·7
Hauver Br.*				195	155	57		24		179		
Fishing Cr. Tr.	1988–1992	47·1	0·7	29·0	44·2	41·3	22·6	37·8	9·5	47·9	41·0	110·5

* Hauver Branch data corrected for deicing salt inputs according to Shanley (in press).

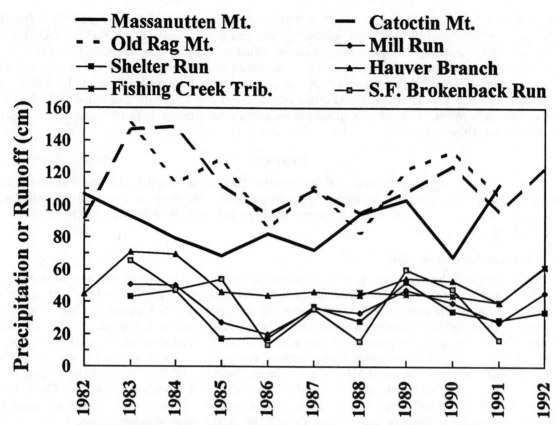

Figure 2. Comparison of 1982–1992 precipitation and runoff (cm) for five mid-Atlantic watersheds

Massanutten Mountain received less precipitation in every year. Consistent with the differences in input volumes, runoff was lower at Mill Run and Shelter Run than Hauver Branch and Fishing Creek Tributary (Figure 2). Runoff volume at S.F. Brokenback Run varied widely over the period of record; runoff measured less than 20 cm in 1986, 1988 and 1991, but was greater than 35 cm in all other years. Total annual inputs and outputs of major anions and cations for each year of record for each of the watersheds are shown in Tables VI–VIII.

Figures 3–11 show the differences in inputs and outputs ($mol^{-1}\ ha^{-1}\ yr^{-1}$) between the watersheds during all years of study. For most solutes, data from each watershed tend to cluster in different areas of the graph. The diagonal line on each figure shows where inputs are equal to outputs. Data that lie above this line indicate an additional atmospheric or watershed source of that solute; data that lie below this line indicate retention or consumption of that constituent in the watershed. Outputs of H^+, Cl^-, NO_3^- and SO_4^{2-}, solutes primarily of atmospheric origin, vary between the watersheds and during the course of the study. Despite high H^+ inputs measured at all sites, substantial H^+ outputs were observed only at Mill Run (Figure 3). Chloride inputs, in contrast, were exceeded by outputs in all years at Fishing Creek Tributary (Figure 4). The greatest Cl^- imbalance was observed consistently at Hauver Branch because of road salt inputs (Table VIII). As discussed previously, these data were corrected for inputs of deicing salts assuming conservation of chloride (Figure 4). In the other watersheds, Cl^{-1} output was greater than inputs in some years and lower than inputs in others. In all watersheds, NO_3^- inputs exceeded outputs (Figure 5). Only very low annual nitrate outputs were observed in the Virginia streams; in the Maryland streams, however, particularly Hauver Branch, NO_3^- outputs were often a substantial fraction of the inputs (Figure 5). In general, SO_4^{2-}

Table VI. Total annual precipitation (PR) or runoff in cm. Mass balances calculated for each year of study at Mill Run (MR) and Shelter Run (SR), Virginia. Annual precipitation inputs and stream outputs in moles ha^{-1} yr^{-1}

		Precip./runoff	% Runoff	H$^+$	Ca^{2+}	Mg^{2+}	Na$^+$	K$^+$	Cl$^-$	NO$_3^-$	SO$_4^{2-}$	HCO$_3^-$	SiO$_2$
1983	PR	106·8		462	33	8	59	29	172	198	235	0	0
	MR	50·7	47	124	172	228	131	139	197	2	522	14	306
	SR	43·1	40	1	748	344	131	135	173	4	328	1617	304
1984	PR	93·0		426	28	5	18	10	53	200	226	0	0
	MR	50·1	54	116	157	246	128	130	147	6	566	10	423
	SR	47·0	51	1	828	399	161	138	133	2	359	1997	69
1985	PR	79·2		404	18	9	63	13	89	174	173	0	0
	MR	27·2	34	74	77	122	86	77	80	4	318	9	250
	SR	17·1	22	0	1011	356	106	50	40	2	113	2627	258
1986	PR	68·5		446	39	7	27	19	45	198	227	0	0
	MR	20·0	29	31	63	116	66	52	61	2	237	20	154
	SR	17·6	26	0	621	250	73	52	45	1	128	1612	216
1987	PR	82·3		409	37	7	63	15	103	219	225	0	0
	MR	36·0	44	60	145	205	99	88	133	4	452	9	273
	SR	37·0	45	0	950	455	126	111	122	4	300	2364	402
1988	PR	72·2		444	48	8	27	13	75	194	206	0	0
	MR	33·2	46	51	119	165	96	94	102	2	415	11	280
	SR	28·0	39	0	810	316	112	92	80	1	206	1941	301
1989	PR	94·4		558	35	10	24	13	68	306	318	0	0
	MR	47·4	50	136	126	204	115	124	145	3	604	6	453
	SR	52·3	55	0	1262	552	182	173	144	6	412	3097	636
1990	PR	103·3		405	21	7	37	19	74	184	193	0	0
	MR	39·5	38	76	102	173	103	111	130	16	416	1	372
	SR	34·0	30	0	786	335	121	110	104	5	243	1948	389
1991	PR	67·9		297	18	4	15	5	40	136	142	0	0
	MR	27·4	40	56	75	125	68	74	83	7	308	8	231
	SR	28·8	41	0	634	276	97	91	79	3	217	1576	295
1992	PR	112·7		394	14	6	24	11	136	189	286	0	0
	MR	45·6	40	127	125	200	113	126	153	5	512	1	373
	SR	33·8	30	0	785	340	118	108	93	2	228	2080	366

outputs were greater than inputs for Mill Run, Shelter Run and Hauver Branch (Figure 6). Outputs were lower than inputs in Fishing Creek Tributary and S.F. Brokenback.

Not surprisingly, outputs of base cations and silica, solutes primarily of watershed origin, exceeded inputs to each of the streams. Annual Na$^+$ outputs in each of the watersheds were fairly similar from year to year; highest Na$^+$ outputs were observed at Hauver Branch and S.F. Brokenback (Figure 7). In general, lower K$^+$ outputs relative to the other watersheds were observed at Hauver Branch and S.F. Brokenback (Figure 8). Ca^{2+} outputs observed at Hauver Branch and Shelter Run were much greater than those observed at Mill Run, Fishing Creek and S.F. Brokenback (Figure 9). A wider range was observed in Mg^{2+} outputs than Ca^{2+} outputs. Greatest Mg^{2+} outputs were observed at Hauver Branch and Shelter Run, followed by Mill Run and Fishing Creek Tributary; lowest outputs were observed at S.F. Brokenback (Figure 10). Highest SiO$_2$ outputs were also observed at Hauver Branch, followed by S.F. Brokenback Run, Fishing Creek Tributary and Shelter Run and Mill Run (Figure 11).

Comparison of geochemical mass balances

We used NETPATH to explore the relative importance of different weathering reactions in determining stream cation loading. Where detailed mineralogical information was available, we compared models based on generalized mineral formulae and those derived from microprobe analyses. NETPATH output is mmoles of mineral per litre of input water; mol ha^{-1} yr^{-1} were calculated by multiplying NETPATH output by

Table VII. Total annual precipitation (PR) or runoff in cm. Mass balances calculated for each year of study at Old Rag, Virginia (OR). Annual precipitation inputs and stream outputs in moles ha^{-1} yr^{-1}

		Precip./runoff	% Runoff	H$^+$	Ca^{2+}	Mg^{2+}	Na$^+$	K$^+$	Cl$^-$	NO$_3^-$	SO$_4^{2-}$	HCO$_3^-$	SiO$_2$
1983	PR	151·2		861	85	15	178	95	380	385	370		
	OR	65·4	43	1	159	75	365	59	223	7	123	388	1007
1984	PR	113·3		550	44	10	92	13	136	263	265		
	OR	47·3	42	2	105	49	239	48	120	4	96	264	733
1985	PR	128·9		579	54	16	93	15	133	278	260		
	OR	54·0	42	1	123	69	355	70	135	11	130	359	945
1986	PR	85·4		543	42	7	42	19	69	249	257		
	OR	13·4	16	0	29	16	91	15	30	6	30	105	173
1987	PR	111·8		565	31	12	115	27	179	316	302		
	OR	35·5	32	1	98	41	198	35	96	7	70	244	588
1988	PR	82·2		422	28	8	40	23	66	202	212		
	OR	15·6	19	1	34	18	92	16	38	4	29	108	229
1989	PR	122·2		798	26	20	25	17	102	347	389		
	OR	60·1	49	4	130	69	355	61	154	9	105	434	956
1990	PR	133·3		539	19	11	77	10	126	233	245		
	OR	47·9	36	1	112	58	278	55	123	15	93	338	731
1991	PR	106·2		501	17	6	32	6	75	227	242		
	OR	16·5	16	0	39	20	97	19	40	14	29	129	258

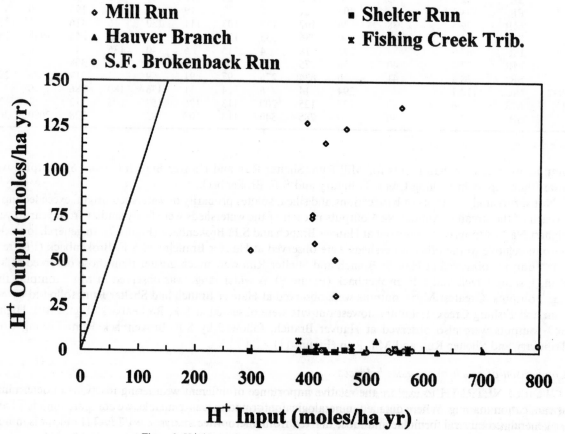

Figure 3. H$^+$ inputs vs outputs in five mid-Atlantic watersheds

Table VIII. Total annual precipitation (PR) or runoff in cm. Mass balances calculated for each year of study at Hauver Branch (HB) and Fishing Creek Tributary (FCT), Maryland. Annual precipitation inputs and stream outputs in moles ha^{-1} yr^{-1}

		Precip./runoff	% Runoff	H$^+$	Ca^{2+}	Mg^{2+}	Na$^+$	K$^+$	Cl$^-$	NO$_3^-$	SO$_4^{2-}$	HCO$_3^-$	SiO$_2$
1982	PR	93.7		441	40	10	92	25	103	225	220	0	0
	HB	44.7	49	0	504	447	496	37	666	199	420	1256	871
	HB*				250	236	296		101		410		
1983	PR	146.8		721	87	25	116	15	245	339	377	0	0
	HB	70.8	48	1	925	653	682	35	754	207	677	1633	1398
	HB*				667	481	475		243		668		
1984	PR	148.5		616	97	18	165	28	158	320	431	0	0
	HB	69.4	47	0	974	703	674	46	814	193	593	1849	1557
	HB*				567	422	529		157		582		
1985	PR	111.9		393	38	19	105	41	119	211	239	0	0
	HB	45.9	42	0	588	492	435	29	520	138	413	1269	756
	HB*				400	345	257		119		406		
1986	PR	94.4		496	51	11	57	17	102	233	259	0	0
	HB	43.8	46	0	624	447	534	37	583	116	407	1228	679
	HB*				477	351	231		102		399		
1987	PR	109.5		517	44	12	78	29	162	301	332	0	0
	HB	46.2	42	6	616	457	562	34	656	67	412	1308	554
	HB*				490	374	188		160		403		
1988	PR	95.1		550	41	12	50	22	95	239	292	0	0
	HB	44.1	46	1	676	509	612	43	662	76	405	1338	744
	HB*				516	401	253		95		395		
	FCT	46.0	48	2	66	99	198	106	201	29	105	203	501
1989	PR	108.2		478	19	6	14	14	52	213	234	0	0
	HB	54.5	50	1	805	608	625	52	763	117	549	1624	819
	HB*				572	447	278		52		537		
	FCT	44.7	41	3	62	99	174	100	155	37	118	156	468
1990	PR	124.3		556	46	17	87	30	134	288	308	0	0
	HB	53.1	43	0	813	637	618	29	836	213	483	1711	889
	HB*				533	433	302		136		471		
	FCT	43.9	35	2	67	102	177	100	164	47	111	177	489
1991	PR	96.4		413	47	21	22	7	94	245	273	0	0
	HB	39.5	41	0	578	455	481	22	652	178	343	1109	621
	HB*				392	321	211		94		333		
	FCT	39.7	41	3	58	85	166	88	147	51	85	165	439
1992	PR	123.2		382	95	36	33	5	94	240	503	0	0
	HB	62.2	50	1	984	762	778	38	882	188	552	2154	975
	HB*				756	603	332		94		538		
	FCT	61.6	50	6	88	133	256	138	217	56	144	263	699

* Data corrected for deicing salt inputs according to Shanley (in press).

runoff volume (Table IX). Although we constrained NETPATH inputs to represent our best understanding of the watershed geochemistry, results were sometimes inconsistent with field observations. The following section presents the weathering scenarios calculated by NETPATH. Discrepancies are addressed in the discussion.

Mill Run, Virginia. Calculations indicate the overall lowest weathering rates in the watersheds studied (Table IX). Weathering reactions include the weathering of biotite, plagioclase and epidote in decreasing order of dissolution rate. Kaolinite and vermiculite are produced during these reactions. NETPATH models indicate that epidote is the main source of Ca^{2+}, while biotite is the main source of dissolved Mg^{2+} and K$^+$ in the stream. Na$^+$ is derived from plagioclase dissolution. Silica precipitates during weathering reactions.

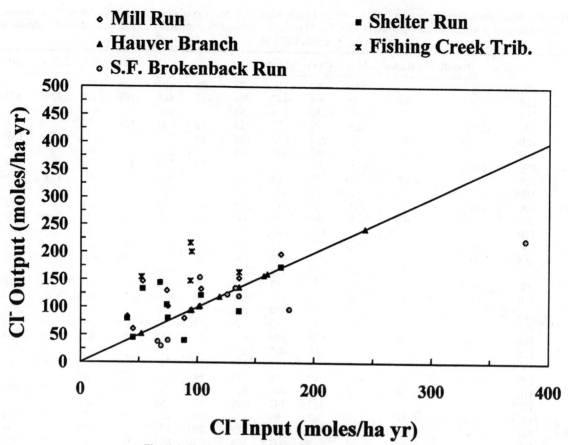

Figure 4. Cl⁻ inputs vs outputs in five mid-Atlantic watersheds

Shelter Run, Virginia. Carbonate minerals apparently are present along the flow path, and calculations indicate the overall highest weathering rates in this watershed (Table IX). NETPATH calculations indicate that more calcite dissolves than any other mineral. Other important weathering reactions include the dissolution of illite, plagioclase and chlorite to produce kaolinite. Calcite is the source of Ca^{2+} in Shelter Run stream water; illite and chlorite are the sources of K^+ and Mg^{2+}, respectively, and plagioclase is the source of Na^+. NETPATH results indicate SiO_2 precipitation in Shelter Run as a product of weathering reactions (Table IX).

S.F. Brokenback Run, Virginia. In both general and specific models, the most important weathering reaction in the S.F. Brokenback Run basin is the dissolution of plagioclase. Plagioclase dissolution is of similar magnitude in both models (Table IX). Also weathering to form kaolinite are actinolite in the specific model and biotite in the general model, although smaller amounts of these minerals are weathered (Table IX). Actinolite is the main source of Ca^{2+} and Mg^{2+} in the specific model and biotite is the main source of Mg^{2+} and K^+ in the general model. Ca^{2+} and Na^{2+} are derived from plagioclase, and K^+ from potassium feldspar in both models (Table IX).

Hauver Branch, Maryland. Models calculated using general mineral formulae differed from those using specific mineral formulae in several ways. The differences are related mainly to the fact that in the specific formulae of amphibole and chlorite some of the magnesium is replaced by other cations, primarily iron and

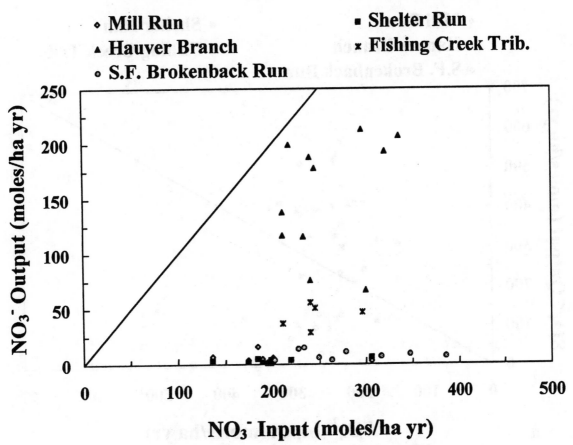

Figure 5. NO$_3^-$ inputs vs outputs in five mid-Atlantic watersheds

calcium (Table III). A small amount of biotite weathered in the general model but not in the specific model, and K-montmarillonite was included as a product in the specific model but not the general model. More amphibole, calcite and chlorite weathered in the specific model than in the general model. The primary weathering reactions in both models of Hauver Branch are the weathering of albite, calcite and chlorite to form kaolinite. Mg^{2+} dissolved in stream water is derived from amphibole (actinolite) and chlorite, and calcite veins present in the Catoctin Formation are a source of Ca^{2+} in both models. In the specific model, amphibole and chlorite are also sources of Ca^{2+}; albite is the source of Na$^+$ in both models. Biotite is the source of K$^+$ in the general model and chlorite is the main source of K$^+$ in the specific model (Table IX).

Fishing Creek Tributary, Maryland. Although reactive minerals are sparse in Fishing Creek Tributary, the most important weathering reactions include the weathering of illite, albite, chlorite and epidote to form kaolinite. In Fishing Creek Tributary watershed, epidote is the source of Ca^{2+}, chlorite and illite the sources of Mg^{2+}, illite the source of K$^+$ and albite the source of Na$^+$. In addition to the SiO$_2$ released by the weathering of silicate minerals, quartz must dissolve in order to match stream silica outputs (Table IX).

DISCUSSION

Comparing and contrasting the solute mass balances of five mid-Atlantic watersheds on different bedrock, shows the relative importance of within-region variations in annual wet deposition and geology on stream

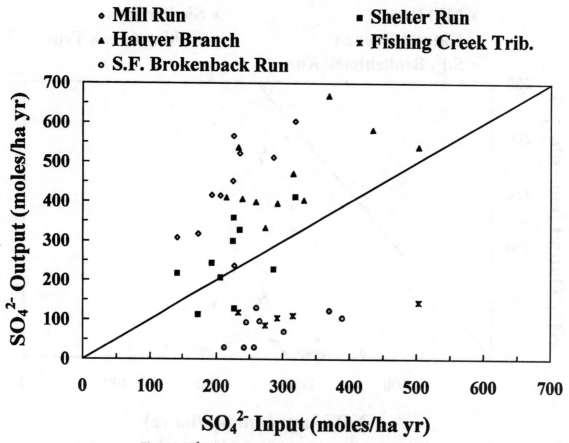

Figure 6. SO_4^{2-} inputs vs outputs in five mid-Atlantic watersheds

solute loadings. Only minor differences were observed in precipitation inputs at the three stations. Higher Ca^{2+} and Mg^{2+} inputs at the Catoctin Mountain station may be explained by the proximity of limestone quarries where particulates may loft into the atmosphere. That Massanutten Mountain typically receives lower amounts of precipitation and Na^+ and Cl^{-1} inputs than the other stations could reflect the different sources of the air masses that contribute to storms at each of the stations. Overall, ten years of data show that similar atmospheric inputs of sulfate and nitrate are observed throughout the mid-Atlantic region of the US. Watershed outputs of these solutes, however, vary over a wide range, even between streams of close proximity.

Very high Cl^- outputs at Hauver Branch are attributed to road salting and data were corrected as described previously (Shanley, 1994). It is not yet understood why Cl^- outputs were always greater than inputs in the Fishing Creek Tributary watershed, or why Cl^- inputs were not consistently related to outputs in the three Virginia watersheds. These data indicate that dry deposition may be a significant source of chloride, as suggested by Juang and Johnson (1967) and Eaton *et al.* (1973), or that chloride is not conservative in these watersheds.

In addition, greater NO_3^- exports were observed from the Maryland watersheds than from the Virginia watersheds. Factors generally thought to control the export of nitrogen species from forested watersheds — forest age, vitality and composition — do not vary significantly between the watersheds. One explanation may be that Fishing Creek Tributary and Hauver Branch watersheds lie closer to frequently travelled, paved roads, and thus are exposed to greater NO_x emissions compared with the Virginia watersheds, which are

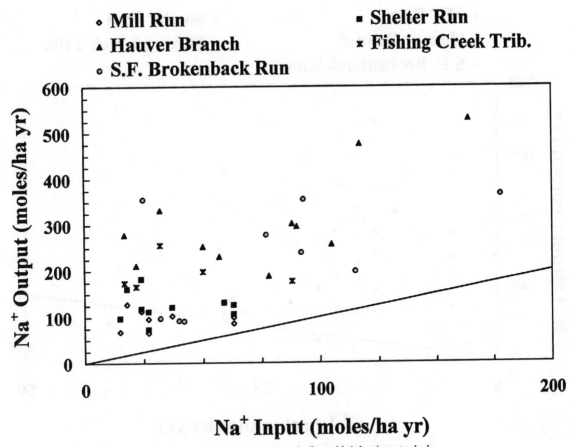

Figure 7. Na$^+$ inputs vs outputs in five mid-Atlantic watersheds

accessed only by non-paved restricted roads. Increased nitrate output as a result of gypsy moth defoliation has been reported for many watersheds in the southern Blue Ridge, such as White Oak Run (Webb *et al.*, 1994). Nitrate outputs at S.F. Brokenback increased in 1990 and 1991, but not to the extent observed in the most damaged watersheds reported by Webb *et al.* (1994). Highest NO$_3^-$ output at S.F. Brokenback was 14 mol ha^{-1} yr^{-1} in 1991, while outputs from White Oak Run were greater than 100 mol ha^{-1} yr^{-1}. It is not known whether the increase at S.F. Brokenback is attributable to gypsy moths.

Dry deposition of sulfur aerosols explains the excess SO$_4^{2-}$ in stream runoff at Hauver Branch, Mill Run and Shelter Run. Because Mill Run and Shelter Run receive the same inputs and neither have a bedrock source of sulfur, the stream water concentrations and exports should be similar. Lower SO$_4^{2-}$ concentrations and exports at Shelter Run indicate possible retention in the Shelter Run watershed. Alternatively, tritium data indicate that Shelter Run is sustained by pre-1950 waters, while Mill Run is sustained by present day waters. The lower sulfate exports may reflect the lower sulfate concentrations in older waters. Lower outputs than wet inputs of SO$_4^{2-}$ at Fishing Creek Tributary and S.F. Brokenback indicate sulfate retention within those watersheds. The difference between outputs and inputs is a conservative estimate of retention, however, because dry deposition is not measured. At S.F. Brokenback watershed, calculations indicate that the amount of sorbed sulfate accumulated below the soil profile is equivalent to nearly 50 years deposition at today's rate (P. Jorgensen, personal communication).

One of the most interesting observations in the solute mass balances is in the comparison of Ca^{2+}, Mg^{2+}, SiO$_2$ and ANC outputs between the acid-sensitive watersheds (ANC < 200) — Mill Run, Fishing Creek

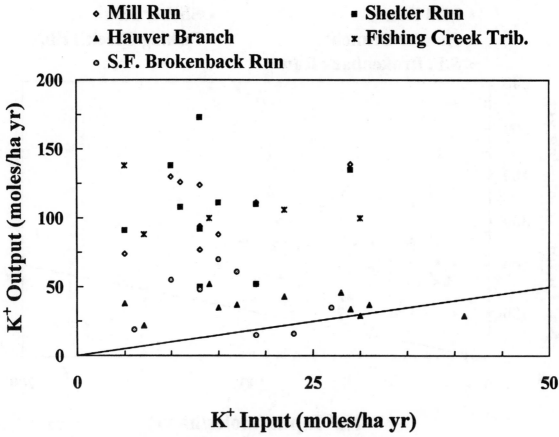

Figure 8. K$^+$ inputs vs outputs in five mid-Atlantic watersheds

Tributary and S.F. Brokenback. Mill Run has the lowest ANC and SiO$_2$ output of the three watersheds. However, of the three watersheds, Mill Run also has the highest Ca^{2+} and Mg^{2+} outputs. These and other differences in watershed outputs of Na$^+$, Mg^{2+}, Ca^{2+} and SiO$_2$ were further explored using NETPATH.

The feasibility of each scenario was first assessed by examining the behaviour of SiO$_2$ in each model. WATEQF calculations indicated saturation of silica in stream water in all of the watersheds included in this study. NETPATH weathering scenarios, however, indicated SiO$_2$ precipitation in only Shelter Run and Mill Run. SiO$_2$ dissolution was indicated in Fishing Creek Tributary and neither precipitation nor dissolution was indicated in S.F. Brokenback Run or Hauver Branch. While SiO$_2$ was not observed to precipitate in any of the streams, secondary silica precipitation was observed in the form of a hard pan layer less than 1 m below the surface in Mill Run (Olson, 1988). During extended dry periods, SiO$_2$ probably precipitated out of concentrated soil solutions. Another explanation for the precipitation of silica in NETPATH weathering scenarios is that not all base cations in the stream are derived from silicate mineral weathering. Alternatively, we may not have accounted for a potential silica sink, such as the silicification of gibbsite to kaolinite. While the presence of gibbsite has not been confirmed in our XRD analysis of subsurface watershed material, gibbsite is difficult to identify unless a very large amount is present. In contrast to Mill Run and Shelter Run models, the NETPATH weathering scenario for Fishing Creek Tributary indicates dissolution of quartz or secondary silica. Quartz is very resistant to weathering and unlikely to contribute a large amount of SiO$_2$ to the stream. A silica-cemented hard pan is not observed at Fishing Creek Tributary watershed. A source of

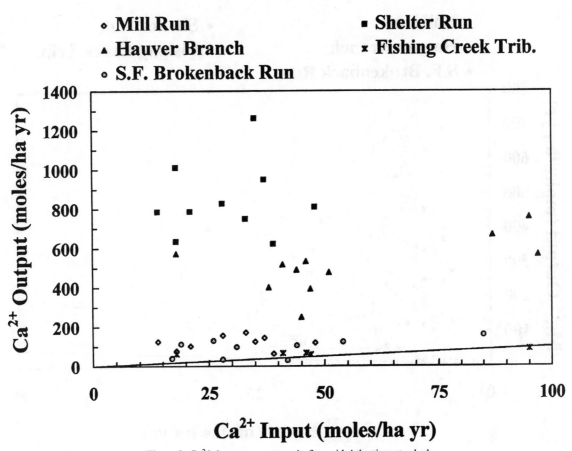

Figure 9. Ca^{2+} inputs vs outputs in five mid-Atlantic watersheds

SiO₂ potentially unaccounted for is the desilicification of kaolinite to gibbsite. As in the other analyses of clay mineralogy, the presence of gibbsite was not confirmed.

The feasibility of the weathering scenario calculated by NETPATH was assessed next, by examining the sources and amounts of minerals weathered to provide the cation loading of each of the streams. The amounts of minerals weathered in each of the watersheds relative to the modal abundance and weatherability was evaluated in the context of the above discussion of silica dissolution and precipitation. Stream water Na^+ in all of the watersheds is attributed to the weathering of plagioclase feldspars of variable composition. A large amount of plagioclase weathering is observed in the S.F. Brokenback Run watershed, consistent with the high modal abundance (7–44%) and calcium content of the plagioclase in this watershed. Plagioclase weathering rates at S.F. Brokenback Run were similar to those measured in Coweeta watersheds underlain by high-grade metasedimentary schists and gneisses (Taylor and Velbel, 1991). Plagioclase is also abundant and important in weathering in the Catoctin Formation underlying Hauver Branch (modal abundance 24–47%). Most of the plagioclase in the Catoctin Formation is albitic; however, some calcic plagioclase is present in the watershed (Badger, 1989; Table IX). The contribution of plagioclase feldspar weathering is much smaller at Mill Run, Shelter Run and Fishing Creek Tributary, consistent with the dominant carbonate mineralogy at Shelter Run and the low abundance of silicate minerals other than quartz at Mill Run and Fishing Creek Tributary.

No consistent patterns in potassium inputs and outputs were observed between the watersheds. Inconsistent outputs may be attributable to the involvement of K^+ in biological reactions.

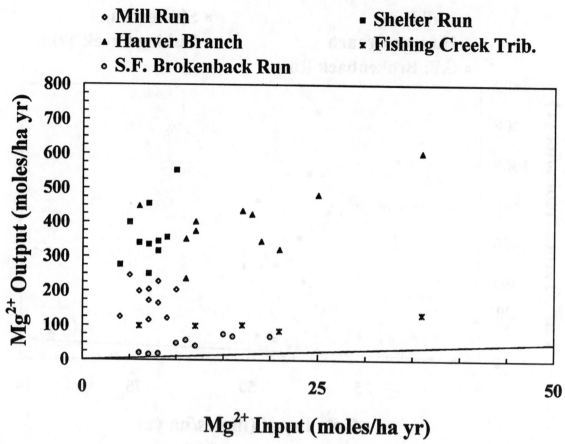

Figure 10. Mg^{2+} inputs vs outputs in five mid-Atlantic watersheds

Calcium outputs were highest in Shelter Run and Hauver Branch (Figure 9), the two watersheds underlain by bedrock containing calcite. Not surprisingly, calculations showed more calcite weathered than any other mineral in all of the weathering scenarios. While Shelter Run is thought to be underlain by calcareous rocks, the Catoctin Formation is only an average of 1·6% calcite. Similarly, 40% of the cations in stream water in the Loch Vale Watershed, Rocky Mountain National Park, Colorado were attributed to the weathering of a small amount of calcite (Mast *et al.*, 1990).

Mill Run, Fishing Creek Tributary and S.F. Brokenback had much lower Ca^{2+} outputs than Shelter Run and Hauver Branch. The potential mineral source of Ca^{2+} in Mill Run and Fishing Creek tributary is epidote; however, the amount of epidote weathering required to produce the amount of Ca^{2+} output in the stream is more than seems reasonable given that the mineral is not very abundant (trace) or reactive. Field observations of abundant epidote associated with the resistant heavy mineral suite and in veins of quartz in the Catoctin Formation suggest that epidote is resistant to weathering. Despite the abundance of the reactive plagioclase in S.F. Brokenback Run, Ca^{2+} outputs from S.F. Brokenback Run are lower than those from Mill Run and Fishing Creek Tributary. This discrepancy may indicate that the calcium-contributing phases in the models of Mill Run and Fishing Creek Tributary are not correct, or that Ca^{2+} is derived from a non-mineral source. The precipitation of SiO_2 in the weathering scenario at Mill Run is consistent with the latter explanation. Other potential sources and sinks of cations in watersheds include biomass and the exchange pools as documented by Paces (1986) and Velbel (1986). The acidic water moving through the Mill Run and Fishing Creek Tributary watersheds may enhance the degradation of organic material or the breakdown of

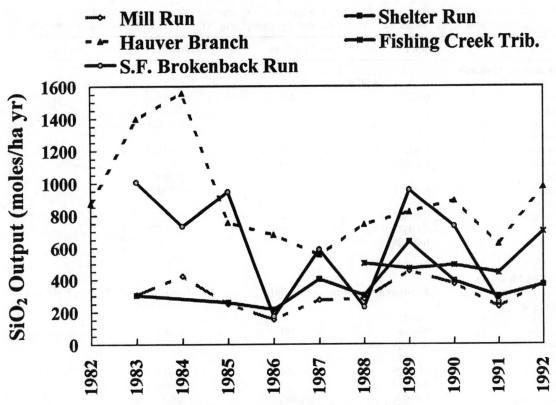

Figure 11. SiO₂ outputs in five mid-Atlantic watersheds

biological cycles, increasing the organic matter transported to the stream. It is possible that the cations in Mill Run and Fishing Creek Tributary are incorporated into or bound to organic matter derived from these processes. This is supported in comparatively higher cation : anion ratios in Mill Run compared with Shelter Run in summer months when cation concentrations are highest. However, the cation : anion ratios in most samples are greater than 0·9 and less than 1·1, or within the precision of the analyses.

Magnesium outputs were highest in Hauver Branch (Figure 10) and are attributed to the weathering of abundant chlorite and some amphibole (actinolite) in the Catoctin Formation. Magnesium outputs were lowest in S.F. Brokenback Run (Figure 10) and are attributed to actinolite in the specific model and biotite in the general model. Consistently, no magnesium-bearing minerals are present in modal abundance greater than 2% in the S.F. Brokenback watershed (Table III). Although magnesium-bearing minerals are not abundant in the Mill Run, Shelter Run and Fishing Creek Tributary watersheds, Mg^{2+} outputs are higher than in S.F. Brokenback Run. Mg^{2+} outputs at Shelter Run and Fishing Creek Tributary are attributed to illite and chlorite dissolution and Mg^{2+} outputs at Mill Run are attributed mainly to biotite dissolution. That illite is resistant to weathering, and chlorite and biotite are not abundant in any of these watersheds, suggests that we do not have the correct Mg^{2+}-bearing phase in the NETPATH input, or that the Mg^{2+} is derived from a non-silicate mineral source. In the Shelter Run watershed waters may contact dolomitic rocks.

Where comparisons were possible, several differences were found between specific and general weathering models of the watersheds. Mineral suites in the S.F. Brokenback Run models were the same except for one reaction. Biotite dissolution was included with no actinolite dissolution in the general model, while actinolite dissolution was included without biotite dissolution in the specific model. The other difference was that more

Table IX. NETPATH results — mmol mineral weathered to produce observed stream chemistry per litre of precipitation input. Moles ha^{-1} yr^{-1} obtained by multiplying NETPATH value by outcome volume

Mineral	mmol l^{-1}	mol ha^{-1} yr^{-1}	mmol l^{-1}	mol ha^{-1} yr^{-1}
Massanutten Mountain				
	Mill Run		Shelter Run	
Biotite	0·02282	200		
Calcite			0·26543	2325
Chlorite			0·02036	178
Clinopyroxene				
CO$_2$ gas	0·0029	25	0·42603	3732
Epidote	0·01075	94		
Illite			0·04407	386
Kaolinite	−0·01776	−156	−0·0854	−748
K-spar				
Plagioclase	0·01802	158	0·02773	243
SiO$_2$	−0·0087	−76	−0·02407	−211
Vermiculite	−0·00488	−43		
Evap. fac	2·4		2·8	
Old Rag Mountain				
	S.F. Brokenback Run — General		S.F. Brokenback Run — PMP data*	
Actinolite			0·00344	40
Biotite	0·00315	36		
CO$_2$ gas	0·0688	791	0·0688	791
Epidote	0·00006	1	0·00004	0
Illite				
Kaolinite	−0·01429	−164	−0·01034	−119
K-spar	0·00284	33	0·00637	73
Plagioclase	0·06034	693	0·05561	639
Vermiculite				
Evap. fac	2·2		2·2	
Catoctin Mountain				
	Hauver Branch — General		Hauver Branch — RLB data†	
Albite	0·04342	494	0·03701	421
Amphibol	0·00894	102	0·01577	179
Biotite	0·00314	36		
Calcite	0·06984	795	0·05942	676
Chlorite	0·00408	46	0·01967	224
CO$_2$ Gas	0·21808	2482	0·2285	2600
Epidote				
Illite				
Kaolinite	−0·02736	−311	−0·03559	−405
K-Mont.			−0·01118	−127
SiO$_2$				
Evap. fac	2·2		2·2	
	Fishing Creek Tributary			
Chlorite	0.00225	26		
CO$_2$ gas	0.041	467		
Epidote	0·00202	23		
Illite	0·03062	348		
Kaolinite	−0·05363	−610		
K-spar				
Albite	0·02626	299		
SiO$_2$	0·01899	216		
Evac. fac	2.35			

* NETPATH run using mineralogical composition data from Piccoli (1987).
† NETPATH run using mineralogical composition data from Badger (1989).

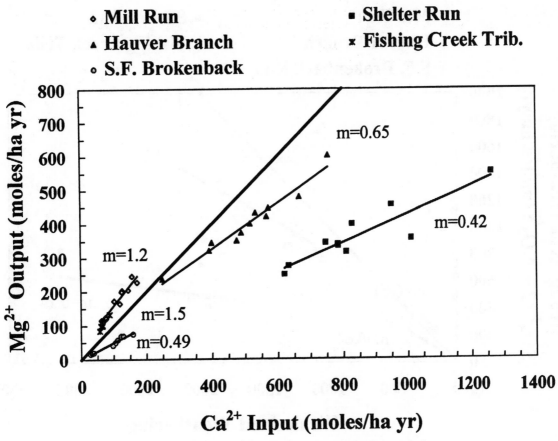

Figure 12. Ca^{2+} vs Mg^{2+} outputs in five mid-Atlantic watersheds

than twice as much K-spar weathered in the specific model than in the general model; however, biotite and K-spar weathering were of relatively minor importance compared with that of plagioclase. Biotite dissolution was included in the general Hauver Branch model but not in the specific model. The main discrepancies observed between the general and specific Hauver Branch models, however, were among the ratios of albite to ferromagnesium mineral (chlorite + amphibole), and chlorite to amphibole weathering. The specific model indicated an albite to ferromagnesium mineral weathering ratio of 1·0, while the general model indicated a ratio of 3·3. Since mineral formulae in the specific model included more iron than the general formulae (Table III), the lower albite to ferromagnesium mineral ratio in the specific model is consistent with observations that mineral stability decreases with increasing iron content (B. Jones, personal communication). In addition, the ratio in the specific model better reflects the actual modal abundances in the Catoctin Formation (1·6). The ratio of chlorite to amphibole in the specific model was 1·2, compared with 0·5 in the general model. The chlorite to amphibole ratio in the Catoctin Formation is about 8·5. Although minerals do not necessarily weather at rates proportional to their relative abundances, this information can be used to assess roughly the NETPATH output. The study indicates more reasonable watershed mass balances may be calculated if specific mineralogic composition and modal abundance are available for the underlying formation. Further work is indicated to assess the feasibility of the general and specific models.

Figure 13 illustrates the watershed sensitivity to acid deposition and the importance of weak versus strong acid weathering in each of the watersheds. The H^+ consumed in strong acid weathering is equal to two times the SO_4^{2-} export plus the NO_3^- export in moles per hectare per year. One mole of HCO_3^- export in stream

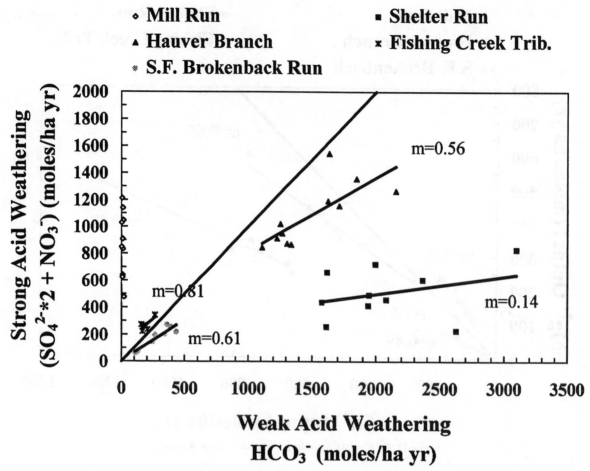

Figure 13. Weak vs strong acid weathering in five mid-Atlantic watersheds

water represents one mole of H^+ consumed in weak acid weathering reactions; therefore, the H^+ consumed in weak acid weathering is equal to the HCO_3^- output. The percentage H^+ consumed in strong acid weathering is represented by the slope of the lines (Figure 13). Only 14% of the H^+ consumed in weathering is attributed to strong acid weathering in Shelter Run. About 56% of the H^+ consumed in weathering is attributed to strong acid weathering at Hauver Branch, and 61 and 81% at S.F. Brokenback and Fishing Creek Tributary, respectively. The relative importance of strong and weak acid weathering is consistent with the ability of each bedrock to buffer incoming acidic deposition. Data plotting close to the 1 : 1 diagonal line, such as S.F. Brokenback Run and Fishing Creek Tributary, indicate that the watershed has little buffering capacity, but neutralizes acidic inputs. Mill Run exports no bicarbonate, indicating that this watershed has no buffering capacity. Hauver Branch may export as much $SO_4^{2-} + NO_3^-$ in some years as Mill Run, but at the same time exports substantially more bicarbonate. Shelter Run $SO_4^{2-} + NO_3^-$ outputs are lower than those at Mill Run, but bicarbonate outputs are the highest of the five watersheds.

CONCLUSIONS

A comparative study of solute and geochemical mass balances shows the relative importance of various watershed processes in determining the stream water chemistry. Results show that concentrations and inputs of SO_4^{2-} and NO_3^- in precipitation are fairly consistent between the three precipitation stations, but that local quarries may explain minor discrepancies in cation inputs. Despite similar concentrations in precipitation and similar inputs, stream solute concentrations and outputs vary greatly between the watersheds. At each watershed, however, mean volume-weighted concentrations and outputs of these constituents are similar from year to year.

NETPATH results indicate the possible combinations of minerals that weather to produce different stream cation concentrations and outputs. Discrepancies in the expected contribution of cations relative to the abundance and reactivities of the minerals between the watersheds suggest other cation inputs to the stream, especially in watersheds underlain by unreactive bedrock. We suggest cations bound to organic material as possible sources that we have not quantified. Weak to strong acid-weathering ratios are consistent with the relative reactivities of each of the bedrock types underlying the watersheds. The results of this study confirm that geology is an excellent predictor of the relative buffering capacity of a watershed and the relative importance of weak versus strong acid weathering (Bricker and Rice, 1988), but indicate that stream chemistry is not quantitatively derived from bedrock geology alone.

ACKNOWLEDGEMENTS

This research was conducted in part through funding from the US Geological Survey under projects 4384-10900, 4384-12800 and 4424-0970, in cooperation with the Maryland Department of the Environment and the Maryland Department of Natural Resources. This document has been subjected to USGS peer and administrative reviews and has been approved for publication. Mention of trade names or commercial products does not constitute endorsement or recommendation for use. The authors thank P. Michael Shackelford, Chris Carter and Amanda Sigillito for their work in the field, Elaine and Jim McGee for their assistance with the scanning electron microscope and microprobe, and Danny Webster for his assistance in analysing clay mineralogy. The manuscript was improved by the suggestions of our reviewers Tanya Furman and Michael Velbel and the insights of Blair Jones.

REFERENCES

Badger, R. L. 1989. 'Geochemistry and petrogenesis of the Catoctin volcanic province', *Ph.D. Dissertation*, Virginia Polytechnic Institute and State University, Blacksburg. 337 p.

Bassett, R. L., Miller, W. R., McHugh, J., and Catts, J. G. 1992. 'Simulation of natural acid sulfate weathering in an alpine watershed', *Wat. Resour. Res.*, **28**, 2197–2209.

Bricker, O. P. and Rice, K. C. 1988. 'Acidic deposition to streams', *Environ. Sci. Technol.*, **23**, 379–385.

Bricker, O. P., Godfrey, A. E., and Cleaves, E. T. 1968. 'Mineral-water interaction during the chemical weathering of silicates', in *Trace Inorganics in Water*, Advances in Chemistry Series, 73. American Chemical Society, Washington, D.C. pp. 128–142.

Busenberg, E. and Plummer, L. N. 1987. 'pH measurements of low-conductivity waters', Water Resources Investigations, *Report 87-4060*. US Geological Survey.

Campbell, D. H. and Turk, J. T. 1988. 'Effects of sulfur dioxide emissions on stream chemistry in the western United States', *Wat. Resour. Res.*, **24**, 871–878.

Campbell, D. H., Turk, J. T., and Spahr, N. E. 1991. 'Response of Ned Wilson Lake Watershed, Colorado, to changes in atmospheric deposition of sulfate', *Wat. Resour. Res.*, **27**, 2047–2060.

Cleaves, E. T., Godfrey, A. E., and Bricker, O. P. 1970. 'Geochemical balance of a small watershed and its geomorphic implications', *Geol. Soc. Am. Bull.*, **81**, 3015–3032.

Durland, P., Lelong, F., and Neal, C. 1992. 'Comparison and significance of annual hydrochemical budgets in three small granitic catchments with contrasting vegetation (Mont-Lozere, France)', *Environ. Pollut.*, **75**, 223–228.

Eaton, J. S., Likens, G. E., and Bormann, F. H. 1973. 'Throughfall and stemflow chemistry in a northern hardwood forest', *J. Ecol.*, **61**, 495–508.

Harris, W. B. 1972. 'High-silica resources of Clarke, Frederick, Page, Rockingham, Shenandoah, and Warren Counties, Virginia', *Mineral Resources Report 11*. Virginia Division of Mineral Resources, Charlottesville. 43 pp.

Garrels, R. M. and Mackenzie, F. T. 1967. 'Origin of the chemical composition of some springs and lakes', in Gould, R. F. (Ed.), *Equilibrium Concepts in Natural Water Systems*, Advances in Chemistry Series, Vol. 67. American Chemical Society, Washington, D.C. pp. 222–242.

Gathright, T. M. 1976. *Geology of the Shenandoah National Park, Virginia*, Bulletin 86. Virginia Division of Natural Resources, Charlottesville. 93 pp.

Gathright, T. M. and Frischmann, P. S. 1986. *Geology of the Harrisonburg and Bridgewater Quadrangles, Virginia*, Publication 60. Virginia Division of Mineral Resources, Charlottesville.

Johnson, P. L. and Swank, W. T. 1973. 'Studies of cation budgets in the southern Appalachians on four experimental watersheds with contrasting vegetation', *Ecology*, **54**, 70–80.

Juang, F. H. T. and Johnson, N. M. 1967. 'Cycling of chlorine through a forested catchment in New England', *J. Geophys. Res.*, **72**, 5641–5647.

Katz, B. G., Bricker, O. P., and Kennedy, M. M. 1985. 'Geochemical mass-balance relationships for selected ions in precipitation and stream water, Catoctin Mountains, Maryland', *Am. J. Sci.*, **285**, 931–962.

Likens, G. E., Bormann, F. H., Pierce, R. S., Eaton, J. S., and Johnson, N. M. 1977. *Biogeochemistry of a Forested Ecosystem*. Springer, New York. 146 pp.

Lukert, M. T. 1982. 'Uranium-lead isotope age of the Old Rag Granite, Northern Virginia', *Am. J. Sci.*, **282**, 391–398.

Mast, M. A., Drever, J. I., and Baron, J. 1990. 'Chemical weathering in the Loch Vale Watershed, Rocky Mountain National Park, Colorado', *Wat. Resour. Res.*, **26**, 2971–2978.

National Atmospheric Deposition Program (NADP), 1990. *NADP/NTN Annual Data Summary: Precipitation Chemistry in the United States — 1990*. Natural Resources Ecological Laboratory, Colorado State University, Fort Collins.

Olson, C. G. 1988. 'Clay-mineral contribution to the weathering mechanisms in two contrasting watersheds', *J. Soil Sci.*, **39**, 457–467.

Olson, C. G. and Hupp, C. R. 1986. 'Coincidence and spatial variability of geology, soils, and vegetation, Mill Run watershed, Virginia', *Earth Surf. Process. Landf.*, **11**, 619–629.

Paces, T. 1986. 'Rates of weathering and erosion derived from mass balance in small drainage basin', in Colman, S. M. and Dethier, D. P. (Eds), *Rates of Chemical Weathering of Rocks and Minerals*. Academic Press, Orlando. pp. 531–550.

Pavich, M. J. 1986. 'Processes and rates of saprolite production and erosion on a foliated granitic rock of the Virginia Piedmont', in Colman, S. M. and Dethier, D. P. (Eds), *Rates of Chemical Weathering of Rocks and Minerals*. Academic Press, Orlando. pp. 551–590.

Peters, N. E. 1991. 'Chloride cycling in two forested lake watersheds in the west-central Adirondack Mountains, New York, USA', *Water Air Soil Pollut.*, **59**, 201–215.

Piccoli, P. M. 1987. 'Petrology and geochemisty of the Old Rag Granite on Old Rag Mountain, Central Virginia', *Masters Thesis*, University of Pittsburgh, Pittsburgh, 227 pp.

Plummer, L. N., Prestemon, E. C., and Parkhurst, D. L. 1991. 'An interactive code (NETPATH) for modeling geochemical reactions along a flowpath', US Geological Survey, Water Resources Investigations. *Report 91-4078*.

Probst, A., Viville, D., Fritz, B., Ambroise, B., and Dambrine, E. 1992. 'Hydrochemical budgets of a small forested granitic catchment, exposed to acid deposition: the Strengbach catchment case study (Vosges Massif, France)', *Water Air Soil Pollut.*, **62**, 337–347.

Rader, E. K. and Biggs, T. H. 1976. *Geology of the Strasburg and Toms Brook Quadrangle, Virginia*, Report of Investigations 45. Virginia Division of Mineral Resources, Charlottesville. 104 pp.

Shaffer, P. W. and Galloway, J. N. 1982. 'Acid precipitation: the impact on two headwater streams in Shenandoah National Park, Virginia', in Herrmann, R. and Johnson, A. I. (Eds), *International Symposium on Hydrometeorology*. American Water Resources Association. pp. 43–53.

Shanley, J. B. 1994. 'Effects of ion exchange on stream solute fluxes in a basin receiving highway deicing salts', *J. Environ. Qual.*, in press.

Stose, G. W. and Stose, A. J. 1946. 'Geology of Carroll and Frederick Counties', in *The Physical Features of Carroll County and Frederick County*. Maryland Department Geology, Mines, Water Resources, Baltimore. pp. 11–131.

Swank, W. and Waide, J. 1988. 'Characterization of baseline precipitation and stream chemistry and nutrient budgets for control watersheds', in Swank, W. and Crossley, D., Jr. (Eds), *Forest Hydrology and Ecology at Coweeta*. Springer-Verrlag, New York. pp. 58–79.

Taylor, A. B. and Velbel, M. A. 1991. 'Geochemical mass balances and weathering rates in forested watersheds of the southern Blue Ridge II. Effects of botanical uptake terms', *Geoderma*, **51**, 29–50.

Velbel, M. A. 1985. 'Geochemical mass balances and weathering rates in forested watersheds of the southern Blue Ridge', *Am. J. Sci.*, **285**, 904–930.

Velbel, M. A. 1986. 'The mathematical basis for determining rates of geochemical and geomorphic processes in small forested watersheds by mass balance: examples and implications', in Colman, S. M. and Dethier, D. P. (Eds), *Rates of Chemical Weathering of Rocks and Minerals*. Academic Press, Orlando. pp. 439–451.

Webb, J. R., Diviney, F. A., Galloway, J. N., Rinehart, C. A., Thompson, P. A., and Wilson, S. 1994. 'The acid-base status of native brook trout streams in the mountains of Virginia: a regional assessment based on the Virginia Trout Stream Sensitivity Study', submitted to Virginia Department of Game and Inland Fisheries, US Fish and Wildlife Service, Shenandoah National Park, George Washington National Forest, Jefferson National Forest, Virginia Council of Trout Unlimited. University of Virginia, Charlottesville.

15

CHEMICAL MASS BALANCE AND RATES OF MINERAL WEATHERING IN A HIGH-ELEVATION CATCHMENT, WEST GLACIER LAKE, WYOMING

JIM B. FINLEY* AND JAMES I. DREVER

Department of Geology & Geophysics, University of Wyoming, Laramie, WY, USA

INTRODUCTION

Mass balance simply means a budget. In the context of catchment studies, it is a budget that describes fluxes of solutes into and out of a catchment, and assigns the solutes to specific sources and sinks such as atmospheric deposition, biomass change or bedrock weathering. Although the concept is simple, identifying and quantifying the various contributions to the individual fluxes is not easy, and the final budget often has ambiguities and uncertainties. We shall discuss here some of the issues associated with assigning the solutes leaving a catchment to specific mineral weathering reactions, a subject that has a long history in the geochemical literature (e.g. Garrels and Mackenzie, 1967; Tardy, 1971; Pačes, 1983; Katz *et al.*, 1985; Clayton, 1986; Drever and Hurcomb, 1986; Velbel, 1986; Baron and Bricker, 1987; Mast *et al.*, 1990; Kirkwood and Nesbitt, 1991; Drever and Zobrist, 1992; Miller *et al.*, 1993; Blum *et al.*, 1994).

A conceptual model of the mass balance equation, calculated from aqueous concentrations of solutes, is:

$$(\text{Solute output})_i - (\text{Solute input})_i = (\text{Solute from weathering})_i \pm \Delta E_i \pm \Delta B_i \tag{1}$$

where (Solute output) is the mass of solute i discharged from the catchment per unit time; (Solute input) is the mass of solute i supplied to the catchment from the atmosphere per unit time; (Solute from weathering) is the mass of solute i generated by chemical weathering of bedrock minerals per unit time; ΔE is the change in the concentration of solute i on the soil exchange complex per unit time; and ΔB is the change in the concentration of solute i associated with biological processes per unit time (Drever, 1988; Mast *et al.*, 1990). The common units of measure are in moles of solute i per hectare per year, or as moles of i per litre of runoff per year. The choice of units depends on how well the water balance is known. The solutes considered in this paper for calculating mineral weathering reactions are the major cations (Ca^{2+}, Mg^{2+}, Na^+, K^+) and dissolved silica.

Terms on the left-hand side of Equation (1) are most readily accessible using field data; thus certain assumptions must first be made in order to differentiate between the biogeochemical processes represented by the three terms on the right-hand side of Equation (1). First, if the chemical composition of precipitation is relatively constant, or varies within a well-defined range of concentration with time, the distribution of cations associated with the exchange complex in the soil should reach a steady state. That is to say, the concentration of cations on the exchange sites does not change with time, which means that the exchange complex cannot be a *net* source of cations. Secondly, if the biological community is in a steady state, ΔB, the biological term, will be zero, at least for non-volatile species. A steady state here implies that new growth each year is exactly balanced by decomposition over an annual cycle. This condition is unlikely to be fulfilled under a productive forest (e.g. Likens *et al.*, 1977; Taylor and Velbel, 1991); however, in an alpine environment with low biomass and low annual productivity, net annual biomass accumulation of solutes is often considered to be negligibly small.

Recent research studies of hydrochemical cycling of solutes in high-elevation catchments have focused on understanding the effects of acid deposition on surface water chemistry (e.g. Turk and Campbell, 1987;

* Presently at: Shepherd Miller Inc., 3801 Automation Way, Suite 100, Fort Collins, CO 80525, USA.

Rochette *et al.*, 1988; Frogner, 1990; Mast *et al.*, 1990; Stauffer, 1990; Denning *et al.*, 1991; Baron, 1992; Drever and Zobrist, 1992; Williams *et al.*, 1993). Since the only geochemical process capable of neutralizing atmospheric acidity over the long term is weathering of soil and bedrock minerals (Likens *et al.*, 1977; Drever, 1988), a major part of these studies has been to use the mass balance method to identify the mineral weathering reactions responsible for the generation of solutes, and to quantify the rates at which they occur. If mineral weathering rates are to be determined in this way, the cation exchange and biological terms in Equation (1) must be small compared with the term for (Solutes from weathering), or they must be known from independent measurement. Ideally, long-term, weighted average concentrations of solutes should be used in order to minimize the possible effects of cation exchange and biotic uptake/ release.

An additional assumption in applying Equation (1) is that 'weathering' represents the chemical alteration of bedrock minerals, with no influence from aeolian dust. Aeolian deposition of sediments in alpine catchments has been identified, but not quantified, in several alpine catchments (e.g. Caine, 1974; Shroba and Birkeland, 1983; Litaor, 1987; Rochette *et al.*, 1988; Finley *et al.*, 1993). At present, there has not been an adequate assessment of long-term deposition of aeolian sediments on calculated rates of mineral weathering.

The final result of a mass balance calculation to determine mineral weathering rates is an overall weathering reaction that 'explains' the observed chemical composition of stream water draining the catchment (e.g. Garrels and Mackenzie, 1967; Stoddard, 1987; Mast *et al.*, 1990; Drever and Zobrist, 1992; Velbel, 1992; Williams *et al.*, 1993). We will, (1) present the results of chemical mass balance calculated for two tributary drainages in a high-elevation catchment located in the Snowy Range of south-eastern Wyoming, and (2) use the results as a vehicle to discuss some of the problems associated with the determination of mineral weathering reactions in high-elevation catchments by mass balance calculations.

SITE DESCRIPTION

West Glacier Lake is located in a small (62 ha), high-elevation headwater catchment located *ca.* 65 km west of Laramie in south-eastern Wyoming (106° 15'W, 40° 23'N). Base elevation is 3275 m, with the spine of the Snowy Range (3500 m) forming an arcuate boundary on the north-west side of the basin (Figure 1). Bedrock consists of Medicine Peak Quartzite (*ca.* 85% by area), cross-cut by amphibolite dikes (*ca.* 15% by area), both of which are Precambrian in age (Karlstrom and Houston, 1984). Medicine Peak Quartzite consists of quartz (60–100%), phyllosilicates (0–38%), kyanite (0–25%) and feldspar (*ca.* 3%; mainly albite, An_7) (Flurkey, 1983). Minerals present in the amphibolite dike include actinolite (44%), epidote (30%), albite (An_3, 7–15%), quartz (0–7%), chlorite (0–5%), sphene (0–4%), biotite (0–2%) and less than 0·5% opaque minerals including chromite and pyrite (Rochette, 1987). Thus, the West Glacier Lake catchment presents an opportunity to compare rates of weathering of ferromagnesian silicates with analogous rates calculated for high-elevation catchments primarily underlain by granitic rocks. The areally dominant, nearly mono-mineralic Medicine Peak Quartzite is the least reactive bedrock, whereas the more spatially heterogeneous amphibolite contains more easily weathered ferromagnesian minerals with a small amount of plagioclase. In addition to the major rock types in the West Glacier Lake catchment, three isolated outcrops of calc-silicate rock are also found, and have been classified as schist or meta-conglomerate lenses (Rochette, 1987) (Figure 1). The calc-silicate rock contains calcite (up to 10 wt.%) and pyrite. Steep, talus-covered slopes dominate the catchment with local accumulation of thin ($\leqslant 1$ m), young soils in topographic depressions. Vegetation is sparse and grades from subalpine fir [*Abies lasiocarpa* (Hook.) Nutt.] and Engelmann spruce (*Picea engelmanni* Parry) at low elevation, to bare rock and talus, interspersed with a krummholz-willow assemblage, higher in the catchment. Grassy meadows form a narrow band along tributary streams.

Four well-defined tributaries, Long Creek (10·84 ha), Cascade Creek (16·55 ha), Meadow Creek (4·25 ha) and Boulder Creek (22·93 ha), feed West Glacier Lake, whereas East Glacier Lake receives diffuse runoff from a smaller area. Mean annual precipitation is 119 cm, based on analysis of five years of data collected at the Snowy Range site as part of the National Atmospheric Deposition Program/National Trends Network program (NADP/NTN, 1992). Of the total precipitation input, snowfall comprises 80–90% (in water equivalence). Snowmelt constitutes the major hydrologic event, with the spring melt beginning, typically in late March and lasting until late July (Sommerfeld *et al.*, 1990). A significant, but unknown, amount of snow is wind deposited into the catchment from adjacent areas. Average wind speed during winter is 6 m s^{-1} (Fox *et al.*, 1994), and the snowpack development within the catchment is controlled by wind redistribution, with interfluve ridges often barren of snow, while tributary drainages are filled with thick (up to 5 m) wedge-shaped deposits.

METHODS

Stream discharge

Discharge is measured by Parshall flumes at Meadow Creek, Cascade Creek and the outlet of West Glacier Lake (Hasfurther *et al.*, 1990) (Figure 1). Stage height data were provided by the Rocky Mountain Forest and Range Experiment Station, US Department of Agriculture, Forest Service. Discharge data for 1989 and 1990 from Meadow Creek, Cascade Creek and West Glacier Lake were included in the present analysis. Discharge measurements are not available for Long Creek and Boulder Creek because the locations where the tributaries discharge into West Glacier Lake are large boulder fields with no well-defined path of flow through the debris. A synthetic hydrograph was developed for Long Creek using discharge data from the Cascade Creek flume. Baseflow discharge in Long Creek is about 0·05 l s^{-1}, which, taken over a year, is equal to about 3% of the lake volume. Average annual discharge from West Glacier Lake is a factor of twenty greater than the lake volume; thus, the groundwater discharge into the lake is negligible compared with the amount of water supplied by snowmelt. Average annual discharge from Meadow Creek is 200 m^3 and from Long Creek is 119 400 m^3.

Periods of discharge vary between the tributaries: Long and Boulder creeks flow year round, whereas Meadow Creek ceases to flow in October as the water level in the bedrock aquifer drops below that of the spring feeding the creek. Only data from samples collected in Long Creek and Meadow Creek are discussed

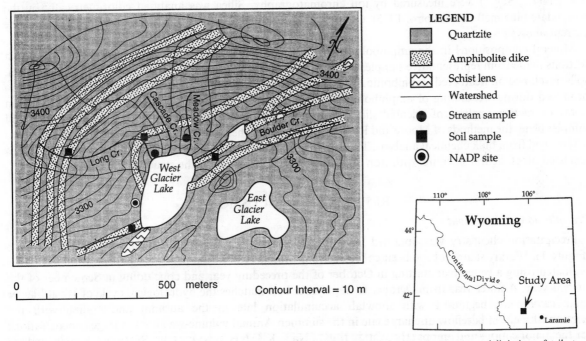

Figure 1. Location map of the West Glacier Lake catchment showing sampling sites, bedrock exposures, deliniation of tributary drainages, and general topography

in this paper as the chemistry of waters discharging from these two tributaries are most indicative of the range of geochemical reactions occurring in the West Glacier Lake catchment.

Precipitation: quality and quantity

Precipitation chemistry and volume are measured with a wet–dry sampler and a Belfort rain gauge, respectively, at the NADP site (NADP/NTN, 1992) (Figure 1). Weekly samples have been collected since the installation of the site in 1986. One concern in using NADP data to characterize the chemical composition of precipitation is the poor capture efficiency of the sampler as a result of high winds that are common for much of the year at West Glacier Lake. In order to apply the chemical data obtained from samples collected at the NADP site, we have assumed that the chemical composition of snow measured at the station is representative of snow throughout the catchment. The total amount of precipitation is calculated from the rain gauge data, and is also affected by strong winds in the catchment.

Stream sampling

Water samples were collected twice weekly from Meadow Creek throughout the snowmelt season (March–July) then bi-weekly until August in 1989 and 1990. The chemical data for Meadow Creek are not greatly different between 1989 and 1990; most of the variation is on short time-scales associated with variations in the timing of the melt season. Additionally, monthly sampling of stream water from Meadow Creek and Long Creek began in July 1991 and continued until October 1992. Subsequent discussions about stream water chemistry are referenced to a water year that begins in October and ends in September of the subsequent year.

All water samples were collected in acid-washed, distilled water-rinsed (3×) polypropylene bottles, and passed through 0·45 µm nylon filters on the same day. Alkalinity was determined by Gran titration to pH 3·5, and pH was measured by specific ion electrode; both were measured within four hours of sample collection. Samples were then stored at 4°C overnight prior to analysis of major ions. Cations (Ca^{2+}, Mg^{2+}, Na^+, K^+) were measured in acidified samples by flame atomic absorption spectrophotometry (FAAS), and anions (Cl^-, NO_3^-, SO_4^{2-}) were measured by ion chromatography. Silica was analysed colorimetrically by the molybdate blue method (Shapiro, 1975). Charge balance was better than $\pm 10\%$ for all samples used in the current analysis.

Mineral compositions in the amphibolite were based on electron microprobe analysis of polished thin sections made from representative samples of rock (Rochette et al., 1988). Four samples were collected from soils developed on weathered amphibolite dikes in an attempt to identify the secondary weathering product(s) produced during weathering of amphibolite minerals. Identification of secondary minerals was based on X-ray diffraction analysis of oriented slides prepared from the less than 0·2 µm size fraction of the soil samples using the methods of Moore and Reynolds (1989). The chemical composition of the clay fraction was determined from bulk chemical analysis. The chemical analysis was used to calculate the stoichiometry of an idealized dioctahedral smectite, with iron allowed to vary in order to fit the structural formula.

RESULTS AND DISCUSSION

Precipitation: Solute Input

Precipitation chemistry was obtained from the database of information generated at the NADP site (Figure 1). Weekly samples have been collected from March 1986 to the present. Concentration data were evaluated using a water year starting in October of the preceding year and concluding in September of the current year. A water year from October to September best matches the hydrological cycle of West Glacier Lake catchment, beginning with snowfall accumulation late in the autumn and ending with the re-establishment of baseflow discharge late in the summer. Annual volume-weighted average concentrations for the major cations and anions (H^+, Ca^{2+}, Mg^{2+}, Na^+, K^+, NH_4^+, NO_3^-, Cl^-, SO_4^{2-}) were calculated for water years 1987 to 1991 (Table I).

The true chemical composition of precipitation can only be estimated since the single point sampling at the NADP site is unlikely to be representative of precipitation over the entire catchment, especially considering the amount of wind reworking and deposition of snow from outside the catchment by wind transport (Reuss et al., 1993). Precipitation samples collected at the NADP site do not provide a measure of solutes from dry deposition. Input of solutes from dry deposition depends strongly on the properties of the surface (Whelpdale and Shaw, 1974; Hicks, 1986), and since the major period of atmospheric input at West Glacier Lake is winter snowfall, the surface characteristics of the snowpack will have the greatest effect on dry deposition of solutes. Most of the recent research has been focused on evaluating depositional velocities to snow surfaces for the acid anion precursors (e.g. SO_2, NO_2, HCl) (e.g. Whelpdale and Shaw, 1974; Bales et al., 1987; Valdez et al., 1987), and few studies have extended the evaluation to include the base cations. Dry deposition of solutes has been shown to be significant in completely forested catchments (e.g. Shepard et al., 1989; Bytnerowicz et al., 1992). Shepard et al. (1989) demonstrated that dry deposition of base cations may constitute greater than 50% of the total amount of Ca^{2+} in a north-eastern US hardwood forest. In contrast, Williams and Melack (1991) found that dry deposition in the Emerald Lake Watershed, Sierra Nevada mountains was not a major contributor to atmospheric input of solutes based on a comparison between cumulative snowfall and snowpack loading; although Clow and Mast (1995) showed that dry deposition of cations to a bare rock surface could be significant. Given the preponderance of precipitation input as snow, and the longevity of snow cover in the West Glacier Lake catchment, we have assumed that contributions of solutes from dry deposition are small compared with wet deposition. The five-year volume-weighted mean solute concentrations listed in Table I will be used for subsequent calculations.

Tributary data: solute output

Meadow Creek.   Intensive sampling during snowmelt in 1989 and 1990 allows the identification of short-term fluctuations in the concentrations of solutes in Meadow Creek. The period of record for Meadow Creek extends only until August in 1989 and 1990. In order to estimate the annual volume-weighted mean concentrations of solutes in Meadow Creek using 1989 and 1990 data, the balance of the water year cycle is constructed using data collected during the autumn of 1991. Time series of chemical data for Meadow Creek are shown as dashed lines in Figure 2 where 1990 data are combined with data from 1991. The two years of data (1989 and 1990) are not significantly different, either in concentration or in discharge, so the data presented in Figure 2 are the results from the 1990 season only. Although stream samples were collected from Meadow Creek during water year 1992, the 1990 water year record for Meadow Creek is much more detailed and is used in the determination of solute flux.

Table I. Volume-weighted mean concentrations of solutes in precipitation for the NADP site located in the West Glacier Lake catchment and at Nash Fork (*ca.* 5 km to the south-east). All concentrations are in µeq l^{-1} except pH

Element	Water year					Grand average 1987–1991	
	1987	1988	1989	1990	1991	Snowy Range	Nash Fork
pH	5·07	5·26	5·51	5·49	5·34	5·32	5·18
H^+	8·47	5·49	3·08	3·22	4·6	4·83	6·67
Ca^{2+}	9·10	8·99	10·2	8·65	8·57	9·02	13·30
Mg^{2+}	2·00	2·15	2·52	2·85	1·93	2·30	2·50
K^+	0·51	0·52	0·36	0·37	0·42	0·43	0·64
Na^+	5·74	5·53	10·16	4·82	3·36	5·62	4·67
NH_4^+	6·72	5·13	10·95	8·55	6·54	7·76	9·15
NO_3^-	11·26	9·19	13·76	12·65	11·06	11·63	13·37
Cl^-	3·70	3·40	4·14	4·35	2·35	3·54	3·11
SO_4^{2-}	13·70	13·49	17·72	12·07	10·49	13·21	15·81
Ppt. (cm)	93·47	112·14	125·14	134·56	128·15	118·69	53·40

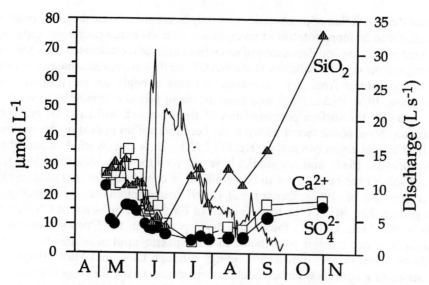

Figure 2. Time series of concentrations for selected solutes in Meadow Creek (symbols with lines) with the annual discharge hydrograph superimposed (solid line). Lines for solutes are dashed where 1990 and 1991 data are combined (see text)

The concentrations of Ca^{2+}, SO_4^{2-}, and SiO_2 during 1990, together with the discharge hydrograph, are shown on Figure 2. Concentrations of the solutes are fairly high during early snowmelt and decrease steadily with time. Minima in the concentration curves correspond to peak discharge. Late season increases in concentration occur after all snow has melted and discharge in Meadow Creek is derived from drainage of the deeper soil zone and the bedrock aquifer. Seasonal differences in the biogeochemical processes controlling the concentrations of solutes are indicated by the change in the ratio of Ca^{2+} : SiO_2 from early season snowmelt to late season baseflow (Figure 2). Finley and Drever (1992) have interpreted the changes through time in terms of the relative influence of soil processes and the preferential elution of solutes from the snowpack during the initial stages of snowmelt; a result consistent with observations in other snowmelt-dominated catchments (Tsiouris *et al.*, 1985; Tranter *et al.*, 1986; Bales *et al.*, 1990). Early season freeze–thaw cycles interrupt the operation of the water level recorder, limiting the amount of discharge data available during the earliest portion of snowmelt. Thus, only that portion of the annual record with discharge data was used to calculate the volume-weighted mean concentrations of solutes for use in the mass balance calculation.

Long Creek. Water samples from Long Creek were collected over the 1992 water year (October 1991–September 1992) yielding a complete set of monthly data. Discharge data from development of the synthetic hydrograph were combined with stream water chemistry to calculate annual volume-weighted mean concentrations of solutes for Long Creek. Baseflow discharge was estimated in the field to be approximately 0.05 l s^{-1}. Errors of even a factor of 10 for baseflow discharge do not make a significant difference, since almost 90% of the total volume of water, and mass of solutes, from any tributary in the West Glacier Lake catchment is discharged during the snowmelt period (March–July).

Concentrations of Ca^{2+}, SO_4^{2-} and SiO_2 are nearly constant during the baseflow period (October–March) (Figure 3). Steady decreases in concentration during the snowmelt season are typical of the influence of preferential elution followed by dilution during peak discharge/snowmelt. Late season increases in concentration coincide with re-establishment of baseflow conditions. In the Long Creek drainage, base flow consists only of discharge from the bedrock aquifer, without contributions from the deeper soil zone, since there is effectively no soil along the reach of the creek where discharge occurs during baseflow. Sulfate concentrations have been discussed in detail elsewhere, and are related to oxidation of pyrite contained in the amphibolite dikes traversing the Long Creek drainage (Finley *et al.*, 1995). Nearly constant concentrations

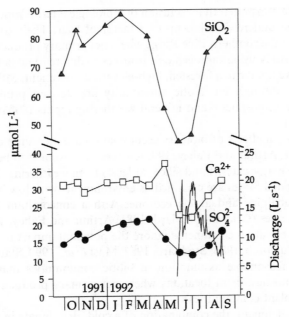

Figure 3. Time series of concentrations for selected solutes in Long Creek with the synthetic (see text) annual discharge hydrograph superimposed. Symbols as in Figure 2

of Ca^{2+} and SiO_2 during baseflow are interpreted to represent the approach to steady-state reactions in the bedrock aquifer. In terms of the annual hydrological cycle, the greatest mass of solutes is discharged during the snowmelt period, when over 90% of the total volume of water passes through the catchment. Thus, although the apparent steady-state concentrations of solutes during baseflow are an enticing measure of mineral weathering, the solute ratios are not the same as in the volume-weighted annual discharge and thus are not representative of discharge over an annual cycle. Water recharging the bedrock aquifer has presumably come into contact with fine-grained material, either through soils or as sediment filling bedrock fractures, and may thus have been influenced by cation exchange or biological processes operating over only part of the annual cycle. Alternatively, the stoichiometry of mineral weathering in the bedrock aquifer may not be identical to that occurring in the soil, so weathering in the bedrock is not representative of weathering in the catchment as a whole. We have used volume-weighted mean concentrations as the best representation of weathering reactions occurring in the catchment as a whole.

Relative rates of processes and mass balance

We have now derived an input–output budget, which represents the left-hand side of Equation (1). Before any inferences can be made regarding the supply of solutes from weathering, the potential influence of biotic uptake/release and cation exchange must be addressed, which brings in the question of time-scale. The first requirement is that the period of data collection must be sufficiently long that changes in water storage can be ignored. The second requirement is that the period of data collection should include at least one annual cycle to average out the effects of the annual cycle of plant growth. The annual cycle is generally associated with much larger fluxes of nutrients than long-term increase or decrease of biomass, and the nutrient uptake may interact with the exchange pool in the soil to influence runoff chemistry. Such effects should average out over an annual cycle, although, obviously, not all years are identical (Likens et al., 1977). The period of record should, ideally, average over drought years and wet years. The next time-scale is that of the cycle of growth and destruction of the plant community which, for forests, is of the order of centuries (Vitousek and Reiners, 1975). There are no data records of this length.

Taylor and Velbel (1991) demonstrated that exclusion of a biological term from the calculation of mineral weathering rates can result in underestimation up to a factor of four. Their study was conducted in the heavily forested catchments of the southern Blue Ridge where net primary productivity is much greater than in alpine environments. In a study using radiogenic Sr isotopes to differentiate between mineral weathering, cation exchange and vegetative uptake in a forested, high-elevation catchment, Miller et al. (1993) concluded that nutrient solutes cycled through the biotic community are derived primarily from the historical accumulation of solutes from atmospheric and mineral weathering inputs (80%), with the balance (20%) from 'new' mineral weathering.

In reality, either the long-term effects of biomass accumulation are measured directly (e.g. Likens et al., 1977; Taylor and Velbel, 1991; Arthur and Fahey, 1993) or they are assumed to be small compared with the rate of release of solutes (the major cations and silica, at least) from weathering. In general, high-elevation catchments in the western United States are classified as environments with low biomass productivity since the sites often span both subalpine and alpine ecotones with a concomitant decrease in biomass and productivity as elevation increases (Grier and Ballard, 1981; Arthur and Fahey, 1990; Arthur, 1992). Most mass balance studies in high-elevation catchments ignore the potential impact of biotic uptake and release (Clayton, 1986; Drever and Hurcomb, 1986; Stoddard, 1987; Mast et al., 1990; Stauffer, 1990; Williams et al., 1993), which is generally a reasonable assumption in biotic communities that have not been recently perturbed (fire, avalanche, grazing), or in locations where net primary productivity, with uptake of base cations, is small (e.g. alpine plant communities).

In the West Glacier Lake catchment, the combination of a cold, dry climate in typical alpine catchments contributes to low, but non-zero, net primary productivity (Grier and Ballard, 1981; Arthur, 1992); the question, then, is whether this non-zero primary productivity is negligible, compared with the supply of solutes by mineral weathering. Arthur and Fahey (1992, 1993) measured the rate of cation accumulation in a spruce fir forest located in the Loch Vale Watershed, Rocky Mountain National Park. The environment at Loch Vale is very similar to West Glacier Lake, and their results provide a reasonable first-order approximation for estimating the potential effect of biotic uptake and release in the West Glacier Lake catchment. The forested portion of the Loch Vale catchment is accumulating cations at a rate of 29 μmol m^{-2} yr^{-1} (Arthur and Fahey, 1993), which, if converted to the total flux of cations on a per litre basis for Long Creek, would represent a sink of 0·026 μmol l^{-1} of runoff. The total cation flux from Long Creek (Table II, column 7) is ca. 50 μmol l^{-1}, almost three orders of magnitude greater than the rate of uptake by vegetation. Thus, the net annual biotic uptake/release of cations by vegetation in the West Glacier Lake catchment should be negligible compared with the supply of cations from mineral weathering reactions.

Cation exchange reactions are generally held to exert minimal influence on time-scales greater than one year, as long as the chemical composition of precipitation is relatively constant (Drever, 1988). In high-elevation catchments where the annual hydrological cycle is dominated by snowmelt, the main period of cation exchange in the soil zone will occur during peak snowmelt, when large volumes of dilute water flush the products of mineral weathering and decomposition of biological material, and drive rapid cation exchange reactions ($\leqslant$ minutes) (Sparks, 1989; Finley and Drever, 1992; Williams et al., 1993). Although soils in the West Glacier Lake catchment are thin and immature, the mass of base cations stored on exchange sites is about two orders of magnitude greater than the mass of base cations eluted each year from the snowpack (J. B. Finley, unpublished data). A major unknown in assessing the potential impact of cation exchange reactions on mass balance is defining just how much soil, or soil-like material, must exist before the effect of cation exchange exerts control on seasonal concentrations of solutes. In snowmelt-dominated hydrological systems, like many of the high-elevation catchments in the western United States, late season flow in streams is produced from discharge of water from bedrock aquifers, and soils are generally unsaturated and are disconnected hydrologically from streams (Williams et al., 1993; J. B. Finley, unpublished data). The best measure of solutes produced by mineral weathering reactions in the soil and bedrock will then be obtained from the longest record of annual inputs and outputs to the catchment. In principle, mineral weathering will change with time as reactive minerals become depleted and soil accumulates, but the time-scale of such processes is much longer than is generally considered in mass balance studies.

Miller *et al.* (1993) demonstrated that 30% of the annual output of Sr^{2+} from a forested catchment was derived from cation exchange sites, but also that, in the catchment studied, annual cation release during mineral weathering is adequate to replenish the cations released from soil exchange sites. Cation exchange was thus not a *net* source of cations. The chemistry of precipitation in the West Glacier Lake catchment has been relatively constant over the last decade, so it is reasonable to assume that the annual export of base cations reflects weathering of minerals in the soil and bedrock and not depletion of the cation exchange pool.

Mineral weathering

Chemical mass balances for Meadow and Long creeks are calculated using the volume-weighted mean concentrations of solutes input by precipitation and output as stream discharge (Table II; Figure 4a–e). Box plots shown in Figure 4 provide a visual comparison between concentrations in the precipitation input and the stream output. Chloride is not a significant element in any of the bedrock minerals, and is not actively sorbed in the soil or taken up significantly by vegetation. Thus, Cl^- should be conservative, with the volume-weighted mean concentration of Cl^- discharged from tributary streams representing the concentration of Cl^- in precipitation modified by the influence of evapotranspiration. As indicated by Figure 4a, there is no significant difference between the concentration of Cl^- input from precipitation and that discharged from either of the tributary drainages ($p \leqslant 0.01$). Nitrate, not shown, is like Cl^-, in that concentrations in both Meadow Creek and Long Creek are not significantly different from the concentration in precipitation; a result consistent with the observations of several investigators (Mast *et al.*, 1990; Campbell *et al.*, 1991; Turk and Spahr, 1991). The data for SO_4^{2-} show clearly the additional source of S in the waters of Long Creek, whereas SO_4^{2-} in Meadow Creek is dominated by atmospheric precipitation (Figure 4b). The concordance between precipitation input and stream output for the major acid anions, especially Cl^-, indicates that evapotranspiration is minor and supports our assumption that dry deposition is negligible.

Table II. Volume-weighted mean concentrations in precipitation (NADP column 2), Meadow Creek (column 3) and Long Creek (column 4). Concentrations listed for Long Creek were calculated using synthetic hydrograph data (see text). Column 5 lists the average concentration of solutes in Long Creek measured using data from September to March (see text). Concentrations are in units of $\mu mol \, l^{-1} \, yr^{-1}$ except pH. Uncertainty is 1σ about the mean. Solute budget (columns 6 and 7) is calculated using volume weighted mean concentrations for Meadow Creek and Long Creek

1 Solute	2 Input NADP	3	4	5	6	7
		Output			Solute budget (Output − Input)	
		Volume-weighted Mean Baseflow				
		Meadow Cr.	Long Cr.	Long Cr.	Meadow Cr.	Long Cr.
Ca^{2+}	4.51 ± 12.28	13.67 ± 12.87	26.15 ± 5.97	31.38 ± 1.15	9.16	21.64
Mg^{2+}	1.15 ± 2.08	4.61 ± 2.24	13.99 ± 3.80	17.90 ± 0.91	3.46	12.84
K^+	0.43 ± 1.20	2.34 ± 1.62	3.55 ± 1.24	4.19 ± 1.12	1.91	3.12
Na^+	5.62 ± 11.50	3.38 ± 5.30	17.19 ± 4.95	17.83 ± 2.27	*	11.57
NH_4^+	7.76 ± 8.16	N.D.†	N.D.	N.D.		
NO_3^-	11.63 ± 14.70	9.83 ± 5.46	$>11.08 \pm 7.67$	13.38 ± 1.13		
Cl^-	3.54 ± 6.37	2.35 ± 1.60	2.91 ± 1.55	4.00 ± 0.71		
SO_4^{2-}	6.61 ± 8.54	7.03 ± 3.64	$>13.84 \pm 4.59$	18.62 ± 2.61	*	7.23
H^+	4.83 ± 12.48	1.84 ± 1.15	0.55 ± 0.40	0.60 ± 0.34		
SiO_2‡	0.00	18.12 ± 8.52	62.46 ± 18.65	74.98 ± 16.36	18.12	62.46

* Statistically indistingushable from zero ($p \leqslant 0.01$).
† Non-detectable.
‡ SiO_2 is not measured in precipitation, but SiO_2 was not detected in samples of snowmelt.

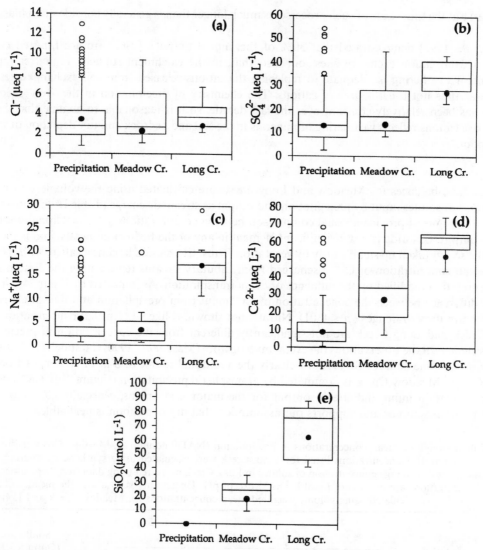

Figure 4. Box plots of (a) Cl^-, (b) SO_4^{2-}, (c) Na^+, (d) Ca^{2+} and (e) SiO_2 for concentrations of solutes measured in samples of (1) precipitation, (2) meadow Creek, and (3) Long Creek. Area of box encompasses 50% of the data, the whiskers are operationally defined to include all data within 1·5 times the interquartile range measured from the upper and lower quartiles; the horizontal line is the median concentration; closed circles (●) are the volume-weighted mean concentration; and open circles (○) are outliers

Calcium and sodium (Figure 4c and d) are derived from both atmospheric deposition and weathering, whereas silica (Figure 4e) is derived almost exclusively from mineral weathering in the soil or bedrock. On short time-scales, for example during snowmelt, cation exchange, if important, could have a large impact on the concentrations of Ca^{2+}, whereas in most high-elevation catchments, where the total concentration of solutes is small, Na^+ should be minimally affected by either cation exchange or biotic uptake (Stauffer, 1990; Appelo and Postma, 1993). The volume-weighted mean concentration of Na^+ in Meadow Creek is not significantly different from the concentration of Na^+ in precipitation, indicating a relatively small input from bedrock weathering, whereas the volume-weighted mean concentration of Na^+ in Long Creek is significantly higher than that of atmospheric input (Figure 4c). In terms of Na^+ released by chemical weathering of primary minerals, albite is the only potential source of sodium. Similarly, Ca^{2+} in Meadow Creek is only slightly higher than Ca^{2+} in precipitation, whereas Ca^{2+} in Long Creek is much greater than Ca^{2+} in precipitation (Figure 4d). The only bedrock source of Ca^{2+} is minerals contained in the amphibolite dike (epidote, actinolite). Silica shows a similar pattern to Ca^{2+}, with the highest concentration in Long Creek and a much lower concentration in Meadow Creek. The difference between the production of Na^+, Ca^{2+} and SiO_2 in Meadow Creek and Long Creek is most likely related to the distribution of amphibolite dikes, boulders of amphibolite as talus and hydrological flow paths in each tributary catchment.

Four amphibolite dikes traverse Long Creek and only one dike crosses the upper portion of Meadow Creek (Figure 1). Circulation of water through each tributary differs markedly; our interpretation of flow paths (Finley and Drever, 1992) is based on chemical data and field observation of water flows. An analysis of chemical data obtained from samples of snowmelt, soil pore water and surface water in Meadow Creek supports the interpretation that much of the discharge through the Meadow Creek basin occurs as: (1) overland flow during snowmelt; (2) soil zone discharge as snowmelt infiltrates the soil and discharges to the stream; and (3) shallow, ephemeral bedrock flow which halts when the water table of the bedrock aquifer falls below the elevation of the only spring (located about one-third of the way up the basin). Although the same general interpretation of hydrological cycling presumably applies to Long Creek, the relative contribution of water from different sources is quite different. Discharge through the bedrock aquifer appears to be much more important in Long Creek.

Concentrations of solutes used in mass balance calculations are listed in Table II. Two compositions are listed for Long Creek to demonstrate the slight difference between volume-weighted mean concentrations and averages of concentrations calculated for the baseflow period. The higher average baseflow concentration compared with the volume-weighted mean concentration indicates the effect of circulation through the bedrock aquifer and discharge from the bedrock. Seasonal variations in the concentrations of Ca^{2+} and SO_4^{2-} are similar for Meadow Creek and Long Creek, but the pattern for SiO_2 is different. Silica concentrations are much higher, both relatively and absolutely, in Long Creek (Figures 2 and 3). Although silica may be, in part, associated with labile silica in the soil zone, the steady-state concentrations during baseflow probably represent mineral weathering.

There are several techniques for extracting rates of mineral weathering from the results of mass balance calculations, but all are ultimately based on the method first presented by Garrels and Mackenzie (1967) and Garrels (1967). In essence, balanced chemical reactions are written in which primary minerals react with water to yield secondary minerals and the observed solute concentrations. Garrels and Mackenzie (1967) wrote a separate equation for each mineral weathering reaction and then summed the results. An alternative method, which we use here, involves writing a single weathering reaction that includes all primary and secondary minerals simultaneously. This approach yields a system of linear equations (one for each solute), which can be solved by matrix methods (Plummer and Back, 1980; Velbel 1986; Bowser and Jones, 1993). In the case of the West Glacier Lake catchment, the bedrock mineralogy is fairly straightforward in that the dominant rock type, Medicine Peak Quartzite, is very unreactive; quartz is the most abundant mineral (60–100%) and the remainder of the minerals are relatively non-reactive (muscovite and related phyllosilicates). The only relatively reactive minerals in the quartzite are the feldspars, albite (An_7) and K-feldspar, which together make up only about 3% of the rock. Amphibolite dikes cross-cutting the quartzite are heterogeneously distributed ribbons of rock that contain the most reactive suite of minerals in the catchment. Thus, the group of bedrock minerals to consider is fairly well defined (epidote, actinolite, albite, K-feldspar, biotite). There is also a possibility that aeolian calcite is contributing solutes. We shall make the initial assumption that the only primary minerals weathering are those found in the amphibolite dikes and quartzite.

Choice of secondary mineral(s). Identification of the secondary mineral(s) to include in the mineral weathering reaction is the least well-constrained step of the procedure, in terms of both the choice of a particular mineral and its chemical composition. In the West Glacier Lake catchment, the clay fraction (<2 μm) of residual soils formed on the amphibolite dikes contains both smectite and kaolinite; although

Rochette *et al.* (1988) reported one sample consisting almost entirely of smectite. Soils in the basin also contain mixed-layer illite–smectite, vermiculite, and chlorite. These are interpreted as being (largely, at least) aeolian (Rochette *et al.*, 1988). Thus, the first-order interpretation is that mineral weathering of primary minerals results in the formation of smectite and/or kaolinite with the additional possibility of amorphous $Al(OH)_3$.

An alternative approach to identifying secondary minerals likely to be forming is to make the assumption that the secondary phase(s) is in equilibrium with the runoff water. Williams *et al.* (1993) plotted their data from the Emerald Lake catchment on a phase diagram depicting the system Na_2O–Al_2O_3–SiO_2–H_2O, which showed that the majority of their data plotted near the gibbsite-kaolinite boundary, mostly in the gibbsite field, and they then used kaolinite as the secondary mineral for mass balance calculations. Drever and Zobrist (1992) demonstrated that a strict interpretation of water-rock interaction based solely on interpretation of phase diagrams may lead to spurious results. According to the distribution of their data on the phase diagram, soil clay minerals should change from $Al(OH)_3$ at high elevations to kaolinite at low elevations, neither of which is supported by mineralogical analysis of soil samples. They suggest that formation of secondary phases may well yield overall stoichiometry similar to minerals like kaolinite, but may consist of more complex mixtures of clay minerals. We do not believe that thermodynamic calculations are a useful constraint on whether kaolinite, or smectite, or both are forming in the West Glacier catchment.

Weathering reactions: Long Creek. A plausible generalized reaction would be:

$$a \text{ Actinolite} + b \text{ Epidote} + c \text{ Albite} + e \text{ Biotite} + f \text{ Quartz}(+83.65CO_2) \rightarrow g \text{ Smectite} +$$

$$21.64Ca^{2+} + 12.84Mg^{2+} + 3.12K^+ + 11.57Na^+ + 62.46H_4SiO_4(+83.65HCO_3^-) \qquad (2)$$

where a–g are the moles of primary mineral altered, or secondary mineral formed, and the coefficients for the base cations and silica are obtained using the solute budgets, and are shown for Long Creek (Table II, column 7). The budget for Long Creek was used because water cycling through the Long Creek sub-catchment most clearly reflects chemical interaction with bedrock minerals. Compositions of the solid phases used in Equation (2) are listed in Table III. The mathematical requirement is that there must be as many mineral phases, reactants or products, as there are chemical components; the number of chemical components is equal to the number of independent solutes plus any other elements (e.g. aluminium) that are explicitly conserved in the solid phases (alternatively, the number of components is equal to the number of independent solutes, with aluminium considered as a solute with a concentration of zero). Bicarbonate is not included as an independent solute (nor CO_2 as a component) because of the requirement of charge balance: if bicarbonate is the only anion present, bicarbonate concentration must equal total cation concentration

Table III. Minerals present in the soils and bedrock derived from amphibolite in the West Glacier Lake catchment

Mineral	Formula
Actinolite*	$(Ca_{1.85}Na_{0.15})(Mg_{3.00}Fe_{1.90})(Si_{7.50}Al_{0.80})O_{22}(OH)_2$
Epidote†	$Ca_{1.95}Al_{2.40}Fe_{0.60}Si_{3.02}O_{12}(OH)$
Biotite‡	$KMg_{1.5}Fe_{1.5}AlSi_3O_{10}(OH)_2$
Albite*	$NaAlSi_3O_8$
Pyrite§	FeS_2
Smectite§	$(Na_{0.01}K_{0.02}Ca_{0.09}Mg_{0.06}Al_{0.04})(Mg_{0.15}Fe_{0.26}Al_{1.59})(Si_{3.70}Al_{0.30})O_{10}(OH)_2$

* Taken from Rochette *et al.* 1988.
† Recalculated from electron microprobe data in Rochette 1987.
‡ Assumed composition.
§ Based on bulk chemical analysis of the clay fraction from soils formed directly on amphibolite dike.

$$\begin{array}{c} \\ \text{Ca} \\ \text{Mg} \\ \text{K} \\ \text{Na} \\ \text{Si} \\ \text{Al} \end{array} \begin{array}{cccccc} \text{Ac} & \text{Bi} & \text{Ep} & \text{Ab} & \text{Qtz} & \text{Smec} \\ \begin{bmatrix} 1.85 & 0.00 & 1.95 & 0.00 & 0.00 & -0.09 \\ 3.00 & 1.50 & 0.00 & 0.00 & 0.00 & -0.21 \\ 0.00 & 1.00 & 0.00 & 0.00 & 0.00 & -0.02 \\ 0.15 & 0.00 & 0.00 & 1.00 & 0.00 & -0.01 \\ 7.50 & 3.00 & 3.02 & 3.00 & 1.00 & -3.70 \\ 0.80 & 1.00 & 2.4 & 1.00 & 0.00 & -1.93 \end{bmatrix} \end{array} \begin{bmatrix} n_{Ac} \\ n_{Bi} \\ n_{Ep} \\ n_{Ab} \\ n_{Qtz} \\ n_{Smec} \end{bmatrix} = \begin{array}{c} \text{Solutes} \\ \begin{bmatrix} 21.64 \\ 12.84 \\ 3.12 \\ 11.57 \\ 62.46 \\ 0.00 \end{bmatrix} \end{array}$$

Figure 5. Matrix representation of Equation (2) showing the system of linear equations with the grand matrix with dimensions of m elements by n phases, a column vector corresponding to the moles of each mineral phase weathered per litre per year, and a column vector of solute concentrations in Long Creek. Conservation of aluminium is achieved by setting the aqueous concentration to zero. Ac, Actinolite; Bi, Biotite; Ep, Epidote; Ab, Albite; Qtz, Quartz; S, Smectite

and is thus not an independent variable. Smectite was chosen as the secondary mineral on the basis of X-ray diffraction (XRD) analysis of the clay mineral fraction (< 2 μm) of samples collected from soils formed on outcrops of the amphibolite dike. There are no particular constraints on the selection of any solid phase except geochemical reasonableness (e.g. presence in bedrock and soils); we shall demonstrate the effect of varying the selection of minerals on the overall interpretation of mineral weathering.

A matrix representation of Equation (2) is shown in Figure 5. Each column represents the stoichiometry of an individual mineral. Substitution of another mineral simply entails listing the stoichiometric coefficient of each element as a column vector and substituting the new mineral into the matrix. The rightmost column vector is the forcing function of the mathematical solution and consists of the chemical composition of stream waters representing solutes derived from mineral weathering based on the input–output budget.

The effects of varying the choice of minerals on the calculated rates of mineral weathering are shown in Figure 6. (Positive values in Figure 6 indicate a mineral being consumed (reactants), and negative values minerals being produced (products)). Six combinations of minerals are shown to demonstrate that, within the constraints imposed by mineral composition and the concentration of solutes in stream water, the mathematical solution to Equation (2) is not unique. Differences in the rates of mineral weathering between sets of minerals are shown as changes in the dimension of the box corresponding to a particular mathematical solution (Figure 6). For example, in the first run, actinolite, albite, K-feldspar, epidote and quartz react to form smectite.

Source of K^+. Identifying the source of K^+ is ambiguous, with two possible minerals: (1) biotite contained in the amphibolite dike or (2) K-feldspar contained in the quartzite. In either case, the modal abundance of biotite and K-feldspar are the same in both rock types (*ca.* 2%), but in terms of total abundance K-feldspar dominates because of the large spatial coverage of quartzite (*ca.* 85% of catchment area). As shown in Figure 6 (a and c), the choice between K-feldspar and biotite has significant implications for the calculated rates of weathering of other minerals, especially the rate at which quartz alters.

In two separate studies of mineral weathering in Sierra Nevada catchments, Stoddard (1987) and Williams *et al.* (1993) used K-feldspar as the source of K^+. Stauffer (1990) applied rigorous statistical analysis to the large set of chemical data generated during the Western Lake Survey (WLS) (Landers *et al.*, 1987) to interpret the influence of mineral weathering reactions over a large, regional scale on the concentrations of solutes found in subalpine and alpine lakes during autumn overturn. He concluded that the primary source of K^+ in lakes from the Sierra Nevada and the northern Cascades was probably the chemical alteration of biotite to hydrobiotite or vermiculite. In a recent study, Blum *et al.* (1994) used radiogenic Sr ($^{87}Sr/^{86}Sr$) as a tracer of base cation sources in streams draining a Sierra Nevada granite batholith. They concluded that biotite weathers preferentially relative to alteration of K-feldspar.

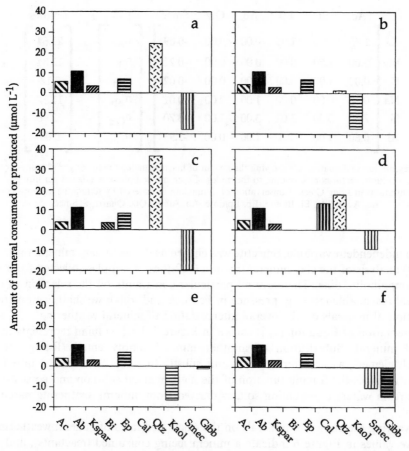

Figure 6. Bar charts showing the effect of mineral composition on the calculated rates of mineral weathering for six (a–f) combinations of possible reactant and product minerals (see text for discussion). Positive values correspond to reactants and negative values are products. Minerals considered are Ac, actinolite; Ab, albite; Kspar, potassium feldspar; Bi, biotite; Ep, epidote; Cal, calcite; Qtz, quartz; Kaol, kaolinite; Smec, smectite and Gibb, gibbsite. Units can be converted to moles per hectare using the total discharge value of approximately $1.1 \times 10^7 \, l \, ha^{-1} \, yr^{-1}$

Thus, even though biotite may represent only a small fraction of the total mass of bedrock, weathering of biotite was apparently more important than K-feldspar as a source of K^+. Use of biotite as a primary mineral for interpreting weathering reactions in the West Glacier Lake catchment requires high rates of weathering for quartz or high rates of precipitation of $Al(OH)_3$ (Figure 6a and b). Quartz is generally considered to be non-reactive in low-temperature environments. There are two reasons for this: the rate of dissolution of quartz is slow (Dove and Elston, 1992), and in many environments infiltrating waters rapidly become supersaturated with respect to quartz as a consequence of dissolution of more reactive silicates (Brantley, unpublished data). It is possible that in the West Glacier Lake catchment, the large area of quartz exposed to meteoric water and the near-absence of reactive silicates in the quartzite result in significant dissolution of quartz. Thus, dissolution of quartz cannot be ruled out as a source of silica. However, a mathematical solution to Equation (2) yielding greater rates of weathering for quartz than for either plagioclase or ferromagnesian minerals seems implausible. If the alternative model, $Al(OH)_3$ precipitation rather than quartz dissolution, is adopted, the amount of $Al(OH)_3$ formed would be 2·6 times the amount of smectite. It seems unlikely that so large an amount would escape detection in the bulk chemical analyses of the soils. According to this logic, K-feldspar is more plausible than biotite as a source of K^+. Both, of course, may contribute, but without some other constraint, such as Sr isotopes, the mass balance calculation cannot determine the relative contributions.

Sources of calcium and the problem of excess calcium. When Garrels and Mackenzie (1967) first calculated rates of mineral weathering using spring waters from the Sierra Nevada, they noted that after back-reacting clays to primary minerals an excess of Ca^{2+} balanced by alkalinity (HCO_3^-) remained. They attributed the excess Ca^{2+} to 'minor amounts of carbonates encountered en route'. They could, mathematically, have explained the excess by weathering of a relatively calcic plagioclase to kaolinite, but there was no geological reason to postulate such a plagioclase. Since the original observations of Garrels and Mackenzie several other investigators have found excess Ca^{2+} when calculating mineral weathering reactions. Drever and Hurcomb (1986) found a large (relative) excess of Ca^{2+} in waters of the South Cascade Glacier catchment in the Cascade Mountains. They argued that there were three possible sources of Ca^{2+} from mineral weathering: (1) plagioclase feldspar; (2) ferromagnesian minerals (e.g. clinopyroxene, amphibole); and (3) calcite. Chemical alteration of the first two minerals would produce not only Ca^{2+}, but also Na^+ in the case of feldspars, Mg^{2+} in the case of ferromagnesian minerals and SiO_2 in either case. Thus, if either group of silicate minerals was the source of excess Ca^{2+}, additional cations or silica should also have been produced. Since there was no 'excess' of silica, they concluded that the most probable source of Ca^{2+} was calcite, which contributes only additional Ca^{2+} and bicarbonate upon dissolution. The only potential problem with the calcite hypothesis was the very low abundance of the mineral in the bedrock; the authors invoked high rates of physical weathering in alpine environments as a mechanism to continuously expose fresh calcite-bearing surfaces to the circulating waters.

Mast *et al.* (1990), in their study of mineral weathering in the Loch Vale catchment, concluded similarly that the most plausible explanation for Ca^{2+} remaining after their back-reaction sequence was the dissolution of calcite. Calcite in bulk rock samples comprised from 0·005 to 0·4 wt.% and was found associated with hydrothermal alteration as well as along grain boundaries and fractures in unaltered bedrock. As noted by the authors, the dissolution rate for calcite is about 6 orders of magnitude faster than for aluminosilicate minerals at circumneutral pH (Lasaga, 1984; Mast *et al.*, 1990), and they again invoked high rates of physical weathering in the alpine environment as the mechanism required to expose fresh, unweathered bedrock. Aeolian dust is another potential source of Ca^{2+} in the western United States, and particulate calcite has been recorded at one site in the Front Range of Colorado (Litaor, 1987). Preliminary evaluation of aeolian dust input to West Glacier Lake has not identified any particulate calcite, even though a nearby, up-wind source exists (J. B. Finley, unpublished data). In the mass balance calculation (Figure 6a and d), dissolution of either calcite or epidote could explain the Ca^{2+} in the waters. Since epidote is present and shows etch features (Rochette *et al.*, 1988), there is no need to postulate calcite as a source of Ca^{2+} in the West Glacier Lakes catchment.

Stauffer (1990), in his review of the WLS data, concluded that Ca^{2+} was the dominant base cation in almost all of the granitic terrains studied, and that the stoichiometric weathering of plagioclase was not the sole source of Ca^{2+}. Sulfate was significantly correlated with Ca^{2+} in all but one region, which Stauffer explained by the oxidation of sulfide minerals in the bedrock with concomitant dissolution of calcite, if present, or accelerated rates of plagioclase weathering.

Although the West Glacier Lake catchment contains calcite-rich bedrock, the outcrops are effectively isolated from the prevailing hydrological cycle. The link between sulfide oxidation and dissolution of calcite inferred by Stauffer (1990) does occur in the localized calcite-rich bedrock, but the small percentage of annual flow passing through these outcrops is not detectable in the water chemistry of either Meadow Creek or Long Creek. The amphibolite dikes crossing the Long Creek drainage do contain pyrite, and hand samples show residual Fe-oxyhydroxide as the remnant of oxidation. Acidity released during oxidation of pyrite is, in all likelihood, consumed by chemical weathering of epidote and actinolite, both of which are calcium bearing. Thus, the strong correlation of Ca^{2+} and SO_4^{2-} in waters discharged from Long Creek are consistent with pyrite oxidation-driven alteration of amphibolite minerals. Weathering of actinolite and epidote can supply sufficient Ca^{2+} to account for all Ca^{2+} discharged in Long Creek. There is no need, on the basis of mass balance arguments, to invoke calcite as a source of Ca^{2+}; however, there is no reason to rule out calcite on the basis of mass balance arguments alone.

Secondary phases. Although smectite is the only unequivocal product of weathering of amphibolite minerals, kaolinite is also observed in XRD patterns of soils formed on the amphibolite dikes. In runs a–d (Figure 6) quartz is weathered, and the rate relative to the other primary minerals is quite high for all runs except b, in which K-feldspar is the source of K^+ and the secondary mineral formed is kaolinite. Thus, of the first four mathematical solutions to Equation 2 the most geochemically reasonable combination of minerals does not contain smectite. Mineral combinations e and f compare the consequences of producing smectite along with gibbsite (or amorphous aluminum hydroxide). Run e is essentially the reaction posed by Rochette *et al.* (1988) with differences in rates a consequence of mineral composition and the concentrations of solutes. When smectite is formed rather than kaolinite, and when the source of potassium is biotite, the amount of gibbsite formed is even greater. Field observation suggests that the reaction should be primary minerals (actinolite, epidote and either biotite or K-feldspar) reacting to form both smectite and kaolinite, since the XRD patterns indicate the presence of both clay minerals in soils developed on amphibolite dikes. Unfortunately, given the constraints of mineral composition and the concentration of SiO_2 in the stream water, a mathematical solution containing both smectite and kaolinite as products does not exist. To provide balance, smectite must be a reactant rather than a product. Conceivably, the smectite we observe could have formed sometime in the past and is now altering to kaolinite, but this does not seem particularly plausible. In summary, balance can be achieved with either smectite or kaolinite as the sole secondary phase provided quartz is allowed as a reactant. Alternatively, if quartz is not allowed as a reactant, balance can be achieved if amorphous $Al(OH)_3$ (or gibbsite) is allowed to precipitate in addition to smectite or kaolinite.

Summary. The general pattern of solutions to the mass balance equation can be summarized as follows:

1. Albite is necessary as a source of Na^+. The amount of albite dissolved is essentially the same in all the permutations, determined by the sodium concentration in the water.

2. Actinolite is necessary as a source of Mg^{2+}. The amount of actinolite dissolving varies somewhat, depending on whether or not Mg^{2+} is present in a secondary phase (smectite) and whether biotite or K-feldspar is included as a source for K^+.

3. Either K-feldspar or biotite must be dissolving to provide a source of K^+.

4. Either epidote or calcite, in addition to actinolite, is contributing Ca^{2+}.

5. If only a single clay mineral is forming, quartz must be dissolving to provide silica. If amorphous $Al(OH)_3$ is postulated to form, quartz dissolution is not necessary.

These conclusions are phrased in terms of the mathematical constraints of the mass balance calculation, that the number of solid phases does not exceed the number of components. In reality, more than that number of phases could be involved (K^+ could come from both K-feldspar *and* biotite; Ca^{2+} could be coming from both epidote and calcite; quartz could be dissolving and amorphous $Al(OH)_3$ precipitating) but there is no way of quantifying these possibilities without data on additional components. This is a serious limitation on the use of the mass balance approach to determine weathering reactions, and underlines the need for additional information on soil mineralogy to supplement solute budgets.

Mineral weathering reactions: Meadow Creek. Meadow Creek illustrates another source of uncertainty in mass balance calculations of mineral weathering, that of uncertainty in the input–output budget. The output concentration of sodium is indistinguishable from that of the precipitation input (Figure 4). The 'best estimate', of sodium derived from weathering is thus zero, but the uncertainty is large. If the boxes containing 50% of the data on Figure 4 are taken as a measure of uncertainty, the Ca^{2+} concentration could range from 17 to 54 $\mu eq\ l^{-1}$ (a factor of three) and the Si concentration from 15 to 29 $\mu eq\ l^{-1}$, a factor of two. With such large uncertainties in the data, detailed mineral weathering budgets would not have much meaning.

CONCLUSIONS

Chemical mass balance has been used in various studies of high-elevation catchments to identify specific mineral weathering reactions and to estimate the rates at which primary minerals react with circulating waters. Such calculations are subject to uncertainties from several sources. We have used the results of mass balance and calculated rates of mineral weathering in the West Glacier Lake catchment to illustrate some of the problems. These include:

1. The accuracy of the chemical mass balance depends on high quality measurements of the mass of solutes input and output, preferably over several annual cycles. Characterizing the amount and chemical composition of precipitation in high-elevation catchments is difficult for several reasons. First, the primary form of precipitation is snow, and strong winds (e.g. an average of 6 m s^{-1} at West Glacier Lake) affect the capture efficiency of standard precipitation gauges. This results in a large uncertainty in the volume of precipitation/snow entering the catchment. Secondly, the composition of precipitation varies with both time and elevation as illustrated by the differences shown in Table I for precipitation at the two NADP sites. The assumption must be made that the volume-weighted average composition from the collector is representative of the composition of precipitation over the catchment as a whole. A third problem that typically affects high-elevation catchments is that weathering rates are low, which means that the flux of solutes derived from weathering is similar in magnitude to the flux from the atmosphere. Since the flux from weathering is calculated from the difference between the measured input flux and the measured output flux, uncertainties in these fluxes result in a large uncertainty in the calculated flux from weathering. Meadow Creek in the West Glacier Lake catchment illustrates this problem well.

2. A quantitative solute budget must be developed, in which the contribution from mineral weathering is separated from the effects of ion exchange and biomass uptake. In general, the argument is made that if long-term average concentrations of solutes are available, then the net effect of cation exchange and biotic uptake/release are small in systems that are not affected by atmospheric pollution and where primary productivity and biomass are small (e.g. Drever, 1988; Mast et al., 1990; Stauffer, 1990). Also, most soils in high-elevation catchments are thin and immature, resulting in low cation exchange capacities.

3. Assigning the solutes derived from weathering to specific mineral weathering reactions is subject to additional uncertainties. For purely mathematical reasons, the number of solid phases (reactants plus products) must equal the number of components used in the calculation (typically six: the solutes Na$^+$, K$^+$, Ca^{2+}, Mg^{2+}, SiO$_2$, and Al, which is conserved). Nature is not constrained in this way, so the actual number of phases reacting or forming may be greater than six, but the relative proportions of a larger number of phases cannot be deduced from the water composition alone. The choice of a set of reactants and products is non-unique. The fact that a set of reactions is found that gives exact mass balance does not mean, even ignoring other uncertainties, that that particular set is a realistic description of what is happening in nature. Identification of primary minerals for inclusion into the mineral weathering reaction is usually straightforward although, in the example discussed, we could not distinguish unequivocally between K-feldspar and biotite as a source of K$^+$, nor between epidote and calcite as a source of Ca^{2+}. The chemical composition of any primary minerals used in the mass balance must be known. For phases of variable composition, particularly for ferromagnesian minerals such as amphibole and biotite, the composition must be determined, which usually requires electron microprobe microanalysis.

Identifying and chemically characterizing the secondary phases formed by weathering is usually difficult. Clay minerals can be identified by X-ray diffraction, but such clay minerals are not necessarily forming at the present time. They could be aeolian, or they could be the products of weathering at sometime in the past. Once a clay mineral has been identified, it must be assigned a chemical composition. For kaolinite or gibbsite, this is straightforward, but for smectite, vermiculite and mixed-layer clays, all of which are common in high-elevation catchments, it is difficult because the compositions are highly variable and it is often impossible to obtain a pure mineral separate for chemical analysis. In the West Glacier Lake catchment we were able to find and analyse soils whose fine fraction was almost pure smectite, but this is not always possible, and there is no way of knowing if this relatively pure smectite (which is found in only a few localities) is chemically representative of smectite forming over the catchment as a whole. There is also an implicit assumption that 'the' secondary product is a well-defined phase. In high-elevation catchments undergoing rapid physical erosion, it is quite possible that primary ferromagnesian phases are partially leached without the formation of well-defined secondary phases (c.f. Miller and Drever, 1977). Also, in our specific example of the Long Creek tributary, the majority of water input is circulated through the bedrock aquifer, and selecting secondary weathering products based on surface samples of soil may not adequately represent the geochemical reaction.

These uncertainties suggest that the exact numerical values for mineral weathering reactions calculated by the mass balance approach should be viewed with a certain amount of scepticism, but they do not mean that the whole approach is worthless: certain conclusions are quite robust, and independent of the details of the calculation. For example, the only source of Na^+ in weathering is commonly plagioclase or albite, and Na^+ is generally not incorporated significantly into secondary phases in high-elevation catchments. The sodium flux can thus be transformed directly into a rate of weathering of the albite component of feldspar. If the composition (or distribution of compositions) of plagioclase in the catchment is known, then the rate of weathering of plagioclase can be estimated. Conversely, if the flux of Na^+ is very low, as in the South Cascade Glacier area (Drever and Hurcomb, 1986), then the rate of plagioclase weathering must be very low and, unless an extremely calcic plagioclase is present, plagioclase weathering cannot be a significant source of Ca^{2+}. Semi-quantitative arguments such as this can usually constrain the rates of alteration of primary minerals fairly well. However, we do not have much confidence in the use of mass balance calculations to identify the nature of secondary phases formed by weathering in high-elevation catchments.

ACKNOWLEDGEMENTS

We thank the US Forest Service, Rocky Mountain Forest and Range Experiment Station for providing logistical support and discharge data for tributaries in the West Glacier Lake catchment. Support for this research was provided by a series of grants from the Wyoming Water Research Center (J.I.D.) and a Geological Society of America Research Grant 4426-90 (J.B.F.). The comments of two anonymous reviewers have helped focus and clarify the manuscript.

REFERENCES

Appelo, C. A. J. and Postma, D. 1993. *Geochemistry, Groundwater and Pollution*. A. A. Balkema.

Arthur, M. A. 1992. 'Vegetation', in Baron, J. (Ed.), *Biogeochemistry of a Subalpine Ecosystem: Loch Vale Watershed*, Ecological Studies V. 90. Springer-Verlag, pp. 76–92.

Arthur, M. A. and Fahey, T. J. 1990. 'Mass and nutrient content of decaying boles in an Englemann Spruce-Subalpine Fir forest, Rocky Mountain National Park, Colorado', *Can. J. For. Res.* **20**, 730–737.

Arthur, M. A. and Fahey, T. J. 1992. 'Biomass and nutrients in an Engelmann spruce-subalpine fir forest in north central Colorado: pools, annual production, and internal cycling', *Can. J. For. Res.*, **22**, 315–325.

Arthur, M. A. and Fahey, T. J. 1993. 'Controls on soil solution chemistry in a subalpine forest in north-central Colorado', *Soil Sci. Soc. Am. J.* **57**, 1122–1130.

Bales, R. C., Valdez, M. P., Dawson, G. A., and Stanley, D. A. 1987. 'Physical and chemical factors controlling gaseous deposition of SO_2 to snow', in Jones, H. G. and Orville-Thomas, W. J. (Eds), *Seasonal Snowshowers: Physics, Chemistry, Hydrology*, NATO ASI Ser. C, V. 211. Reidel, pp. 289–297.

Bales, R. C., Sommerfeld, R. A., and Kebler, D. G. 1990. 'Ionic tracer movement through a Wyoming snowpack', *Atm. Environ.*, **24A**, 2749–2785.

Baron, J. 1992. 'Biochemical Fluxes', in Baron, J. (Ed.), *Biogeochemistry of a Subalpine Ecosystem: Loch Vale Watershed*, Ecological Studies V. 90. Springer-Verlag, pp. 218–231.

Baron, J. and Bricker, O. P. 1987, 'Hydrologic and chemical flux in Loch Vale Watershed, Rocky Mountain National Park', in McKnight, D. and Averett, R. C. (Eds), *Chemical Quality of Water and the Hydrologic Cycle*. Lewis Publishers. pp. 141–156.

Blum, J. D., Erel, Y., and Brown, K. 1994. '$^{87}Sr/^{86}Sr$ ratios of Sierra Nevada stream waters: implications for relative mineral weathering rates', *Geochim. Cosmochim. Acta*, **57**, 5019–5025.

Bowser, C. J. and Jones, B. F. 1993. 'Mass balances of natural waters: silicate dissolution, clays, and the calcium problem. *Biogeomon Symposium on Ecosystem Behavior: Evaluation of Integrated Monitoring in Small Catchments*, Prague, Czech Republic, Abstracts, pp. 30–31.

Bytnerowicz, A., Dawson, P. J., Morrison, C. L., and Poe, M. P. 1992. 'Atmospheric dry deposition on pines in the Eastern Brook Lake watershed, Sierra Nevada, California', *Atm. Environ.*, **26A**, 3195–3201.

Caine, N. 1974. 'The geomorphic processes of the alpine environment', in Ives, J. D. and Barry, R. G. (Eds), *Arctic and Alpine Environments*. Methuen. pp. 722–748.

Campbell, D. H., Turk, J. T., and Spahr, N. E. 1991. 'Response of Ned Wilson Lake watershed, Colorado, to changes in atmospheric deposition of sulfate', *Wat. Resour. Res.*, **27**, 2047–2060.

Clayton, J. L. 1986. 'An estimate of plagioclase weathering rate in the Idaho Batholith based upon geochemical transport rates', in Colman, S. M. and Dethier, D. P. (Eds), *Rates of Chemical Weathering of Rocks and Minerals*. Academic Press. pp. 453–467.

Clow, D. W. and Mast, M. A. 1995. 'Composition of precipitation, bulk deposition, and runoff at a granitic bedrock catchment in the Loch Vale watershed, Colorado, USA', in Tonnessen, K. A., Williams, M. W., and Tranter, M. (Eds), *Biogeochemistry of Seasonally Snow-covered Catchments*. IAHS Publ. No. 228. IAHS, Wallingford. pp. 235–242.

Denning, A. S., Baron, J., Mast, M. A., and Arthur, M. 1991. 'Hydrologic pathways and chemical composition of runoff during snowmelt in Loch Vale watershed, Rocky Mountain National Park, Colorado, USA', *Water Air Soil Pollut.* **59**, 107–123.

Dove, P. M. and Elston, S. F. 1992. 'Dissolution kinetics of quartz in sodium chloride solutions: analysis of existing data and a rate model for 25°C', *Geochim. Cosmochim. Acta*, **56**, 4147–4156.

Drever, J. I. 1988. *The Geochemistry of Natural Waters*, 2nd Edn. Prentice-Hall.

Drever, J. I. and Hurcomb, D. R. 1986, 'Neutralization of atmospheric acidity by chemical weathering in an alpine drainage basin in the North Cascade Mountains', *Geology*, **14**, 221–224.

Drever, J. I. and Zobrist, J. 1992. 'Chemical weathering of silicate rocks as a function of elevation in the southern Swiss Alps', *Geochim. Cosmochim. Acta*, **56**, 3209–3216.

Finley, J. B. and Drever, J. I. 1992. 'Chemical hydrograph separation using field and experimental data with implications for solute cycling in an alpine catchment', *Proc. 7th Int. Symp. Water-rock Interaction*, Vol. I, pp. 553–556.

Finley, J. B., Drever, J. I., and Frost, C. D. 1993. 'Eolian flux to an alpine catchment, Snowy Range, Wyoming', *Geol. Soc. Am. Abstracts with Programs*, **25**, A318.

Finley, J. B., Drever, J. I., and Turk, J. T. 1995. 'Sulfur dynamics in the Glacier Lakes catchment, Snowy Range, Wyoming', *Water Air Soil Pollut.*, **79**, 227–241.

Flurkey, A. J. 1983. 'Depositional environment and petrology of the Medicine Peak Quartzite Early (Proterozoic), southern Wyoming', *Ph.D. Dissertation*, University of Wyoming, Laramie.

Fox, D. G., Humphries, H. C., Zeller, K. F., Connell, B. H., and Wooldridge, G. L. 1994. 'Meteorology', in Musselmann, R. C. (Ed.), *The Glacier Lakes Ecosystem Experiments Site*, Rocky mountain Forest and Range Experiment Station, General Tech. Rept. RM–249, pp. 42–47.

Frogner, T. 1990. 'The effect of acid deposition on cation fluxes in artificially acidified catchments in western Norway', *Geochim. Cosmochim. Acta*, **54**, 769–780.

Garrels, R. M. 1967. 'Genesis of some ground waters from igneous rocks', in Abelson, P. H. (Ed.), *Researches in Geochemistry*, Vol. 2. J. Wiley & Sons. pp. 405–420.

Garrels, R. M. and Mackenzie, F. T. 1967. 'Origin of the chemical compositions o some springs and lakes', *Am. Chem. Soc. Adv. Chem. Ser.* **67**, 222–242.

Grier, C. C. and Ballard, T. M. 1981, 'Biomass, nutrient distribution, and net production in alpine communities of the Kluane mountains, Yukon Territory, Canada', *Can. J. Bot.*, **59**, 2635–2649.

Hasfurther, G., Kerr, G., Parks, G., and Wetstein, J. 1990. Glacier Lakes Hydrological Balance — final report. Wyoming Water Research Center, University of Wyoming, Laramie.

Hicks, B. B. 1986. 'Measuring dry deposition: a re-assessment of the state of the art', *Water Air Soil Pollut.*, **30**, 75–90.

Karlstrom, K. E. and Houston R. S. 1984. 'The Cheyenne Belt: analysis of a Proterozoic suture in southern Wyoming', *Precambrian Res.*, **25**, 415–446.

Katz, B. G., Bricker, O. P., and Kennedy, M. M. 1985. "Geochemical mass-balance relationships for selected ions in precipitation and stream water, Cotactin Mountains, Maryland', *Am. J. Sci.*, **285**, 931–962.

Kirkwood, D. E. and Nesbitt, H. W. 1991. 'Formation and evolution of soils from an acidified watershed: Plastic Lake, Ontario, Canada', *Geochim. Cosmochim. Acta*, **55**, 1295–1308.

Landers, D. H. *et al.* 1987, 'Characteristics of lakes in the western United States. Vol. 1. Population descriptions and physico-chemical relationship', US EPA/600/3–86/054a.

Lasaga, A. C. 1984. 'Chemical kinetics of water–rock interaction', *J. Geophys. Res.*, **89**, 4009–4025.

Likens, G. E., Bormann, F. H., Pierce, R. S., Eaton, J. S., and Johnson, N. M. 1977. *Biogeochemistry of a Forested Ecosystem*. Springer-Verlag.

Litaor, M. I. 1987. 'The influence of eolian dust on the genesis of alpine soils in the Front Range, Colorado', *Soil Sci. Soc. Am. J.*, **51**, 142–147.

Mast, M. A., Drever, J. I., and Baron, J. 1990. 'Chemical weathering in the Loch Vale Watershed, Rocky Mountain National Park, Colorado', *Wat. Resour. Res.*, **26**, 2971–2978.

Miller, W. R. and Drever, J. I. 1977. 'Chemical weathering and related controls on water chemistry in the Absarota Mountains, Wyoming', *Geochim. Cosmochim. Acta*, **41**, 1693–1702.

Miller, E. K., Friedland, A. J., and Blum, J. D. 1993. 'Determination of soil exchange-cation loss and weathering rates using Sr isotopes', *Nature*, **362**, 439–411.

Moore, D. M. and Reynolds, R. C. 1989. *X-ray Diffraction and the Identification and Analysis of Clay Minerals*. Oxford University Press, Oxford.

National Atmospheric Deposition Program (NRSP-3)/National Trends Network, 1992. NADP/NTN Coordination Office, Natural Resource Ecology Laboratory, Colorado State University, Fort Collins.

Pačes, T. 1983, 'Rate constants of dissolution derived from the measurements of mass balance in hydrological catchments', *Geochim. Cosmochim. Acta*, **36**, 537–544.

Plummer, L. N. and Back, W. 1980. 'The mass balance approach: application to interpreting the chemical evolution of hydrologic systems', *Am. J. Sci.*, **280**, 130–142.

Reuss, J. O., Vertucci, F. A., Musselman, R. C., and Sommerfeld, R. A. 1993. *Biogeochemical fluxes in the Glacier Lakes catchments', Rocky Mountain Forest and Range Experiment Station, Research Paper RM-314*. 27 p.

Rochette, E. A. 1987, 'Chemical weathering in the West Glacier Lake drainage basin, Snowy Range, Wyoming', *M.S. Thesis*, University of Wyoming, Laramie.

Rochette, E. A., Drever, J. I., and Sanders, F. S. 1988. 'Chemical weathering in the West Glacier Lake drainage basin, Snowy Range, Wyoming: implications for future acid deposition', *Contr. Geol.*, **26**, 29–44.

Shapiro, L. 1975. 'Rapid analysis of silicate, carbonate, and phosphate rocks — revised edition', *US Geol. Surv. Bull.*, **1401**, 76.

Shepard, J. P., Mitchell, M. J., Scott, T. J., Zhang, Y. M., and Raynal, D. J. 1989. 'Measurements of wet and dry deposition in a northern hardwood forest', *Water Air Soil Pollut.*, **48**, 225–238.

Shroba, R. R. and Birkeland, P. W. 1983. 'Trends in late-Quaternary soil development in the Rocky mountains and Sierra Nevada of the western United States', in Porter, S. C. (Ed.), in *Late Quaternary Environments of the United States*, Vol. 1. pp. 145–156.

Sommerfeld, R. A., Musselman, R. C., and Wooldridge, G. L. 1990. 'Comparison of estimates of snow input with a small alpine catchment', *J. Hydrol.*, **120**, 295–307.

Sparks, D. L. 1989. *Kinetics of Soil Chemical Processes*. Academic Press.

Stauffer, R. E. 1990. 'Granite weathering and the sensitivity of alpine lakes to acid deposition', *Limnol. Oceanogr.*, **35**, 1112–1134.

Stoddard, J. L. 1987. 'Alkalinity dynamics in an unacidified alpine lake, Sierra Nevada, California', *Limnol. Oceanogr.*, **32**, 825–839.

Tardy, Y. 1971. 'Characterization of the principal weathering types by the geochemistry of waters from some European and African crystalline massifs', *Chem. Geol.*, **7**, 253–271.

Taylor, A. B. and Velbel, M. A. 1991. 'Geochemical mass balances and weathering rates in forested watersheds of the southern Blue Ridge II. Effects of botanical uptake terms', *Geoderma*, **51**, 29–50.

Tranter, M., Brimblecombe, P., Davies, T. D., Vincent, C. E., Abrahams, P. W., and Blackwood, I. 1986. 'The composition of snowfall, snowpack, and meltwater in the Scottish Highlands: evidence for preferential elution', *Atm. Environ.*, **20**, 517–525.

Tsiouris, S., Vincent, C. E., Davies, R. D., and Brimblecombe, P. 1985. 'The elution of ions through field and laboratory snowpacks', *Ann. Glaciol.*, **7**, 196–201.

Turk, J. T. and Campbell, D. H. 1987. 'Estimates of acidification of lakes in the Mt. Zirkel Wilderness Area, Colorado', *Wat. Resour. Res.*, **23**, 1757–1761.

Turk, J. T. and Spahr, N. E. 1991. 'Rocky Mountains', in Charles D. and Christie, S. (Eds), *Acid Deposition and Aquatic Ecosystems*. Springer-Verlag. pp. 471–501.

Valdez, M. P., Bales, R. C., Stanley, D. A., and Dawson, G. A. 1987. 'Gaseous deposition to snow 1. Experimental study of SO_2 and NO_2 deposition', *J. Geophys. Res.* **92**, 9779–9787.

Velbel, M. A. 1986. 'The mathematical basis for determining rates of geochemical and geomorphic processes in small forested watersheds by mass balance: Examples and implications', in Colman, S. P. and Dethier, D. P. (Eds), *Rates of Chemical Weathering of Rocks and Minerals*. Academic Press. pp. 439–451.

Velbel, M. A. 1992. 'Geochemical mass balances and weathering rates in forested watersheds of the southern Blue Ridge. III. Cation budgets and the weathering rate of amphibole', *Am. J. Sci.*, **292**, 58–78.

Vitousek, P. M. and Reiners, W. A. 1975. 'Ecosystem succession and nutrient retention: a hypothesis', *Bioscience*, **25**, 376–381.

Whelpdale, D. M. and Shaw, R. W. 1974. 'Sulfur dioxide removal by turbulent transfer over grass, snow, and water surfaces', *Tellus*, **26**, 196–204.

Williams, M. W. and Melack, J. M. 1991. 'Precipitation chemisty in and ionic loading to an alpine basin, Sierra Nevada', *Wat. Resour. Res.*, **27**, 1563–1574.

Williams, M. W., Brown, A. D., and Melack, J. M. 1993. 'Geochemical and hydrologic controls on the composition of surface water in a high-elevation basin, Sierra Nevada, California', *Limnol. Oceanogr.*, **38**, 775–797.

16

THE USE OF MASS BALANCE INVESTIGATIONS IN THE STUDY OF THE BIOGEOCHEMICAL CYCLE OF SULFUR

H. E. EVANS[1], P. J. DILLON[2] AND L. A. MOLOT[3]

[1] *RODA Environmental Research, PO Box 447, Lakefield, Ontario K0L 2H0, Canada*
[2] *Ontario Ministry of Environment and Energy, Dorset Research Centre, PO Box 39, Bellwood Acres Road, Dorset, Ontario P0A 1E0, Canada*
[3] *Faculty of Environmental Studies, York University, 4700 Keele Street, Toronto, Ontario M3J 1P3, Canada*

ABSTRACT

The use of mass balances in the investigation of the biogeochemical cycle of sulfur is reviewed for three systems: 1) upland catchments, 2) wetlands, and 3) lakes. In upland catchments, the major inputs of sulfur are via wet and dry atmospheric deposition, whereas outputs or losses occur primarily through volatilization and/or runoff. In addition, sulfur may be stored in vegetation and in the forest floor. In wetlands (particularly peatlands), a large proportion of the sulfur inputs are derived from surface and groundwater originating in the upland system. Because of the fluctuating water table in wetlands, they can act as a source or sink for sulfate, depending on the redox conditions. Wetlands, therefore, can significantly affect input-output budgets for lakes. In most lakes, only a small portion of the sulfate input is retained, (i.e. not lost from the lake via outflow), indicating that there is an excess of sulfate relative to biological needs. Seepage lakes are exceptions to this generalization. Although the reactivity of the sulfate input to many lakes is low, sulfate levels, especially in regions receiving substantial atmospheric sulfur deposition, are high enough that the portion reduced results in substantial in-lake alkalinity production; in fact, in many cases, alkalinity production from sulfate reduction is greater than that resulting from not only other in-lake processes but from external sources (the catchment) as well.

The importance of mass balance investigations in elucidating the biogeochemical cycling of sulfur is stressed and the need for additional studies on a whole-system basis stressed.

INTRODUCTION

The use of mass balances (also called input–output budgets) for lakes and catchments has become an increasingly important means of assessing the sources, fate (i.e. loss to sinks or via transformations) and cycling of major ions (Dillon *et al.*, 1982; Wright, 1983; Bergstrom and Gustafson, 1985; Jeffries *et al.*, 1988), nutrients such as phosphorus (Vollenweider, 1969; Dillon and Rigler, 1975; Dillon and Evans, 1993; Dillon *et al.*, 1993; Dillon and Molot, 1996), nitrogen (Dillon and Molot, 1990; Dillon *et al.*, 1991; Molot and Dillon, 1993) and carbon (Dillon and Molot, 1997) and metals (Cross and Rigler, 1983; Schut *et al.*, 1985; White and Driscoll, 1985, 1987a,b; Dillon *et al.*, 1988; Nurnberg and Dillon, 1993). The advantages of using mass balance studies are numerous (Brezonik *et al.*, 1987): (1) the method is holistic and readily integrates information over large areas and extended periods of time; (2) the calculations are conceptually simple; (3) the method is based on a fundamental precept of science, that of conservation of mass; and (4) in the case of some whole system studies, the data needed for the calculations can be obtained from routine monitoring programmes. In addition, whole system input–output budgets are useful for predicting concentrations as a function of inputs and losses, and for evaluating the causes and effects of temporal patterns in concentrations.

There has been a continual increase in research interest in sulfur, and in particular, sulfate (SO_4^{2-}), for many years because of its importance as a plant nutrient (Wainwright, 1990), a major component of acidic precipitation and a control in determining phosphorus availability in lakes (Hasler and Einsele, 1948; Caraco *et al.*, 1993). Several authors have used SO_4^{2-} budgets to determine the relative importance of in-lake (Schindler, 1986; Schindler *et al.*, 1986; Schafran and Driscoll, 1987; Brezonik *et al.*, 1987; Lin *et al.*, 1987; Eshleman and Hemond, 1988; Psenner, 1988) versus watershed alkalinity generation (Likens *et al.*, 1977; Wright and Johannessen, 1980; Wright, 1983; Schindler *et al.*, 1986; Eshleman and Hemond, 1988; see Shaffer *et al.*, 1988 for summary), whereas others (Rudd *et al.*, 1990) have used S and nitrogen (N) budgets to determine the relative 'acidification' efficiencies of nitric and sulfuric acid in an experimentally acidified, soft water, oligotrophic lake (Lake 302) in the Experimental Lakes Area (ELA) of Ontario, Canada. In these studies, SO_4^{2-} retention is equated with generation of alkalinity.

The S cycle is relatively complex in that it involves several gaseous species (e.g. H_2S, SO_2), ions (e.g. SO_4^{2-}, HS^-), solid phases (S, metal sulfides), including poorly soluble minerals such as pyrite (FeS_2), and organic compounds (C-bonded and ester sulfate forms). Because S can have several oxidation states (i.e. -2, -1, 0, $+2$, $+4$ and $+6$) and because it is closely linked with the oxygen cycle, the S cycle involves many complex reactions and processes. These include abiotic redox reactions such as the conversion of SO_2 to sulfuric acid in the atmosphere, hydrolysis and organic substitution reactions [e.g. HS^- acts as a nucleophile and displaces a halide ion (e.g. Br, Cl) on an alkyl halide (RX) to produce a thiol (RSH) or a thiol thiolate ion, RS^-)], precipitation and dissolution of solid phases (e.g. precipitation of FeS_2 under anoxic conditions and solubilization in the presence of oxygen) and microbial redox processes (e.g. the microbial reduction of SO_4^{2-} to H_2S under anoxic conditions — termed dissimilatory reduction). In addition to these processes, S cycling rates in the environment are affected by mass transfer processes such as diffusion across the sediment–water, plant–water, or soil–water interface and volatilization, gas exchange and adsorption at the air–water interface.

Mass balance studies of S, or indeed any other substance, can be conducted at many physical scales. For example, microcosm experiments wherein sediment cores (Davison and Woof, 1990; Kling *et al.*, 1991) or soil/peat samples (Spratt and Morgan, 1990; Ghani *et al.*, 1993) were manipulated in the lab (e.g. radio-labelled S was added, or the cores aerated), and after suitable incubation periods, the various S species were determined. While these sorts of experiments are useful for elucidating reaction rates and end-products, the results apply only to the experimental conditions used and, thus, cannot be extrapolated necessarily to the natural environment.

Mesocosm experiments involve the examination and/or experimental manipulation of part of the system *in situ*. The experimental area or study site is generally referred to as a 'plot' if the terrestrial catchment is involved, or a 'limnocorral' if the lake is involved. Schiff and Anderson (1987) measured mass balances of SO_4^{2-} and other ions in enclosures in two lakes at the Experimental Lakes Area (ELA). The rate of SO_4^{2-} consumption in the lakes increased in response to increased SO_4^{2-} loading, suggesting that SO_4^{2-}-reducing bacteria were limited by SO_4^{2-}. In mesocosm experiments conducted in two forested watersheds in Mont-Lozere, France, Vannier *et al.* (1993) addressed the relationship between the distribution of the various sulfur forms and the SO_4^{2-} fluxes within three soil profiles. Unfortunately, as in microcosm experiments, the information derived from soil plots may apply only to the area studied and cannot be extrapolated necessarily to the entire catchment. Likewise, while input–output budgets using limnocorrals are suitable in the short term, Levine (1983, as cited by Schindler, 1987) found that for experiments lasting more than a few weeks, even very large enclosures developed abnormally high populations of periphyton (phytoplankton attached to the limnocorral wall) affecting the cycling of most elements substantially.

The most effective scale of mass balance studies for assessing the fate and behaviour of S in the environment is at the whole system level. In several areas in the northern hemisphere in which whole lakes and/or catchments are being, or have been, manipulated, mass balances have been measured so that the S cycle might be further elucidated. Some of these studies include: (1) the addition of H_2SO_4 or alum to Lakes 223, 114 and 302 at the ELA (Schindler, 1991); (2) the acidification of the north basin of Little Rock Lake, Wisconsin with H_2SO_4 (Baker *et al.*, 1989); (3) the RAIN project in Norway (Wright, 1985; Wright and Gjessing, 1986; Wright *et al.*, 1986, 1988a) which comprises two parallel experiments representing both addition (with H_2SO_4 and HNO_3) experiments and also exclusion experiments; (4) the Gardsjon Project in

Sweden in which acid has been added to two subcatchments of Lake Gardsjon (5) the addition of dry $(NH_4)_2SO_4$ by helicopter to the West Bear Brook, Maine catchment (Sampson *et al.*, 1994) as part of the Watershed Manipulation Project (WMP) in the United States; and (6) the EXperimental MANipulation of Forest Ecosystems Project (EXMAN) in Europe (Rasmussen *et al.*, 1992). Experimental additions of sulfuric acid also have been attempted in wetlands (Bayley *et al.*, 1986, 1987).

The information derived from these and other mass balance studies is essential for the construction of predictive models that are based on mass balance concepts and equations (Lee and Schnoor, 1988). Examples of models used to predict both the spatial and temporal response of lakes and catchments to changes in, for example, sulfur deposition, include the Birkenes model (Christophersen and Wright, 1981), the Wisconsin lake model (Baker and Brezonik, 1988), the ILWAS model (Integrated Lake-Watershed Acidification Study; Gherini *et al.*, 1985), MAGIC (Model of Acidification of Groundwater in Catchments; Cosby *et al.*, 1985) and the TAME model (Terrestrial Aquatic Model for Ecosystems; Rees and Schnoor, 1994), an outgrowth of the ETD (Enhanced Trickle-Down model; Nikolaides *et al.*, 1988). Mass balance studies are even more informative when used in conjunction with parallel or published reductionist studies because the latter can assist in the interpretation of mass balance results and in the design of models.

Conceptually, the use of whole system mass balances in the study of the biogeochemical cycle of sulfur can be divided into studies of three components: (1) the 'upland' or terrestrial component of the catchment; (2) the 'lowland' or wetland component of the catchment; and (3) the lake component. Using this framework, the use of input–output budgets in elucidating the fate and behaviour of S in the environment is discussed in the following sections.

UPLAND CATCHMENTS

We define 'upland' here as the terrestrial portion of a catchment where oxic processes predominate at all times of the year. A schematic representation of the S cycle in an upland forest is shown in Figure 1 (from Johnson, 1984; Reuss and Johnson, 1986; Van Stempvoort *et al.*, 1991). While this review will address only forested upland catchments, similar processes occur in other types of upland catchments. The reader is referred to Maynard *et al.* (1985) for a review of S cycling in grassland and parkland soils.

In those catchments where minerological sources of S are rare or absent, the major input of S to the forest canopy is by wet and dry atmospheric deposition. While measurements of both past and current sulfur deposition are abundant, a variety of methods have been used to collect atmospheric deposition. In the past, scientists have collected or measured wet deposition, bulk deposition and, in some instances, dry deposition. Wet deposition may be defined operationally as deposition measured using collectors open only during periods of precipitation; bulk deposition is deposition measured using collectors which are open at all times (Vet *et al.*, 1988). Dry deposition, on the other hand, is deposited through or by mechanisms other than precipitation and fog water deposition. These mechanisms include the uptake of gases and particulates by surface elements such as water, vegetation and soil, i.e. scavenging by vegetation and wash-out (Rochelle and Church, 1987). Dry deposition can occur during both precipitation and non-precipitation events.

Although bulk deposition samplers have been used in the past, and continue to be used to estimate the total atmospheric deposition of S to terrestrial catchments (Likens *et al.*, 1977), most bulk deposition collectors collect wet deposition, little or no gaseous dry S deposition and some unknown, variable fraction of the dry particulate S deposition (Vet *et al.*, 1988). Input–output budgets measured at the Hubbard Brook Experimental Forest in New Hampshire first demonstrated the deficiency in bulk deposition collectors. When Likens *et al.* (1977), using bulk deposition collectors, found that the output of SO_4^{2-} from the catchment via stream flow exceeded the measured input from the atmosphere by about 40%, they attributed the difference to unmeasured inputs of S from atmospheric gaseous and aerosol sources. Similarly, Dillon *et al.* (1982), using the mass balance approach, reported that the output of SO_4^{2-} from 11 catchments in the Sudbury region of Ontario, Canada exceeded bulk deposition inputs in all cases.

The discrepancy between bulk deposition and total (wet + dry) deposition of S appears to be a function of distance from the point source. Because SO_2 is oxidized rapidly to SO_4^{2-} in the atmosphere, bulk deposition collectors may be suitable in areas remote from the S source because the SO_4^{2-} is washed out efficiently during precipitation events. However, in areas that are close to point sources, gaseous inputs of SO_2 or deposition of dry particulate SO_4^{2-} can be significant. For example, bulk deposition collectors were unsuitable in Sudbury, Ontario because of the presence of large point sources (smelters; Dillon *et al.*, 1982). At the Turkey Lakes Watershed, Ontario, bulk deposition measurements of SO_4^{2-} were only 15 and 20% lower, respectively, than estimates of total (wet + dry) deposition (Vet *et al.*, 1988). On the other hand, Johnson *et al.* (1985) determined that wet + estimated dry deposition (160 ± 10 meq m^{-2} yr^{-1}) to two deciduous forests in eastern Tennessee was about twice that measured using bulk deposition collectors (78 ± 7 meq m^{-2} yr^{-1}).

Because there is no routine, direct method suitable for all conditions, estimates of dry deposition are difficult to obtain (see summary by Lovett and Lindberg, 1984). One approach for estimating the dry deposition flux of SO_2 is to multiply the air concentration of SO_2 above the deposition surface by the estimated dry deposition velocity (typically about 0.5 cm s^{-1}; Baker *et al.*, 1989). Aerosol deposition of SO_4^{2-} can also be estimated from dry bucket and snow core measurements.

A second approach is to calculate the difference between throughfall SO_4^{2-} input and measured SO_4^{2-} input in bulk precipitation (Eshlemann and Hemond, 1988), although stemflow should be added to the throughfall measurement because it can represent a significant (10–15%) input of SO_4^{2-} (Johnson *et al.*, 1986; Neary and Gizyn, 1994). Unfortunately, there is no agreement regarding definitions or terminology. For example, throughfall minus bulk deposition has been referred to as 'interception' rather than dry deposition by Hambuckers and Ramacle (1990). Part of the uncertainty in these terms is that the magnitude and nature of the dry component of throughfall varies (e.g. gaseous impact on to the leaf surface or dry particle deposition on to the leaf surface followed by wash-out), and, thus, the discrepancy between throughfall and bulk deposition differs according to the distance from the source of the SO_2 as discussed above.

Alternatively, total dry deposition has sometimes been assumed to be simply a fraction of the wet deposition. For example, in attempting to estimate retention of S in watersheds in the eastern US, Rochelle and Church (1987) included a 'low' deposition and a 'high' deposition scenario in which, depending on the region, dry deposition ranged between 15 and 100% of the wet deposition.

The uncertainty in the actual wet + dry, as well as bulk, deposition measurements is further complicated by the type, size and shape of the collection funnel and bottle or bucket, by the location of the collector and by the frequency and/or timing of sample collection (Vet *et al.*, 1988). Recently, Padro (1994) found that the dry deposition of SO_2 was greater over a forest canopy wetted by both dew and rain, than over a dry forest. Furthermore, SO_2 uptake by the trees was enhanced when the forest was wetted.

Thus, the actual deposition of SO_4^{2-} to upland soils, although largely controlled by wet and dry deposition to the forest canopy, is influenced, to a significant extent, by the above-ground vegetation (Figure 1). Processes such as interception, throughfall, foliar leaching, stemflow and litterfall all modify atmospheric SO_4^{2-} deposition and thus affect soil SO_4^{2-} pools. The construction of mass balance budgets within the forest ecosystem has elucidated to some extent, the magnitude of some of these processes. For example, Van Stempvoort *et al.* (1991) determined that up to 60% of the SO_4^{2-} in throughfall and stemflow at an upland (primarily) coniferous forest at Plastic Lake, Ontario was derived from above-ground vegetation, including dry deposited aerosols and SO_2, and mineralized plant organic S.

A common approach to the problem of assessing the importance of dry deposition is to determine the ratio between throughfall SO_4^{2-} deposition rate and bulk deposition rate. This approach can also be used to evaluate differences between various forest canopies as sources of S. Johnson *et al.* (1985), working in two deciduous forests in eastern Tennessee, found that the ratio of throughfall flux to the flux of SO_4^{2-} in bulk deposition was 1·64 and 1·97 at a chestnut oak and yellow poplar site, respectively. Foster and Nicolson (1988) reported that the ratio of throughfall flux to bulk deposition flux was 1·2–1·5 (depending on the year)

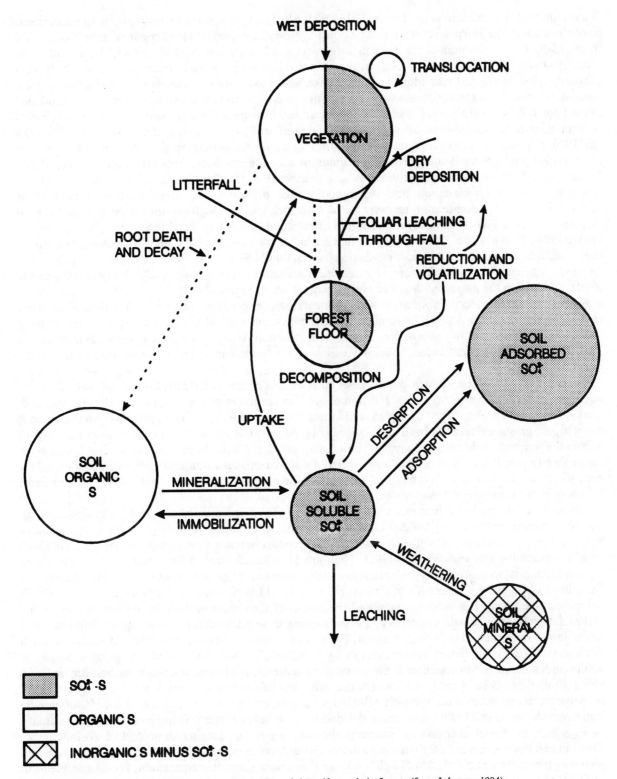

Figure 1. Schematic representation of the sulfur cycle in forests (from Johnson, 1984)

in a deciduous forest situated in the Turkey Lakes Watershed (TLW), Ontario. At a beech–maple hardwood forest located in the Harp Lake watershed, Ontario, Neary and Gizyn (1994) reported that the ratio of throughfall SO_4^{2-} deposition rate to bulk deposition rate, annually, was 1·11, 1·16 and 1·00 for the years 1983, 1984 and 1985, respectively. However, the ratio was higher (i.e. 1·38, 1·59 and 1·50, respectively) in a primarily coniferous forest located in the Plastic Lake watershed. Likewise, the ratio of throughfall to bulk precipitation was higher in coniferous versus deciduous forest sites in Germany (Bredemeier, 1988) and also in the French Ardennes (Nys *et al.*, 1990). Hambuckers and Ramacle (1990) measured very high mean ratios of throughfall SO_4^{2-} deposition rate to bulk deposition rate (i.e. 4·24, 1·79, 2·72 and 2·20 for 1981, 1982, 1983 and 1986, respectively) in a coniferous watershed in Belgium for the period April–September, as did Probst *et al.* (1992) who observed that throughfall inputs to a coniferous forest (primarily Norway spruce) in France, were 4–5 times greater than those in bulk precipitation. On the other hand, DeWalle *et al.* (1988) reported that bulk precipitation and throughfall fluxes of SO_4^{2-} were approximately equal (about 85 kg S ha^{-1} yr^{-1}) at deciduous forest sites in the Appalachians, although the annual throughfall fluxes in their study were similar to those reported for other hardwood forests (e.g. Lindberg *et al.*, 1979; Neary and Gizyn, 1994). DeWalle *et al.* (1988) suggested that the bulk deposition of SO_4^{2-} that they measured may have been unusually high as a result of post-depositional oxidation of SO_2 in the collectors and/or their ridge-top locations. The fact that the collectors were remote from point source inputs of SO_2, may also affect the similarity between the measured fluxes in bulk precipitation and throughfall.

Because of the problems inherent in measurements of S deposition, budgets constructed on upland ecosystems are particularly sensitive to measurement errors, since all of the inputs of S to many of these systems are via wet and dry deposition. However, in the long term, mineral weathering (Figure 1) can contribute SO_4^{2-} to the catchment soils, but this source is significant only in areas with sulfur-bearing minerals.

Soil SO_4^{2-} pools can be affected by processes such as adsorption and desorption of soil-sorbed SO_4^{2-}, microbial mineralization of organic S and immobilization (assimilation) of organic S, SO_4^{2-} or SO_2 and, finally, plant root uptake of SO_4^{2-} and plant exudation of S (see Figure 1). Of these processes, soil sorption is probably the most significant and will be discussed later. As shown in Figure 1, the major output pathways of S from the forested catchment are by volatilization and runoff, the latter being either as surface (outflow) or subsurface (groundwater) flow. In addition, S may be stored in vegetation and in the forest floor as, for example, FeS, although these were reported to be minor fluxes of SO_4^{2-} at the Hubbard Brook Experimental Forest in New Hampshire (Driscoll *et al.*, 1989).

Provided that the problems inherent in the measurement of atmospheric deposition of S are addressed, the use of long-term input–output budgets can provide crucial information regarding catchment retention of SO_4^{2-}. There are many examples of such studies in the literature; a few will be mentioned here. Driscoll *et al.* (1989), Christopherson and Wright (1981), Nicolson (1988) and Hambuckers and Ramacle (1990) all reported that S input–output budgets approximately balanced in watersheds located in the Hubbard Brook Experimental Forest, the Birkenes Catchment, Norway, the TLW, Ontario and a Norway spruce forest in Belgium, respectively. The latter authors used the results of their mass balance study to compare the net accumulation of SO_4^{2-}-S with estimates of S accumulating in wood, in order to evaluate the importance of sulfur losses from forest harvesting. Johnson (1984) found that a chestnut oak forest at Walker Branch Watershed (WBW), Tennessee was accumulating 21 kg SO_4^{2-}-S ha^{-1} yr^{-1} (75% of inputs). Similarly, Kallio and Kauppi (1990) found that five forested catchments in Finland accumulated between <1 and about 10 kg SO_4^{2-}-S ha^{-1} yr^{-1} (<11% to more than 60% of inputs), whereas in a Norway spruce forest in Belgium, Hambuckers and Ramacle (1990) found that 0·3 kg SO_4^{-2}-S ha^{-1} yr^{-1} (<1% of inputs) was retained. Nys *et al.* (1990) found that a deciduous forest in the French Ardennes was losing about 20–24 kg S ha^{-1} yr^{-1} while a coniferous system in the same region was gaining about 5–13 kg S ha^{-1} yr^{-1}. They suggest that the gain of SO_4^{2-} in the coniferous forest may be attributable to a combination of factors including lower pH (about 0·5 of a pH unit) and higher free acidity and free aluminium. The higher retention of SO_4^{2-} at the more acidic site is consistent with the findings of Johnson *et al.* (1986), who observed that lowered soil solution pH reduced the leaching of SO_4^{2-} as a result of protonation and thus increased positive charges on hydrous iron and aluminium oxide and hydroxy surfaces (Turner *et al.*, 1990). On the other hand, Probst *et al.* (1992) found that from 1986 to 1989 a coniferous forest in France was losing about 13 kg S ha^{-1} yr^{-1}.

At the RAIN (Reversing Acidification in Norway) project, parallel manipulation experiments have been conducted involving the exclusion of ambient acidic precipitation at an acidified catchment in southernmost Norway (Risdalsheia site) and the artificial acidification with H_2SO_4 and HNO_3 of two pristine catchments in western Norway (Sogndal site). The results of input–output budgets determined during the first four years of the study (Wright *et al.*, 1988) indicated that at both control sites at Sogndal, inputs of SO_4^{2-} were approximately equal to outputs. However, at one Sogndal site, where H_2SO_4 was artificially added, about 62% of the total inputs of SO_4^{2-} were retained, equivalent to about 80% of the SO_4^{2-} experimentally added to the catchment, and were apparently stored in the soil. In fact, soil samples collected during the period, 1984–1986 revealed that the pool of readily available SO_4^{2-} (adsorbed + water-soluble SO_4^{2-}) had doubled.

In one small catchment at the Ridalsheia site, KIM where the ambient acidic precipitation was excluded by means of a roof and clean precipitation was added beneath, there was a net export of SO_4^{2-} representing 58% of the total inputs from the catchment. In contrast, at the control site EGIL, less than 10% of the incoming SO_4^{2-} was exported (Wright *et al.*, 1988). Wright and Gjessing (1986) report that during the first one and a half years of the study, the input–output budget for SO_4^{2-} actually balanced at EGIL. The net loss of SO_4^{2-} at KIM can be explained by the desorption of water-soluble SO_4^{2-} in the soils, although it is possible that other processes such as mineralization and transformation of organic S occurred (Seip *et al.*, 1979). However, the observed 43% decrease in the pool of readily available SO_4^{2-} in the soil during the period 1984–1986 (Wright *et al.*, 1988) would tend to support desorption. In addition, the response of the catchment to the treatments (i.e. addition or exclusion) was more rapid at Risdalsheia, the site of exclusion experiments, than at Sogndal, where addition experiments were conducted. This is consistent with the response expected if the process of SO_4^{2-} adsorption-desorption dominates (Wright *et al.*, 1988). Theoretically, an increase in SO_4^{2-} deposition (Sogndal) should result in retention followed by a rapid 'breakthrough' in SO_4^{2-} export as sites for SO_4^{2-} adsorption become saturated, with the lag time before the elevated SO_4^{2-} concentrations appear in soil solution and runoff, following the increase in deposition being dependent on the SO_4^{2-} adsorption properties of the soil (Reuss and Johnson, 1986). Ultimately, this would result in an increased proportion of the atmospheric input of SO_4^{2-} being leached to the surface waters and thus a decrease in the net retention of S by the catchments. Alternatively, a decrease in deposition (Risdalsheia) should result in a more rapid response (decrease) in SO_4^{2-} runoff provided the chemical changes brought about initially by acidic precipitation are largely reversible.

In an extensive survey of hundreds of watersheds in the eastern USA, Rochelle and Church (1987) used input–output budgets to determine that the retention of S by the catchments varied from around 0 to 78%, depending on the region and the estimated dry deposition scenario. Their findings supported the hypothesis (Galloway *et al.*, 1983) that regional patterns exist in the retention of S and that these patterns can be related to the spatial extent of the Lake Wisconsin glaciation. Soils of the south-eastern United States (primarily Utisols) adsorb greater amounts of SO_4^{2-} than do soils of the north-eastern United States (primarily Spodisols), probably because of the higher Fe and Al oxide content and lower organic content of the former (Reuss and Johnson, 1986; Turner *et al.*, 1990). Thus, catchments in the north-east typically show little S retention, whereas those in the south-east typically retain a large percentage of the incoming S (Rochelle and Church, 1987; Rochelle *et al.*, 1987).

Sulfate mass balance budgets are also useful in determining temporal (i.e. past, current and future) trends in catchment S retention. In the short term, seasonal or monthly input–output budgets can elucidate when pulses or peaks in SO_4^{2-} export might occur (Christopherson and Wright, 1981; Galloway *et al.*, 1987; Stottlemeyer, 1987), whereas, as discussed above, annual input–output budgets are useful in evaluating the status of watersheds in a region with respect to their retention of deposited S (Herlihy *et al.*, 1993). From these budgets, Herlihy *et al.* (1993) have suggested that a response curve of percentage sulfur retention versus time (i.e. cumulative sulfur loading) can be derived for a given watershed so that the 'breakthrough' sulfur loading, i.e. the total loading at which catchment retention begins to decrease, can be determined. This curve, potentially, could be applied to other watersheds having similar characteristics so that a critical SO_4^{2-} loading, with respect to catchment leaching of sulfur, might be predicted.

LOWLAND CATCHMENTS (WETLANDS)

Turner *et al.* (1990) describe wetlands as 'poorly drained areas transitional between terrestrial and aquatic systems, where the water table is periodically at or near the surface or the land is covered by water'. When the thickness of the organic soils within the wetland is greater than 30 cm, it is referred to as a peatland. Peatlands are a common feature of the landscape, encompassing approximately 420–450 million ha in area globally, including about 150 million ha in Canada and 60 million ha in the United States (Galloway *et al.*, 1984). Peatlands, which will be the focus of this section, can be classified according to the source and chemistry of their water supply. Bogs are peatlands that receive their water solely from atmospheric precipitation. Typically, they have low dissolved mineral content, nutrient availability and pH and are therefore described as ombrotrophic (Galloway *et al.*, 1984; Turner *et al.*, 1990). Fens receive their water from precipitation, surface flow, interflow and groundwater (Turner *et al.*, 1990). Because most of this water has percolated through mineral soils, fens are described as minerotrophic (Galloway *et al.*, 1984). Reversals of hydraulic gradient and flow have recently been demonstrated in some peatlands (Devito *et al.*, 1996), complicating these definitions.

A conceptual model of the S cycle in an idealized peatland is shown in Figure 2 (modified from Hemond, 1980; Gorham *et al.*, 1984). This figure illustrates that SO_4^{2-} is involved in similar reactions in both peatland and upland forested catchments (cf. Figure 1). For example, SO_4^{2-} can be assimilated by plants and micro-organisms resulting in ester sulfate formation or in the production of organic sulfides (i.e. C-bonded S compounds such as proteins) in both types of systems. Sulfate can also be reduced by dissimilatory processes (anaerobic microbial respiration) or by the decay and mineralization of C-bonded S, resulting in the production of H_2S. Dihydrogen sulfide (H_2S) can react with organic matter or with reduced metals to form FeS, or it can be reoxidized to SO_4^{2-} by microorganisms. Dihydrogen sulfide can also be lost to the atmosphere along with other reduced S gases such as dimethyl sulfide [$(CH_3)_2S$]. In addition, abiotic processes such as ion exchange, and adsorption and desorption of soil-sorbed SO_4^{2-} with soil-soluble SO_4^{2-}, can occur.

As is the case with upland catchments, the major output pathways of SO_4^{2-} from peatlands are as surface (outflow) or groundwater flow. While the importance of gaseous loss of reduced S species is still uncertain, Brown and Macqueen (1985 as cited by Bayley *et al.*, 1986) concluded that volatilization of H_2S from peat cores was negligible. Similarly, other studies comparing rates of gas exchange from peatlands suggest that this is not an important loss mechanism for the peatland, although it may be an important source of S to the atmosphere, particularly in areas remote from point source inputs (Nriagu *et al.*, 1987).

In direct contrast to upland terrestrial catchments, peatlands do not receive all of their S from direct atmospheric deposition. Studies conducted in three minerotrophic fens located in the Canadian Shield region of Ontario have shown that from 33–63% (Bayley *et al.*, 1986) to over 93% (Devito, 1994) of S inputs were derived from surface and groundwater flow originating in the upland system. Ombrotrophic bogs are an exception because they receive all, or most, of their water, and thus chemical inputs, from the atmosphere.

Microcosm mass balance experiments conducted on peat cores have been useful in elucidating the relative magnitude of the various S reactions and pool sizes in peatlands. Total S in peatlands appears to be dominated by organic S, primarily C-bonded S, whereas ester sulfate concentrations are generally lower (Morgan, 1992). Wieder and Lang (1988) found that 81% of the total S in Big Run Bog, West Virgina was C-bonded S and 10·4% was ester sulfate. Likewise, Spratt and Morgan (1990) reported that C-bonded S and ester sulfate represented 74% and approximately 13%, respectively, of the total S in a cedar swamp located in New Jersey. In all of these studies, reduced inorganic sulfur pools (e.g. H_2S, FeS, FeS_2) were generally

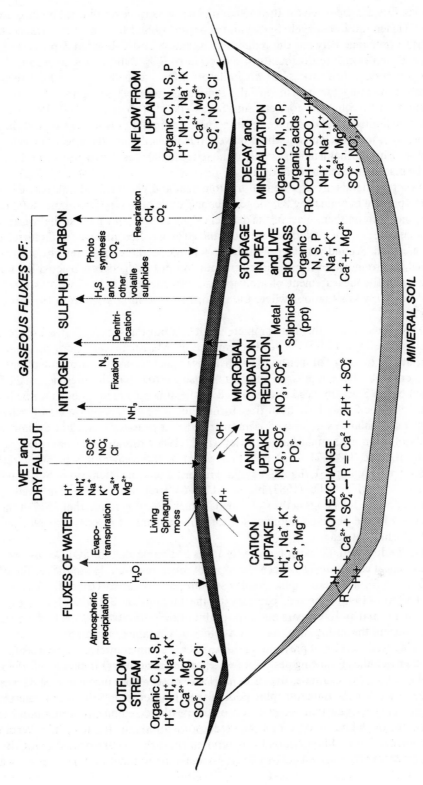

Figure 2. Biogeochemical processes and flow paths for solutes in an idealized peatland (source: Gorham *et al.*, 1984, reprinted with permission from the *Canadian Journal of Fisheries and Aquatic Sciences*). Inputs and outputs to and from the groundwater pool are ignored here

< 10% of the total S. Despite these results, the cycling of S appears to be dominated by fluxes through these inorganic S pools (Wieder and Lang, 1988; Spratt and Morgan, 1990; Morgan, 1992). It has been suggested that there is extensive exchange between the reduced inorganic S and C-bonded S pools with the reduced inorganic sulfur produced from the dissimilatory reduction of SO_4^{2-} and/or the anaerobic breakdown of C-bonded S (Morgan, 1992). The process of dissimilatory SO_4^{2-} reduction, while long recognized as an important alkalinity generating mechanism in lake sediments (see next section), has only recently been shown to be an important mechanism for retaining S in wetlands (Morgan, 1992). Because dissimilatory SO_4^{2-} reduction can be limited by SO_4^{2-} concentration, increases in S inputs to the wetland should stimulate S storage and thus provide a mechanism whereby wetlands can mitigate the effects of acid precipitation (Gorham *et al.*, 1984). This mechanism works only if the reduced SO_4^{2-} is stored permanently in the wetland, a factor that will be discussed later in more detail.

Despite the fact that peatlands are an essential hydrological and geochemical link between upland areas and lakes, and are important in regulating the biogeochemical cycle of S, very few whole system mass balance studies have been conducted on them. Part of the deficiency in this area may be that the constantly changing size of the wetland, as a result of periods of wetting and drying, often makes it difficult to delineate the boundary of the watershed. Secondly, there is a problem with the accurate measurement of hydrology that is primarily controlled by groundwater fluxes. Finally, input–output budgets constructed for wetlands suffer from the same errors in the measurement of atmospheric deposition of SO_4^{2-} as do forested catchments, although this is critical only in situations where the majority of the inputs are from the atmosphere, as in ombrotrophic bogs.

Hemond (1990) presented one of the first complete input–output budgets for S in an ombrotrophic bog. He determined that while inputs of SO_4^{2-} to Thoreau's Bog, Massachusetts, from precipitation totalled 87 meq m^{-2}, outputs were only 20 meq m^{-2}, indicating a net transformation/retention of SO_4^{2-} of 67 meq m^{-2} yr^{-1}, or 77% of the inputs, in the bog. He suggested that the presence of H_2S in the bog indicated that much of the S was reduced and returned to the atmosphere in gaseous form. In addition, as much as 34 meq m^{-2} of the S may have been transformed/retained in the bog by biological uptake.

Since then, other mass balance studies have demonstrated that peatlands can, but do not always, retain significant amounts of S on an annual basis. Bayley *et al.* (1986) found that from 1981 to 1984, SO_4^{2-} retention ranged from 22 to 73% (21 to 51 meq m^{-2} yr^{-1}) of the total inputs of SO_4^{2-} to a fen at the Experimental Lakes Area. Similarly, Urban *et al.* (1986) reported that the retention of SO_4^{2-} in Marcell Bog, Minnesota, averaged 56%, while Calles (1983) measured 25–64% retention of SO_4^{2-} in Swedish catchments that contained between 8 and 33% bogs in the watersheds. Braekke (1981) found that 35% of the total inputs of SO_4^{2-} were retained in the peatland-dominated Storgama catchment in southern Norway during dry years, but only 4% during wet years.

The fact that an individual wetland can alternatively act as a variable sink for SO_4^{2-} or as a source of SO_4^{2-} depending on hydrological conditions, has been the topic of much research. From the results of seven years of input–output budgets for a minerotrophic conifer swamp (i.e. a forested wetland) located in the Plastic Lake catchment, LaZerte (1993) reported, contrary to the findings of Braekke (1981), that the swamp retained SO_4^{2-} during normal and wet years and exported it during dry years. While high concentrations of SO_4^{2-} were released during the spring melt, SO_4^{2-} was efficiently retained during the dry summer (LaZerte and Dillon, 1984). This cyclic release of previously retained SO_4^{2-} on an annual and also a seasonal basis, has been observed in other wetlands following periods of water table drawdown (Kerekes *et al.*, 1986; Urban and Bayley, 1986; Devito, 1994). For example, Bayley *et al.* (1986) observed an increase in SO_4^{2-} concentrations from 17 to over 700 µeq l^{-1} in the minerotrophic pool of their fen (located in ELA) in September following the summer drought. They suggest that desaturation of the peatland sediments during the drought period may have resulted in the oxidation to SO_4^{2-} of stored reduced S. Subsequently, water flow resulted in export of the SO_4^{2-} from the peatland. Dillon *et al.* (1997) reported that the occurrence of droughts followed by massive SO_4^{2-} export from wetlands was controlled by large-scale weather patterns, specifically following El Niño events.

The fluctuating water table in wetlands therefore results in constantly changing redox conditions in the near-surface layer; if hydrological conditions favour the oxidation of reduced S, then any temporary gains in acid-neutralizing capacity that resulted from the S-reducing reactions would be reduced. Peatlands can thus lose a portion of their previously stored S and subsequently act as a source of SO_4^{2-} rather than a sink. This has serious implications with respect to the reduction in atmospheric SO_2 emissions. The fact that wetlands can store S during certain years, or during periods of the year, and then release it at a later time, means that inputs of SO_4^{2-} to surface waters may continue after the atmospheric source of S to the system has been discontinued (Dillon et al., 1997). Since weather is unpredictable, one cannot predict when, or even if, a wetland might release its stored sulfur. Changes in temperature and rainfall that are expected to occur as a result of global warming further complicate predictions of wetland behaviour. None the less, using input–output budgets, it may be possible to extrapolate the effects of short-term drought or flooding conditions if it is assumed that rapid responses observed during short-term studies (seasonal and short-term annual observations) would be permanent features under a new climatic regime. Long-term mass balance studies on wetlands are necessary for verification of this assumption.

LAKES

Interest in the sulfur cycle in lakes has increased over the past decade, in part because of the realization that the reduction of SO_4^{2-} within a lake can provide an important mechanism for in-lake alkalinity generation (IAG) (Cook et al., 1986, Schindler et al., 1986; Lin et al., 1987) and thus mitigate the effects of increased inputs of SO_4^{2-} to the lake. Baker et al. (1985) and Brezonik et al. (1987) have summarized the S cycle reactions that can produce or consume alkalinity within lakes. These reactions/processes are the same as those that were described previously for both upland catchments (Figure 1) and wetlands (Figure 2).

Although whole lake mass balance estimates for SO_4^{2-} are abundant in the literature, there are problems in the determination of these input–output budgets. A drawback with some lake mass balance studies is that there is no accurate method for measuring groundwater inputs. This is particularly important in closed lakes, or seepage lakes, which have no surface inflows or outflows. Such lakes may have seepage out of the lake in all directions (groundwater recharge lakes) or they may have seepage into the lake on the 'upstream' side and seepage out on the 'downstream' side (groundwater flow-through lakes) (Turner et al., 1990). While water balance measurements, i.e. input from precipitation versus output from evaporation, should provide a hydrological estimate of the magnitude of these water sources and losses, it is difficult to measure S concentrations in all the seepage flows and thus determine an accurate S mass balance.

Since atmospheric deposition provides the only measurable input of S to a seepage lake, it is critical that these measurements be accurate. As discussed previously, there are problems associated with the measurement of total S deposition. Fortunately, the problems are minor compared with those in forested catchments because processes such as throughflow, stemflow, interception, etc. do not need to be considered.

Another problem in the determination of input–output budgets for lakes is in assessing atmospheric losses and inputs of S through gaseous exchange of H_2S, $(CH_3)_2S$ and SO_2. Generally this component is not included in mass balance budgets for lakes, although it may be important.

Nonetheless, the results from mass balance studies on lakes have shown that non-seepage lakes are relatively conservative with respect to S, i.e. there is very little retention, indicating an excess of S relative to biological needs. In seepage lakes, however, net SO_4^{2-} retention is quite high. For example, Marnette et al. (1993) found that retention of S in two Dutch seepage pools averaged 74 and 83% for the period 1982–1990. Baker et al. (1989) reported that 49% of the total inputs of S to Little Rock Lake, Wisconsin were retained. Similarly, retention rates, i.e., percentage of the total inputs, of 40% were reported in nearby Vandercook Lake, Wisconsin (Lin et al., 1987) and also in acidic McCloud Lake, Florida (43–46%; Baker et al., 1986). The Dutch pools, however, become desiccated in summer and so may act as intermediate between wetland and lake. Perhaps retention is high in seepage lakes because of extended contact time with sediments.

Artificial impoundments and reservoirs may behave differently from natural lakes because of their complex internal dynamics and variable outflow control (Turner *et al.*, 1990). For example, in Bickford Reservoir, Massachusetts, Eshleman and Hemond (1988) reported that 340 meq m^{-2} of SO_4^{2-} was produced by the reservoir annually based on two years of mass balance data; although they attributed the apparent export of SO_4^{2-} to unmeasured dry deposition of S to the reservoir surface. Retention of S was 58 and 47% (1983 and 1984, respectively) in the Contrary Creek Arm of Lake Anna, Virginia ($\tau = 0.27$ yr), an impoundment that receives acid mine drainage (Herlihy *et al.*, 1987).

The retention of S in drainage lakes, i.e. those having channelized inflows and outflow with negligible seepage, can be described with a steady-state model first used to describe phosphorus fluxes (Dillon and Rigler, 1975). The model assumes that lakes are continuously stirred, homogeneous reactors and predicts whole lake, annual SO_4^{2-} concentration ($[SO_4^{2-}]$) as a function of SO_4^{2-} loading rate, internal loss rates and areal water discharge rate. The mass balance equation is,

$$Ln = Lo - Ls \tag{1}$$

where Ln is the SO_4^{2-} load from streams, precipitation and ungauged portions, Lo is the SO_4^{2-} loss via outflow and Ls is the SO_4^{2-} net loss to sediments, to other dissolved S species and to the atmosphere. Units are in mass per lake area per yr. If it is assumed that loss and discharge of SO_4^{2-} are first-order processes, then at steady state,

$$Ln = q[SO_4^{2-}] + v[SO_4^{2-}] \tag{2}$$

where v is a mass transfer or generalized loss coefficient (m yr^{-1}) representing the net rate of SO_4^{2-} loss to sediments, to other dissolved S species and to the atmosphere, and q is the areal water discharge rate (m yr^{-1}) given by the total water discharge rate divided by the lake surface area. The lake residence time, τ, is equal to $\bar{z}/q$, where $\bar{z}$ is the mean depth. At steady state,

$$[SO_4^{2-}] = Ln/(q + v) \tag{3}$$

SO_4^{2-} retention, R, is defined relative to total inputs,

$$R = (Ln - Lo)/Ln \tag{4}$$

where $Ln - Lo$ is the net gain. Equations (2) and (4) together give,

$$R = v/(q + v) \tag{5}$$

Equation (5) predicts that R should decrease as q increases at constant v. Rearranging Equation (5),

$$1/R = (1/v)q + 1 \tag{6}$$

In lakes near Dorset in central Ontario, R was small, ranging from -0.06 in Dickie Lake (i.e. the lake was a source of SO_4^{2-}) to 0.11 in Plastic Lake (P. J. Dillon, unpublished data) for the period 1980–1992. Correspondingly, v was also very low, ranging from -0.09 to 0.36 m yr^{-1}, with a mean of 0.11 ± 0.16 m yr^{-1}. Although two of the lakes had negative v, none of the lakes was significantly different from zero. Between-year variation within a lake was much larger than variation in between-lake means. Similarly, Kelly *et al.* (1987) reported R values for nine lakes, some of which received experimental additions of SO_4^{2-}, in north-western Ontario, New York State and Norway to be 0–0.50 with a mean v of 0.36 ± 0.30 m yr^{-1}. For comparison, mean loss coefficients in the Dorset study lakes were 7.9 m yr^{-1} for total phosphorus (TP) (Dillon and Molot 1996a), 3.6 m yr^{-1} for dissolved organic C (Dillon and Molot 1996b), 3.5 m yr^{-1} for total N and 16 m yr^{-1} for ammonium (Molot and Dillon, 1993).

These results demonstrate that hydrology controls SO_4^{2-} retention and that significant retention can still occur in lakes with very low v when q is very low. Of the nine lakes studied by Kelly *et al.* (1987), R exceeded 0·20 in the lakes with $q < 2$ m yr^{-1}, whereas R in the remaining four lakes, with $q > 5$ m yr^{-1} was less than 0·05. Similarly, in the seven Dorset study lakes, R was highest in the lakes with lowest q (P. J. Dillon, unpublished data). Presumably, low q increases contact time between SO_4^{2-} and utilization sites such as anoxic sediments, thereby promoting retention. The model also predicts that the absolute amount of SO_4^{2-} retained will increase with increasing SO_4^{2-} concentration, even though R may remain constant because of first-order loss kinetics.

A major conclusion drawn from S mass balance studies of lakes is that in-lake alkalinity generation (IAG) produced by SO_4^{2-} reduction is more important than IAG produced from other processes, including NO_3^- consumption (assimilation during primary production, denitrification and mineralization of organic N), and than alkalinity produced from terrestrial sources (terrestrial alkalinity generation or TAG). Dillon and Evans (unpublished data) ascertained that the 12-year average (1980–1992) rates of IAG resulting from SO_4^{2-} retention in eight lakes in the Dorset area of Ontario ranged from about 0 to 54 meq m^{-2} yr^{-1}, and represented as much as 33% of the total alkalinity generated within the lakes. By comparison, NO_3^- consumption represented 10–20% of the total IAG. Their results are comparable to those obtained from shorter term S mass balance studies conducted on other soft water lakes in North America. For example, Baker *et al.* (1986), using one year of mass balance data, found that SO_4^{2-} retention was 26–34 meq m^{-2} yr^{-1} (54–61% of the net IAG compared with 17–20 meq m^{-2} yr^{-1} or 31–33% generated from NO_3^- immobilization) in McCloud Lake, a seepage lake located in Florida. In Vandercook Lake, Wisconsin, SO_4^{2-} removal determined from over two years of mass balance data was 18 meq m^{-2} yr^{-1} and accounted for 54% of the alkalinity generated (Lin *et al.*, 1987). However, that value may have been underestimated because only wet deposition inputs of S were included in the atmospheric component of the budget. In 16 lakes in northern Wisconsin, SO_4^{2-} retention determined by Stauffer (1993) averaged 35 ± 8 meq m^{-2} over the 161-day period from early May to mid-October 1984, while in a seepage lake in northern Wisconsin (Little Rock Lake), retention of SO_4^{2-} was 20·4 meq m^{-2} yr^{-1} prior to the experimental acidification of the lake (Baker *et al.*, 1989).

At ELA, reported whole lake SO_4^{2-} removal rates from mass balance calculations were 39 meq m^{-2} yr^{-1} (53% of the IAG; 1981–1983 average) in Lake 239 (Schindler *et al.*, 1986), 50 meq m^{-2} yr^{-1} (16% of the IAG; 1982–1986 average) in Lake 302N (Rudd *et al.*, 1990), 245 meq m^{-2} yr^{-1} (about 40% of the net IAG; 1976–1983 average) for Lake 223 (Cook *et al.*, 1986) and 180 meq m^{-2} yr^{-1} (1982–1986 average) in Lake 302S (Rudd *et al.*, 1990). The high values in Lakes 223 and 302S can be attributed to the fact that the lakes were experimentally acidified with H_2SO_4 during the study. Evidence that SO_4^{2-} reduction rates increase with increasing lake water SO_4^{2-} concentrations has been documented in both experimentally acidified lakes (Baker *et al.*, 1989) and limnocorrals (Schiff and Anderson, 1987), as well as in laboratory experiments using sediment cores (Kelly and Rudd, 1984). As a consequence, it can be postulated that increasing inputs of SO_4^{2-} to a lake might, potentially, increase rates of IAG and thus mitigate, to some extent, the effects of acid precipitation.

In Europe, SO_4^{2-} mass balance data collected by Psenner (1988) from Piburger See, Austria, demonstrated that S reactions produced 20% of the alkalinity generated (234 meq m^{-2} yr^{-1} for the period 1975–1976), whereas in Langtjern, southern Norway, Wright (1983) found that, on average, 56·5 meq m^{-2} yr^{-1} of SO_4^{2-} was retained during the period 1974–1980. Wright (1983) and Baker *et al.* (1985) point out that there are a number of possible processes that could contribute to SO_4^{2-} retention in lakes. These include (modified from Wright, 1983): (1) reduction of SO_4^{2-} to S and release as H_2S gas to the atmosphere; (2) reduction of SO_4^{2-} to sulfide and deposition as, for example, FeS in anaerobic lake sediments; (3) assimilation of sulfur by macrophytes and phytoplankton and (4) diffusion/sorption of SO_4^{2-} into lake sediments. Unfortunately, while whole lake input–output budgets can ascertain the magnitude of S retention, the quantitative importance of each of these processes cannot be determined without parallel process-oriented studies.

Experiments conducted using lake sediment cores and limnocorrals, in conjunction with whole lake mass balance studies, have demonstrated that dissimilatory SO_4^{2-} reduction is a major process in many if not all lake sediments (e.g. Baker *et al.*, 1986; Cook *et al.*, 1986; Schiff and Anderson, 1987). Similar to wetlands, then, the magnitude of SO_4^{2-} reduction and the 'permanency' of the alkalinity generated by this process is dependent on redox conditions within the lake, since SO_4^{2-} reduction generates permanent alkalinity only if the reduced, uncharged products are isolated from reoxidation by burial, outflow or volatilization (Cook *et al.*, 1986; Rudd *et al.*, 1986; Giblin *et al.*, 1990). Moreover, while long-term mass balance studies on lakes are necessary in order to determine annual variability and trends in S retention in lakes, short-term studies are crucial in order to establish seasonal or monthly cycles in SO_4^{2-} reduction.

CONCLUSIONS

Mass balance studies have been a useful means of elucidating factors influencing S fluxes on a landscape scale. Most of these studies have focused on the single S species, SO_4^{2-}, partly because of its mobility in the environment, and partly because of its role in the acid rain phenomenon. The regional effect of SO_4^{2-} fluxes in areas lacking carbonate formations depends on landscape composition, i.e. the relative densities of uplands, wetlands and lakes, because of the different ways in which each affects S cycling. Since lakes and wetlands are typically 'down gradient' from uplands, the ability of uplands to process SO_4^{2-} is particularly important to SO_4^{2-} input to wetlands and lakes. Similarly, reactions of SO_4^{2-} wetlands can affect input to lakes.

Landscapes dominated by aggrading forests with acidic soils and ample SO_4^{2-} adsorption sites, wetlands resistant to drought and lakes with long residence times would be expected to retain significant quantities of SO_4^{2-} via several different mechanisms and, hence, generate significant quantities of alkalinity. Stored SO_4^{2-} in soils and wetlands may, however, be released if soils become markedly less acidic or if wetlands become drier. Elevated SO_4^{2-} export may also occur in soils that become saturated after extensive periods of deposition. Long-term temporal trends are potentially useful in assessing 'breakthrough' sulfur loading (i.e. the cumulative load beyond which retention begins to decrease).

Conversely, landscapes dominated by thin, non-carbonate soils having few adsorption sites. Wetlands subject to frequent drying because of limited recharge from uplands and lakes with short residence times are most affected by S deposition because these landscape forms have the least potential for reactions involving SO_4^{2-} retention.

REFERENCES

Baker, L. A. and Brezonik, P. L. 1988. 'Dynamic model of in-lake alkalinity generation', *Wat. Resour. Res.*, **24**, 65–74.

Baker, L. A., Brezonik, P. L., Edgerton, E. S., and Ogburn, R. W., III, 1985. 'Sediment acid neutralization in softwater lakes', *Wat. Air Soil Pollut.*, **25**, 215–230.

Baker, L. A., Brezonik, P. L., and Edgerton, E. S. 1986. 'Sources and sinks of ions in a soft water, acidic lake in Florida', *Wat. Resour. Res.*, **22**, 715–722.

Baker, L. A., Urban, N. R., Brezonik, P. L., and Sherman, L. A. 1989. 'Sulfur cycling in an experimentally acidified seepage lake', in Saltzman, E. S. and Cooper, W. J. (Eds), *Biogenic Sulfur in the Environment*, ACS Symposium Series 393. American Chemical Society, Washington, D.C. pp. 79–100.

Bayley, S. E., Behr, R. S., and Kelly, C. A. 1986. 'Retention and release of S from a freshwater wetland', *Wat. Air Soil Pollut.*, **31**, 101–114.

Bayley, S. E., Vitt, D. H., Newbury, R. W., Beaty, K. G., Behr, R., and Miller, C. 1987. 'Experimental acidification of a Sphagnum-dominated peatland: first year results', *Can. J. Fish. Aquat. Sci.*, **44**, 194–205.

Bergstrom, L. and Gustafson, A. 1985. 'Hydrogen ion budgets of four small runoff basins in Sweden', *Ambio*, **14**, 346–348.

Braekke, F. H. 1981. 'Hydrochemistry of high altitude catchments in southern Norway. 1. Effects of summer droughts and soil-vegetation characteristics', *Report of the Norwegian Forestry Research Institute No. 36.8*. 26 pp.

Bredemeier, M. 1988. 'Forest canopy transformation of atmospheric deposition', *Water Air Soil Pollut.*, **40**, 121–138.

Brezonik, P. L., Baker, L. A., and Perry, T. E. 1987. 'Mechanisms of alkalinity generation in acid-sensitive soft water lakes', in Hites, R. A. and Eisenreich, S. J. (Eds), *Sources and Fates of Aquatic Pollutants*. American Chemical Society, Washington, D.C. pp. 229–260.

Calles, U. M. 1983. 'Dissolved inorganic substances: a study of mass balance in three small drainage basins', *Hydrobiologia*, **101**, 13–18.

Caraco, N. F., Cole, J. J., and Likens, G. E. 1993. 'Sulfate control of phosphorus availability in lakes', *Hydrobiologia*, **253**, 275–280.

Christophersen, N. and Wright, R. F. 1981. 'Sulfate budget and a model for sulfate concentrations in stream water at Birkenes, a small forested catchment in southernmost Norway', *Wat. Resour. Res.*, **17**, 377–389.

Cook, R. B., Kelly, C. A., Schindler, D. W., and Turner, M. A. 1986. 'Mechanisms of hydrogen ion neutralization in an experimentally acidified lake', *Limnol. Oceanogr.*, **31**, 134–148.

Cosby, B. J., Wright, R. F., Hornberger, G. M., and Galloway, J. N. 1985. 'Modelling the effects of acid deposition: estimation of long-term water quality response in a small forested catchment', *Wat. Resour. Res.*, **21**, 1591–1601.

Cross, P. M. and Rigler, F. H. 1983. 'Phosphorus and iron retention in sediments measured by mass budget calculations and directly', *Can. J. Fish. Aquat. Sci.*, **40**, 1589–1597.

Davison, W. and Woof, C. 1990. 'The dynamics of alkalinity generation by an anoxic sediment exposed to acid water', *Water Res.*, **24**, 1537–1545.

Devito, K. J. 1994. 'Hydrologic control of sulfur dynamics in headwater wetlands of the Canadian Shield', *Ph.D. Thesis*, York University, Toronto, Ontario.

Devito, K. J., Waddington, J. M., and Fowle, B. A. 1997. 'Flow reversals in peatlands influenced by local groundwater systems', *Hydrol. Process.*, **11**, 103–110.

DeWalle, D. R., Sharpe, W. E., and Edwards, P. J. 1988. 'Biogeochemistry of two Appalachian deciduous forest sites in relation to episodic stream acidification', *Water Air Soil Pollut.*, **40**, 143–156.

Dillon, P. J. and Evans, H. E. 1993. 'A comparison of phosphorus retention in lakes determined from mass balance and sediment core calculation', *Water Res.*, **27**, 659–668.

Dillon, P. J. and Molot, L. A. 1990. 'The role of ammonium and nitrate retention in the acidification of lakes and forested catchments', *Biogeochemistry*, **11**, 23–43.

Dillon, P. J. and Molot, L. A. 1996. 'Long-term phosphorus budgets and an examination of the steady state mass balance model for central Ontario lakes', *Water Res.*, **30**, 2273–2280.

Dillon, P. J. and Molot, L. A. 1997. 'Dissolved organic and inorganic carbon mass balances in central Ontario lakes', *Biogeochemistry*, **36**, 29–42.

Dillon, P. J. and Rigler, F. H. 1975. 'A simple method for predicting the capacity of a lake for development based on lake trophic status', *J. Fish. Res. Board Can.*, **32**, 1519–1531.

Dillon, P. J., Jeffries, D. S., and Scheider, W. A. 1982. 'The use of calibrated lakes and watersheds for estimating atmospheric deposition near a large point source', *Water Air Soil Pollut.*, **18**, 241–258.

Dillon, P. J., Evans, H. E., and Scholer, P. J. 1988. 'The effects of acidification on metal budgets of lakes and catchments', *Biogeochemistry*, **5**, 201–220.

Dillon, P. J., Molot, L. A., and Scheider, W. A. 1991. 'Phosphorus and nitrogen export from forested stream catchments in central Ontario', *J. Envion. Qual.*, **20**, 857–864.

Dillon, P. J., Reid, R. A., and Evans, H. E. 1993. 'The relative magnitude of phosphorus sources for small, oligotrophic lakes in Ontario, Canada', *Verh. Int. Verein. Limnol.*, **25**, 355–358.

Dillon, P. J., Molot, L. A., and Futter, M. 1997. 'A note on the effect of El Niño-related drought on the recovery of acidified lakes. *Int. J. Environ. Mon. Assess.*, in press.

Driscoll, C. T., Fuller, R. D., and Schecher, W. D. 1989. 'The role of organic acids in the acidification of surface waters in the eastern U.S.', *Water Air Soil Pollut.*, **43**, 21–40.

Eshleman, K. N. and Hemond, H. F. 1988. 'Alkalinity and major ion budgets for a Massachusetts reservoir and watershed', *Limnol. Oceanogr.*, **33**, 174–185.

Foster, N. W. and Nicolson, J. A. 1988. 'Acid deposition and nutrient leaching from deciduous vegetation and podzolic soils at the Turkey Lakes watershed', *Can. J. Fish. Aquat. Sci.*, **45**, 96–100.

Galloway, J. N., Schofield, C. L., Peters, N. E., Hendrey, G. R., and Altwicker, E. R. 1983. 'Effect of atmospheric sulfur on the composition of three Adirondack lakes', *Can. J. Fish. Aquat. Sci.*, **40**, 799–806.

Galloway, J. N., Likens, G. E., and Hawley, M. E. 1984. 'Acid precipitation: natural versus anthropogenic components', *Science*, **226**, 829–831.

Galloway, J. N., Hendrey, G. R., Schofield, C. L., Peters, N. E., and Johannes, A. H. 1987. 'Processes and causes of lake acidification during spring snowmelt in the west-central Adirondack Mountains, New York', *Can. J. Fish. Aquat. Sci.*, **44**, 1591–1602.

Ghani, A., McLaren, R. G., and Swift, R. S. 1993. 'Mobilization of recently-formed soil organic sulfur', *Soil Biol. Biochem.*, **25**, 1739–1744.

Gherini, S. A., Mok, L., Hudson, R. J. M., Davis, G. P., Chen, C. W., and Goldstein, R. A. 1985. 'The ILWAS model: formulation and application', *Water Air Soil Pollut.*, **26**, 425–459.

Giblin, A. E., Likens, G. E., White, D., and Howarth, R. W. 1990. 'Sulfur storage and alkalinity generation in New England lake sediments', *Limnol. Oceanogr.*, **35**, 852–869.

Gorham, E., Bayley, S. E., and Schindler, D. W. 1984. 'Ecological effects of acid deposition upon peatlands: a neglected field in "acid-rain" research', *Can. J. Fish. Aquat. Sci.*, **41**, 1256–1268.

Hambuckers, A. and Remacle, J. 1990. 'A six-year nutrient balance for a coniferous watershed receiving acidic rain inputs', in Harrison, A. F., Ineson, P., and Heal, O. W. (Eds), *Nutrient Cycling in Terrestrial Ecosystems. Field Methods, Application and Interpretation*. Elsevier, New York. pp. 130–138.

Hasler, A. C. and Einsele, W. G. 1948. 'Fertilization for increasing productivity of natural inland waters', *Trans. North Am. Wildr. Conf.*, **13**, 527–555.

Hemond, H. F. 1980. 'Biogeochemistry of Thoreau's Bog, Concord, Massachusetts', *Ecol. Monogr.* **50**, 507–526.

Hemond, H. F. 1990. 'Acid neutralizing capacity, alkalinity, and acid-base status of natural waters containing organic acids', *Environ. Sci. Technol.*, **24**, 1486–1489.

Herlihy, A. T., Mills, A. L., Hornberger, G. M., and Bruckner, A. E. 1987. 'The importance of sediment sulfate reduction to the sulfate budget of an impoundment receiving acid mine drainage', *Water Resour. Res.*, **23**, 287–292.

Herlihy, A. T., Kaufmann, P. R., Church, M. R., Wigington, P. J., Jr., Webb, J. R., and Sale, M. J. 1993. 'The effects of acidic deposition on streams in the Appalachian Mountain and Piedmont Region of the mid-Atlantic United States', *Water Resour. Res.*, **29**, 2687–2703.

Jeffries, D. S., Semkin, R. G., Neureuther, R., and Seymour, M. 1988. 'Ion mass budgets for lakes in the Turkey Lakes Watershed, June 1981–May 1983', *Can. J. Fish. Aquat. Sci.*, **45**, 47–58.

Johnson, D. W. 1984. 'Sulfur cycling in forests', *Biogeochemistry*, **1**, 29–43.

Johnson, D. W., Richter, D. D., Lovett, G. M., and Lindberg, S. E. 1985. 'The effects of atmospheric deposition on potassium, calcium, and magnesium cycling in two deciduous forests', *Can. J. For. Res.*, **15**, 773–782.

Johnson D. W., van Miegroet, H., and Kelly, J. M. 1986. 'Sulfur cycling in five forest ecosystems', *Water Air Soil Pollut.*, **30**, 965–979.

Kallio, K. and Kauppi, L. 1990. 'Ion budgets of small forested basins', in Kauppi, K. *et al.* (Eds), *Acidification in Finland*. Springer-Verlag, Berlin. pp. 811–823.

Kelly, C. A. and Rudd, J. W. M. 1984. 'Epilimnetic sulfate reduction and its relationship to lake acidification', *Biogeochemistry*, **1**, 63–77.

Kelly, C. A., Rudd, J. W. M., Hesslein, R. H., Schindler, D. W., Dillon, P. J., Driscoll, C. T., Gherini, S. A., and Hecky, R. E. 1987. 'Prediction of biological acid neutralization in acid-sensitive lakes', *Biogeochemistry*, **3**, 129–140.

Kerekes, J., Beauchamp, S., Tordon, R., and Pollock, T. 1986. 'Sources of sulphate and acidity in wetlands and lakes in Nova Scotia', *Water Soil Pollut.*, **31**, 207–214.

Kling, G. W., Giblin, A. E., Fry, B., and Freeman. B. J. 1991. 'The role of seasonal turnover in lake alkalinity dynamics', *Limnol. Oceanogr.*, **36**, 106–122.

LaZerte, B. D. 1993. 'The impact of drought and acidification on the chemical exports from a minerotrophic conifer swamp', *Biogeochemistry*, **18**, 153–172.

LaZerte, B. D. and Dillon, P. J. 1984. 'Relative importance of anthropogenic versus natural sources of acidity in lakes and streams of central Ontario', *Can. J. Fish. Aquat. Sci.*, **41**, 1664–1677.

Lee, S. and Schnoor, J. L. 1988. 'Reactions that modify chemistry in lakes of the National Surface Water Survey', *Environ. Sci. Technol.*, **22**, 190–195.

Likens, G. E., Bormann, F. H., Pierce, R. S., Eaton, J. S., and Johnson, N. M. 1977. *Biogeochemistry of a Forested Catchment*. Springer-Verlag, New York. 146 pp.

Lin, J. C., Schnoor, J. L., and Glass, G. E. 1987. 'Ion budgets in a seepage lake', in Hites, R. A. and Eisenreich, S. J. (Eds), *Sources and Fates of Aquatic Pollutants*. American Chemical Society, Washington, D.C. pp. 209–227.

Lindberg, S. E., Harriss, R. C., Turner, R. R., Shriner, D. S., and Huff, D. D. 1979. 'Atmospheric deposition to a deciduous forest watershed', *Report ORNL/TM-6674*. Oak Ridge National Laboratory, Oak Ridge, TN.

Lovett, G. M. and Lindberg, S. E. 1984. 'Dry deposition and canopy exchange in a mixed oak forest determined from analysis of throughfall', *J. Appl. Ecol.*, **21**, 1013–1028.

Marnette, E. C. L., Houweling, H., van Dam, H. and Erisman, J. W. 1993. 'Effects of decreased atmospheric deposition on the sulfur budgets of two Dutch moorland pools', *Biogeochemistry*, **23**, 119–144.

Maynard, D. G., Stewart, J. W. B., and Bettany, J. R. 1985. 'The effects of plants on soil sulfur transformations', *Soil Biol. Biochem.*, **17**, 127–134.

Molot, L. A. and Dillon, P. J. 1993. 'Nitrogen mass balances and denitrification rates in central Ontario lakes', *Biogeochemistry*, **20**, 195–212.

Morgan, M. D. 1992. 'Sulfur pool sizes and stable isotope ratios in humex peat before and immediately after the onset of acidification', *Environ. Int.*, **18**, 545–553.

Neary, A. J. and Gizyn, W. I. 1994. 'Throughfall and stemflow chemistry under deciduous and coniferous forest canopies in south-central Ontario', *Can. J. For. Res.*, **24**, 1089–1100.

Nicolson J. A. 1988. 'Water and chemical budgets for terrestrial basins at the Turkey Lakes Watershed', *Can. J. Fish. Aquat. Sci.*, **45**, 88–95.

Nikolaides, N. P., Rajaram, H., Schnoor, J. L., and Georgakakos, K. P. 1988. 'A generalized soft water acidification model', *Wat. Resour. Res.*, **4**, 1983–1996.

Nriagu, J. O., Holdway, D. A., and Coker, R. D. 1987. 'Biogenic sulfur and the acidity of rainfall in remote areas of Canada', *Science*, **237**, 1189–1192.

Nurnberg, G. K. and Dillon, P. J. 1993. 'Iron budgets in temperate lakes', *Can. J. Fish. Aquat. Sci.*, **50**, 1728–1737.

Nys, C., Stevens, P., and Ranger, J. 1990. 'Sulfur nutrition of forests examined using a sulfur budget approach', in Harrison, A. F., Ineson, P., and Heal, O. W. (Eds), *Nutrient Cycling in Terrestrial Ecosystems. Field Methods, Application and Interpretation*. Elsevier, New York. pp. 356–372.

Padro, J. 1994. 'Observed characteristics of the dry deposition velocity of O_3 and SO_2 above a wet deciduous forest', *Sci. Total Environ.*, **147**, 395–400.

Probst, A., Viville, D., Fritz, B., Ambroise, B., and Dambrine, E. 1992. 'Hydrochemical budgets of a small forested granitic catchment exposed to acid deposition: the Strengbach Catchment Case Study (Vosges Massif, France).

Psenner, R. 1988. 'Akalinity generation in a soft-water lake: watershed and in-lake processes', *Limnol. Oceanogr.*, **33**, 1463–1475.

Rasmussen, L., Beier, C., deVisser, P., van Breeman, N., Kreutzer, K., Schierl, R., Bredemeier, M., Raben, G., and Farrell, E. P. 1992. 'The "EXMAN" Project — EXperimental MANipulations of Forest Ecosytems', in Teller, A., Matty, P., and Jeffers, J. N. R. (Eds), *Responses of Forest Ecosystems to Environmental Changes*. Elsevier Science Publishers, London. pp. 325–334.

Rees, T. H. and Schnoor, J. L. 1994. 'Long-term simulation of decreased acid loading on forested watershed', *J. Environ. Engng — ASCE*, **120**, 291–312.

Reuss, J. O. and Johnson, D. W. 1986. *Acid Deposition and the Acidification of Soils and Waters*, Ecological Studies Volume 59. Springer-Verlag, New York. 119 pp.

Rochelle, B. P. and Church, M. R. 1987. 'Regional patterns of sulfur retention in watersheds of the eastern U.S.', *Water Air Soil Pollut.*, **36**, 61–73.

Rochelle, B. P., Church, M. R., and David, M. B. 1987. 'Sulfur retention at intensively studied sites in the U.S. and Canada', *Water Air Soil Pollut.*, **33**, 73–84.

Judd, J. W. M., Kelly, C. A., and Furutani, A. 1986. 'The role of sulphate reduction on long term accumulation or organic and inorganic sulfur in lake sediments', *Limnol. Oceanogr.*, **31**, 1281–1291.

Rudd, J. W. M., Kelly, C. A., Schindler, D. W., and Turner, M. A. 1990. 'A comparison of the acidification efficiencies of nitric and sulfuric acids by two whole-lake addition experiments', *Limnol. Oceanogr.*, **35**, 663–679.

Sampson, C. J., Brezonik, P. L., and Weir, E. P. 1994. 'Effects of acidification on chemical composition and chemical cycles in a seepage lake: inferences from a whole-lake experiment', in Baker, L. A. (Ed.), *Environmental Chemistry of Lakes and Reservoirs*, Advances in Chemistry Series #237. ACS, Washington, D.C. pp. 121–159.

Schafran, G. C. and Driscoll, C. T. 1987. 'Comparison of terrestrial and hypolimnetic sediment generation of acid neutralizing capacity for an acid Adirondack lake', *Environ. Sci. Technol.*, **21**, 988–993.

Schiff, S. L. and Anderson, R. F. 1987. 'Limnocorral studies of chemical and biological acid neutralization in two freshwater lakes', *Can. J. Fish. Aquat. Sci.*, **44**, 173–187.

Schindler, D. W. 1986. 'The significance of in-lake production of alkalinity', *Water Air Soil Pollut.*, **30**, 931–944.

Schindler, D. W. 1987. 'Evolution of phosphorus limitation in lakes', *Science*, **195**, 260–262.

Schindler, D. W. 1991. 'Whole lake experiments at the Experimental Lakes Area', in Mooney, H. A., Schindler, D. W., and Schutz, E. D. (Eds), *Ecosystem Experiments*. John Wiley & Sons, Chichester. 304 pp.

Schindler, D. W., Turner, M. A., Stainton, M. P., and Linsey, G. A. 1986. 'Natural sources of acid neutralizing capacity in low alkalinity lakes of the Precambrian Shield', *Science*, **232**, 844–847.

Schut, P. H., Evans, R. D., and Scheider, W. A. 1985. 'Variations in trace metal exports from small Canadian Shield watersheds', *Water Air Soil Pollut.*, **28**, 225–237.

Seip, H. M., Abrahemsen, G., Gjessing, T., and Stuanes, A. 1979. 'Studies of soil-, precipitation- and runoff-chemistry in six small natual plots (mini-catchments)', *SNSF Project IR 46/79. Ås, Oslo, Norway*.

Shaffer, P. W., Hooper, R. P., Eshleman, K. N., and Church, M. R. 1988. 'Watershed vs in-lake alkalinity generation: a comparison of rates using input-output studies', *Water Air Soil Pollut.*, **39**, 263–273.

Spratt, H. G., Jr. and Morgan, M. D. 1990. 'Sulfur cycling in a cedar-dominated, freshwater wetland', *Limnol. Oceanogr.*, **35**, 1586–1593.

Stauffer, R. E. 1993. 'Chemistry of soft-water seepage lakes in the upper midwest: lacustrine alkalinity production in summer', *Water Air Soil Pollut.*, **71**, 1–12.

Stottlemeyer, R. 1987. 'Snowpack ion accumulation and loss in a basin draining to Lake Superior', *Can. J. Fish. Aquat. Sci.*, **44**, 1812–1819.

Turner, R. S., Cook, R. B., Van Miegrot, H., Johnson, D. W., Elwood, J. W., Bricker, O. P., Lindberg, S. E., and Hornberger, G. M. 1990. 'Watershed and lake processes affecting surface water acid-base chemistry', NAPAP Report 10, in *Acidic Deposition: State of Science and Technology*. National Acid Precipitation Asessment Program, Washington, D.C.

Urban, N. R. and Bayley, S. E. 1986. 'The acid-base balance of peatlands: a short-term perspective', *Water Air Soil Pollut.*, **30**, 791–800.

Urban, N. R., Eisenreich, S. J., and Gorham, E. 1986. 'Proton cycling in bogs: geographic variation in Northeastern North America', in Hutchinson, T. C. and Meema, K. M. (Eds), *Effects of Air Pollutants on Forests, Agriculture and Wetlands*. Wiley and Sons, New York.

Van Stempvoort, D. R., Wills, J. J., and Fritz, P. 1991. 'Aboveground vegetation effects on the deposition and cycling of atmospheric sulfur: chemical and stable isotope evidence', *Water Air Soil Pollut.*, **60**, 55–82.

Vannier, C., Didonlescot, J. F., Lelong, F., and Guillet, B. 1993. 'Distribution of sulfur forms in soils from beech and spruce forests of Mont-Lozere (France)', *Plant Soil*, **154**, 197–209.

Vet, R. J., Sirois, A., Jeffries, D. S., Semkin, R. G., Foster, N. W., Hazlett, P., and Chan, C. H. 1988. 'Comparison of bulk, wet-only, and wet-plus-dry deposition measurements at the Turkey Lakes Watershed', *Can. J. Fish. Aquat. Sci.*, **45**, 26–37.

Vollenweider, R. A. 1969. 'Secchi disc readings as an auxiliary measure to characterize the optical conditions', in Vollenweider, A. (Ed.), *A Manual for Measuring Primary Production in Aquatic Environments*. IBP Handbook 12. Blackwell, Oxford. p. 171.

Wainwright, M. 1990. 'Field methods used in the determination of sulfur transformations in soil', in Harrison, A. F., Ineson, P., Heal, O. W. (Eds), *Nutrient Cycling in Terrestrial Ecosystems. Field Methods, Application and Interpretation*. Elsevier, New York. pp. 218–232.

White, J. R. and Driscoll, C. T. 1985. 'Lead cycling in an acidic Adirondack lake', *Environ. Sci. Technol.*, **19**, 1182–1187.

White, J. R. and Driscoll, C. T. 1987a. 'Zinc cycling in an acidic Adirondack lake', *Environ. Sci. Technol.*, **21**, 211–216.

White, J. R. and Driscoll, C. T. 1987b. 'Manganese cycling in an acidic Adirondack lake', *Biogeochemistry*, **3**, 87–103.

Wieder, R. K. and Lang, G. E. 1988. 'Cycling of inorganic and organic sulfur in peat from Big Run Bog, West Virginia', *Biogeochemistry*, **5**, 221–242.

Wright, R. F. 1983. 'Input–output budgets at Langtjern, a small acidified lake in southern Norway', *Hydrobiologia*, **101**, 1–12.

Wright, R. F. 1985. 'Chemistry of Lake Hovvatn, Norway, following liming and reacidification', *Can. J. Fish. Aquat. Sci.*, **42**, 1103–1113.

Wright, R. F. and Johannessen, M. 1980. 'Input–output budgets of major ions at gauged catchments in Norway', in Drablos, D. and Tollan, A. (Eds), *Ecological Impact of Acid Precipitation*. pp. 250–253.

Wright, R. F. and Gjessing, E. T. 1986. 'RAIN Project Annual Report for 1985', *Acid Rain Research Report 9/1986*. NIVA, Oslo. 33 pp.

Wright, R. F., Gjessing, E. T., Christophersen, N., Lotse, E., Seip, H. M., Semb, A., Sletaune, B., Storhaug, R., and Wedum, K. 1986. 'Project RAIN: changing acid deposition to whole catchments. The first year of treatment', *Water Air Soil Pollut.*, **30**, 47–64.

Wright, R. F., Lotse, E., and Semb, A. 1988. 'Reversibility of acidification shown by whole-catchment experiments', *Nature*, **334**, 670–675.

17

NITROGEN FLUXES IN A HIGH ELEVATION COLORADO ROCKY MOUNTAIN BASIN

JILL S. BARON[1] AND DONALD H. CAMPBELL[2]

[1] US Geological Survey, Natural Resource Ecology Laboratory, Colorado State University,
Fort Collins, CO 80523, USA
[2] US Geological Survey, Mail Stop 415, Denver Federal Center, Lakewood, CO 80225, USA

ABSTRACT

Measured, calculated and simulated values were combined to develop an annual nitrogen budget for Loch Vale Watershed (LVWS) in the Colorado Front Range. Nine-year average wet nitrogen deposition values were 1·6 ($s = 0.36$) kg NO_3-N ha^{-1}, and 1·0 ($s = 0.3$) kg NH_4-N ha^{-1}. Assuming dry nitrogen deposition to be half that of measured wet deposition, this high elevation watershed receives 3·9 kg N ha^{-1}. Although deposition values fluctuated with precipitation, measured stream nitrogen outputs were less variable. Of the total N input to the watershed (3·9 kg N ha^{-1} wet plus dry deposition), 49% of the total N input was immobilized. Stream losses were 2·0 kg N ha^{-1} (1125 kg measured dissolved inorganic N in 1992, 1–2 kg calculated dissolved organic N, plus an average of 203 kg algal N from the entire 660 ha watershed). Tundra and aquatic algae were the largest reservoirs for incoming N, at approximately 18% and 15% of the total 2574 kg N deposition, respectively. Rocky areas and forest stored the remaining 11% and 5%, respectively. Fully 80% of N losses from the watershed came from the 68% of LVWS that is alpine.

INTRODUCTION

Historically, nitrogen has been among the nutrients most often limiting to the growth of aquatic and terrestrial organisms (Aber *et al.*, 1989; Stoddard, 1992). Recently, however, there is increasing evidence that N limitation has been overcome in a number of diverse environments (Likens *et al.*, 1977; Ågren and Bosatta, 1988; Aber *et al.*, 1989; Schelske, 1991; Tietema and Verstraeten, 1991; Stoddard, 1992; Stoermer *et al.*, 1993; Galloway *et al.*, 1994). Globally, increased nitrogen availability is due in part to large regional increases in nitrogen oxide emissions (Brimblecombe and Stedman, 1982; Schindler and Bayley, 1993; Galloway *et al.*, 1994), agricultural ammonia emissions (van Breeman *et al.*, 1987; Schlesinger and Hartley, 1991), and waste water and agricultural effluent (Brezonik, 1972; Valiela *et al.*, 1990).

In the Front Range of the Colorado Rocky Mountains a number of alpine and subalpine lakes have high concentrations of nitrate (6·3–16·8 µeq NO_3 l^{-1}; Eilers *et al.*, 1986), including Loch Vale Watershed (LVWS) lakes in Rocky Mountain National Park, with an annual mean 1982–1992 concentration of 16·2 µeq NO_3 l^{-1} (Baron, 1992). Mean deposition rates of N in wetfall in Loch Vale Watershed are 1·6 ($s = 0.36$) kg NO_3-N ha^{-1}, and 1·0 ($s = 0.3$) kg NH_4-N ha^{-1} for the nine years 1984–1992 (NADP/NTN Data Base, 1993). These values are greater than those that are considered background N deposition values (about 0·2 kg N ha^{-1}) typical of remote, non-industrialized parts of the world (Galloway *et al.*, 1982; Hedin *et al.*, 1995). The plains adjacent to the Colorado Front Range include Denver and other metropolitan areas and a major agricultural region, and, frequently, surface winds flow up into the mountains, drawing agricultural and urban air to high elevations (Parrish *et al.*, 1990; Langford and Fehsenfeld, 1992; Baron and Denning, 1993). Additionally, the predominant westerly winds contribute nitrate to snowpacks from western sources (Turk *et al.*, 1992). For this paper, results from a variety of published and original data are synthesized in order to develop a model of nitrogen flux in Loch Vale Watershed, and to explore the fate of atmospherically deposited nitrogen in alpine and subalpine terrestrial and aquatic ecosystems.

Site description

Loch Vale Watershed (LVWS) is a 6·6 km^2 north-east-facing basin located in Rocky Mountain National Park, Colorado. Streams and four lakes are located in a narrow glaciated valley 500–1000 m below surrounding granitic cliffs (Figure 1). Three lakes, Sky Pond, Glass Lake and Andrews Tarn, are alpine, with a combined volume of 173 000 m^3. The Loch is a shallow 61 100 m^3 lake below treeline. A fourth-order stream drains the basin at The Loch outlet, where discharge has been measured since 1983. Two alpine subbasins, Icy Brook (2·8 km^2) and Andrews Creek (1·5 km^2), drain into subalpine Icy Brook, with a watershed area of 2·3 km^2. Although the stream below the confluence is bordered by meadow and forest, the slopes of this subalpine portion are either unvegetated or covered by sparse tundra. The percentage cover by landform for the entire 6·6 km^2 is 81% rock, 11% tundra, 6% forest, 1% meadow and 1% surface waters. The forest is a mature stand of Englemann spruce and subalpine fir, and trees are greater than 500 years in age (Arthur and Fahey, 1992).

The climate is continental and characterized by strong westerly winds, particularly during the winter months. Mean annual temperature is 1·5°C, with January minimum and maximum temperatures of −10·9 and −2·9°C, respectively. July minimum and maximum temperatures are 7·7 and 19·5°C, respectively. Of the approximately 110 cm precipitation per year, 65–80% occurs as snow. Much of the remainder occurs as summertime cloudbursts that originate from up-valley funnelling of warm air from lower elevations. As the uplifted air rises, cools and condenses, thunderstorms often develop (Baron and Denning, 1993).

METHODS

Precipitation

Precipitation is measured with two Belfort rain gauges located at 3160 m elevation adjacent to a solar-powered remote area weather station. One rain gauge is equipped with an alter wind shield, and the other with a nipher wind shield. A comparison of performance of the two shields showed an insignificant difference, so the two have been used interchangeably to create the most complete record (Bigelow *et al.*, 1990).

Precipitation chemistry is measured with an Aerochemetrics collector located adjacent to a meteorological station. Precipitation has been collected weekly since September 1983, as part of the National Atmospheric Deposition Program/National Trends Network (NADP/NTN Data Base, 1993). NADP procedures are described in Peden (1986). Wet deposition is presented as kg per watershed and is calculated as the product of the volume-weighted mean concentration and the total measured precipitation for each time sequence. Precipitation collected in summer 1986 was contaminated; mean summer values from the other nine years were substituted in order to develop a full 1986 deposition year.

Stream discharge and chemical composition

Stream discharge is monitored at two alpine tributaries and at the outlet to the entire basin in LVWS (Figure 1). Stage height at the tributaries was recorded at 15 minute intervals beginning in April 1992. Discharge was measured once to twice per week during spring runoff to develop discharge rating curves. The uncertainty associated with this method of calculation is 4% (Winter, 1981). The discharge record was extended through the winter months using a logarithmic extrapolation of stream flow recession. Because the estimates of winter flow were 0·5% of the annual total for Icy Brook and 7·0% for Andrews Creek, the overall uncertainty introduced by the extrapolation was considered insignificant to overall budget estimates.

Stream discharge at The Loch outlet was calculated from continuous measurements of stream stage in a Parshall flume since 1983. The flume was calibrated for stream stages of 6 cm or more; very low flows during the winter were unmeasurable. Stage heights at low flows were assigned, instead, as follows: in October the

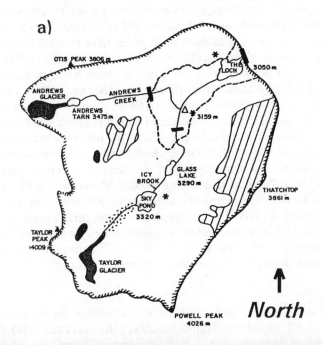

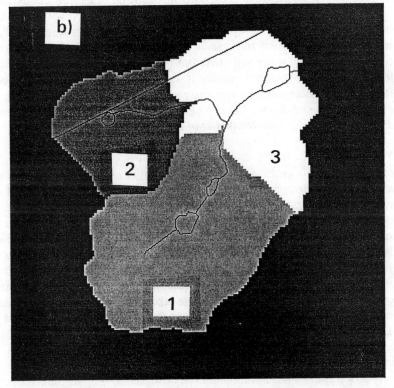

Figure 1. Map of Loch Vale Watershed in Rocky Mountain National Park, Colorado. (a) Major landforms. Hatched areas are alpine tundra, the area within the dotted line is forest, and remaining areas are bedrock or talus. Stream gauging stations are shown by dark rectangles; the meteorological station is shown with an open triangle; (b) Loch Vale Watershed subbasins: Icy Brook (alpine) (1), Andrews, Creek (alpine) (2) and Icy Brook (subalpine) (3). The line running through (b) is the Alva Adams tunnel. This figure was generated with GRASS 4.1

last valid recorded reading was carried through to the end of the month; in November the stage was set to 0·036 m; in December 0·019 m; January and February 0·002 m; March 0·019 m; and April 0·036 m, until the stage increased to measurable values. The uncertainty assigned to Parshall flumes is 5%; another 5% was added to account for slight leakage around the flume (Winter, 1981; Baron, 1992).

Seasons were defined for this paper as winter, spring and summer, based on examination of the measured hydrograph and observed weather parameters (Baron and Bricker, 1987). Winter, the period from 1 October–14 April, corresponds to minimum surface water drainage from the basin. All precipitation that occurred during this time is assumed to have accumulated within Loch Vale as cold snowpack. Spring, from 15 April–14 July, is when that snowpack is released. This is the period of maximum discharge. Summer is defined as the rest of the year, 15 July–30 September, and corresponds to declining flow.

Stream water samples were collected manually and by autosampler at varying time intervals, ranging from daily to weekly. Samples were collected in acid-washed bottles, filtered with 0·45 μm filters and analysed by ion chromatography. The mean relative percentage difference was less than 15% for 80 NO_3 duplicate samples, and less than 42% for 65 NH_4 paired samples over the period 1983–1993. As expected, major constituents of water, such as NO_3, are analysed more precisely than those present in trace amounts, such as NH_4. Ammonium concentrations were close to the 0·166 μeq l^{-1} detection limit, and NO_3 concentrations were often much greater than the 0·161 μeq l^{-1} detection limit.

Terrestrial ecosystem processes

Nitrogen imports and exports from alpine tundra and subalpine forest were taken from previously published results of CENTURY model simulations for LVWS (Baron *et al.*, 1994). CENTURY is a general model of plant–soil ecosystems that represents carbon, nitrogen and other nutrient dynamics (Sanford *et al.*, 1991; Parton *et al.*, 1993). Climate data from LVWS for forest, and from Niwot Ridge, Colorado, for alpine tundra, were used to run the model. The separate forest and tundra runs were initialized with previously published measurements from the two sites, such as above- and below-ground plant tissue C and N, percentage soil organic matter and other soil attributes (Webber and May, 1977; Arthur and Fahey, 1992; Baron *et al.*, 1992, 1994). The simulated results at steady state (1066 year runs) compared well with additional, previously unused data from the two sites, as well as measured values from other similar forest and tundra ecosystems (Cole and Rapp, 1981; Vogt *et al.*, 1986; Prescott *et al.*, 1989; Bowman, 1992; Bowman *et al.*, 1993; Baron *et al.*, 1994; Walker *et al.*, 1994).

Phytoplankton biomass N

Chlorophyll *a* measurements from 1984 to 1986 (McKnight *et al.*, 1986, 1988) from surface, middle and bottom waters were averaged for each season for The Loch and Sky Pond. Ratios of C:chlorophyll *a* in phytoplankton vary depending on the availability of P. A ratio of C:chlorophyll *a* of 8 μmol C μg chl *a*$^{-1}$ was used, based on the finding of Morris and Lewis (1992) that Rocky Mountain lake biota have severe P limitations and on C:chlorophyll *a* ratios reported by Hecky *et al.* (1993) for P-deficient phytoplankton. Phytoplankton N was then bracketed using C:N ratios of 7–12, which are similar to C:N ranges found from N-rich, P-poor subarctic or temperate lakes (Hecky *et al.*, 1993). The resulting values were summed by season for the volume times seasonal turnover rates in total water volume of The Loch, and the volume times seasonal turnover rates in total water volume of alpine lakes (Sky Pond plus Glass Lake), to yield phytoplankton N contents. Turnover rates were estimated from Baron (1992) to be 15, 5 and 1 per season, for Sky Pond and Glass Lake, and 60, 22 and 1 per season for The Loch for spring, summer and winter, respectively.

RESULTS

Measured water fluxes

Measured precipitation ranged from 85·1 cm in the driest year (1989) to 125·8 and 125·3 cm in 1984 and 1986, the two wettest years (Table I). The highest discharge was during the years of greatest precipitation,

Table I. Water fluxes (cm) for Loch Vale Watershed, 1984–1993. Water years are 1 October–30 September

Year	Precipitation (P)	Discharge (Q)	$P - Q$	% Evaporation*
1984	125·8	91·3	34·5	0·27
1985	110·7	67·5	43·2	0·39
1986	125·3	93·5	31·8	0·25
1987	102·7	74·8	27·9	0·27
1988	94·3	73·5	20·8	0·22
1989	85·1	65·4	19·6	0·23
1990	119·0	74·6	44·4	0·37
1991	105·9	74·0	31·9	0·30
1992	96·5	58·9	37·6	0·39
1993	120·4	76·5	43·9	0·36

* Evaporation, transpiration and sublimation, calculated by difference between precipitation and discharge.

91·3 cm in 1984 and 93·5 cm in 1986; the lowest discharge of 65·4 cm was during 1989. The difference between measured precipitation and measured discharge ranged between 19·6 and 44·4 cm, or 22–39% of precipitation. This is consistent with the calculated evaporative water losses (evaporation, evapotranspiration, and sublimation) rates of Baron (1992), using a combination of values depending on season and land cover type.

Discharge hydrographs for 1992 were representative of other years (Figure 2). Discharge occurs year round at The Loch outlet and Andrews Creek; in Icy Brook, flow ceases during the winter months and resumes during spring snowmelt. Approximately half of the total discharge of LVWS was from the 290 ha of Icy Brook basin that is above the treeline, and another third came from the 160 ha Andrews Creek sub-basin.

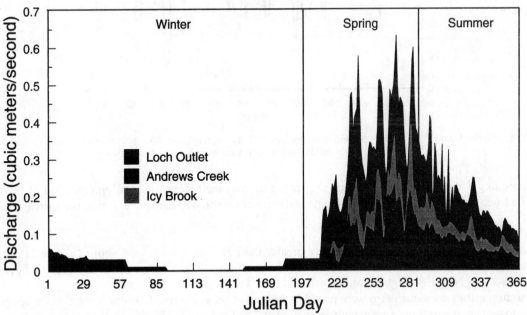

Figure 2. Discharge for water year 1992. Icy Brook and Andrews Creek are alpine tributaries in Loch Vale Watershed. The Loch outlet is the furthest downstream extent of the watershed. Julian days from 1 October 1991 are shown on the *x*-axis

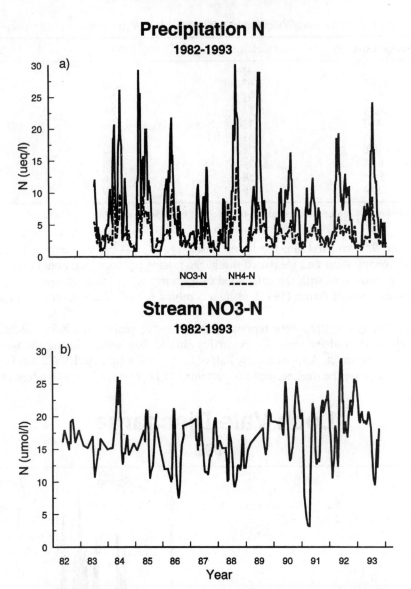

Figure 3. Time series, from 1982 to 1993, of N concentrations (µeq l^{-1}) of (a) precipitation NH$_4$-N and NO$_3$-N and (b) stream NO$_3$-N in Loch Vale Watershed. Stream values are from The Loch outlet

The remaining water was supplied during the spring by snowmelt from the subalpine portion of the basin into The Loch or its tributaries. The subalpine water contribution was negligible during the summer season.

Nitrogen concentrations

Nitrate concentrations in precipitation were greater than ammonium concentrations (Figure 3a). Stream nitrogen was almost exclusively nitrate and, with the exception of low values observed in 1991, fluctuated around 16·2 µeq NO$_3$ l^{-1} (Figure 3b).

Maximum nitrate concentrations were found in Andrews Creek, intermediate concentrations in Icy Brook and the lowest concentrations were usually measured at The Loch outlet (Figure 4). Nitrate concentrations at all sites were seasonally dynamic; maximum concentrations were measured in early spring just after the

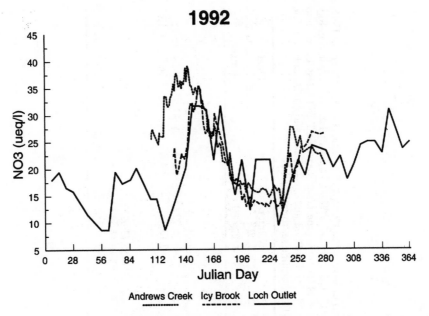

Figure 4. Nitrate concentrations (μeq/l) in Icy Brook, Andrews Creek and at The Loch outlet for water year 1992

beginning of snowmelt. Concentrations declined to their lowest levels during midsummer. Nitrate concentrations increased again at the end of the summer and remained between 8 and 20 μeq l^{-1} for The Loch outlet until the beginning of the spring snowmelt.

Calculated algal N

Algal biomass varied considerably in LVWS from year to year, over the seasons (Spaulding *et al.*, 1993) and from lake to lake (McKnight *et al.*, 1986, 1988). Cell numbers were much greater in Sky Pond (ave. 59 900, $s = 55\,570$, $n = 46$) than in The Loch (ave. 5880 cells ml^{-1}, $s = 4460$, $n = 57$). Mean chlorophyll *a* was more abundant in Sky Pond than The Loch during all seasons; the highest mean value of 5·18 ($s = 4·20$) μg l^{-1} occurred in the winter, and the lowest mean value of 3·75 ($s = 1·41$) μg l^{-1} was found during the summer (Table II). Winter chlorophyll *a* was also highest for The Loch, with an average of 4·65 ($s = 3·92$) μg l^{-1}; but spring season chlorophyll *a* was lower than summer values, with a mean of 1·28 ($s = 0·73$) μg l^{-1}, possibly because of rapid lake water turnover during snowmelt (Baron, 1992; McKnight *et al.*, 1990). Chlorophyll *a* values were highly variable, reflecting the dynamic nature of algal populations. The alpine lakes have significantly greater lake volumes, and they turn over less frequently, and this influenced the total amount of N calculated in algal biomass for alpine versus subalpine lakes. We calculated a range of 36–56 kg N ha^{-1} yr^{-1} for Sky Pond and Glass Lake, compared with a range of 13–25 kg N ha^{-1} yr^{-1} for The Loch.

Nitrogen fluxes

Loch Vale Watershed. In general, more N was deposited in wetter years, but there was not a one-to-one relationship between measured amounts of precipitation and annual wet N deposition (Table I, Figure 5). Somewhat less than half of the total annual N in precipitation was deposited during the winter season; the remainder was distributed evenly between spring and summer.

Measured efflux from the watershed did not fluctuate as greatly as influx, although there was some indication that more N left during high precipitation years (Figure 5). The bulk of the N efflux was during spring, although a significant fraction was also lost during the summer season. An average of 43% of the total N entering the basin from wet deposition flushed out of LVWS in any given year.

Table II. Ranges of N in algal biomass calculated from the average (and standard deviation) of measured chlorophyll a values (McKnight et al., 1986, 1988). Values are derived assuming a C:chlorophyll a ratio of 8, and a range of C:N of 7–12 (Hecky et al., 1993). Chlorophyll a was measured in Sky Pond and The Loch. Volumetric calculations were made using a total of 147 374 m³ and 4·04 ha for alpine lakes (Sky Pond and Glass Lake), and 61 100 m³ and 4·98 ha for The Loch. Seasonal turnover rates from Baron (1992) of 15, 5 and 1, for Sky Pond and Glass Lake, and 60, 22 and 1 for The Loch for spring, summer and winter, respectively, were used to determine seasonal algal N

	Spring		Summer		Winter		Total	
	n	Average (s)	n	Average (s)	n	Average (s)	n	Range
Sky Pond and Glass Lake								
chlor a	45	4·85 (2·97)	34	3·75 (1·41)	10	5·18 (4·20)	89	
µmol N l⁻¹		3·23–5·50 (1·98–3·39)		2·50–4·29 (0·94–1·61)		3·45–5·92 (2·80–4·80)		9·18–15·71
kg N ha⁻¹		27·38–42·15		7·30–10·95		1·83–3·03 (1·43–2·45)		36·50–56·13
The Loch								
chlor a	67	1·28 (0·73)	37	2·22 (2·00)	10	4·65 (3·92)	114	
µmol N l⁻¹		0·85–1·46 (0·38–0·65)		1·48–2·54 (1·33–2·29)		3·10–5·31 (2·61–4·48)		5·43–9·30
kg N ha⁻¹		7·26–14·52		5·44–9·79		0·49–0·91		13·19–25·21

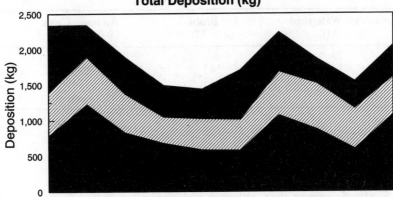

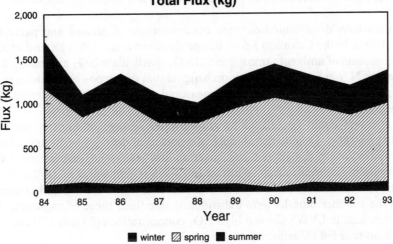

Figure 5. Measured nitrogen in wet deposition and stream efflux from The Loch outlet, water years 1984–1993. Fluxes are separated by seasons of the water year, with winter being 1 October to 14 April, spring 15 April to 14 July and summer 15 July to 30 September

Sub-basins. Most (62%) of the N efflux from the two alpine basins in 1992 occurred during snowmelt. For Icy Brook the efflux was 383 kg N out of the total yearly N flux of 618 kg N, and for Andrews Creek, 289 kg N out of the total flux of 464 kg N. The remaining 38% efflux occurred during the summer. For the whole LVWS basin, 55% of the total 1125 kg N loss was from Icy Brook, and an additional 41% came from Andrews Creek (Table III).

Only 4% of the measured N leaving the basin was contributed by the area below the treeline (calculated by difference between loss from the alpine sub-basins and loss at The Loch outlet), and this subalpine contribution of 43 kg N occurred only during the spring. During the summer there was net consumption of N in the aquatic system below the treeline, since more N passed from the two alpine basins into the subalpine basin than left at The Loch outlet.

Table III. Dissolved inorganic nitrogen losses (in kg) measured from 660 ha Loch Vale Watershed and its alpine sub-basins Icy Brook (290 ha) and Andrews Creek (160 ha) for water year 1992. The subalpine N loss was calculated by the difference between the alpine losses (Icy Brook and Andrews Creek) and the watershed as a whole. A negative value denotes N retention

	Loch Vale Watershed ($n = 53$)	Icy Brook ($n = 87$)	Andrews Creek ($n = 107$)	Subalpine
Winter	79	—	—	—
Spring	730	383	289	59
Summer	315	235	175	−95
Annual	1124	(618)	(464)	−36

DISCUSSION

Dynamics of inputs and outputs

Nitrogen fluxes (Figure 5) suggest that, although wet deposition values fluctuated with precipitation, measured stream N outputs were less variable. The stability of N outputs could be a result of one factor, or a combination of several: dry-deposited N species that may make overall deposition less variable than observed only in wet deposition; a large groundwater reservoir that releases N throughout the year; or biological (vegetative, algal or microbial) N demand that serves to dampen the amplitude of surface water N dynamics.

Several investigators have documented elevated concentrations of aerosol and particulate N species in ambient air above 3000 m in the Colorado Front Range (Heubert et al., 1983; Parrish et al., 1990; Sievering et al., 1994). Measurements of ambient atmospheric HNO_3, particulate NO_3 and particulate NH_4, coupled with elevated snowpack N over wet-only deposition, suggest that dry-deposited N along the Colorado Front Range may be significant, although less so than measured wet deposition (Sievering et al., 1992, 1994; Campbell et al., 1995).

Smoothed stream fluxes could be related to groundwater storage of high N snowmelt waters, and subsequent release as baseflow. Despite the absence of a source of deep groundwater along the Continental Divide, and thin and sparsely distributed soil cover in the watershed, subsurface water may play an important role in regulating concentrations of nitrate in LVWS (Campbell et al., 1995; Kendall et al., 1995). Evidence supporting this idea comes from Back (1994), who found that the ^{18}O and D isotopic signatures of lake waters during the summer months were indistinguishable from those of snowmelt. Measurements of glacial meltwater from high in LVWS showed high NO_3 concentrations (30 µeq l^{-1}) even after a prolonged dry period in which no rain fell (Martin, 1994).

Examination of hysteresis patterns for nitrate and other solutes in concentration–discharge relationships in Andrews Creek and Icy Brook showed that the hysteresis patterns were consistent with piston-type flow through a reservoir of soil or groundwater (Campbell et al., 1995). During late spring snowmelt, large day to day variations in stream flow were accompanied by little or no change in solute concentrations. The most likely reservoir of groundwater storage is in the interstices of talus fields that make up a substantial portion of the alpine zone of LVWS. In the talus, vegetation to take up nitrogen is sparse, allowing nitrate concentrations to remain high throughout the growing season. A large supply of N to streams from groundwater could serve to dampen interannual variability. This is an area of active research, as we do not know the residence times of water in talus.

Several investigators attributed the lack of a relation between N deposition and stream losses to terrestrial N demand (Johnson, 1992; MacDonald et al., 1992). In the systems studied by these investigators, N from deposition was apportioned first to terrestrial vegetation and microbial processes, with the remainder available for export. Lajtha et al. (1995) suggest that this applies particularly to aggrading forests or those that are chronically N limited.

Nitrogen fluxes

If wet-only deposition is considered, as in Figure 5, net N retention is 25% of inputs in LVWS. Of the mean annual wet N deposition of 1716 kg N ($s = 354$; or 2·6 kg N ha^{-1}, $s = 0·5$) to the watershed, 1125 kg inorganic N ($s = 200$; or 1·7 kg N ha^{-1}, $s = 0.3$) was lost via stream flow. This loss is very similar to reported fluxes from an alpine watershed in the southern Sierra Nevada Mountains of California, where an influx of 4·1 kg N ha^{-1} versus an efflux of 2·9 kg N ha^{-1} resulted in 29% accumulation (Williams *et al.*, 1995). Using the wet-only deposition values, Baron (1992) previously postulated that steep mountain basins behaved as 'flow-through' systems where most atmospheric inputs were flushed rapidly downstream. The reasons given for this included the low ratio of vegetative cover to bedrock, the short growing season and a snowmelt-dominated hydrological regime that rapidly removed 70–80% of water and total solutes from the watershed during a three-month period each year. However, a more realistic assessment of total deposition, including dry as well as wet, brings deposition closer to the range of 2574 kg N total, or 3·9 kg N ha^{-1}. Nearly half of the total N input is immobilized within the watershed at this level of deposition.

Immobilization of N can occur within tundra, forest, rock or aquatic areas. Using previously published work, the results described above and an aerial weighting of ecosystem components, an annual N mass balance was calculated (Table IV, Figure 6). In order to compare the calculations with direct measured values presented in this paper, the watershed was somewhat arbitrarily divided into alpine and subalpine units discharging into an aquatic unit. By constraining the N dynamics of these three units with measured, modelled or calculated values, we balanced the budget by calculating the flux through a fourth unit, bedrock, by difference. Given the smoother efflux than influx (Figure 5), and the growing evidence of dynamic subsurface storage reservoirs for water and N, an annual budget may be too short a time to capture LVWS N dynamics. However, calculation of the annual budget as a first approximation is a useful exercise to identify the relative importance of the different nitrogen cycling components, and to recognize that the relative importance of some processes may change as a function of the time-scale considered.

Tundra and meadow. Tundra and wet meadows comprise about 76 ha of the total 660 ha in LVWS; therefore, 12% of total N deposition, or 296 kg N, was deposited directly on tundra. Additionally, 12 kg N were added to account for direct deposition to surface water area that is above the treeline. Tundra systems are highly organic (Bowman *et al.*, 1993), and are the largest pools of both C and N in LVWS, storing 98 400 kg N in biomass and 824 000 kg N in soil organic matter. According to the CENTURY model results, 3600 kg N is taken up yearly, 3120 of which comes from mineralization. Because the model predicted 160 kg N leached from tundra into surface waters, 316 kg N was taken as runoff from bedrock on to tundra to meet the tundra N requirement. The measured N flux from Icy Brook and Andrews Creek in 1992 was 1080 kg N. To account for this difference an additional 908 kg N was routed from bedrock directly into alpine streams (Table IV, Figure 6).

Forests. Only 6%, or 40 ha, of LVWS is forested, so 156 kg N was routed on to forest directly from deposition. Another 11 kg N was added to account for deposition to surface water area that is below the treeline. The CENTURY model predicted 36 kg N volatilized from forest soils as N_2O (Baron *et al.*, 1994). Forest biomass, according to the model, held 16 000 kg N, while soil organic matter was a repository for 312 800 kg N. Soil organic matter values reported here are much higher than those previously reported, because earlier values were for forest floor N only, instead of for total soil to 60 cm (Arthur and Fahey, 1992; Baron, 1992). Forest N inputs and outputs were well balanced. The simulated amount of forest leachate to streams was 30 kg N, from Baron *et al.* (1994).

Aquatic systems. The inputs to aquatic systems from direct deposition (23 kg N), and measured from the alpine unit (1080 kg N) was 1103 kg N. Inorganic outputs measured in 1992 at The Loch outlet were 1125 kg N. The pool of inorganic N in surface waters was calculated to be 30 kg N based on 1982 concentrations. Calculated phytoplankton N ranged from 280 to 490 kg after conversion of values per ha

Table IV. Summary of nitrogen (N) values used to develop The Loch Vale Watershed N budget depicted in Figure 6

	Area of landform (ha)	N (kg ha^{-1})	N (kg watershed^{-1})	Reference
Wet deposition	660	2·6	1716	NADP, 1993
Dry deposition	660	1·3	858	Sievering *et al.*, 1994
Tundra and meadow				
Inputs				
Deposition	76	4·0	308	NADP, 1993; Sievering *et al.*, 1994
From bedrock	76	4·2	316	Calculated by difference, this study
Pools				
Biomass	76	1295·0	98 400 (6400)	Baron *et al.*, 1994
Soil organic matter	76	10 842·0	824 000 (8000)	Baron *et al.*, 1994
Internal fluxes				
Plant uptake	76	47·0	3600 (1040)	Baron *et al.*, 1994
Mineralization	76	41·0	3120 (640)	Baron *et al.*, 1994
Net loss				
Inorganic N	76	2·1	160 (160)	Baron *et al.*, 1994
Forest				
Inputs				
Deposition	40	4·2	167	NADP, 1993; Sievering *et al.*, 1994
Pools				
Biomass	40	400·0	16 000 (4)	Baron *et al.*, 1994
Soil organic matter	40	7820·0	312 800 (4)	Baron *et al.*, 1994
Internal fluxes				
Plant uptake	40	31·0	1240 (360)	Baron *et al.*, 1994
Mineralization	40	28·0	1120 (320)	Baron *et al.*, 1994
Volitilization	40	0·9	36	Baron *et al.*, 1994
Net loss				
Inorganic N	40	0·8	30 (20)	Baron *et al.*, 1994
Bedrock				
Inputs				
Deposition	540	3·9	2100	NADP, 1993; Sievering *et al.*, 1994
Retention	540	0·5	279	Calculated by difference, this study
Net loss				
Via alpine	540	2·3	1236	Calculated by difference, this study
Via subalpine	540	1·1	585	Calculated by difference, this study
Lakes and streams				
Inputs				
From alpine	450	2·4	1080	Measured, this study
From subalpine	210	2·8	585	Calculated by difference, this study
Pools				
Inorganic N	6	5·0	30	Baron, 1992
Algal biomass	6	46·6–81·6	280–490	Calculated using McKnight *et al.*, 1988
Internal fluxes				
Algal uptake	6	73·0–123·0	438–738	Calculated, this study
Denitrification	6	0·2–0·3	1–2	Calculated, this study
Net loss				
Dissolved inorganic N	660	1·7	1125	Measured, this study
Dissolved organic N	6	0·1–0·3	1–2	Calculated, after Hedin *et al.*, 1995
Algal N flushed	6	26·3–41·3	158–248	Calculated, after McKnight *et al.*, 1990

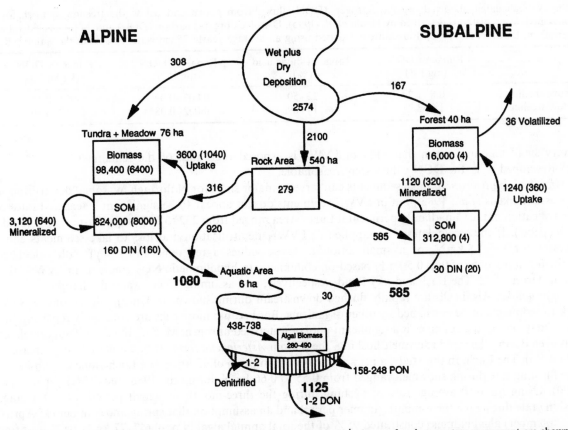

Figure 6. Nitrogen budget for LVWS. Values are estimates, in kg watershed^{-1}. Areas of each ecosystem component are shown in hectares. The budget is divided into an alpine side (tundra and meadows) and a subalpine side (forest). The vegetation is divided into biomass and soil organic matter (SOM). The 'Rock Area' is made up of bedrock and talus fields; runoff from this unit supplies alpine or subalpine areas. Stream fluxes are divided into dissolved inorganic nitrogen (DIN), which is mostly NO$_3$, dissolved organic nitrogen (DON) and particulate organic nitrogen (PON), calculated from estimates of phytoplankton flushed downstream. Measured values are shown in bold type; all others are either calculated or the results of CENTURY model simulations (Baron *et al.*, 1994). Values in parentheses are standard deviations about the mean of the last ten years of a 1060 year simulation. See text and Table V for further explanation

from Table II to total surface water area (6 ha). Additional aquatic N processes were conversion to organic N, microbial denitrification in lake sediments or uptake into algae and subsequent sedimentation, flushing or consumption by zooplankton.

Dissolved organic nitrogen. The contribution of dissolved organic nitrogen (DON) to total stream N flux was calculated from known organic carbon measurements (Hedin *et al.*, 1995). In many waters organic anions are dominated by fulvic acids, and N concentrations can be derived using 1·6–3·0% N relative to fulvic acid carbon (Aiken *et al.*, 1990; McKnight *et al.*, 1992). In LVWS fulvic acid comprised 25–50% of dissolved organic carbon above the treeline, and 14–69% in stream and lake samples below the treeline (Baron *et al.*, 1992). The DON associated with fulvic acids therefore ranged 0·02–0·15 and 0·02–0·60 kg ha^{-1} for alpine and subalpine waters, respectively (Table V). This quantity is lower than the 1·9 kg ha^{-1} inorganic N losses from LVWS, and is much lower than DON terms reported from other studies. Hedin *et al.* (1995) found DON to be the major form of stream nitrogen in old-growth temperate forest of southern Chile, where inorganic N deposition is extremely low. Even if all the dissolved organic carbon were fulvic acids, maximum DON losses from 6 ha of surface waters would be 2·3 kg watershed^{-1},

Table V. Calculated dissolved organic nitrogen (DON) fluxes from above and below the treeline in Loch Vale Watershed, using methods described by Hedin *et al.* (1995). Dissolved organic carbon (DOC) and % fulvic acid values are from Baron *et al.* (1992). Fluxes are calculated using an average yearly 75 cm water loss from the watershed

	Range of DOC (mg l^{-1})	Range of fluvic acid (%)	Range of DON (mg l^{-1})	Range of DON (kg ha^{-1})
Above treeline	0·8–1·2	25–50	0·003–0·02	0·023–0·15
Below treeline	1·0–3·7	14–69	0·002–0·08	0·015–0·60

a very small amount. Given that 81% of LVWS is exposed rock and, of the remainder, only 7% has well-developed soils, the DON values seem reasonable.

Microbial denitrification. Lake sediments can serve as major sinks for N through microbial denitrification processes (Likens *et al.*, 1985), but in LVWS sediments do not appear to be significant. Measured rates of denitrification in the Canadian Experimental Lakes Area ranged from 232 to 602 µeq NO$_3$ m^{-2} d^{-1} (Rudd *et al.*, 1986). If these calculations are applied to LVWS, denitrification and loss to lake sediments could account for 0·7–1·8 kg N, or a small amount. These values agree remarkably well with calculated denitrification values of 0·2–0·5 kg N based on epilimnetic to hypolimnetic NO$_3$ gradients in LVWS lakes (from Baron, 1992, The Loch and Sky Pond, respectively, and assuming 75 cm water discharge).

Algal uptake. Algal cells are rapidly flushed downstream during snowmelt. During all seasons cells can sink to sediments or be consumed by other organisms. But because more cells are constantly replacing this loss, the standing stock of algae is a significant pool of watershed nitrogen. McKnight *et al.* (1990) calculated that even during the rapid snowmelt flushing, the diatom *Asterionella formosa* had a net rate of increase of 0.34 d^{-1} in The Loch. In this study, a net aquatic N consumption of 95 kg in The Loch alone was observed. The flushing rate during snowmelt ranged from 0·55 to 0·66 kg N during the 1985 study (McKnight *et al.*, 1990). Using the 0.60 average rate of flushing during the three-month snowmelt period, a 0.30 average flushing rate during the three-month summer period and an assumption that spring and summer (15 April–30 September) algal biomass constituted 95% of the total annual algal N pool (47–77 kg N ha^{-1} in spring and summer algal biomass, Table II), algal N uptake was calculated. For simplicity the algal population was assumed to be in steady state. By increasing the spring and summer algal pool by 0·6 and 0·3, respectively, and summing up the rest of the year, algal uptake of N is 438–738 kg ha^{-1}; flushing of cells accounts for 144–246 kg N ha^{-1}. Consumption of phytoplankton by higher organisms, sedimentation of cells or microbial activity in the water column was not taken into account. Nevertheless, at 73–123 kg N uptake ha^{-1}, phytoplankton constitute the largest biological consumer of N within LVWS.

Bedrock and talus. Eighty-two percent of LVWS (540 ha) is exposed bedrock and boulder fields, so 82% of total N deposition (2110 kg N) was routed directly from precipitation into this rock unit (Table IV, Figure 6). Nitrogen residing in the bedrock pool was then used to balance the rest of the watershed N budget. As previously stated, 316 kg N was used to balance the internal tundra and meadow N budget. An addition 920 kg N was routed into streams for alpine losses to match measured N fluxes from streams draining tundra. Finally, 585 kg N was taken from bedrock to satisfy the average algal N demand and the measured N losses from the watershed outlet. The remainder of 279 kg N within rock could be utilized by lichens or microorganisms, or stored in subsurface reservoirs.

SUMMARY

The amount of dry deposited nitrogen to LVWS is unknown, but if it is slightly less than measured wet deposition, approximately 52% of the N was flushed downstream in dissolved inorganic or organic form. Dissolved organic nitrogen appeared to be insignificant to the overall N budget, making up less than 1% of the total efflux of N. Based on calculations, particulate organic N as phytoplankton flushed downstream could make up as much as 18% of the total efflux.

Alpine tundra immobilized 18% of the N inputs. Only 5% was incorporated into forest vegetation or soils, in part owing to the low aereal coverage of forests in LVWS, and in part because these forests, at 500 years of age, are not aggrading. Algal uptake processes consume about 22% of total inputs. Short-lived algae do not remain in the water column; they settle to the bottom, are consumed by other organisms, or are flushed downstream as particulate organic matter. Roughly 7% of nitrogen fixed by phytoplankton is washed out of the watershed because of hydrological flushing, leaving 15% within the LVWS aquatic system. Denitrification at the sediment–water interface accounted for up to 2 kg N stored: or an insignificant amount.

We cannot place a high degree of confidence in the bedrock N retention values, because they were derived by difference. The budget calculation exercise suggests, however, that biological processes or subsurface storage within the large bedrock area is significant, accounting for about 10% of total annual N inputs. This finding is consistent with the findings of Kendall *et al.* (1995) and Campbell *et al.* (1995) that suggest both water retention and biological activity in talus significantly influences stream water chemical and isotopic composition.

This study points out that algal N uptake is important to the overall watershed N budget, despite large watershed N fluxes during spring and summer growing seasons. Better measurements are necessary, but the study results suggest that the range of algal consumption of N, at 45–75 kg N ha^{-1}, is equal or greater per unit area than that for alpine tundra and subalpine meadows. The study also suggests large areas of bedrock are biologically active.

ACKNOWLEDGEMENTS

Many people have helped to collect data from LVWS, and to them we are grateful. We specifically wish to thank Brian Newkirk, data manager of the LVWS project at Colorado State University. The comments of several reviewers, particularly Diane McKnight, USGS, greatly improved this manuscript. This work was supported by the National Park Service, the US Geological Survey Water Resources Division, the US Geological Survey Water, Energy, and Biogeochemical Budgets Program and the National Biological Service.

REFERENCES

Aber, J. D., Nadelhoffer, K. J., Steudler, P., and Melillo, J. M. 1989. 'Nitrogen saturation in northern forest ecosystems', *BioScience*, **39**, 378–386.

Ågren, G. I. and Bosatta, E. 1988. 'Nitrogen saturation of terrestrial ecosystems', *Environ. Pollut.*, **54**, 185–197.

Aiken, G., McKnight, D., Wershaw, R., and Miller, L. 1990. 'Evidence for diffusion of aquatic fulvic acid from the sediments of Lake Fryxell, Antarctica', in Baker, R. L. (Ed.), *Organic Substances and Sediments and Water*, Vol. 1: *Humics and soils*. Lewis Publishers, Ann Arbor, MI. pp. 75–88.

Arthur, M. A. and Fahey, T. J. 1992. 'Biomass and nutrients in an Englemann spruce-subalpine fir forest in north-central Colorado: pools, annual production, and internal cycling', *Can. J. For. Res.*, **22**, 315–325.

Back, J. 1994. 'Stable isotopes as tracers of hydrologic sources to three alpine lakes, Rocky Mountain National Park', *M.S. Thesis*, Colorado State University.

Baron, J. 1992. 'Biogeochemistry of a subalpine ecosystem: Loch Vale Watershed', *Ecology Series 90*. Springer-Verlag, New York.

Baron, J., and Bricker, O. P. 1987. 'Hydrologic and chemical flux in Loch Vale Watershed, Rocky Mountain National Park', in Averett, R. C. and McKnight, D. M. (Eds), *Chemical Quality of Water and the Hydrologic Cycle*. Lewis Publishers, Ann Arbor, MI. pp. 141–157.

Baron, J. and Denning, A. S. 1993. 'The influence of mountain meteorology on precipitation chemistry at low and high elevations of the Colorado Front Range, USA', *Atm. Environ.*, **27A**, 2337–2349.

Baron, J., McKnight, D. M., and Denning, A. S. 1992. 'Sources of dissolved and particulate organic material in Loch Vale Watershed, Rocky Mountain National Park, Colorado, USA', *Biogeochemistry*, **15**, 89–110.

Baron, J., Ojima, D. S., Holland, E. A., and Parton, W. J. 1994. 'Analysis of nitrogen saturation potential in Rocky Mountain tundra and forest: implications for aquatic systems', *Biogeochemistry*, **27**, 61–82.

Bigelow, D. S., Denning, A. S., and Baron, J. S. 1990. 'Differences between nipher and alter-shielded Universal Belfort precipitation gages at two Colorado deposition monitoring sites', *Environ. Sci. Technol.*, **24**, 758–760.

Bowman, W. D. 1992. 'Inputs and storage of nitrogen in winter snowpack in an alpine ecosystem', *Arc. Alp. Res.*, **24**, 211–215.

Bowman, W. D., Theodose, T. A., Schardt, J. C., and Conant, R. T. 1993. 'Constraints of nutrient availability on primary production in two alpine tundra communities', *Ecology*, **74**, 2085–2097.

Brezonik, P. L. 1972. 'Nitrogen: sources and transformations in natural waters', in Allen, H. and Kramer, J. (Eds), *Nutrients in Natural Waters*. John Wiley and Sons, New York. pp. 1–50.

Brimblecombe, P. and Stedman, D. H. 1982. 'Historical evidence for a dramatic increase in the nitrate component of acid rain', *Nature*, **298**, 460–462.

Campbell, D. H., Clow, D. W., Ingersoll, G. P., Mast, M. A., Spahr, N. E., and Turk, J. T. 1995. 'Temporal variation in the chemistry of 2 snowmelt-dominated streams in the Rocky Mountains', *Wat. Resour. Res.*, **31**, 2811–2822.

Cole, D. W. and Rapp, M. 1981. 'Elemental cycling in forest ecosystems', in Raichle, R. E. (Ed.), *International Biome Programme 23*. Cambridge University Press. pp. 341–409.

Eilers, J. M., Kanciruk P., McCord, R. A., Overton, W. S., Hook, L., Blick, D. J., Brakke, D. F., Kellar, P., Silverstein, M. E., and Landers, D. H. 1986. *Characteristics of Lakes in the Western United States, Vol. II, Data compendium for selected physical and chemical variables*. EPA-600/3-86/054B. USEPA, Washington, D.C.

Galloway, J. N., Likens, G. E., Keene, W. C., and Miller, J. M. 1982. 'The composition of precipitation in remote areas of the world', *J. Geophys. Res.*, **87**, 8771–8776.

Galloway, J. N., Levy, H., II, and Kasibhatia, P. S. 1994. 'Year 2020: consequences of population growth and development on deposition of oxidized nitrogen', *Ambio*, **23**, 120–123.

Hecky, R.E., Campbell, P., and Hendzel, L. L. 1993. 'The stoichiometry of carbon, nitrogen, and phosphorous in particulate matter of lakes and oceans', *Limnol. Oceanogr.*, **38**, 709–724.

Hedin, L. O., Armesto, J. J., and Johnson, A. H. 1995. 'Patterns of nutrient loss from unpolluted, old-growth temperate forests: evaluation of biogeochemical theory', *Ecology*, **76**, 493–509.

Heubert, B. J., Norton, R. B., Bollinger, M. J., Parrish, D. D., Hahn, C., Bush, Y. A., Murphy, P. C., Fehsenfeld, F. C., and Albritton, D. L. (1983). 'Gas phase and precipitation acidities in the Colorado mountains', in Herrmann, R. and Johnson, A. I. (Eds), *International Symposium on Hydrometeorology*, Proceedings American Water Resources Association, Bethesda. pp. 17–24.

Johnson, D. W. 1992. 'Nitrogen retention in forest soils', *J. Environ. Qual.*, **21**, 1–12.

Kendall, C., Campbell, D. H., Burns, D. A., Shanley, J. B., Silva, S. R., and Chang, C. C. Y. 1995. 'Tracing sources of nitrate in snowmelt runoff using the oxygen and nitrogen isotopic composition of nitrate', in Tonnesson, K. A., Williams, M. W., and Tranter, M. (Eds), *Biogeochemistry of Seasonally Snow-Covered Catchments*. IAHS Publication No. 228. IAHS Press, Institute of Hydrology, Wallingford, UK. pp. 339–348.

Lajtha, K., Seely, B., and Valiela, I. 1995. 'Retention and leaching losses of atmospherically-derived nitrogen in the aggrading coastal watershed of Waquoit Bay, MA', *Biogeochemistry*, **28**, 33–54.

Langford, A. O. and Fehsenfeld, F. E. 1992. 'Natural vegetation as a source or sink for atmospheric ammonia: a case study', *Science*, **255**, 581–583.

Likens, G. E., Bormann, F. H., Pierce, R. S., and Eaton, J. S. 1977. *Biogeochemistry of a Forested Ecosystem*. Springer-Verlag, New York.

Likens, G. E., Eaton, J. S., Johnson, N. M. and Pierce, R. S. 1985. 'Flux and balance of water and chemicals', in Likens, G. E. (Ed.) *An Ecosystem Approach to Aquatic Ecology: Mirror Lake and its Environment*. Springer-Verlag, New York. pp. 135–155.

MacDonald, N. W., Burton, A. J., Leichty, H. O., Witter, J. A., Pregitzer, K. S., Mroz, G. D., and Richter, D. D. 1992. 'Ion leaching in forest ecosystems along a Great Lakes air pollution gradient', *J. Environ. Qual.*, **21**, 614–623.

Martin, M. M. 1994. 'Variability of inorganic nitrogen in a small alpine watershed, Rocky Mountain National Park', *M.S. Thesis*, Colorado State University, Fort Collins, CO.

McKnight, D. M., Brenner, M., Smith, R., and Baron, J. 1986. 'Seasonal changes in phytoplankton populations and related chemical and physical characteristics in lakes in Loch Vale, Rocky Mountain National Park, Colorado', *USGS Water-Resources Investigations Report 86-4101*. USGS Denver, CO. 64 pp.

McKnight, D. M., Miller, C., Smith, R., Baron, J., and Spaulding, S. 1988. 'Phytoplankton populations in lakes in Loch Vale, Rocky Mountain National Park, Colorado: sensitivity to acidic conditions and nitrate enrichment', *USGS Water-Resources Investigations Report 88-4115*. USGS Denver, CO. 102 pp.

McKnight, D. M., Smith, R. L., Bradbury, J. P., Baron, J. S., and Spaulding, S. A. 1990. 'Phytoplankton dynamics in three Rocky Mountain Lakes, Colorado, USA', *Arct. Alp. Res.*, **22**, 264–274.

McKnight, D. M., Bencala, K. E., Zellweger, G. W., Aiken, G. R., Feder, G. L., and Thorn, K. A., 1992. 'Sorption of dissolved organic carbon by hydrous aluminum and iron oxides occurring at the confluence of Deer Creek with the Snake River, Summit County, Colorado', *Environ. Sci. Technol.*, **26**, 1388–1396.

Morris, D. P. and Lewis, W. M., Jr. 1992. 'Nutrient limitations of bacterioplankton growth in Lake Dillon, Colorado', *Limnol. Oceanogr.*, **37**, 1179–1192.

NADP/NTN Data Base (1993) National Atmospheric Deposition Program Tape of Weekly Data. National Atmospheric Deposition Program (IR-7)/National Trends Network. July 1978–December 1993. [Magnetic tape, 9 track, 1600 cpi, ASCII.] NADP/NTN Coordination Office, Natural Resource Ecology Laboratory, Colorado State University, Fort Collins, CO.

Parrish, D. D., Hahn, C. H., Fahey, D. W., Williams, E. J., Bollinger, M. J., Hubler, G., Buhr, M. P., Murphy, P. C., Trainer, M., Hsie, E. Y., Liu, S. C., and Fehsenfeld, F. C. 1990. 'Systematic variations in the concentrations of NO_x (NO plus NO_2) at Niwot Ridge, Colorado', *J. Geophys. Res.*, **95**(D2) 1817–1836.

Parton, W. J., Scurlock, J. M. O., Ojima, D. S., Gilmanov, T. G., Scholes, R. J., Schimel, D. S., Kirchner, T., Menaut, J-C., Seastedt, T., Garcia Moya, E., Kamnalrut, A., and Kinyamario, J. I. 1993. 'Observations and modeling of biomass on soil organic matter dynamics for the grassland biome worldwide', *Glob. Biogeochem. Cyc.*, **7**, 785–810.

Peden, M. E. 1986. 'Methods of collection and analysis of wet deposition', *Illinois State Water Survey, Report No. 381*. Champaign, IL.

Prescott, C. E., Corbin, J. P., and Parkinson, D. 1989. 'Biomass, productivity, and nutrient use efficiency of aboveground vegetation in four Rocky Mountain coniferous forests', *Can. J. For. Res.*, **19**, 309–317.

Rudd, J. W. M., Kelly, C. A., St. Louis, V., Hesslein, R. H., Furutani, A., and Holoka, M. H. 1986. 'Microbial, consumption of nitric and sulfuric acids in acidified north temperature lakes', *Limnol. Oceanogr.*, **31**, 1267–1280.

Sanford, R. L., Jr., Parton, W. J., Ojima, D. S., and Lodge, D. J., 1991. 'Hurricane effects on soil organic matter dynamics and forest production in the Luquillo Experimental Forest, Puerto Rico: results of simulation modeling', *BioTropica*, **23**, 364–372.

Schelske, C. L. 1991. 'Historical nutrient enrichment of Lake Ontario: paleolimnological evidence', *Can. J. Fish. Aquat. Sci.*, **48**, 1529–1538.

Schindler, D. W. and Bayley, S. E. 1993. 'The biosphere as an increasing sink for atmospheric carbon: estimates from increased nitrogen deposition', *Glob. Biogeochem. Cyc.*, **7**, 717–733.

Schlesinger, W. H. and Hartley, A. E. 1991. 'A global budget for atmospheric NH_3', *Biogeochemistry*, **15**, 191–212.

Sievering, H. D., Burton, D., and Caine, N. 1992. 'Atmospheric loading of nitrogen to alpine tundra in the Colorado Front Range', *Glob. Biogeochem. Cyc.*, **6**, 339–346.

Sievering, H. D., Rusch, D., Marquez, L., Williams, M. W., Bardsley, T. J., Sannes, M., and Siebold, C. 1994. 'Nitrogen loading at the Saddle Site and a comparison with previous N loading data from Niwot Ridge', Abstract from paper presented at Niwot Ridge/ Green Lakes Valley Research Workshop, 23 August 1994. University of Colorado Mountain Research Station.

Spaulding, S. A., Ward, J. V., and Baron, J. S. 1993. 'Winter phytoplankton dynamics in a subalpine lake, Colorado, USA', *Arch. Hydrobiol.*, **129**, 179–198.

Stoddard, J. L. 1994. 'Long-term changes in the watershed retention of nitrogen: its causes and consequences', in Baker, L. A. (Ed.), *Environmental Chemistry of Lakes and Reservoirs*, Advances in Chemistry Series No. 237. American Chemical Society, Washington, D.C.

Stoermer, E. F., Wolin, J. A., and Schelske, C. L. 1993. 'Paleolimnological comparison of the Laurentian Great Lakes based on diatoms', *Limnol. Oceanogr.*, **38**, 1311–1316.

Tietema, A. and Verstraeten, J. M. 1991. 'Nitrogen cycling in an acid forest ecosystem in the Netherlands under increased atmospheric nitrogen input: the nitrogen budget and the effect of nitrogen transformations on the proton budget', *Biogeochemistry*, **15**, 21–46.

Turk, J. T., Campbell, D. H., Ingersoll, G. P., and Clow, D. A., 1992. 'Initial findings of synoptic snowpack sampling in the Colorado Rocky Mountains', *US Geological Survey Open-File report 92-645*. USGS Denver, Colorado. 6 pp.

Valiela, I., Costa, J., Foreman, K., Teal, J. M., Howes, B., and Aubrey, D. 1990. 'Transport of groundwater-borne nutrients from watersheds and their effects on coastal waters', *Biogeochemistry*, **10**, 177–197.

van Breeman, N., Mulder, J., and Van Grinsven, J. J. M. 1987. 'Impacts of acid atmospheric deposition on woodland soils in the Netherlands. II: nitrogen transformations', *Soil Sci. Soc. Am. J.*, **51**, 1634–1640.

Vogt, K. A., Grier, C. C., and Vogt, D. J. 1986. 'Production, turnover, and nutrient dynamics of above- and belowground detritus of world forests', *Adv. Ecol. Res.*, **15**, 303–377.

Walker, M. D., Webber, P. J., Arnold, E. H., and Ebert-May, D. 1994. 'Effects of interannual climate variation on aboveground phytomass in alpine vegetation', *Ecology*, **75**, 393–408.

Webber, P. J. and May, D. E. 1977. 'The magnitude and distribution of belowground plant structures in the alpine tundra of Niwot Ridge, Colorado', *Arc. Alp. Res.*, **9**, 157–174.

Williams, M. W., Bales, R. C., Brown, A. D., and Melack, J. M. 1995. 'Fluxes and transformations of nitrogen in a high-elevation catchment. Sierra Nevada', *Biogeochemistry*, **28**, 1–31.

Winter, T. C. 1981. 'Uncertainties in estimating the water balance of lakes', *Wat. Resour. Bull.*, **17**, 82–115.

18

SULFUR AND NITROGEN BUDGETS FOR FIVE FORESTED APPALACHIAN PLATEAU BASINS

CHARLES L. DOW[1] AND DAVID R. DeWALLE[2]

[1] *The Pinelands Commission, PO Box 7, New Lisbon, NJ 08064, USA*
[2] *School of Forest Resources and Environmental Resources Research Institute, The Pennsylvania State University, 132 Land and Water Research Building, University Park, PA 16802, USA*

ABSTRACT

Sulfur and nitrogen input–output budgets were estimated for five forested Appalachian Plateau basins in Pennsylvania for the period October 1988 to March 1990. Wet and dry deposition inputs were determined on a weekly basis from data collected at atmospheric deposition monitoring stations located near the study sites. Stream export was estimated from intensively sampled stream chemistry and continuous discharge data collected on all five basins. On four of the five basins, deposited sulfur was essentially in balance with stream flow export of sulfur (92–120% exported) for the 1989 water year. The fifth basin had net retention of deposited sulfur, with only 42% exported. All five basins retained the vast majority of deposited nitrogen (only 3–18% exported). The fraction of atmospherically deposited sulfur exported in stream flow was greater by a mean factor of 14 versus nitrogen, implying that sulfur dominates base cation leaching processes on these non-carbonate-based catchments. Although basins in the study were relatively homogeneous in terms of topography, climate, geology and land use, local basin conditions caused significant differences in input–output budgets, pointing to the need for replicated basin studies in a region.

INTRODUCTION

Sulfur and nitrogen have long been implicated as major chemical species involved in the episodic and chronic acidification of surface waters (Reuss and Johnson, 1986; Wigington *et al.*, 1993). Recent attention has also been focused on the question of nitrogen saturation in forested basins and the effects this would have on water quality and forest health (Aber *et al.*, 1989; Kahl *et al.*, 1993; Sullivan, 1993; Stevens *et al.*, 1994). Chemical input–output budgets as pioneered through the research at the Hubbard Brook Experimental Forest (Likens *et al.*, 1977) would aid in determining the influence of sulfur and nitrogen on these processes. Assessment of the retention capacity of sulfur and nitrogen within watersheds (Rochelle and Church, 1987; Rochelle *et al.*, 1987; Johnson, 1992), would provide clues to the chemical and nutrient dynamics of each element on the watershed scale. Policy decisions related to trade-offs in the reduction of sulfur versus nitrogen emissions from industry (under consideration by the US EPA as part of the Clean Air Act Amendments of 1990, Title IV, Section 403) would also be enhanced by examining the input–output budgets of these species in relation to watershed acidification processes.

Often, sulfur and nitrogen budgets are computed for single basins in a particular region, which does not permit intraregional examination of budget variation. Data generated by the EPA-sponsored Episodic Response Project (ERP) provided an opportunity to compare the nitrogen and sulfur budgets for five, relatively undisturbed forested basins in the Northern Appalachian Plateau region of Pennsylvania for the 18-month duration of the study (Wigington *et al.*, 1993). For the period 1979–1984, the Appalachian Plateau region received wet deposition with the highest concentrations of H^+, NO_3^-, and seventh highest SO_4^{2-} concentrations in the United States (Knapp *et al.*, 1988) and at the time of the ERP, the highest

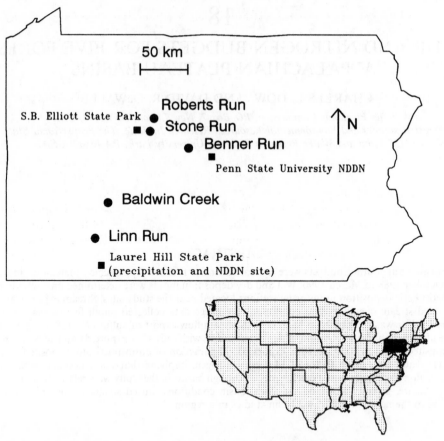

Figure 1. Locations of the five study watersheds along with the two precipitation monitoring sites (state parks) and the two air quality monitoring sites (NDDN) on the Northern Appalachian Plateau of Pennsylvania

dry deposition of sulfur in the eastern United States (Edgerton *et al.*, 1992). Budget results could therefore have important implications for anthropogenic nitrogen and sulfur emission policy in the Appalachian Plateau region.

STUDY SITES

The five study watersheds are located on the Northern Appalachian Plateau in the state of Pennsylvania (Figure 1). These watersheds can be characterized as relatively undisturbed, forested basins with poorly buffered, second-order streams that were not glaciated during the last major period of glaciation. Four of the basins are approximately 10 km^2 in area, while the fifth basin, Baldwin Creek, has an area of 5 km^2. Maximum elevations on these basins range from 701 to 893 m. Basin slopes for four of the five streams are similar (70 m km^{-1}), while the fifth, Linn Run, is much more steep (150 m km^{-1}). A complete description of the study basins is available elsewhere (DeWalle *et al.*, 1993; Wigington *et al.*, 1993).

Average precipitation between 1949 and 1990 for the north-central Pennsylvania region (Benner, Roberts and Stone Run basins) on a water year basis (October to September) was 101 cm at the Philipsburg 8E National Oceanic and Atmospheric Administration (NOAA) climatological station. For the south-west region of Pennsylvania (location of Baldwin Creek and Linn Run basins), average precipitation from 1972 to 1990 was 133 cm on a water year basis at the Laurel Mountain NOAA climatological station. Precipitation amounts for the 1989 water year at the north-central and south-western Pennsylvania sites were 114 and 142 cm, respectively, indicating wetter than normal conditions for the 1989 study period.

Benner Run and Baldwin Creek are dominated by Pocono Group sandstone strata, which generally produces higher ANC (acid neutralizing capacity) groundwater than the Pottsville and Allegheny Group sandstone found on Linn, Stone and Roberts runs (DeWalle *et al.*, 1993). The soils found on these basins are mostly sandy loams and are extremely stony. Well-drained Hazelton or Dekalb soils (Typic Dystrochrepts) dominate on all catchments.

These basins are forested with a mature deciduous stand resulting from regrowth following extensive cutting at the turn of the century. Defoliation by oak leaf roller (*Archips semiferanus*) and gypsy moth caterpillars (*Porthetria dispar*) affected the north-central basins (Benner, Stone, and Roberts runs) during three periods: oak leaf roller in the early 1970s, gypsy moth ca. 1980 and gypsy moth again in the late 1980s. Baldwin Creek basin was slightly affected in 1988 and 1989 by defoliating insects, while Linn Run was unaffected by either defoliation or stand harvesting for the period considered in this paper. Mortality of trees from defoliation was not assessed on these basins, but some salvage logging of trees killed by defoliation was conducted on the Roberts Run basin at the beginning of the ERP study. Clear cuts exist on Benner (12% basin area cut in 1973–1974) and Roberts Runs (6% basin area cut in 1985–1986). A thinning cut, where approximately one-third of the trees were cut over 6% of the basin area, took place on Baldwin Creek in 1982.

METHODS

The approach used in this research was to compare sulfur and nitrogen inputs from wet and dry atmospheric deposition with outputs in stream flow on each of the five basins. Deposition data were collected at sites near the study basins where possible. Export was computed from stream discharge and chemistry data collected as part of the Episodic Response Project (Wigington *et al.*, 1993). The total period of study was for 18 months beginning in October 1988 and ending in March 1990. Monthly grab sampling has continued on these streams since November 1991 as part of the EPA-sponsored Long-Term Monitoring (LTM) project. This continued sampling has provided additional stream chemistry data which can be used to infer longer term stream export differences between basins.

Wet deposition

Weekly deposition was computed from precipitation amounts and chemistry collected at two sites that are part of a state-wide wet deposition monitoring network (Lynch *et al.*, 1989, 1990, 1991). The data collection protocol for the wet deposition samples followed the guidelines set forth by the National Atmospheric Deposition Program/National Trends Network (NADP/NTN). The S. B. Elliott State Park site used to represent the north-central basins was 6 to 48 km from the Stone, Roberts and Benner Run basins, respectively (Figure 1). The Laurel Hill State Park site used to represent the south-west Pennsylvania sites was located 18 to 42 km from the Linn Run and Baldwin Creek basins, respectively (Figure 1).

The wet deposition contribution of sulfur (SO_4^{2-}) and nitrogen (NO_3^-, NH_4^+) was corrected for each watershed to account for differences between precipitation measured on the watershed and that measured at the specific wet deposition monitoring site. Precipitation amounts measured at the S. B. Elliott State Park site were reduced by 9, 13 and 12% for the Benner, Roberts and Stone Run basins, respectively. A correction was not needed for the south-western sites.

Dry deposition

Atmospheric chemistry data collected at two EPA National Dry Deposition Network (NDDN) sites (obtained from the US EPA through Environmental Science and Engineering, Inc., Gainesville, Florida); Laurel Hill State Park and Pennsylvania State University (Figure 1), were used in conjunction with deposition velocities to estimate the dry component of sulfur (SO_2 gas, SO_4^{2-} particulates) and nitrogen

Table I. Average deposition velocities for three deciduous forest sites, in cm s^{-1} (± 1 standard deviation between sites), used in dry deposition computations (from data given by Meyers *et al.*, 1991)

	Winter (Dec. to Feb.)	Spring (Mar. to May)	Summer (Jun. to Aug.)	Autumn (Sept. to Nov.)
SO_2 *	0.17 ± 0.05	0.33 ± 0.13	0.51 ± 0.10	0.27 ± 0.11
SO_4^{2-} †	0.19 ± 0.05	0.41 ± 0.08	0.33 ± 0.09	0.28 ± 0.06
HNO_3 *	1.56 ± 0.19	2.05 ± 0.42	2.11 ± 0.86	1.77 ± 0.37
NO_3^- †	0.53 ± 0.22	0.51 ± 0.40	0.55 ± 0.36	0.51 ± 0.13

* Gas.
† Particulate.

(NO_3^- particulates, HNO_3 gas) inputs to the south-west and north-central watersheds, respectively. Gaseous NO_x as well as particulate and gaseous NH_4^+ were not included in the dry nitrogen estimation owing to the inability to produce reasonably accurate deposition estimates given data collection inadequacies for these components (Hicks *et al.*, 1991; Meyers *et al.*, 1991). The NDDN Laboratory Operations and Field Operations manuals offer a more detailed description of the collection and analysis for these data (US EPA, July 1990, August 1990). Weekly dry deposition flux was computed as the product of the deposition velocity and the mean weekly concentration of sulfur or nitrogen species in the atmosphere (Hicks and Meyers, 1988). Deposition velocities (V_d) were calculated from published average seasonal concentration (C_d) data and deposition rates (F_d) for three NDDN deciduous forest sites (Meyers *et al.*, 1991). The three sites are Oak Ridge, Tennessee; Whiteface Mountain, New York; and Shenandoah, Virginia. The sites had from one to four years of data each (1984 to 1987) and a mean deposition velocity was calculated ($V_d = F_d/C_d$) for each season of available data. Values from each site were then averaged for each season (Table I). Puckett (1990) (SO_2), Lovett and Lindberg (1984) (SO_4^{2-}), Hicks *et al.* [in Shanley (1989)] (SO_4^{2-}), Meyers *et al.* (1989) (HNO_3) and Sirois and Summers (1989) (NO_3^-) show similar deposition velocity values for deciduous forest basins.

Stream export

Stream NO_3^-, NH_4^+ and SO_4^{2-} concentrations and hourly flow rates needed to compute stream export of SO_4^{2-}, NO_3^- and NH_4^+ from the basins were collected for the ERP (DeWalle *et al.*, 1993). The EPA ion chromatography method was used to analyse samples for SO_4^{2-}, the EPA phenate method was used for NH_4^+ analysis, and NO_3^- was analysed using the EPA cadmium reduction method (US EPA, 1983). Stream samples were filtered before analysis of dissolved chemistry using a 0.45 µm pore size filter. Continuous stream stage height data were collected with water level chart/electronic recorders at each of the study streams. Discharge was computed based on rating equations established through stream gauging with natural channel controls. Baseflow and intensive storm flow chemistry sampling were conducted throughout the 18-month period of the ERP, with approximately 300 samples collected per stream. Storm flow sampling was accomplished with automatic water samplers actuated by stage changes so that samples could be obtained on the rising and falling portions of the hydrographs and at peak flows. Regression, period-weighted and discharge-weighted methods (Dann *et al.*, 1986) were used to determine the most appropriate method for computing export. The regression approach was chosen since it is more responsive to weekly and seasonal fluctuations in export owing to intensive and uneven sampling. Hourly SO_4^{2-} and NO_3^- concentrations were predicted for flux calculations from stepwise, forward-selected multiple regression equations for each basin from discharge, time of year harmonic variables (up to the 3rd harmonic terms) and interaction terms (Table II) (Bliss, 1970).

NH_4^+ could not be predicted with the regression approach owing to extremely low correlation (R^2 range of 0.06 to 0.14) between the NH_4^+ concentration and flow, or the harmonic time variables. Consequently, median NH_4^+ concentration, calculated over the entire study period, for each stream was multiplied by the weekly flow total to determine the total NH_4^+ export.

Table II. Significant ($\alpha = 0.10$ for retention, marked with 'x') independent variables in the prediction equations for stream flow concentrations of sulfate and nitrate on each basin

Basin	Dependent variable	Significant independent variables*															R^2
		lQ	$t1$	$t2$	$t3$	$t4$	$t5$	$t6$	$i1$	$i2$	$i3$	$i4$	$i5$	$i6$	n	
Benner	(SO_4^{2-})	x	x	x	x			x		x	x	x			x	277	0·51
	(NO_3^-)	x		x					x	x	x	x	x	x	274	0·31
Roberts	(SO_4^{2-})	x	x	x			x	x	x				x	x	319	0.53
	(NO_3^-)			x			x	x	x			x	x		305	0·40
Stone	(SO_4^{2-})	x	x	x		x			x	x					362	0·61
	(NO_3^-)		x	x	x	x	x		x						362	0·25
Baldwin	(SO_4^{2-})	x	x	x		x			x	x					300	0·28
	(NO_3^-)	x		x			x	x	x	x	x	x			295	0·62
Linn	(SO_4^{2-})		x	x	x	x		x				x	x		256	0·27
	(NO_3^-)		x	x		x		x		x	x	x	x		251	0·47

* Independent variables defined:
$lQ = \log_{10}$ of stream flow in m^3 s^{-1} km^{-2}
$t1 = \cos^*ct$; $t2 = \sin^*ct$; $t3 = \cos^*2ct$; $t4 = \sin^*2ct$; $t5 = \cos^*3ct$; $t6 = \sin^*3ct$
$i1 = \log Q^*t1$; $i2 = \log Q^*t2$; $i3 = \log Q^*t3$; $i4 = \log Q^*t4$; $i5 = \log Q^*t5$; $i6 = \log Q^*t6$
$c = 2\pi/k$ radian day^{-1} where $k = 365$ days
t = Julian date

Particulate and dissolved organic nitrogen and sulfur analysis was not conducted during the ERP. The particulate flux of these elements in stream-suspended solids was assumed to be negligible after Likens *et al.* (1977). Crude estimates of dissolved organic sulfur and nitrogen exported in stream flow were obtained as the product of appropriate nitrogen:carbon and sulfur:carbon ratios and median DOC concentrations for each stream (1·93 to 2·73 mg l^{-1}). Ratios of C:S = 71 and C:N = 17 for sulfur and nitrogen, respectively, were used in these calculations and were based on average chemical composition data of soil humic acid substances world-wide (Sposito, 1989).

RESULTS AND DISCUSSION

Sulfur budgets

Wet deposition of sulfur (as SO_4^{2-}) contributed 54–58% of the total annual atmospheric input of sulfur to these basins (Table III). The south-western basins had a larger amount of wet sulfur deposition even

Table III. Sulfur input–output fluxes and total differences (kg S ha^{-1}) for the 1989 water year (October 1988 to September 1989). Budgets only include dissolved, inorganic chemical transport

Basin	Atmospheric deposition inputs				Outputs	Input–output	% Retained
	Wet SO₄-S	Dry SO₂-S	Dry SO₄-S	Total	SO₄-S		
Benner Run	11·7	6·3	2·0	20·0	8·4	11·6	58
Roberts Run	11·1	6·3	2·0	19·4	19·6	−0·2	−1
Stone Run	11·4	6·3	2·0	19·7	18·1	1·6	8
Baldwin Creek	13·6	9·3	2·3	25·2	24·3	0·9	4
Linn Run	13·6	9·3	2·3	25·2	30·2	−5·0	−20
MEAN	12·3	7·5	2·1	21·9	20·1	1·8	10

though the precipitation-weighted mean SO_4^{2-} concentration was less than that found in the north-central region (0·86 mg S l^{-1} and 0·90 mg S l^{-1}, respectively) owing to the greater precipitation in south-west Pennsylvania (142 cm) versus north-central Pennsylvania (114 cm) .

The majority of dry sulfur deposition (76–80% for the two deposition sites) came in the form of SO_2 gas. This was attributed to the overall greater average atmospheric concentration of sulfur from SO_2 gas (7·4–10·5 µg S m^{-3}) than from SO_4^{2-} particulates (1·9–2·3 µg S m^{-3}) since the deposition velocities for SO_2 and SO_4^{2-} particulates were roughly equal (Table I).

Dry deposition of sulfur is a major contributor to the total atmospheric input of sulfur to the basins and is among the highest reported anywhere in the United States. The dry component ranged from 42 to 46% of the total input of sulfur for the north-central and south-western basins, respectively, for the 1989 water year (Table III). These percentages of total sulfur input by dry deposition are within the range 22–53% reported for several other forested North American sites (Sirois and Summers, 1989; Meyers et al., 1991; Lindberg and Lovett, 1992). Rustad et al. (1994) reported a dry deposition percentage of 63% for SO_4^{2-} on the Bear Brook watershed in Maine using a dry deposition estimation methodology appropriate for a mixed, uneven-aged stand.

On four of the basins (Roberts, Stone, Linn Runs and Baldwin Creek) the input of sulfur was essentially balanced by the sulfur exported in stream flow (92–120% exported, with a mean of 102% exported) (Table III). The median SO_4^{2-} concentrations in stream flow ranged from 104 (Benner Run) to 220 µeq l^{-1} (Linn Run). Export of sulfur for the 1989 water year averaged 20·1 kg S ha^{-1} with a low of 8·4 kg S ha^{-1} on Benner Run and a high of 30·2 kg S ha^{-1} on Linn Run (Table III). The Benner Run basin was retaining the majority of input sulfur (only 42% exported).

Given the very high C:S ratio of 71 for soil humic acid substances, the dissolved organic sulfur exported from these basins would appear to be very low. Based upon this ratio and dissolved organic carbon (DOC) measured in each stream, dissolved organic sulfur was computed to be less than 1% of the inorganic sulfur export for the 1989 water year on all basins. It is likely that export of dissolved organic sulfur in stream flow can be ignored in such budget calculations.

It is hypothesized that the geology/soils unique to the Benner Run basin may be the cause of greater SO_4^{2-} retention. The Benner Run basin does contain a significant amount of Andover (Typic Fragiaguults) soil series which has higher clay content than soils on the other basins (DeWalle et al., 1993). Drilling on Benner Run basin has also revealed extensive fragipan layers with high clay content. High iron and aluminium oxide content associated with higher clay content soils tends to generate greater SO_4^{2-} adsorption capacity (Johnson and Todd 1983; Kleckner-Polk, 1991) and is believed to be the cause of higher sulfur retention on Benner Run basin.

Rochelle et al. (1987) found, in examining sulfur input–output budgets for sites in the US and Canada, that the extent of the last glaciation (Wisconsinian) could be used to define the boundary between sulfur loss and retention. They found that sites south of this line were retaining 20–90% of incoming sulfur, while those to the north had zero retention or net loss of sulfur. The budgets for four of the five basins presented here fall below the 20–90% sulfur retention range, which does not support this hypothesis. This result, however, may be somewhat fortuitous given uncertainties in dry sulfur deposition input calculations, and the assumption of no internal watershed sources of sulfur, such as those from weathering. Overall, results are taken to suggest that on four of five basins, atmospheric sulfur inputs are roughly balanced by stream flow export with strong SO_4^{2-} adsorption still active on Benner Run basin owing to higher clay-content soil.

Monthly grab samples collected from November 1991 to February 1995 as part of the LTM project lend support to the sulfur budget results. Median stream SO_4^{2-} concentrations on Benner Run continue to be about half of that found on the other study streams (median stream SO_4^{2-} concentrations on Benner Run = 90·5 µeq l^{-1}; average of the median stream SO_4^{2-} concentrations for the remaining streams = 190·1 µeq l^{-1} for November 1991 to February 1995). This similarity in the relative SO_4^{2-} concentrations between Benner Run and the other four study streams generally confirms differences in sulfur retention found during the 1989 budget period.

Table IV. Nitrogen input–output fluxes and total differences (kg N ha^{-1}) for the 1989 water year (October 1988 to September 1989). Budgets only include dissolved, inorganic chemical transport

Basin	Atmospheric deposition inputs					Stream flow outputs		Input–output	% Retained
	Wet NO$_3$-N	Wet NH$_4$-N	Dry HNO$_3$-N	Dry NO$_3$-N	Total	NO$_3$-N	NH$_4$-N		
Benner Run	5·4	3·2	3·0	0·5	12·1	0·68	0·05	11·4	94
Roberts Run	5·1	3·1	3·0	0·5	11·7	0·58	0·06	11·1	95
Stone Run	5·2	3·1	3·0	0·5	11·8	0·21	0·05	11·5	97
Baldwin Creek	5·2	3·4	3·3	0·2	12·1	2·12	0·07	9·9	82
Linn Run	5·2	3·4	3·3	0·2	12·1	1·97	0·07	10·1	83
Mean	5·2	3·2	3·1	0·4	12·0	1·11	0·06	10·8	90

Nitrogen budgets

The wet nitrogen deposition component comprised 70–71% of the total annual atmospheric nitrogen input (Table IV). NO$_3^-$ explained approximately 62% of the nitrogen wet deposition on each basin while NH$_4^+$ accounted for the remaining 38%. Unlike sulfur, the wet nitrogen deposition total was essentially the same for all basins. Although the precipitation-weighted mean NO$_3^-$ concentrations were greater in north-central Pennsylvania (0·47 mg N l^{-1}) than in south-west Pennsylvania (0·36 mg N l^{-1}), the precipitation-weighted mean NH$_4^+$ concentrations were essentially equal (0·26 and 0·27 mg N l^{-1} for north-central and south-west Pennsylvania, respectively).

Nearly all (86–94%) of the dry deposited nitrogen was in the form of HNO$_3$. This can be attributed to the greater average concentrations of gaseous HNO$_3$ (0·49–0·54 µg N m^{-3}) versus particulate NO$_3^-$ (0·11–0·32 µg N m^{-3}) coupled with the much larger deposition velocities for HNO$_3$ than for NO$_3^-$ (Table I). Dry nitrogen deposition was 29–30% of the total input for the five study basins (Table IV). These values are less than those calculated at other sites (Lindberg et al., 1987; Shepard et al., 1989; Meyers et al., 1991). Edgerton et al. (1992) found that the two Pennsylvania dry deposition monitoring sites included in this study were in the intermediate range for measured dry nitrogen deposition in the eastern United States in 1989.

Median NO$_3^-$ stream concentrations were between 9 (Benner Run) and 21 µeq l^{-1} (Linn Run). NO$_3^-$ accounted for the vast majority of nitrogen exported from the basins with the range being 0·21 (Stone Run) to 2·12 kg N ha^{-1} y^{-1} (Baldwin Creek) (Table IV). The NH$_4^+$ cation was a very minor component in stream flow for these basins, with maximum concentrations ranging from 0·13 to 0·22 µeq l^{-1}. For four of the basins (Benner, Roberts, Baldwin and Linn) NH$_4^+$-N accounted for only 3–9% of the total nitrogen export (Table IV). On Stone Run it accounted for 19% of the total nitrogen export owing to lower NO$_3^-$ in that stream.

Nitrogen budgets for the five basins indicate an annual retention of nitrogen (Table IV) ranging from 82 to 97% of the deposited nitrogen. Nitrogen retention percentages are similar to values reported for other forested North American watersheds (Johnson, 1992). Results are also consistent with those of DeWalle and Swistock (1994), who found that stream NO$_3^-$ concentrations were less responsive to episodic changes in flow compared with SO$_4^{2-}$ concentrations.

Despite the overall similarity in nitrogen retention between basins, the two south-west Pennsylvania basins (Baldwin Creek and Linn Run) retain approximately 13% less nitrogen than the three north-central Pennsylvania basins. Differences in nitrogen export may be attributed to forest conditions found on the basins. The north-central basins were subjected to defoliation and subsequent stand mortality over the past several decades, whereas the south-west Pennsylvania basins experienced these stresses to a much lesser degree. Although insect defoliation stresses may have caused immediate increases in nitrogen export from a basin (Swank *et al.*, 1981), a subsequently greater demand for nitrogen by regrowth following tree mortality may be occurring on the north-central basins. Thus, a disturbance that originally caused a greater nitrogen export may lead to a greater nitrogen demand in subsequent years (Johnson, 1992). Significant differences in nitrogen retention resulting from past forest stand management practices have also been found in two Black Forest stands in Germany (Feger *et al.*, 1990). Stevens *et al.* (1994) found greater inorganic-N stream water concentrations in older versus younger Sitka spruce stands in Wales, suggesting greater uptake of nitrogen in younger stands.

Other site factors may also explain differences in nitrogen retention between these basins, but these results suggest that important differences in nitrogen retention can occur between relatively similar forested basins located within a region. DeWalle and Pionke (1995) have also found that variations in nitrogen retention between the study basins may be part of a more widespread regional pattern. The forested watersheds from this study in south-west Pennsylvania, along with watersheds located in western Maryland and north-central West Virginia, show higher nitrogen export than north-central Pennsylvania basins, including Benner, Roberts and Stone Runs (DeWalle and Pionke, 1995).

Nitrogen input–output budgets raise the issue of what stage these forested watersheds are at, in terms of response to chronic levels of increased nitrogen deposition (Aber *et al.*, 1989). Clearly, nitrogen export is much less than nitrogen deposition, suggesting that nitrogen saturation (stage 2, as defined by Aber *et al.*, 1989) has not been reached, and since some nitrogen export is occurring these watersheds are also probably not at stage 0 (Aber *et al.*, 1989). However, nitrogen export from the two south-west Pennsylvania basins is about four times greater than the export from the three north-central basins for nearly identical nitrogen inputs, indicating that the initial effects of chronic nitrogen deposition (stage 1) have been reached in south-west Pennsylvania. The three north-central Pennsylvania basins, possibly as a result of increased nitrogen uptake in response to regrowth following past insect defoliation, appear to be in an earlier phase of stage 1 response.

Export of dissolved organic nitrogen in stream flow may be an important budget component, which is often overlooked. Estimated organic nitrogen exported by the streams ranged from 0·68 to 1·01 kg N ha^{-1} for the study basins for the 1989 water year. Organic nitrogen export amounted to less than half of the inorganic nitrogen export in the south-western Pennsylvania streams, but was one to three times the inorganic nitrogen export on the three north-central Pennsylvania streams. Although inclusion of organic nitrogen would not change the general pattern of nitrogen retention among basins, it is clear that budgets based only upon inorganic nitrogen can significantly underestimate nitrogen export.

As was the case for the sulfur budgets, stream chemistry data collected as part of the more recent LTM project also lend support to the nitrogen budgets results. The ratio of stream NO_3^- concentrations in the south-west streams to the north-central streams during the 1989 budget period was 5·3 in this study. This ratio has increased to 6·7 (average median stream NO_3^- concentrations for Benner, Roberts and Stone runs = 6·9 μeq l^{-1}; for Baldwin Creek and Linn Run = 46·2 μeq l^{-1}) during the period November 1991 to February 1995, strengthening the position that the two regions are not similar in terms of nitrogen retention.

Seasonal variations in the sulfur and nitrogen budgets

Monthly sulfur and nitrogen budgets demonstrate a pronounced seasonality, with inputs peaking in the growing season and outputs peaking in the late dormant to early spring seasons (Figures 2 and 3). Mean lag time between inputs and exports was about 10 months for both nitrogen and sulfur as determined from sine waves fitted to monthly input and output data (Bliss, 1970). Growing season input peaks were primarily controlled by wet deposition peaks in late spring and early summer, and peak outputs appear to be controlled by high discharge in late winter and early spring.

An example of the detailed variation in wet and dry deposition inputs of sulfur and nitrogen for Benner Run is presented in Figures 4 and 5, respectively. Precipitation amounts peak in May and June and appear largely to control seasonality in computed wet deposition inputs of both sulfur and nitrogen. Peak wet deposition sulfur concentrations were somewhat more pronounced in the July to August period, but little

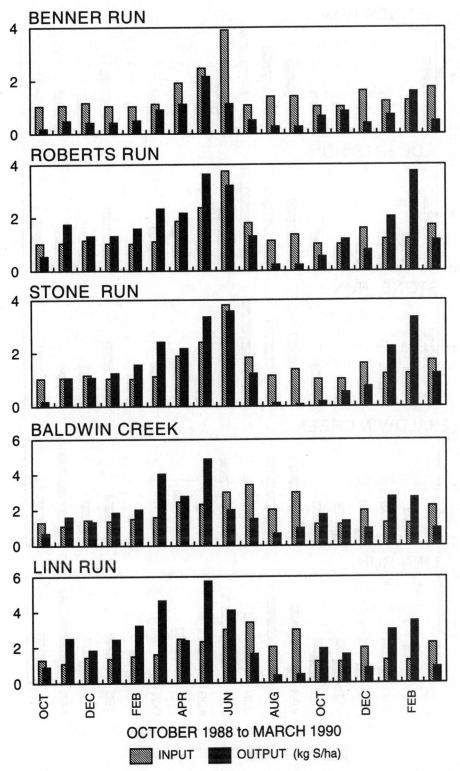

Figure 2. Monthly input–output sulfur budgets (October 1988 to March 1990) in kg S ha^{-1} for the five study basins

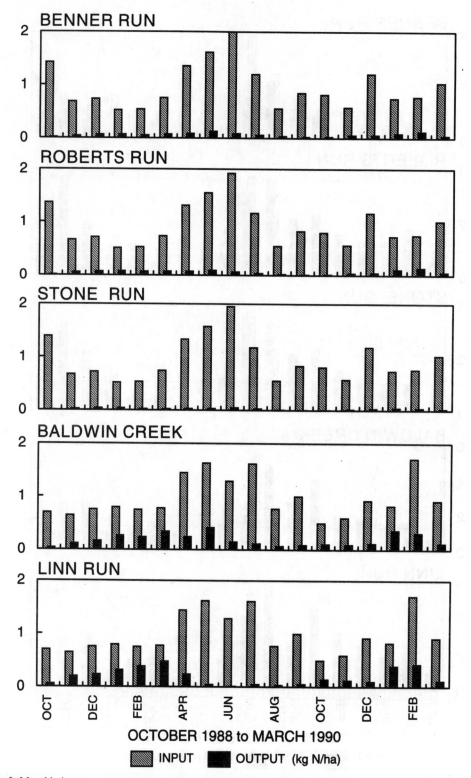

Figure 3. Monthly input–output nitrogen budgets (October 1988 to March 1990) in kg N ha^{-1} for the five study basins

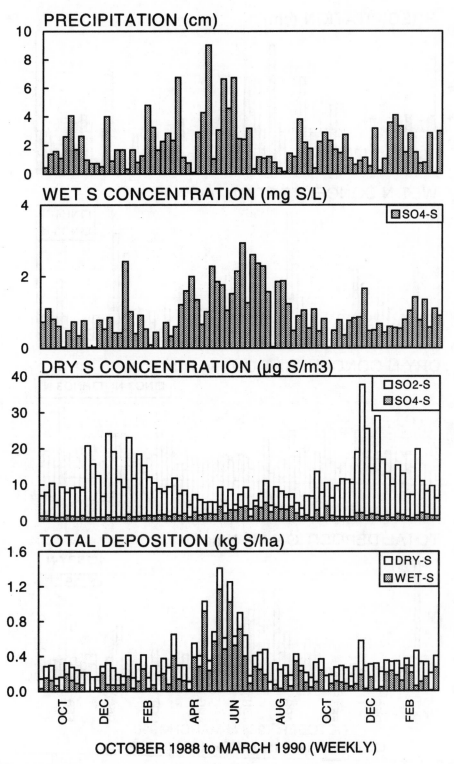

Figure 4. Weekly precipitation, concentrations of sulfur species included in deposition calculations and computed sulfur inputs for the Benner Run basin from October 1988 to March 1990

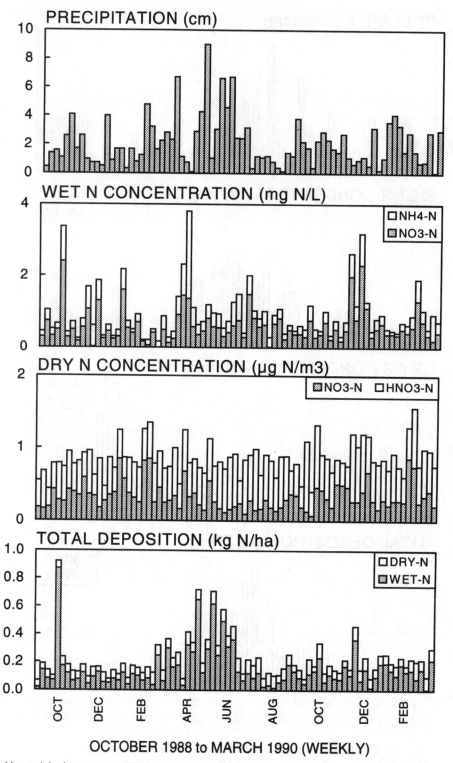

Figure 5. Weekly precipitation, concentrations of nitrogen species included in deposition calculations and computed nitrogen inputs for the Benner Run basin from October 1988 to March 1990

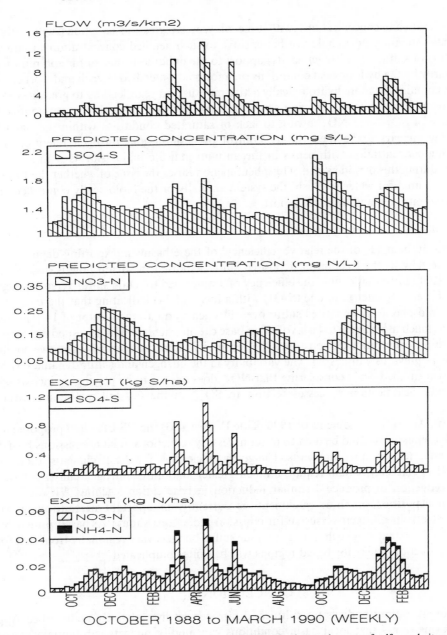

Figure 6. Weekly stream flows, predicted stream nitrogen and sulfur concentrations and export of sulfur and nitrogen from the Benner Run basin from October 1988 to 1990

seasonal variations occurred in nitrogen concentrations of wet deposition. Although seasonal variations in atmospheric concentrations of nitrogen and sulfur did occur (Figures 4 and 5), computed dry deposition with seasonally varying deposition velocities (Table I) showed little seasonality.

Although discharge and export for both sulfur and nitrogen peaked in spring, peak stream concentrations for both SO_4^{2-} and NO_3^- occurred in autumn (Figure 6). This caused secondary peaks in export in autumn. Seasonal harmonic terms in the concentration prediction equations (Table II) are needed to account for such variations in stream concentrations with time of year. Regression equations with only discharge variables to predict stream export could not be sensitive to such detailed concentration variations.

Seasonal cycling of sulfur and nitrogen lends support to the hypothesis that sulfur and nitrogen export is controlled primarily by the hydrological conditions on a forested watershed (Lynch and Corbett, 1989). The dormant season contains the more hydrologically active months of a year, leading to greater export of SO_4^{2-} and NO_3^- in stream flow, while the growing season has the less hydrologically active months, leading to temporary storage of SO_4^{2-} and NO_3^- owing to lack of saturated conditions within the watershed. Even though hydrology appears to be the primary control over sulfur and nitrogen retention and export, seasonally varying concentration with peaks in the autumn indicate major storms with subsequent high export can occur during this period as well. This phenomenon raises the issue of whether the 10-month lag is simply the time required for water to 'flush' the system annually or the biologically controlled lag between input of sulfur and nitrogen and eventual export.

Efficiency in exporting sulfur versus nitrogen

An assessment can be made of the relative 'efficiency' of these basins in exporting deposited sulfur and nitrogen in stream flow. If the fraction of deposited sulfur exported in stream flow is divided by that for nitrogen a value is obtained that defines the efficiency of a watershed to retain sulfur versus nitrogen. All the basins have an efficiency greater than one (5–31, with a mean of 14) indicating that these basins are more likely to export atmospherically deposited sulfur over nitrogen by an average factor of 14. This implies that deposited sulfur is much more likely to be involved in base cation leaching than deposited nitrogen. The wide range in sulfur to nitrogen ratios demonstrates the sensitivity of the budgets to vegetative conditions and sulfur adsorption as previously discussed. The variability in the nitrogen and sulfur dynamics within a given region is further exemplified when considering that NO_3^- does appear to be playing an increasingly greater role in the episodic acidification processes relative to SO_4^{2-} in the Catskill mountains (Murdoch and Stoddard, 1992).

Under the Clean Air Act Amendments of 1990 (Title IV, Sec. 403) the US EPA had proposed a provision where reduced SO_x emissions could be used to offset necessary reductions in NO_x emissions by industries to meet overall emission reduction requirements. The given 'efficiencies' for the catchments studied here show that this trade-off in emissions would require a much greater amount of nitrogen emission reduction that sulfur emission reduction to produce a similar reduction in base cation leaching. Viewed in a different context, nitrogen deposition on these catchments is currently unimportant in controlling base cation leaching and trade-offs in nitrogen versus sulfur emission reductions should be based primarily on other considerations. To the extent that such nitrogen versus sulfur budgets vary regionally from that found here, setting emission reduction policy for broad regions will be quite complicated.

CONCLUSIONS

Comparison of input–output budgets for sulfur and nitrogen on five relatively undisturbed Appalachian Plateau forest basins shows that local basin conditions can modify budgets substantially, especially for sulfur. The deposited sulfur was essentially balanced by the sulfur exported via stream flow (92–120% exported) on four basins, while one basin showed only 42% sulfur export, probably owing to higher soil sulfate adsorption. The nitrogen budgets indicated that the majority (82–97%) of deposited nitrogen was being retained within all basins, but two basins in south-west Pennsylvania showed four times greater nitrogen export than three north-central Pennsylvania basins. Calculations indicated that stream flow export of dissolved organic nitrogen was important relative to inorganic export, but that organic sulfur export in stream flow was negligible. Overall, basins in this study were much more likely to leach atmospherically deposited sulfur than nitrogen (14 : 1 mean efficiency ratio), which has implications for substitutions of SO_x or NO_x in atmospheric emissions reduction policies. Additional years of budget calculations are needed to verify trends on each basin; however, subsequent years of monthly stream sampling from 1991 to 1995 suggest future budget results would be consistent with the 1989 budget period.

ACKNOWLEDGEMENTS

Although the research described in this paper has been funded in part by the US Environmental Protection Agency (agreement CR814566-02 to The Pennsylvania State University), it has not been subjected to the Agency's review and therefore does not necessarily reflect the views of the Agency, and no official endorsement should be inferred. The authors wish to acknowledge the cooperation of the following agencies in providing site access and data: US EPA Environmental Response Project; Pennsylvania State University state-wide precipitation monitoring network; the Pennsylvania Department of Environmental Resources, Bureau of Forestry; Pennsylvania Game Commission; and the US EPA National Dry Deposition Network.

REFERENCES

Aber, J. D., Nadelhoffer, K. J., Steudler, P. and Melillo, J. M. 1989. 'Nitrogen saturation in northern forest ecosystems', *BioScience*, **39**, 378–386.

Bliss, C. I. 1970. *Statistics in Biology*, Vol. 2. McGraw-Hill Book Co., New York. pp. 219–287.

Clean Air Act Amendments, 1990; Title IV, Sec. 403(c). *Environmental Reporter-Federal Laws*. Bureau of National Affairs, Inc., Washington, DC, **71**, 1249–1252.

Dann, M. S., Lynch, J. A. and Corbett, E. S. 1986. 'Comparison of methods for estimating sulfate export from a forested watershed', *J. Environ. Qual.*, **15**, 140–145.

DeWalle, D. R. and Pionke, H. B. 1995. 'Nitrogen export from forest land in the Chesapeake Bay Region', *Proc., 1994 Chesapeake Bay Research Conference*, June 1–3, 1994, Norfolk, VA, pp. 649–655.

DeWalle, D. R. and Swistock, B. R. 1994. 'Causes of episodic acidification in five Pennsylvania streams on the northern Appalachian Plateau', *Wat. Resour. Res.*, **30**, 1955–1963.

DeWalle, D. R., Swistock, B. R., Dow, C. L., Sharpe, W. E., and Carline, R. F. 1993. 'Episodic response project–Northern Appalachian Plateau: site description and methodology', *EPA/600/R-93/023*. US/EPA, Washington, D.C. 55 pp.

Edgerton, E. S., Lavery, T. F., and Boksleitner, R. P. 1992. 'Preliminary data from the USEPA dry deposition network: 1989', *Environ. Pollut.*, **75**, 145–156.

Feger, K. H., Brahmer, G., and Zottl, H. W. 1990. 'Element budgets of two contrasting catchments in the Black Forest (Federal Republic of Germany)', *J. Hydrol.*, **116**, 85–99.

Hicks, B. B. and Meyers, T. P. 1988. 'Measuring and modeling dry deposition in mountainous areas' in Unsworth, M. H. and Fowler, D. (Eds), *Acid Deposition at High Elevation Sites*. Kluwer Academic Publishers, London. pp. 541–552.

Hicks, B. B., Hosker, R. P., Jr, Meyers, T. P., and Womack, J. D. 1991. 'Dry deposition inferential measurement techniques-I. Design and tests of a prototype meteorological and chemical system for determining dry deposition', *Atm. Environ.*, **25A**, 2345–2359.

Johnson, D. W. 1992. 'Nitrogen retention in forest soils', *J. Environ. Qual.*, **21**, 1–12.

Johnson, D. W. and Todd, D. E. 1983. 'Relationships among iron, aluminum, carbon, and sulfate in a variety of forest soils', *Soil Sci. Soc. Am. J.*, **47**, 792–800.

Kahl, J. S., Norton, S. A., Fernandez, I. J., Nadelhoffer, K. J., Driscoll, C. T., and Aber, J. D. 1993. 'Experimental inducement of nitrogen saturation at the watershed scale', *Environ. Sci. Technol.*, **27**, 565–568.

Kleckner-Polk, D. E. 1991. 'Soil sulfate retention properties at three Appalachian forest sites', *Masters of EPC paper*, Pennsylvania State University, 73 pp.

Knapp, W. W., Bowersox, V. C., Chevone, B. I., Krupa, S. V., Lynch, J. A., and McFee, W. W. 1988. 'Precipitation chemistry in the United States, 1, summary on ion concentration variability 1979–1984', *Continuum 3*, Cent. Environ. Res., Water Resour. Inst., Cornell University, Ithaca, New York. 239 pp.

Likens, G. E., Bormann, F. H., Pierce, R. S., Eaton, J. S., and Johnson, N. M. 1977. *Biogeochemistry of a Forested Ecosystem*. Springer-Verlag, New York. 146 pp.

Lindberg, S. E. and Lovett, G. M. 1992. 'Deposition and forest canopy interactions of airborne sulfur: results from the integrated forest study', *Atm. Environ.*, **26A**, 1477–1492.

Lindberg, S. E., Lovett, G. M., and Meiwes, K.-J. 1987. 'Deposition and forest canopy interactions of airborne nitrate' in Hutchinson, T. C. and Meema, K. M. (Eds), *Effects of Atmospheric Pollutants on Forests, Wetlands and Agricultural Ecosystems*, NATO ASI Series, Vol. G16. Springer-Verlag, Berlin, Heidelberg. pp. 117–130.

Lovett, G. M. and Linberg, S. E. 1984. 'Dry deposition and canopy exchange in a mixed oak forest as determined by analysis of throughfall', *J. Appl. Ecol.*, **21**, 1013–1027.

Lynch, J. A. and Corbett, E. S., 1989. 'Hydrologic control of sulfate mobility in a forested watershed', *Wat. Resour. Res.*, **25**, 1695–1703.

Lynch, J. A., Grimm, J. W. and Corbett, E. S. 1989. 'Atmospheric deposition: spatial and temporal variations in Pennsylvania 1988', *ER8909*. Pennsylvania State University, Environmental Resources Research Institute. 105 pp.

Lynch, J. A., Grimm, J. W., and Corbett, E. S. 1990. 'Atmospheric deposition: spatial and temporal variations in Pennsylvania 1989', *ER9010*. Pennsylvania State University, Environmental Resources Research Institute. 98 pp.

Lynch, J. A., Grimm, J. W., and Corbett, E. S. 1991. 'Atmospheric deposition: spatial and temporal variations in Pennsylvania 1990', *ER9103*. Pennsylvania State University, Environmental Resources Research Institute. 99 pp.

Meyers, T. P., Huebert, B. J. and Hicks, B. B. 1989. 'HNO_3 deposition to a deciduous forest', *Boundary-Layer Meteorol.*, **49**, 395–410.

Meyers, T. P., Hicks, B. B., Hosker, R. P., Jr, Womack, J. D., and Satterfield, L. C. 1991. 'Dry deposition inferential measurement techniques-II. Seasonal and annual deposition rates of sulfur and nitrate', *Atm. Environ.*, **25A**, 2361–2370.

Murdoch, P. S. and Stoddard, J. L. 1992. 'The role of nitrate in the acidification of streams in the Catskill mountains of New York', *Wat. Resour. Res.*, **28**, 2707–2720.

Puckett, J. L. 1990. 'Estimate of ion sources in deciduous and coniferous throughfall', *Atm. Environ.*, **24A**, 545–555.

Reuss, J. O. and Johnson, D. W. 1986. *Acid Deposition and the Acidification of Soils and Waters*. Springer-Verlag, New York. 119 pp.

Rochelle, B. P. and Church, M. R. 1987. 'Regional patterns of sulfur retention in watersheds of the eastern US', *Water Air Soil Pollut.*, **36**, 61–73.

Rochelle, B. P., Church, M. R., and David, M. B. 1987. 'Sulfur retention at intensively studied sites in the U.S. and Canada', *Water Air Soil Pollut.*, **33**, 73–83.

Rustad, L. E., Kahl, J. S., Norton, S. A., and Fernandez, I. J. 1994. 'Underestimation of dry deposition by throughfall in mixed northern hardwood forests', *J. Hydrolo.*, **162**, 319–336.

Shanley, J. B. 1989. 'Field measurements of dry deposition to spruce foliage and petri dishes in the Black Forest, F.R.G.', *Atm. Environ.*, **23**, 403–414.

Shepard, J. P., Mitchell, M. J., Scott, T. J., Zhang, Y. M., and Raynal, D. J. 1989. 'Measurements of wet and dry deposition in a northern hardwood forest', *Water Air Soil Pollut.*, **48**, 225–238.

Sirois, A. and Summers, P. W. 1989. 'An estimation of atmospheric deposition input of sulphur and nitrogen oxides to the Kejimkujik watershed: 1979–1987', *Water Air Soil Pollut.*, **46**, 29–43.

Sposito, G. 1989. *The Chemistry of Soils*. Oxford University Press, New York. 277 pp.

Stevens, P. A., Norris, D. A., Sparks, T. H., and Hodgson, A. L. 1994. 'The impacts of atmospheric N inputs on throughfall, soil and stream water interactions for different aged forest and moorland catchments in Wales', *Water Air Soil Pollut.*, **73**, 297–317.

Sullivan, T. J. 1993. 'Whole-ecosystem nitrogen effects research in Europe', *Environ. Sci. Technol.*, **27**, 1482–1486.

Swank, W. T., Waide, J. B., Crossley, D. A., Jr, and Todd, R. L. 1981. 'Insect defoliation enhances nitrate export from forest ecosystems', *Oecologia*, **51**, 297–299.

US Environmental Protection Agency, 1983. *Methods for Chemical Analysis of Water and Wastes*. US EPA, Environmental Monitoring Support Laboratory, Cincinnati, OH. EPA 600/4–79–020.

US Environmental Protection Agency, July 1990. *National Dry Deposition Network Laboratory Operations Manual*. US EPA, Atmospheric Research and Environmental Assessment Laboratory, Research Triangle Park, NC. EPA Contract No. 68-02-4451.

US Environmental Protection Agency, August 1990. *National Dry Deposition Network Laboratory Operations Manual*. US EPA, Atmospheric Research and Environmental Assessment Laboratory, Research Triangle Park, NC. EPA Contract No. 68-02-4451.

Wigington, P. J., Jr, Baker, J. P., DeWalle, D. R., Krester, W. A., Murdoch, P. S., Simonin, H. A., Van Sickle, J., McDowell, M. K., Peck, D. V., and Barchet, W. R. 1993. *Episodic acidification of streams in northeastern United States: chemical and biological results of the Episodic Response Project*. US EPA, Environmental Research Laboratory, Corvallis, OR. EPA/600/R-93/190, 336 pp.

INDEX